T0220767

Lecture Notes in Artificial Intelligence   12869

Subseries of Lecture Notes in Computer Science

Sheng Li · Maosong Sun ·
Yang Liu · Hua Wu · Liu Kang ·
Wanxiang Che · Shizhu He ·
Gaoqi Rao (Eds.)

Chinese Computational Linguistics

20th China National Conference, CCL 2021
Hohhot, China, August 13–15, 2021
Proceedings

Springer

Editors
Sheng Li
Harbin Institute of Technology
Harbin, China

Maosong Sun
Tsinghua University
Beijing, China

Yang Liu
Tsinghua University
Beijing, China

Hua Wu
Baidu (China)
Beijing, China

Liu Kang
Chinese Academy of Sciences
Beijing, China

Wanxiang Che
Harbin Institute of Technology
Harbin, China

Shizhu He
Chinese Academy of Sciences
Beijing, China

Gaoqi Rao
Beijing Language and Culture University
Beijing, China

ISSN 0302-9743   ISSN 1611-3349   (electronic)
Lecture Notes in Artificial Intelligence
ISBN 978-3-030-84185-0   ISBN 978-3-030-84186-7   (eBook)
https://doi.org/10.1007/978-3-030-84186-7

LNCS Sublibrary: SL7 – Artificial Intelligence

This Springer imprint is published by the registered company Springer Nature Switzerland AG
The registered company address is: Gewerbestrasse 11, 6330 Cham, Switzerland

Preface

Welcome to the proceedings of the twentieth China National Conference on Computational Linguistics. The conference and symposium were hosted and co-organized by Inner Mongolia University, China.

CCL is an annual conference (bi-annual before 2013) that started in 1991. It is the flagship conference of the Chinese Information Processing Society of China (CIPS), which is the largest NLP scholar and expert community in China. CCL is a premier nationwide forum for disseminating new scholarly and technological work in computational linguistics, with a major emphasis on computer processing of the languages in China such as Mandarin, Tibetan, Mongolian, and Uyghur.

The Program Committee selected 108 papers (77 Chinese papers and 31 English papers) out of 303 submissions for publication. The acceptance rate was 35.64%. The 31 English papers cover the following topics:

- Machine Translation and Multilingual Information Processing (2)
- Minority Language Information Processing (2)
- Social Computing and Sentiment Analysis (2)
- Text Generation and Summarization (3)
- Information Retrieval, Dialogue, and Question Answering (6)
- Linguistics and Cognitive Science (1)
- Language Resource and Evaluation (3)
- Knowledge Graph and Information Extraction (6)
- NLP Applications (6)

The final program for CCL 2021 was the result of intense work by many dedicated colleagues. We want to thank, first of all, the authors who submitted their papers, contributing to the creation of the high-quality program. We are deeply indebted to all the Program Committee members for providing high-quality and insightful reviews under a tight schedule, and we are extremely grateful to the sponsors of the conference. Finally, we extend a special word of thanks to all the colleagues of the Organizing Committee and secretariat for their hard work in organizing the conference, and to Springer for their assistance in publishing the proceedings in due time.

We thank the Program and Organizing Committees for helping to make the conference successful, and we hope all the participants enjoyed the first online CCL conference.

July 2021

Sheng Li
Maosong Sun
Yang Liu
Hua Wu
Kang Liu
Wanxiang Che

Organization

Conference Chairs

Sheng Li   Harbin Institute of Technology, China
Maosong Sun   Tsinghua University, China
Yang Liu   Tsinghua University, China

Program Chairs

Hua Wu   Baidu, China
Kang Liu   Institute of Automation, CAS, China
Wanxiang Che   Harbin Institute of Technology, China

Program Committee

Renfen Hu   Beijing Normal University, China
Xiaofei Lu   Pennsylvania State University, USA
Zhenghua Li   Soochow University, China
Pengfei Liu   Carnegie Mellon University, USA
Jiafeng Guo   Institute of Computing Technology, CAS, China
Qingyao Ai   The university of Utah, USA
Rui Yan   Renmin University of China, China
Jing Li   The Hong Kong Polytechnic University, China
Wenliang Chen   Soochow University, China
Bowei Zou   Institute for Infocomm Research, Singapore
Shujian Huang   Nanjing University, China
Hui Huang   University of Macau, China
Yating Yang   Xinjiang Physics and Chemistry Institute, CAS, China
Chenhui Chu   Kyoto University, Japan
Junsheng Zhou   Nanjing Normal University, China
Wei Lu   Singapore University of Technology and Design, Singapore
Rui Xia   Nanjing University of Science and Technology, China
Lin Gui   University of Warwick, UK
Chenliang Li   Wuhan University, China
Jiashu Zhao   Wilfrid Laurier University, Canada

Local Arrangement Chairs

Guanglai Gao   Inner Mongolia University, China
Hongxu Hou   Inner Mongolia University, China

Evaluation Chairs

Hongfei Lin   Dalian University of Technology, China
Wei Song   Capital Normal University, China

Publications Chairs

Shizhu He   Institute of Automation, CAS, China
Gaoqi Rao   Beijing Language and Culture University, China

Workshop Chairs

Xianpei Han   Institute of Software, CAS, China
Yanyan Lan   Tsinghua University, China

Sponsorship Chairs

Zhongyu Wei   Fudan University, China
Yang Feng   Institute of Computing Technology, CAS, China

Publicity Chair

Xipeng Qiu   Fudan University, China

Website Chair

Liang Yang   Dalian University of Technology, China

System Demonstration Chairs

Jiajun Zhang   Institute of Automation, CAS, China
Haofen Wang   Tongji University, China

Student Counseling Chair

Zhiyuan Liu   Tsinghua University, China

Student Seminar Chairs

Xin Zhao   Renmin University of China, China
Libo Qin   Harbin Institute of Technology, China

Finance Chair

Yuxing Wang   Tsinghua University, China

Organizers

Chinese Information Processing Society of China

Tsinghua University, China

Inner Mongolia University, China

Publishers

Lecture Notes in Artificial Intelligence,
Springer

Science China

Journal of Chinese Information Processing

清华大学学报（自然科学版）
Journal of Tsinghua University (Science and Technology)

Journal of Tsinghua University
(Science and Technology)

Sponsoring Institutions

Gold

Silver

Bronze

Contents

Information Retrieval, Dialogue and Question Answering

Linguistics and Cognitive Science

Language Resource and Evaluation

Knowledge Graph and Information Extraction

NLP Applications

Machine Translation and Multilingual Information Processing

Reducing Length Bias in Scoring Neural Machine Translation via a Causal Inference Method

Xuewen Shi[1,2], Heyan Huang[1,2], Ping Jian[1,2(✉)], and Yi-Kun Tang[1,2]

[1] School of Computer Science and Technology, Beijing Institute of Technology, Beijing 100081, China
[2] Beijing Engineering Research Center of High Volume Language Information Processing and Cloud Computing Applications, Beijing 100081, China
{xwshi,hhy63,pjian,tangyk}@bit.edu.cn

Abstract. Neural machine translation (NMT) usually employs beam search to expand the searching space and obtain more translation candidates. However, the increase of the beam size often suffers from plenty of short translations, resulting in dramatical decrease in translation quality. In this paper, we handle the length bias problem through a perspective of causal inference. Specifically, we regard the model generated translation score S as a degraded true translation quality affected by some noise, and one of the confounders is the translation length. We apply a Half-Sibling Regression method to remove the length effect on S, and then we can obtain a debiased translation score without length information. The proposed method is model agnostic and unsupervised, which is adaptive to any NMT model and test dataset. We conduct the experiments on three translation tasks with different scales of datasets. Experimental results and further analyses show that our approaches gain comparable performance with the empirical baseline methods.

Keywords: Machine translation · Causal inference · Half-sibling regression

1 Introduction

Recently, with the renaissance of deep learning, end-to-end neural machine translation (NMT) [2,26] has gained remarkable performances [6,27,28]. NMT models are usually built upon an encoder-decoder framework [5]: the encoder reads an input sequence $\mathbf{x} = \{x_1, ..., x_{T_x}\}$ into a hidden memory H, and the decoder is designed to model a probability over the translation $\mathbf{y} = \{y_1, ..., y_{T_{\hat{y}}}\}$ by:

$$P(\mathbf{y}|\mathbf{x}) = \prod_{t=1}^{T_y} P(y_t|y_{<t}, H). \tag{1}$$

Most existing NMT approaches employ beam search to obtain more translation candidates and then gain a better translation hypothesis $\hat{\mathbf{y}} = \{\hat{y}_1, \cdots, \hat{y}_{T_{\hat{y}}}\}$

© Springer Nature Switzerland AG 2021
S. Li et al. (Eds.): CCL 2021, LNAI 12869, pp. 3–15, 2021.
https://doi.org/10.1007/978-3-030-84186-7_1

by ranking the translation candidates set $\hat{\mathbf{Y}} = \{\hat{\mathbf{y}}_1, \cdots, \hat{\mathbf{y}}_b\}$ across a score function $s(\hat{\mathbf{y}})$:

$$s(\hat{\mathbf{y}}) = \sum_{t=1}^{T_{\hat{\mathbf{y}}}} \log P(\hat{y}_t | \mathbf{x}; \theta), \tag{2}$$

where b is the beam size and θ is the parameter set of the NMT model.

However, continuously increasing the beam size has been shown to degrade performances and lead to short translations [13]. One decisive reason is that the large search space is easy to introduce more short $\hat{\mathbf{y}}$, and the shorter $\hat{\mathbf{y}}$ tends to be scored higher under $s(\hat{\mathbf{y}})$ in Eq. (2). Previous efforts usually deal with the above length bias problem by two mechanisms: i) performing length normalization on $s(\hat{\mathbf{y}})$ via dividing $s(\hat{\mathbf{y}})$ by the length penalty lp, i.e. $s'(\hat{\mathbf{y}}) \leftarrow s(\hat{\mathbf{y}})/lp$ [3,9,13,16,30], and ii) adding an additional length-related reward r to $s(\hat{\mathbf{y}})$, i.e. $s'(\hat{\mathbf{y}}) \leftarrow s(\hat{\mathbf{y}})+\gamma \cdot r$ [7,8,14,17,30]. For the second strategy, the correcting ratio γ of the reward is usually determined by supervised training [7,17] or manually fine-tuning [8] before the testing stage, which lacks the ability of self-adapting to the unseen data.

In this paper, we introduce a causal motivated model agnostic and unsupervised method to solve the length bias problem for NMT. As shown in Fig. 1, for a translation hypothesis $\hat{\mathbf{y}}$, suppose that Q is an unobservable true translation quality of $\hat{\mathbf{y}}$, and the model generated score S can be seen as an observed degraded version of Q which is affected by some noise N. Generally, S equals $s(\hat{\mathbf{y}})$ in conventional NMT approaches, and it can be viewed as one of the measurement methods of Q with systematic errors. As mentioned above, one kind of systematic errors has a strong correlation with the translation length, therefore, the noise caused by length will be eliminated if we subtract the length effect from S. Specifically, we utilize the Half-Sibling Regression (HSR) [22] method to perform the noise elimination operation for NMT. The method first apply a regression model to appraise the effect of the translation length on the model generated score, i.e. $E[S|T_{\hat{\mathbf{y}}}]$. Then, the denoised score is obtained by removing $E[S|T_{\hat{\mathbf{y}}}]$ from S:

$$S' := S - E[S|T_{\hat{\mathbf{y}}}]. \tag{3}$$

We propose two branches of the framework, corpus based (C-HSR) and single source sentence based (S-HSR) re-scoring method. The difference is that C-HSR performs the estimation of $E[S|T_{\hat{\mathbf{y}}}]$ on the whole test set, while S-HSR uses the translation candidates in a beam of the NMT inference process to predict $E[S|T_{\hat{\mathbf{y}}}]$. The operation of approximating $E[S|T_{\hat{\mathbf{y}}}]$ for both C-HSR and S-HSR entirely rely on the current testing data of NMT without fine-tuning or any supervised information. In this work, we regard the NMT model as a black-box and apply the HSR-based denoised method to the re-scoring procedure for NMT.

We conduct the experiments on three translation tasks: Uyghur→Chinese, Chinese→ English and English→French, which represent low-resource, medium-resource and high-resource NMT, respectively. The experimental results show the proposed approaches achieve comparable performances with empirical length normalization methods. Further analyses show the flexibility of the proposed methods and the assumptions that our approaches rely on are reliable.

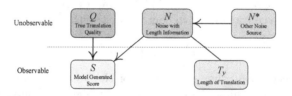

Fig. 1. A causal directed acyclic graph shows the relations among the true translation quality Q, the model generated score S and the translation length $T_{\hat{y}}$. See Sect. 1 and Sect. 3.1 for more details.

2   Related Work

The length bias reduction methods can be mainly divided into two categories: i) dividing the log probability by the length penalty lp:

$$s'(\hat{\mathbf{y}}) \leftarrow s(\hat{\mathbf{y}})/lp, \tag{4}$$

and ii) adding an additive length-related reward to the log probability of the hypothesis:

$$s'(\hat{\mathbf{y}}) \leftarrow s(\hat{\mathbf{y}}) + \gamma \cdot r. \tag{5}$$

For the first branch, the predominant form of the length penalty lp is the length of the hypothesis [3,9,13,16]. Google's NMT system [28] employ an empirical length penalty that is computed as:

$$s'(\hat{\mathbf{y}}) \leftarrow s(\hat{\mathbf{y}})/\frac{(5+T_{\hat{\mathbf{y}}})^\alpha}{(5+1)^\alpha}, \tag{6}$$

where the parameter α is used to control the strength of the length normalization. Stahlberg and Byrne [25] apply another variant of lp, which introduces the information of the length ratio of the hypotheses over the source sentence. Yang et al. [30] propose a brevity penalty normalization which adds the log brevity penalty bp to the normalized score:

$$s'(\hat{\mathbf{y}}) \leftarrow s(\hat{\mathbf{y}})/T_{\hat{\mathbf{y}}} + \log bp, \tag{7}$$

where bp is same as the form of brevity penalty in BLEU [20]:

$$bp = \begin{cases} 1 & gr \cdot T_{\mathbf{x}} < T_{\hat{\mathbf{y}}} \\ e^{(1-T_{\mathbf{y}}/T_{\hat{\mathbf{y}}})} & gr \cdot T_{\mathbf{x}} \geq T_{\hat{\mathbf{y}}} \end{cases}, \tag{8}$$

where gr is the generation ratio i.e. $T_{\mathbf{y}}/T_{\mathbf{x}}$. Since $T_{\mathbf{y}}$ is unknown in the inference step, Yang et al. [30] apply a 2-layer multi layer perceptron (MLP) to predict the gr by taking the mean of the hidden states of the NMT encoder as the input.

The second branch is similar to the word penalty in statistical machine translation [11,19]. The parameter γ can be automatically optimized with supervised learning [7,17] or manually assignment [8].

He et al. [7] propose a log-linear NMT framework which incorporates a word reward feature to the framework to control the length of the translation:

$$s'(\hat{\mathbf{y}}) \leftarrow s(\hat{\mathbf{y}}) + \gamma \cdot T_{\hat{\mathbf{y}}}, \tag{9}$$

where γ is trained with other parameters of the log-linear NMT model using minimum error rate training [7,18]. Murray and Chiang [17] make the optimization process of γ independent to the NMT training process, so that the γ can be trained on a relatively small dataset. Huang et al. [8] introduce a Bounded Length Reward that includes the prior knowledge of the generation ratio gr of reference translation length over source sentence length:

$$s'(\hat{\mathbf{y}}) \leftarrow s(\hat{\mathbf{y}}) + \gamma \cdot \min(gr \cdot T_{\mathbf{x}}, T_{\hat{\mathbf{y}}}), \tag{10}$$

where the length reward γ is fine-tuned manually. All the above methods [7,8,17] fine-tune the correcting ratio γ by a supervised data, which may lead to less optimal results on unseen test datasets. Yang et al. [30] propose a Bounded Adaptive-Reward to remove the hyperparameter γ: $s'(\hat{\mathbf{y}}) \leftarrow s(\hat{\mathbf{y}}) + \sum_{t=1}^{T^*} r_t$, where b is the beam size and r_t is the average negative log-probability of the words in the beam at time step t. $T^* = \min\{T_{\hat{\mathbf{y}}}, T_{pred}(x)\}$, where $T_{pred}(x)$ is predicted with a 2-layer MLP instead of using the constant gr [8] as Eq. (10) does.

The proposed HSR-based debiasing method is motivated entirely by a causal structure shown in Fig. 1, although the form of the approach is same as the reward-based length normalization in Eq. (5). Formally, we can regard $E[S|T_{\hat{\mathbf{y}}}]$ in Eq. (3) as an instance of $(\gamma \cdot r)$ in Eq. (9) with very few prior assumptions or handcrafted designs. The leaning process of $E[S|T_{\hat{\mathbf{y}}}]$ is entirely model agnostic and unsupervised, which makes the proposed method more competitive to the previous supervised approaches [7,8,17] in real practical applications.

3 Approach

3.1 Correcting Length Bias via Half-Sibling Regression

In this paper, we apply a debiasing framework of Half-Sibling Regression (HSR) [22] to subtract the NMT scoring bias caused by the length of the translation. For a translation hypothesis $\hat{\mathbf{y}}$, suppose that Q is the true translation quality that we cannot observe directly, and we regard S as an observable degraded version of Q which is affected by Q and some noise N, simultaneously. Considering a conventional NMT decoder, S is usually calculated by $s(\hat{\mathbf{y}})$ in Eq. (2). As discussed in Sect. 1, $T_{\hat{\mathbf{y}}}$, as the length of $\hat{\mathbf{y}}$, has undesired crucial impacts on S. We refer $s(\hat{\mathbf{y}})$ as a measurement of Q with systemic errors N, then $T_{\hat{\mathbf{y}}}$ is the correlative variable of N that satisfies $N \not\perp T_{\hat{\mathbf{y}}}$. At the same time, we assume

---

Algorithm 1. HSR in translation re-scoring for correcting length bias. See Sect. 3.2 for more details.

Input: m translation candidates: $\hat{Y} = \{\hat{\mathbf{y}}_1, \cdots, \hat{\mathbf{y}}_m\}$, the lengths set of the translation candidates: $T(\hat{Y}) = \{T_{\hat{\mathbf{y}}_1}, \cdots, T_{\hat{\mathbf{y}}_m}\}$, NMT model scores for the m translation candidates: $s(\hat{Y}) = \{s(\hat{\mathbf{y}}_1), \cdots, s(\hat{\mathbf{y}}_m)\}$ and a hyperparameter $\alpha \in [0,1]$.

1: Find the optimal parameters θ_R^* for a regression model $\mathrm{R}(T_{\hat{\mathbf{y}}}; \theta_R)$ by minimize the mean square error:

$$\theta_R^* = \arg\min_{\theta_R} \frac{1}{m} \sum_{\hat{\mathbf{y}} \in \hat{Y}} |\mathrm{R}(T_{\hat{\mathbf{y}}}; \theta_R) - s(\hat{\mathbf{y}})|^2$$

2: Subtract length information from the model estimated score:

$$s'(\hat{Y}) \leftarrow s(\hat{Y}) - \alpha \times \mathrm{R}^*(T(\hat{Y}); \theta_R^*) \tag{12}$$

Output: The debiased translation scores $s'(\hat{Y}) = \{s'(\hat{\mathbf{y}}_1), \cdots, s'(\hat{\mathbf{y}}_m)\}$.

---

that $Q \perp\!\!\!\perp T_{\hat{\mathbf{y}}}$, therefore, we can subtract the effects of $T_{\hat{\mathbf{y}}}$ on S, i.e. $\mathrm{E}[S|T_{\hat{\mathbf{y}}}]$, from S to eliminate length bias without affect the connection between S and Q:

$$S' \leftarrow S - \mathrm{E}[S|T_{\hat{\mathbf{y}}}]. \tag{11}$$

In practice, the value of $\mathrm{E}[S|T_{\hat{\mathbf{y}}}]$ can be estimated by a regression model that is trained on the observed $(S, T_{\hat{\mathbf{y}}})$ pairs.

Figure 1 shows the causal directed acyclic graph (DAG) that illustrates the causalities between Q, S, N, N^* and $T_{\hat{\mathbf{y}}}$, where N^* is other noise source that satisfies $N^* \perp\!\!\!\perp T_{\hat{\mathbf{y}}}$. We set up an undirected connection between N and $T_{\hat{\mathbf{y}}}$ to represent $N \not\perp\!\!\!\perp T_{\hat{\mathbf{y}}}$ since the causal direction between the two variables is not important in this paper. It is worth noting that $Q \perp\!\!\!\perp T_{\hat{\mathbf{y}}}$ is a strong assumption when we don't know the specific form of Q. The possible forms of Q and the assumption of $Q \perp\!\!\!\perp T_{\hat{\mathbf{y}}}$ will be discussed in more detail in Sect. 3.3.

3.2 Re-scoring Translation Candidates

The HSR-based length debiasing method is model agnostic and it views the NMT model as a black-box. Therefore, we simply apply the HSR-based approach to the translation re-scoring process to verify its effectiveness. Algorithm 1 shows a sketch of the proposed re-scoring framework. As described in Algorithm 1, we first optimize a regression model $\mathrm{R}(T_{\hat{\mathbf{y}}}; \theta_R)$ that parameterized by θ_R to estimate the length effect on $s(\hat{\mathbf{y}})$ by using the data $(T(\hat{Y}), s(T_{\hat{\mathbf{y}}})) = \{(T_{\hat{\mathbf{y}}_i}, s(\hat{\mathbf{y}}_i))\}_{i=1}^m$. Then, we adopt the optimal $\mathrm{R}^*(T_{\hat{\mathbf{y}}}; \theta_R^*)$ as an approximate to $\mathrm{E}[S|T_{\hat{\mathbf{y}}}]$ in Eq. (11) to eliminate the length information from $s(\hat{\mathbf{y}})$:

$$s'(\hat{\mathbf{y}}) \leftarrow s(\hat{\mathbf{y}}) - \alpha \times \mathrm{R}^*(T_{\hat{\mathbf{y}}}; \theta_R^*). \tag{13}$$

Following [28], we introduce a hyperparameter $\alpha \in [0,1]$ to control the strength of the debiasing operation. $\alpha = 0$ means no debiasing operation is conducted and empirical studies show that setting $\alpha = 1$ usually gains better performances for $b \geq 8$. (Note that Eq. (12) in Algorithm 1 is in a set form while Eq. (13) is in a single value form.)

We propose two branches of implementations for the proposed re-scoring framework in practice: i) a corpus based re-scoring method (C-HSR) and ii) a single source sentence based re-scoring method (S-HSR). For C-HSR, we perform the regression over the translations and their model scores of the whole test dataset, in other words, it needs the NMT model to finish translating the whole test set. For S-HSR, the regression model is optimized on the translation candidates and their model scores of a single input source sentence. Therefore, the size of $\hat{Y}$ in Algorithm 1, i.e. m, equals the beam size b and $b \times |X_{test}|$ (the size of test set) for S-HSR and C-HSR, respectively.

3.3   Discussion

The Assumption of Q is Independent of $T_\mathbf{y}$. Considering one of ideally forms of Q that is straightforward defined as a conditional probability:

$$Q := P(\mathbf{y}|\mathbf{x}) = P(\{y_1, ..., y_{T_\mathbf{y}}\}|\mathbf{x}). \tag{14}$$

In Eq. (14), $T_{\hat{\mathbf{y}}}$ is an inherent feature of $\mathbf{y}$, so it is also involved in Q. Therefore, executing the calculation of Eq. (11) will inevitably eliminate parts of Q itself.

However, the condition where $T_\mathbf{y}$ is almost independent of Q is also sufficient for HSR in practice, according to [22]. Hence, we should verify the correlation between Q and $T_{\hat{\mathbf{y}}}$ before employing our approach to specific applications. Since, Q as well as $P(\mathbf{y}|\mathbf{x})$ is theoretic and unobservable, we adopt a more precise and pricey observable variable, the professional translators' direct assessment (DA) score, as an approximation to the Q[1]. We use the datasets from WMT 2020 Quality Estimation Share Task 1[2]: Sentence-Level Direct Assessment [24] to analyze the Pearson's and Spearman's correlation scores between the length of translation and the DA score, and the results are presented in Table 1.

As Table 1 shows, for most conditions, the absolute values of the correlation scores are less than 0.20, which indicates that Q is almost independent of the translation length in a linear 2-dimensional space. However, there are multiple possible variables that influence the human DA score such as the number of the rare words in the source sentence and the translation hypothesis. Although partial correlation [1] might be effective for analyzing multiple correlative variables, the information about the other observable variables is unavailable. In general, we believe that removing $E[S|T_{\hat{\mathbf{y}}}]$ will not harm the information of Q too much, and the debiasing ratio α is also a conservative design to avoid punishing the length information overly.

---

[1] Note that, $P(\mathbf{y}|\mathbf{x})$ is one of the formal definitions of Q, and it is not the essence of Q. On the other hand, the human generated DA score is the currently available best approximation of Q to our best knowledge.

[2] http://www.statmt.org/wmt20/quality-estimation-task.html.

Table 1. The Pearson's and Spearman's correlation scores between the DA score and $T_{\hat{y}}$.

Language pair	Train		Valid		Test	
	Pearson	Spearman	Pearson	Spearman	Pearson	Spearman
English-German	−0.06	−0.11	−0.15	−0.18	−0.18	−0.18
English-Chinese	−0.07	−0.12	−0.08	−0.09	−0.00	−0.02
Romanian-English	−0.20	−0.15	−0.20	−0.14	−0.25	−0.18
Estonian-English	−0.09	−0.13	−0.09	−0.10	−0.11	−0.11
Nepalese-English	−0.12	−0.02	−0.12	−0.05	−0.09	−0.01
Sinhala-English	−0.14	−0.06	−0.11	−0.05	−0.17	−0.07
Russian-English	0.07	−0.07	0.00	−0.10	−0.01	−0.16

The Connection to the Word Reward. The proposed HSR-based debiasing method is motivated by a causal structure, although the formalized form of our proposed approach is same as adding length-related reward in Eq. (5), by regarding $E[S|T_{\hat{y}}]$ as a special instance of $(\gamma \cdot r)$. In particular, if we only consider the linear effects, i.e. $R(T_{\hat{y}}; \theta_R) = \theta_1 T_{\hat{y}} + \theta_2$, then Eq. (13) is expand as:

$$s'(\hat{y}) \leftarrow s(\hat{y}) - \alpha \times (\theta_1^* T_{\hat{y}} + \theta_2^*) = s(\hat{y}) - \alpha\theta_1^* T_{\hat{y}} - \alpha\theta_2^*, \tag{15}$$

which is similar to the word reward in Eq. (9). The $\theta_1^* \in \mathbb{R}$ and $\theta_2^* \in \mathbb{R}$ in Eq. (15) are optimal parameters of the linear regression. Therefore, under the above linear assumption, the proposed method can be seen as a simple and effective unsupervised strategy to optimize γ for the word penalty [7,17]. Since most of the previous word penalty efforts determine γ through a supervised procedure [7,8,17] before the testing stage, they may fall into less optimal results on unseen datasets.

However, if we do not apply the linear regression, the form will be different to the word penalty. In this paper, we study the performances of various typical regression models including linear regression, support vector regression, k-neighbors regression, multi-layer perceptron (MLP) regression and random forest regression. We find that applying linear regression and MLP regression to C-HSR and S-HSR respectively gain better performances.

4 Experiments

4.1 Datasets and Evaluation Metric

We evaluate the proposed approaches on three translation tasks: Uyghur→Chinese (Ug→Zh), Chinese→English (Zh→En) and English→French (En→Fr). For each of the translation task, the corpus is tokenized by the Moses [12] *tokenizer.perl*[3] before encoded with byte-pair encoding [23]. For

---

[3] Moses scripts: https://github.com/moses-smt/mosesdecoder/blob/master/scripts/.

Zh→En and Ug→Zh translation tasks, the Chinese parts are segmented by the LTP segmentor [4] before tokenizing.

Ug→Zh. For Uyghur→Chinese translation, the training corpus is from Uyghur to Chinese News Translation Task in CCMT2019 Machine Translation Evaluation [29]. Apart from the Moses [12] tokenizer, we do not use any other tools to segment Uyghur. The training set contains 0.17M parallel sentence pairs, and the vocabularies are 30K for both Uyghur and Chinese corpus. The official validation set and the test set are applied in our experiments.

Zh→En. For Chinese→English translation, the training data is extracted from four LDC corpora[4]. The training set finally contains 1.3M parallel sentence pairs in total. After preprocessing, we get a Chinese vocabulary of about 39K tokens, and an English vocabulary of about 30K tokens. We use NIST2002 dataset for validation and NIST 2003–2006 datasets for test.

En→Fr. For English→French translation, we conduct our experiments on the publicly available WMT'14 En→Fr datasets which consist of 18M sentences pairs. Both source and target vocabulary contains 30K tokens after preprocessing. We report results on newstest2014 dataset, and newstest2013 dataset is used as the validation set.

Evaluation. Following [27], we report the results of a single model by averaging the 5 checkpoints around the best model selected on the development set. The translation results are measured in case-insensitive BLEU [20] by *multi-bleu.perl* (see footnote 3). For the Ug→Zh translation task, the BLEU scores are reported at character-level.

4.2 Length Normalization Baselines

We adopt two popular empirical length normalization strategies ((i), (ii)) and a complicated MLP-based method ((iii)) as the comparison baseline methods: i) Length Norm: directly dividing the translation score by the length of the translation [3,9,13] as shown in Eq. (4), ii) GNMT: the length normalization method of Google NMT [28], as shown in Eq. (6), and iii) BP Norm: the length normalization method that applies a model predicted *bp* constraint [30] as shown in Eq. (7) and Eq. (8). We average the outputs of the Transformer encoder instead of the LSTM hidden layers as the input of the 2-layers MLP used in [30]. For fairness considerations, those methods are all unsupervised[5], since our proposed methods do not rely on any human reference.

---

[4] LDC2005T10, LDC2003E14, LDC2004T08 and LDC2002E18. Since LDC2003E14 is a document-level alignment comparable corpus, we use Champollion Tool Kit [15] to extract parallel sentence pairs from it.

[5] "unsupervised" means that the method is not trained on the dataset that consists the pairs of translation hypothesis and human reference.

4.3   Model Setups

We apply the base model of Transformer [27] as the specific implement of the NMT baseline in our work, and we build up the NMT models based on OpenNMT-py [10]. We analyze different regression models for both C-HSR and S-HSR, and finally select linear regression for C-HSR and one-hidden layer MLP regression S-HSR, denoted by "C-HSR$_{LR}$" and "S-HSR$_{MLP}$", respectively. The regression models used in our work are implemented by using scikit-learn [21]. Following [28], we use α to control the strength of length bias correcting. The α is selected according to the performance on the validation set and the detail selections of α for different model setups are shown in Table 2.

Table 2. Correcting ratio α for different model setups. "-" means same as the left value.

Language pair	Method	$b = 4$	$b = 8$	$b = 16$	$b = 32$	$b = 64$	$b = 100$	$b = 200$
Ug→Zh	$GNMT$	1.0	–	–	–	–	–	–
	C-HSR$_{LR}$	0.9	1.0	–	–	–	–	–
	S-HSR$_{MLP}$	1.0	–	–	–	–	–	–
Zh→En	$GNMT$	0.5	0.9	1.0	–	–	–	–
	C-HSR$_{LR}$	0.7	1.0	–	–	–	–	–
	S-HSR$_{MLP}$	0.7	0.9	0.9	0.9	1.0	–	–
En→Fr	$GNMT$	0.9	–	–	–	–	–	–
	C-HSR$_{LR}$	0.8	–	–	–	–	0.9	1.0
	S-HSR$_{MLP}$	0.8	–	–	–	–	–	–

4.4   Main Results

Table 3. BLEU scores on En→Fr and Ug→Zh translation tasks. "b" represents the beam size.

Method	En→Fr		Ug→Zh	
	b=4	b=200	b=4	b=200
Transformer	39.61	30.66	37.52	36.00
+*Length Norm*	39.41	39.13	37.85	37.96
+*GNMT*	39.77	**39.35**	37.76	37.83
+*BP Norm*	38.36	37.35	37.87	**38.14**
+C-HSR$_{LR}$	39.73	39.13	**37.88**	37.87
+S-HSR$_{MLP}$	**39.80**	39.28	37.81	38.02

Table 4. BLEU scores on NIST 2003∼2006 Zh→En translation task.

Method	03		04		05		06		Average	
	$b{=}4$	=200	$b{=}4$	=200	$b{=}4$	=200	$b{=}4$	=200	$b{=}4$	=200
Transformer	40.10	33.55	42.09	35.31	40.33	33.46	39.94	32.71	40.62	33.76
+*Length Norm*	39.99	40.13	42.05	42.23	39.67	40.10	**40.42**	**40.14**	40.53	40.65
+*GNMT*	40.13	40.08	42.18	42.18	**40.39**	**40.59**	40.24	39.89	40.74	40.69
+*BP Norm*	39.46	39.25	41.50	41.22	39.19	39.15	39.84	39.91	40.00	39.88
+C-HSR$_{LR}$	40.35	39.58	**42.60**	42.00	40.32	40.22	40.25	39.34	**40.88**	40.29
+S-HSR$_{MLP}$	**40.40**	**40.25**	42.42	**42.44**	40.33	40.40	40.25	40.04	40.85	**40.78**

We conduct experiments on three translation tasks with disparate corpora scales: low-resource Ug→Zh, medium-resource Zh→En and high-resource En→Fr. We present BLEU scores on translations with two different decoding beam sizes: $b = 4$ and $b = 200$, in order to compare the model performances on small and large beam sizes. The experimental results are shown in Table 3 and Table 4.

The overall results show that all the length debiasing approaches obtain better BLEU scores than the baseline NMT model for large beam size. For the condition of smaller beam size, "*Length Norm*" tends to disrupt the model performance on En→Fr and Zh→En datasets, which is contrary to the case of a larger search space.

Our proposed C-HSR$_{LR}$ and S-HSR$_{MLP}$ seem to produce stable BLEU scores across multiple datasets and beam sizes. The results show that S-HSR$_{MLP}$ usually gains better BLEU scores than C-HSR$_{LR}$ on the large beam size (Ug→Zh, Zh→En and En→Fr), while C-HSR$_{LR}$ performs better on the small beam size (Zh→En and En→Fr). We consider the reason is that S-HSR$_{MLP}$ is trained better on $b = 200$ than that on small dataset. On the other hand, the requirements for training a linear regression model is not as strict as it for MLP, although the accuracy of the linear model may be lower than the MLP-based model when both of them are well trained.

The performance of BP Norm is unsatisfactory, which we consider the reason is that the MLP-based generation ratio predictor does not work well. If our hypothesis is correct, the length of the translation will be too long or too short under the rescore method of BP Norm. Further analyses about the performances of those method on various beam sizes are shown in Sect. 4.5.

4.5 Performance on Wider Beam Size

As a supplement to Sect. 4.4, we analyze the performances of the proposed approaches on different decoding beam sizes. Figure 2 shows the trend of the BLEU scores with respect to the beam sizes of $[4, 8, 16, 32, 64, 100, 200]$ for the three translation tasks. From Fig. 2 we can observe that all the length debiasing methods achieve stable and comparable performances when the beam size increases.

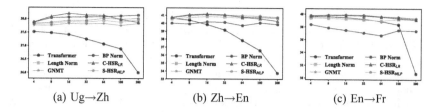

(a) Ug→Zh   (b) Zh→En   (c) En→Fr

Fig. 2. BLEU scores of different methods with respect to different beam sizes [4–200]. The y-axis is the BLEU score, and the x-axis is the decoding beam size. For Zh→En task, we present the averaged the BLEU score of NIST 2003–2006. See Sect. 4.5 for more details.

5   Conclusion and Future Work

In this paper, we introduce a causal motivated method to reduce the length bias problem in NMT. We employ a Half-Sibling Regression [22] method to handle this task and corroborate the task satisfies the independence assumption of HSR. Experimental results on three language pairs with distinct data scales show the effectiveness of the proposed method. In the future, we will complete our experiments on the task of Quality Estimation. Since the proposed approaches are model agnostic and unsupervised, we will verify the effectiveness of our approaches on other natural language generation tasks, such as dialogue system and summarization.

Acknowledgments. We thank all anonymous reviewers for their valuable comments. This work is supported by the National Key Research and Development Program of China (Grant No. 2017YFB1002103) and the National Natural Science Foundation of China (No. 61732005).

References

1. Baba, K., Shibata, R., Sibuya, M.: Partial correlation and conditional correlation as measures of conditional independence. Austr. New Zealand J. Stat. **46**(4), 657–664 (2004)
2. Bahdanau, D., Cho, K., Bengio, Y.: Neural machine translation by jointly learning to align and translate. In: Bengio, Y., LeCun, Y. (eds.) 3rd International Conference on Learning Representations, ICLR 2015, San Diego, CA, USA, 7–9 May 2015, Conference Track Proceedings (2015)
3. Boulanger-Lewandowski, N., Bengio, Y., Vincent, P.: Audio chord recognition with recurrent neural networks. In: de Souza Britto, A., Jr., Gouyon, F., Dixon, S. (eds.) Proceedings of the 14th International Society for Music Information Retrieval Conference, ISMIR 2013, Curitiba, Brazil, 4–8 November 2013, pp. 335–340 (2013)
4. Che, W., Li, Z., Liu, T.: LTP: a Chinese language technology platform. In: Coling 2010: Demonstrations, pp. 13–16. Coling 2010 Organizing Committee, Beijing, August 2010

5. Cho, K., van Merriënboer, B., Bahdanau, D., Bengio, Y.: On the properties of neural machine translation: encoder-decoder approaches. In: Proceedings of SSST-8, Eighth Workshop on Syntax, Semantics and Structure in Statistical Translation, pp. 103–111. Association for Computational Linguistics, Doha, October 2014

6. Gehring, J., Auli, M., Grangier, D., Yarats, D., Dauphin, Y.N.: Convolutional sequence to sequence learning. In: Precup, D., Teh, Y.W. (eds.) Proceedings of the 34th International Conference on Machine Learning, ICML 2017, Sydney, NSW, Australia, 6–11 August 2017. Proceedings of Machine Learning Research, vol. 70, pp. 1243–1252. PMLR (2017)

7. He, W., He, Z., Wu, H., Wang, H.: Improved neural machine translation with SMT features. In: Schuurmans, D., Wellman, M.P. (eds.) Proceedings of the Thirtieth AAAI Conference on Artificial Intelligence, Phoenix, Arizona, USA, 12–17 February 2016, pp. 151–157. AAAI Press (2016)

8. Huang, L., Zhao, K., Ma, M.: When to finish? Optimal beam search for neural text generation (modulo beam size). In: Proceedings of the 2017 Conference on Empirical Methods in Natural Language Processing, pp. 2134–2139. Association for Computational Linguistics, Copenhagen, September 2017

9. Jean, S., Firat, O., Cho, K., Memisevic, R., Bengio, Y.: Montreal neural machine translation systems for WMT 2015. In: Proceedings of the Tenth Workshop on Statistical Machine Translation, pp. 134–140. Association for Computational Linguistics, Lisbon, September 2015

10. Klein, G., Kim, Y., Deng, Y., Senellart, J., Rush, A.: OpenNMT: open-source toolkit for neural machine translation. In: Proceedings of ACL 2017, System Demonstrations, pp. 67–72. Association for Computational Linguistics, Vancouver, July 2017

11. Koehn, P.: Statistical Machine Translation. Cambridge University Press, New York (2010)

12. Koehn, P., et al.: Moses: open source toolkit for statistical machine translation. In: Proceedings of the 45th Annual Meeting of the Association for Computational Linguistics Companion Volume Proceedings of the Demo and Poster Sessions, pp. 177–180. Association for Computational Linguistics, Prague, June 2007

13. Koehn, P., Knowles, R.: Six challenges for neural machine translation. In: Proceedings of the First Workshop on Neural Machine Translation, pp. 28–39. Association for Computational Linguistics, Vancouver, August 2017

14. Li, J., Jurafsky, D.: Mutual information and diverse decoding improve neural machine translation. CoRR abs/1601.00372 (2016)

15. Ma, X.: Champollion: a robust parallel text sentence aligner. In: Proceedings of the Fifth International Conference on Language Resources and Evaluation (LREC 2006). European Language Resources Association (ELRA), Genoa, Italy, May 2006

16. Meister, C., Cotterell, R., Vieira, T.: If beam search is the answer, what was the question? In: Proceedings of the 2020 Conference on Empirical Methods in Natural Language Processing (EMNLP), pp. 2173–2185. Association for Computational Linguistics, Online, November 2020

17. Murray, K., Chiang, D.: Correcting length bias in neural machine translation. In: Proceedings of the Third Conference on Machine Translation: Research Papers, pp. 212–223. Association for Computational Linguistics, Brussels, October 2018

18. Och, F.J.: Minimum error rate training in statistical machine translation. In: Proceedings of the 41st Annual Meeting of the Association for Computational Linguistics, pp. 160–167. Association for Computational Linguistics, Sapporo, July 2003

19. Och, F.J., Ney, H.: Discriminative training and maximum entropy models for statistical machine translation. In: Proceedings of the 40th Annual Meeting of the Association for Computational Linguistics, pp. 295–302. Association for Computational Linguistics, Philadelphia, July 2002
20. Papineni, K., Roukos, S., Ward, T., Zhu, W.J.: Bleu: a method for automatic evaluation of machine translation. In: Proceedings of the 40th Annual Meeting of the Association for Computational Linguistics, pp. 311–318. Association for Computational Linguistics, Philadelphia, July 2002
21. Pedregosa, F., et al.: Scikit-learn: machine learning in Python. J. Mach. Learn. Res. **12**, 2825–2830 (2011)
22. Schölkopf, B., et al.: Modeling confounding by half-sibling regression. Proc. Natl. Acad. Sci. U.S.A. **113**(27), 7391–7398 (2016)
23. Sennrich, R., Haddow, B., Birch, A.: Neural machine translation of rare words with subword units. In: Proceedings of the 54th Annual Meeting of the Association for Computational Linguistics (Volume 1: Long Papers), pp. 1715–1725. Association for Computational Linguistics, Berlin, August 2016
24. Specia, L., et al.: Findings of the WMT 2020 shared task on quality estimation. In: Proceedings of the Fifth Conference on Machine Translation, pp. 743–764. Association for Computational Linguistics, Online, November 2020
25. Stahlberg, F., Byrne, B.: On NMT search errors and model errors: cat got your tongue? In: Proceedings of the 2019 Conference on Empirical Methods in Natural Language Processing and the 9th International Joint Conference on Natural Language Processing (EMNLP-IJCNLP), pp. 3356–3362. Association for Computational Linguistics, Hong Kong, November 2019
26. Sutskever, I., Vinyals, O., Le, Q.V.: Sequence to sequence learning with neural networks. In: Ghahramani, Z., Welling, M., Cortes, C., Lawrence, N.D., Weinberger, K.Q. (eds.) Advances in Neural Information Processing Systems 27: Annual Conference on Neural Information Processing Systems 2014, Montreal, Quebec, Canada, 8–13 December 2014, pp. 3104–3112 (2014)
27. Vaswani, A., et al.: Attention is all you need. In: Guyon, I., et al. (eds.) Advances in Neural Information Processing Systems 30: Annual Conference on Neural Information Processing Systems 2017, Long Beach, CA, USA, 4–9 December 2017, pp. 5998–6008 (2017)
28. Wu, Y., et al.: Google's neural machine translation system: bridging the gap between human and machine translation. CoRR abs/1609.08144 (2016)
29. Yang, M., et al.: CCMT 2019 machine translation evaluation report. In: Huang, S., Knight, K. (eds.) CCMT 2019. CCIS, vol. 1104, pp. 105–128. Springer, Singapore (2019). https://doi.org/10.1007/978-981-15-1721-1_11
30. Yang, Y., Huang, L., Ma, M.: Breaking the beam search curse: a study of (re-)scoring methods and stopping criteria for neural machine translation. In: Proceedings of the 2018 Conference on Empirical Methods in Natural Language Processing, pp. 3054–3059. Association for Computational Linguistics, Brussels, October–November 2018

Low-Resource Machine Translation Based on Asynchronous Dynamic Programming

Xiaoning Jia, Hongxu Hou$^{(\boxtimes)}$, Nier Wu, Haoran Li, and Xin Chang

College of Computer Science-College of Software, Inner Mongolia University, Hohhot, China
cshhx@imu.edu.cn

Abstract. Reinforcement learning has been proved to be effective in handling low resource machine translation tasks and different sampling methods of reinforcement learning affect the performance of the model. The reward for generating translation is determined by the scalability and iteration of the sampling strategy, so it is difficult for the model to achieve bias-variance trade-off. Therefore, according to the poor ability of the model to analyze the structure of the sequence in low-resource tasks, this paper proposes a neural machine translation model parameter optimization method for asynchronous dynamic programming training strategies. In view of the experience priority situation under the current strategy, each selective sampling experience not only improves the value of the experience state, but also avoids the high computational resource consumption inherent in traditional valuation methods (such as dynamic programming). We verify the Mongolian-Chinese and Uyghur-Chinese tasks on CCMT2019. The result shows that our method has improved the quality of low-resource neural machine translation model compared with general reinforcement learning methods, which fully demonstrates the effectiveness of our method.

Keywords: Low-resource · Machine translation · Asynchronous Dynamic Programming

1 Introduction

Traditional machine translation [1] uses cross entropy to measure the entropy of the generated token and reference token during the translation process. Its training process is usually to maximize the logarithmic likelihood of each token in the target sentence by taking the source sentence and the translated target token as input. This training method is called maximum likelihood estimation (MLE). Although it is easy to infer, the objective function at the token level in the training is not consistent with the evaluation indicators at the sequence level such as BLEU [6] and the model uses a given real sample word as input to predict the next word. However, since there is no real distribution in the inference stage, the model can only generate the whole sequence by predicting one word

© Springer Nature Switzerland AG 2021
S. Li et al. (Eds.): CCL 2021, LNAI 12869, pp. 16–28, 2021.
https://doi.org/10.1007/978-3-030-84186-7_2

at a time based on its own prediction output. The inconsistent methods of the prediction and inference stages are easy to cause the accumulation of errors. In order to solve the problem of inconsistency and accumulation of errors, a Minimum Risk Training (MRT) [9] algorithm is proposed to alleviate the problem. This training algorithm introduced evaluation index as the loss function, which aims to minimize the expected loss of training data and combines sentence-level BLEU into the loss function, so that the maximum likelihood estimation result of the machine translation model can be significantly improved. However, this function is not necessarily differentiable, which hinders the model from applying the BLEU score directly to the training process.

Aiming at the above problems, reinforcement learning (RL) [7] method has been applied to target optimization at the sequence level. This method directly optimises the BLEU value through Reinforcement training, which is a successful attempt of RL in strategy optimization at the sequence level. BR-CSGAN [13] is the first application of adversarial learning training to machine translation. The idea of reinforcement learning is cited in the generative adversarial network, which enables the neural machine translation model to obtain the BLEU value corresponding to each possible sequence by the bootstrap method, and use this as the basis for optimizing the model parameters. In order to stabilize the training, the loss gradient of the reward is obtained by the reinforcement learning strategy through backpropagation, and finally the translation effect is steadily improved, it uses Monte Carlo for sampling and the sampling efficiency is improved to a certain extent. But due to the independent scalability of the Monte Carlo method, the sampled words are relatively independent of each other and the estimated state value does not depend on the previous state value, thereby reducing the learning accuracy.

In a study [12], it is the first time that reinforcement learning methods were applied to machine translation, and reward shaping is used to improve the effectiveness of monolingual data in translation quality. Nevertheless, the training strategy using reinforcement learning is still restricted by the setting of rewards and the efficiency of sample sampling. As mentioned above, the traditional Monte Carlo method has independent malleability. Compared with the Monte Carlo method, the state value of each iteration of the agent training is determined by the previous state value and it is difficult to achieve the bias-variance trade-off. In order to achieve the trade-off, we hope that the model has the characteristics of low bias and low variance. However, the most serious challenge in the field of deep learning is the problem of high variance. The general solutions to solve the high variance problems are as follows: reduce the complexity of the model, reduce dimensionality of the data and denoise, increase the number of samples, use validation sets and regularization methods. Because of the scarcity of low-resource corpus data, our method involves the use of validation sets and regularization methods to reduce variance. That is, use the validation set to observe the generalization of the training parameters and update the new state through the latest K-step state difference. The Dynamic Programming method further solves the problem of correlation between words, but each updated action reward needs to refer to the previous round of experience, so all the learned experience will be stored. Based on the above analysis, we combined with the Dynamic Programming to propose an asynchronous training strategy suitable for the training of low-resource neural machine translation model. The bias-variance trade-off is achieved by selectively sampling the experience each time through the experience priority.

In this paper, we used a reinforcement learning strategy combined with Asynchronous Dynamic Programming (ADP) to train machine translation models. The Priortised Sweeping (PS) algorithm is used to determine the experience priority. The higher the priority, the higher the priority. Ultimately, we maximize the priority to generate rewards to update the strategy and use the ADP method for sampling. Compared with the Monte Carlo method, we can pay more attention to the estimation Q of non-terminal state. Compared with the general Dynamic Programming method, the K-step update method avoid the calculation disaster and the real-time supervision of model training can also avoid the accumulation of errors. It improves the accuracy rate and effectively alleviates the sparse reward problem and has a higher universal computing capability.

We map the reinforcement learning algorithm to machine translation according to the Markov characteristics of reinforcement learning. Similar to the traditional machine translation model, we input the source language sentence into the translation model to generate the target sentence. Here we use the LSTM [3] and Transformer [10] frameworks of the machine translation model based on soft attention. The second chapter of this article is about the introduction of background knowledge. The third chapter is divided into framework, asynchronous strategy and training. The fourth chapter is the experiment and analysis part, and the fifth chapter is the conclusion.

2   Background

2.1   Neural Machine Translation

The typical Neural machine translation (NMT) [11] model is an encoder-decoder framework based on attention mechanism and is usually using a cross-entropy function for training. The encoder first maps the source sentence to $X = (x_1, ..., x_n)$ vector form and then the decoder converts the set of vectors to generate the corresponding target sentence $Y = (y_1, ..., y_n)$. And mark the words of the target statement one by one. In the time step t, the model maximizes the likelihood of the current time step t by using the source statement and the target statement before the time step t as input in the training stage. The goal of the translation model is to maximize the maximum likelihood function, MLE is usually used to optimize the model:

$$L_{MLE} = \sum_{i=1}^{S} logp(\hat{y}^i|x^i) = \sum_{i=1}^{S}\sum_{t=1}^{m} logp(\hat{y}_t^i|\hat{y}_1^i...\hat{y}_{t-1}^i, x^i) \tag{1}$$

where S represents the number of sentences in the training set, m is the length of sentence $\hat{y}^i$.

Among all the frameworks based on the encoder-decoder model, the recently popular Transformer architecture has achieved the best translation quality so far. The main difference between Transformer and the previous RNNSearch [2] or Conv2Seq [4] is that it completely relies on the self-attention mechanism [14], using its own structure to calculate the representation of the source and target sentences without recursion or convolution operations.

2.2 RL Based NMT

Reinforcement learning solves the decision-making problem in complex state space by mapping the environment state to the action process to obtain the maximum cumulative reward and provides a new idea for natural language processing tasks. It reduces the gap between NMT training and inference by directly optimizing evaluation indicators (such as BLEU) during training. Specifically, the NMT model can be regarded as an agent interacting with the environment and the parameter definition strategy is the conditional probability $p\left(y_t|x, y_{<t}\right)$. The agent will select a word from the vocabulary according to the strategy, that is, a candidate word. Once the agent generates a complete sequence, it will get a final reward. Reward is introduced to encourage the model to produce an output with a higher reward. In practice, the reward in machine translation is the BLEU value. Obtaining a higher reward means having a higher BLEU score. Here we denote the sentence-level reward as R, where $\hat{y}$ represents the prediction result of the model, y^i represents the reference translation of sentence i, $R\left(\hat{y}, y^i\right)$ represents the reward value of sentence i and by sampling y from the strategy $p\left(\hat{y}|x^i\right)$. At this time, the goal of reinforcement learning training is to maximize the expected reward:

$$L_{RL} = \sum_{i=1}^{S} R\left(\hat{y}^i, y^i\right), \hat{y}^i \sim p\left(\hat{y}|x^i\right), \forall_i \in [S]. \tag{2}$$

3   Approach

In this section, we describe in detail the asynchronous strategy of NMT model training, apply the reinforcement learning algorithm to the neural machine translation model, use the ADP strategy to train the gradient and update the network with parameterization.

3.1   Priority Acquisition of Experience

We abstract machine translation into a Markov process, where s represents the network hidden layer of a specific state of the model, a represents the model decoding prediction, P represents the state transition probability, *reward* represents the BLEU value reward that can be obtained when transferring from one state s to another state s', *strategy* π represents the probability distribution of any action that the agent may take next in a state s, it can also be

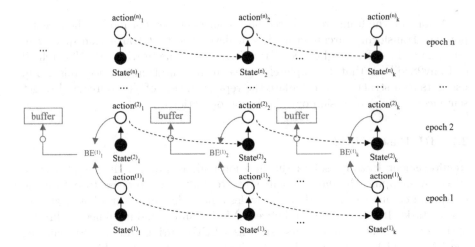

Fig. 1. The illustration of the experience priority.

said that our strategy is to directly select a certain action next in each state. In any case, the reinforcement learning method aims to learn an efficient strategy from the environment to enable agents to have certain model generalization. We can continuously improve our strategy so that we can finally get the maximum cumulative reward. In order to ensure the consistency and rationality of sampling in the training and inference stages and to avoid the problem of strictly traversing each state in Dynamic Programming. Different from the previous work, this paper uses an ADP to sampling, namely the important state priority update(PS) algorithm. The algorithm learns strategy during sampling and reinforcement training. Bellman error is used to calculate the experience priority based on the experience learned from the strategy. The experience that will be collected in the next round of training is determined by priority.

As shown in formula (3), we use the idea of Bellman Error to determine the experience priority situation and use the absolute value of the difference between the state value of the previous round and the current state value to judge. If the Bellman Error is large, it means that the state is unstable and it is necessary to give priority to updating. The experience priority is shown in Fig. 1. We use translation prediction probability *action* instead of state value to judge and record the Bellman Error of each *action* in the first round of epoch and the second round of epoch as $BE^{(1)}_{1...k}$, and so on. After that, $BE^{(1)}_{1...k}$ is stored in the buffer for the optimized call of the asynchronous strategy later. Compared with the sumtree method to extract data priority, the difference data obtained by our method is more dependent on the information content of the previous stage and has more correlation between contexts.

$$\text{Bellman Error} = |\max_{a \in A}(R_s^a + \gamma \sum_{s' \in S} P_{ss'}^a v(s') - v(s))| \qquad (3)$$

We store the sorted experience priorities in a specific experience pool, so that the subsequent asynchronous training strategy can extract experience from the experience pool in turn to optimize the model. This method is called experience playback technology. Experience playback is to store the experience of past agents in an experience playback pool and then repeatedly sample from the experience playback pool to optimize the strategy. By using past experience many times to train the current strategy, in order to improve the sample efficiency and training stability of the reinforcement learning algorithm.

3.2   Asynchronous Strategy

The action space of machine translation tasks based on reinforcement learning is huge and discrete and its size is the whole word list. In the previous algorithm, we will completely store and update all the learned experiences in each iteration, thus requiring more program resources. Therefore, we use the important state priority update algorithm in ADP, and trains the machine translation model through the reinforcement learning method. The experience determined by Bellman Error is stored in the experience pool, and the experience *value* with large error is extracted from the experience pool through the K-step update method for optimization. K-step to select the difference of all states accumulated and $R\left(\sum_k\right)$ as the final reward. According to the reward guidance next state $State_{1,...,k}^{(2)}$ update prediction and update until repeated Bellman Error tends to stabilize. This method not only improves the value of the experience state, but also avoids the problem of high computing resource consumption inherent in traditional Dynamic Programming. As shown in Fig. 2. We verified the effectiveness of ADP in the experimental part.

We know that establishing an appropriate reward mechanism is crucial to obtaining high-quality translation. In the process of machine translation, the NMT system acts as an agent of reinforcement learning, obtains the status information of the current moment through continuous interaction with the environment. After each action prediction probability step, the translation model uses the difference between the translation prediction probability of the previous round and the current time as the empirical priority evaluation standard in the current environment. In the K-step, the optimization is updated to the model convergence by selecting the priority to maximize.

Compared with the general Dynamic Programming method, our method uses ADP method for sampling and uses the K-step update method to select the priority to maximize the update to avoid the consumption of computing resources. Moreover, the real-time supervision of model training also avoids error accumulation, improves the accuracy rate while effectively alleviating the problem of reward sparseness and has higher ubiquitous computing capabilities. Drawing on Bellman Error idea, the sum of rewards for all updates in the K-step is used as the final feedback. The formula is as follows:

$$R\left(\hat{y}_k\right) = \sum_{k=1}^{K} Bellman_{1...k}value \tag{4}$$

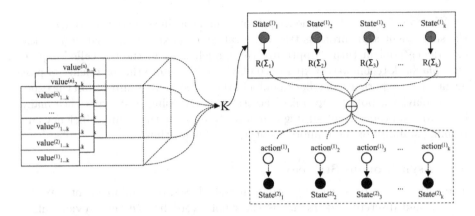

Fig. 2. The illustration of the Asynchronous Dynamic Programming.

3.3   Training

In this paper, ADP strategy is used as a sampling strategy for low-resource reinforcement learning to improve the translation model. The important state with
higher experience priority is iteratively updated within K-step, the operation is
selected by using the current strategy and the reward is determined by comparing the value of the last moment with the current moment. In the training
process, the best sequence comes from the distribution of standard translations.
The goal of the training is to find the model parameters that maximize the
expected reward, so the objective function is set to:

$$L_K = \sum_{i=1}^{S} R\left(\hat{y}_k^i, y^i\right), \hat{y}_k^i \sim p\left(\hat{y}|x^i\right), \forall_i \in [S]. \quad (5)$$

Finally, in order to further stabilize the RL training process and alleviate the
large variance caused by reinforcement learning, we combine the MLE training
target with the RL target. The cross-entropy loss function of traditional machine
translation is retained in the loss function and it is linearly combined with the
reinforcement learning training target. The loss function after mixing is:

$$L_{COM} = \lambda \times L_{MLE} + (1 - \lambda) \times L_K \quad (6)$$

where λ is the hyperparameter to control the balance between MLE and RL.
L_{COM} is the strategy to stabilize RL training progress.

4   Experiment and Analysis

We verified the effectiveness of our method on two low-resource NMT tasks,
including Mongolian-Chinese (Mo-Zh) and Uyghur-Chinese (Ug-Zh).

4.1 Datasets and Preprocessing

For the Mongolian-Chinese (Mo-Zh) experiment, the data resources used to train the translation model and the translation quality evaluation model come from the Mongolian Intelligent Information Processing Key Laboratory of the Inner Mongolia Autonomous Region. Nearly 250,000 bilingual parallel corpora provided by CCMT2019 are used to train translation models. And 1,000 parallel corpora in CWMT2017 are used as the test set and 1,000 as the verification set. For Uyghur-Chinese (Ug-Zh), 330,000 Uyghur-Chinese machine translation data published by CWMT2017 are used for experiments and 1,000 pieces of data were selected as the test set and the verification set. To avoid the model allocating too much training time to long sentences, all sentence pairs that are longer than 50 at the source or target end will be discarded.

The Stanford Word Segmentation Tool[1] is used to segment the words in Chinese. Because Mongolian and Uyghur are low-resource languages and too many low-frequency words in the vocabulary. So Byte Pair Encoder(BPE) [8] is used to process to alleviate the data sparse problem of low-resource languages to a certain extent in this paper.

4.2 Setting

Our model is improved based on the self-attention Transformer[2] model, use the traditional LSTM[3] and the latest Transformer model as the baseline model. For LSTM, we follow the basic settings and embed the word with a dimension of 512, the learning rate to 0.1, and the beam search size to 4. The settings for Transformer are exactly the same as the original paper, dropout = 0.1, word embedding dimension is 512, header count is 8, encoder and decoder both have 6-layer stack. The quality evaluation is based on the sentence level evaluation standard BLEU. The gradient optimization algorithm uses Adam [5]. During reinforcement learning training, MLE model is used to initialize parameter. The learning rate is set to 0.0001 and the beam search width is set to 6.

4.3 Main Results and Analysis

Comparison Results of Benchmark Experiments. In order to verify the translation performance of the model, we conduct low-resource machine translation comparison experiments with several common representative neural machine translation models under the same hardware conditions and corpus scale. Calculating the BLEU value of

Table 1. The BLEU scores of different NMT systems on Mongolian-Chinese and Uyghur-Chinese.

Model	Mo-Zh	Ug-Zh
Transformer	27.58	36.83
Transformer+RL	31.37	39.71
Our Method	**34.52**	**44.35**
LSTM	25.28	35.67
LSTM+RL	30.95	38.00
Our Method	**33.27**	**42.97**

---

[1] https://nlp.stanford.edu/software/segmenter.html.
[2] https://github.com/tensorflow/tensor2tensor.
[3] https://github.com/xwgeng/RNNSearch.

Table 2. Asynchronous strategy performance verification.

Priority	Strategy	BLEU		Latency
		Mo-Zh	Ug-Zh	
Random	MC	-	-	-
	DP	-	-	-
	ADP	26.18	33.82	15.7 ms
KD-tree	MC	-	-	-
	DP	-	-	-
	ADP	33.47	42.05	21.3 ms
Bellman	MC	34.36	43.86	382 ms
	DP	34.21	43.75	401 ms
	ADP	**34.52**	**44.35**	**20.5 ms**

each model on the test set separately. It can be seen from the table that our method can surpass the baseline model in both Mongolian-Chinese and Uyghur-Chinese translation tasks. As shown in Table 1.

As can be seen from the translation results, the baseline method of Mongolian-Chinese Machine Translation Transformer is 27.58, Transformer+RL[4] translation method is 31.37, up 3.79. The LSTM baseline method is 25.28, the LSTM+RL[5] is 30.95, and the addition of RL is also 5.67 higher than the baseline, indicating that the introduction of reinforcement learning plays an important role in improving the accuracy of machine translation. Our method is 3.15 BLEU and 2.32 BLEU higher than Transformer+RL and LSTM+RL, respectively. Compared with general reinforcement learning method, our method alleviates the problem that its training strategy is restricted by reward setting and sample sampling efficiency. We avoid computational disasters through the asynchronous update method and the real-time supervision of model training also avoids the accumulation of errors, which can effectively alleviate the problem of sparse rewards while improving the accuracy. The same conclusion applies to Uyghur-Chinese translation.

Asynchronous Strategy Performance Verification. We compare the three priority judgment methods (Random, KD-tree and Bellman error) through different training strategies to verify the effectiveness of the asynchronous strategy and use the latency of each round as the performance verification basis. It can be seen from Table 2 that the priority translation effect of Bellman Error is optimal. Although the translation performance of Monte Carlo and Dynamic Programming strategies are similar to that of ADP, according to the comparison of the duration of each round, it can be found that ADP is much less time-consuming than other strategies.

---

[4] https://github.com/apeterswu/RL4NMT.
[5] https://github.com/facebookarchive/MIXER.

Due to the nature of asynchronous update based on experience priority, there is no need to wait for the update after each round of traversal is completed. Thereby, reducing the consumption of computing resources, which proves the effectiveness of our method.

Translation Analysis Example. In this paper, Mongolian-Chinese and Uygur-Chinese machine translation examples are explained and the Transformer + RL translation model with better translation performance is used as a comparative example.

From the Fig. 3, we can see that our method is closer to the reference translation than the traditional Transformer+RL translation model. And the translation obtained on the two low-resource corpora of Mongolian-Chinese and Uyghur-Chinese is relatively accurate. The fluency and fidelity of the translation conform to the target language specification, and the quality of the translation is significantly better than its baseline model, which proves that our method can

	Mo-Zh
Source	ᠪᠢ ᠲᠡᠭᠦᠨ ᠤ ᠭᠡᠷ ᠤᠨ ᠬᠠᠶᠢᠭ ᠢ ᠮᠡᠳᠡᠪᠡᠯ ᠪᠢ ᠲᠡᠭᠦᠨ ᠳᠤ ᠵᠠᠬᠢᠳᠠᠯ ᠪᠢᠴᠢᠨᠡ᠃
Reference	如果我知道他的住址，我会写信给他。
Transformer+RL	我知道住址，写信给他。
Ours	**如果**我知道他的住址，我**就**会写信给他。
	Ug-Zh
Source	‫. چىدىمىغۇدەكمەن ، نېغىركەن بەك بىسمى خىزمەت‬
Reference	工作压力太大，我真受不了了。
Transformer+RL	压力大，我受不了了。
Ours	工作压力太大，令**我难以接受**。

Fig. 3. The example of translation results.

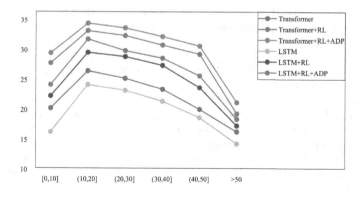

Fig. 4. The length of source sentence.

effectively improve the performance of the low-resource neural machine translation model.

Results of Sentence Length Analysis. We conduct a set of typical experiments using Transformer on the Mongolian-Chinese task to verify the performance of this method on long sentences. We divided the development set data and test set data of the Mongolian-Chinese task according to the sentence length. Figure 4 shows the BLEU scores for different sentence lengths. No matter on LSTM, Transformer or compared to joining RL with the best baseline performance, our work have outstanding behaviors continuously. It is due to our method trains machine translation models based on ADP strategies and increases the ability to analyze low-resource sentence structures through parameter optimization, which makes our method perform better on both long and short sentences.

5   Conclusion

Since different reinforcement learning sampling methods affect the performance of the model and in order to alleviate the quality problem of generated translations in low-resource machine translation tasks, this paper proposes an Asynchronous Dynamic sampling strategy based on reinforcement learning. This strategy selectively samples and stores each time according to the experience priority situation, thereby reducing the consumption of computing resources and improving the value of the experience state. Experimental results show that this method can effectively improve the performance of low-resource machine translation. In the next step, we will combine more low-resource language features to explore more suitable for low-resource machine translation, so as to further improve the performance of low-resource machine translation systems.

References

1. Ahmadnia, B., Dorr, B.J.: Enhancing phrase-based statistical machine translation by learning phrase representations using long short-term memory network. In: Mitkov, R., Angelova, G. (eds.) Proceedings of the International Conference on Recent Advances in Natural Language Processing, RANLP 2019, Varna, Bulgaria, 2–4 September, 2019, pp. 25–32. INCOMA Ltd. (2019). https://doi.org/10.26615/978-954-452-056-4_004
2. Bahdanau, D., Cho, K., Bengio, Y.: Neural machine translation by jointly learning to align and translate. In: Bengio, Y., LeCun, Y. (eds.) 3rd International Conference on Learning Representations, ICLR 2015, San Diego, CA, USA, 7–9 May, 2015, Conference Track Proceedings (2015). http://arxiv.org/abs/1409.0473
3. Cui, Y., Wang, S., Li, J.: LSTM neural reordering feature for statistical machine translation. In: Knight, K., Nenkova, A., Rambow, O. (eds.) NAACL HLT 2016, The 2016 Conference of the North American Chapter of the Association for Computational Linguistics: Human Language Technologies, San Diego California, USA, 12–17 June, 2016. pp. 977–982. The Association for Computational Linguistics (2016). https://doi.org/10.18653/v1/n16-1112

4. Gehring, J., Auli, M., Grangier, D., Yarats, D., Dauphin, Y.N.: Convolutional sequence to sequence learning. In: Precup, D., Teh, Y.W. (eds.) Proceedings of the 34th International Conference on Machine Learning, ICML 2017, Sydney, NSW, Australia, 6–11 August 2017. Proceedings of Machine Learning Research, vol. 70, pp. 1243–1252. PMLR (2017). http://proceedings.mlr.press/v70/gehring17a.html

5. Kingma, D.P., Ba, J.: Adam: a method for stochastic optimization. In: Bengio, Y., LeCun, Y. (eds.) 3rd International Conference on Learning Representations, ICLR 2015, San Diego, CA, USA, 7–9 May, 2015, Conference Track Proceedings (2015). http://arxiv.org/abs/1412.6980

6. Papineni, K., Roukos, S., Ward, T., Zhu, W.: BLEU: a method for automatic evaluation of machine translation. In: Proceedings of the 40th Annual Meeting of the Association for Computational Linguistics, 6–12 July, 2002, Philadelphia, PA, USA, pp. 311–318. ACL (2002). https://doi.org/10.3115/1073083.1073135, https://www.aclweb.org/anthology/P02-1040/

7. Ranzato, M., Chopra, S., Auli, M., Zaremba, W.: Sequence level training with recurrent neural networks. In: Bengio, Y., LeCun, Y. (eds.) 4th International Conference on Learning Representations, ICLR 2016, San Juan, Puerto Rico, 2–4 May, 2016, Conference Track Proceedings (2016). http://arxiv.org/abs/1511.06732

8. Sennrich, R., Haddow, B., Birch, A.: Neural machine translation of rare words with subword units. In: Proceedings of the 54th Annual Meeting of the Association for Computational Linguistics, ACL 2016, 7–12 August, 2016, Berlin, Germany, Volume 1: Long Papers. The Association for Computer Linguistics (2016). https://doi.org/10.18653/v1/p16-1162

9. Shen, S., et al.: Minimum risk training for neural machine translation. In: Proceedings of the 54th Annual Meeting of the Association for Computational Linguistics, ACL 2016, 7–12 August, 2016, Berlin, Germany, Volume 1: Long Papers. The Association for Computer Linguistics (2016). https://doi.org/10.18653/v1/p16-1159

10. Vaswani, A., et al.: Attention is all you need. In: Guyon, I., et al. (eds.) Advances in Neural Information Processing Systems 30: Annual Conference on Neural Information Processing Systems 2017, 4–9 December, 2017, Long Beach, CA, USA, pp. 5998–6008 (2017). https://proceedings.neurips.cc/paper/2017/hash/3f5ee243547dee91fbd053c1c4a845aa-Abstract.html

11. Wu, L., et al.: Beyond error propagation in neural machine translation: Characteristics of language also matter. In: Proceedings of the 2018 Conference on Empirical Methods in Natural Language Processing, Brussels, Belgium, 31 October – 4 November, 2018, pp. 3602–3611 (2018). https://www.aclweb.org/anthology/D18-1396/

12. Wu, L., Tian, F., Qin, T., Lai, J., Liu, T.: A study of reinforcement learning for neural machine translation. In: Riloff, E., Chiang, D., Hockenmaier, J., Tsujii, J. (eds.) Proceedings of the 2018 Conference on Empirical Methods in Natural Language Processing, Brussels, Belgium, 31 October – 4 November, 2018, pp. 3612–3621. Association for Computational Linguistics (2018). https://doi.org/10.18653/v1/d18-1397

13. Yang, Z., Chen, W., Wang, F., Xu, B.: Improving neural machine translation with conditional sequence generative adversarial nets. In: Walker, M.A., Ji, H., Stent, A. (eds.) Proceedings of the 2018 Conference of the North American Chapter of the Association for Computational Linguistics: Human Language Technologies, NAACL-HLT 2018, New Orleans, Louisiana, USA, 1–6 June, 2018, Volume 1 (Long Papers), pp. 1346–1355. Association for Computational Linguistics (2018). https://doi.org/10.18653/v1/n18-1122

14. Yun, H., Hwang, Y., Jung, K.: Improving context-aware neural machine translation using self-attentive sentence embedding. In: The Thirty-Fourth AAAI Conference on Artificial Intelligence, AAAI 2020, The Thirty-Second Innovative Applications of Artificial Intelligence Conference, IAAI 2020, The Tenth AAAI Symposium on Educational Advances in Artificial Intelligence, EAAI 2020, New York, NY, USA, 7–12 February, 2020, pp. 9498–9506. AAAI Press (2020). https://aaai.org/ojs/index.php/AAAI/article/view/6494

Minority Language Information Processing

Uyghur Metaphor Detection via Considering Emotional Consistency

Qimeng Yang, Long Yu$^{(\boxtimes)}$, Shengwei Tian, and Jinmiao Song

Xinjiang University, Urumqi 830046, China
`yul@xju.edu.cn`

Abstract. Metaphor detection plays an important role in tasks such as machine translation and human-machine dialogue. As more users express their opinions on products or other topics on social media through metaphorical expressions, this task is particularly especially topical. Most of the research in this field focuses on English, and there are few studies on minority languages that lack language resources and tools. Moreover, metaphorical expressions have different meanings in different language environments. We therefore established a deep neural network (DNN) framework for Uyghur metaphor detection tasks. The proposed method can focus on the multi-level semantic information of the text from word embedding, part of speech and location, which makes the feature representation more complete. We also use the emotional information of words to learn the emotional consistency features of metaphorical words and their context. A qualitative analysis further confirms the need for broader emotional information in metaphor detection. Our results indicate the performance of Uyghur metaphor detection can be effectively improved with the help of multi-attention and emotional information.

Keywords: Metaphor detection · Uyghur · Deep learning

1 Introduction

With the rapid development of digital technology and the Internet, social media has become a powerful platform where people can express their opinions on various topics such as politics, finance, education, and other social issues. It is worth noting that more users use a lot of metaphors in online texts to express their thoughts and emotions. According to statistical research, metaphors appear in every three sentences in natural language [1–3]. Metaphor involves not only language expression, but also the cognitive process of conceptual knowledge [4]. According to Lakoff and Johnson, metaphor is a conceptual mapping. More specifically, metaphor is a concept used to describe another concept. They are widely used in oral and written language to convey rich linguistic and emotional information. For instance, in the metaphorical utterance: "knowledge is treasure.," we use "treasure" to describe "knowledge" to emphasize that "knowledge" can be valuable. To take another metaphorical instance as an example:

© Springer Nature Switzerland AG 2021
S. Li et al. (Eds.): CCL 2021, LNAI 12869, pp. 31–44, 2021.
https://doi.org/10.1007/978-3-030-84186-7_3

"this is an ocean of flowers." "Ocean" has broad characteristics, which means that the flower area is large. Metaphor detection is an important subtask of natural language processing, which provides a more complete representation for semantic analysis [4]. Moreover, interpreting metaphors helps to improve the performance of tasks such as machine translation and human-machine dialogue analysis [5].

The existing computational models of metaphor detection are mainly based on lexicons [6,7] and supervision methods [8–10]. Lexicon-based methods do not require data annotation, but they cannot detect new metaphoric usage and capture contextual information. The supervised method can obtain text context information from the sentence level, thereby obtaining broader semantic information. However, it requires a complete annotated corpus.

Current metaphor detection mostly focuses on English because of its abundant annotation data. The popular metaphor detection corpus in current research includes VU Amsterdam Metaphor Corpus (VUA) [11], MOH-X [12] and TroFi [13]. In addition, the task of metaphor detection relies on specific semantic resources, and corpora of other languages are gradually established, such as Chinese Metaphor Corpus [14], Russian Metaphor Corpus [15], and Arabic Metaphor Corpus [16]. The establishment of these corpora provides a data foundation for metaphor detection tasks in specific languages.

In recent years, more Uyghur-speaking users have expressed their thoughts and opinions on political, economic, and cultural topics through Internet channels. As more people like to express their opinions in metaphorical language, Uyghur metaphor data is also growing rapidly. The expression and meaning of metaphor differ greatly in different language environments. Due to the influence of the special language environment, the task of metaphor detection in Uyghur language has not yet started. Therefore, it is imperative to establish Uyghur metaphor detection resources and verify the effectiveness of metaphor detection models.

In this paper, we have collected Uyghur metaphor data from multiple fields and proposed to use multi-attention mechanism for metaphor detection. More specifically, we use word embedding, part-of-speech (POS) and position as model inputs, which reveal the semantic information of the text from three levels to make the feature expression more complete. In addition, due to the important relationship between metaphor and emotional expression, we extract emotional features from both the word level and sentence level, which enables the model to learn emotional consistency information. The experimental results also verify that metaphor detection requires extensive emotional information. In this task, we treat the Uyghur metaphor detection as a sequence tagging problem, and sequence tagging tasks such as POS tagging and named entity recognition (NER) have always existed in natural language processing.

To summarize, this paper makes the following contributions:

- We are the first to study and apply deep neural networks to the task of Uyghur metaphor detection.

- The proposed multi-attention model leverages POS and position to reveal semantic information from multiple aspects, which makes the feature expression more complete.
- Emotional embedding allows the model to learn emotional consistency information from the level of words and its context.

The rest of this paper is organized as follows. The next section introduces related work. Section 3 describes our proposed model for Uyghur metaphor detection. Section 4 presents our experiments, including the dataset, experimental details, result analysis. The Sect. 5 is about the conclusion.

2   Related Work

2.1   Metaphor Detection

Current metaphor detection models include supervised machine learning methods combined with hand-designed features, unsupervised representation learning methods, and deep learning models based on sequence tagging. Shutova et al. [17] proposed a method of metaphor detection and recognition through interpretation. The performance of this method is better than previous metaphor detection methods, and it has important significance for metaphor modeling. Dunn et al. [18] proposed a new language independent ensemble-based approach to identifying linguistic metaphors in natural language text. This strategy achieved state-of-the-art results in multiple languages and made significant improvements to existing methods. Tsvetkov et al. [19] proposed a general semantic feature method for metaphor detection. This method verifies that metaphors can be detected at the conceptual level. Specifically, metaphors are part of the universal conceptual system.

Some metaphor detection methods utilize unsupervised learning. Shutova et al. [20] proposed the first metaphor detection method that simultaneously extracts knowledge from language and visual data. Experimental results show that it performs better than language and visual models in isolation, and it is also better than the best-performing metaphor detection method. Mao et al. [21] proposed a method that can recognize and interpret metaphors at the word level without any preprocessing. This model is extended to explain the recognized metaphors and translate them into corresponding literal meanings so that the machine can better translate them.

Recently, with the application of deep learning in natural language processing, a wide variety of techniques for deep learning models for metaphor detection have been proposed. Do et al.[22] proposed a method combining word embedding and neural network for metaphor detection. This method shows that only relying on word embeddings trained on a large corpus can achieve better classification and eliminate the need for additional resources. Swarnkar et al.[23] proposed a metaphor detection using contrasting deep neural structures. The model uses contrast features generated by pre-trained word embeddings to achieve considerable performance. They also verified that using additional features and adjusting the weight of examples can significantly improve performance. Pramanick et al. [24] used a hybrid model of Bi-LSTM and CRF, which uses word2vec to embed tag words and their lemmas. In addition, they used a 20d vector to indicate POS and a heat vector to indicate whether the lemma is the same as the mark, and whether there is a lemma in the mark.

More recently, the metaphor detection task is modeled as a sequence tagging task. More specifically, the word is predicted as a literal or metaphor at each time step. Wu et al. [25] proposed to use the CNN-LSTM model to complete the task of metaphor detection. This model combines two layers of CNN and LSTM, and uses local and remote context information to identify metaphorical information. In addition, they also compared the performance of softmax classifier and conditional random field (CRF) in metaphor detection tasks. Mao et al. [26] proposed a method based on MIP [27] and SVP [28,29] linguistic theories, and surpassed existing models to obtain the best classification performance on datasets VUA, MOH, and TroFi.

2.2 Uyghur Metaphor

Uyghur is a kind of agglutinative language with rich forms, which has special configuration morphemes to express various syntactic relations between words. Uyghur metaphors are mainly affected by the language environment and the grid grammar. More specifically, nouns have three important grammatical categories: number, person, and grid. The grid category includes nominative, accusative, subordinate, directional, time, subordinate, range, boundary, and similar. Adjectives have prototype, comparative and superlative forms. Verbs have grammatical categories such as voice, form, tense, and aspect. Metaphorical expressions in Uyghur use rhetorical methods extensively, and are closely related to the grid grammar. For example, metaphorical demonstratives in Uyghur are often reflected in the grammatical phenomenon of grid. More specifically, it is to add special additional elements (تەك/دەك or خۇددى, etc.) after the vocabulary of the vehicle. For example sentence : Saphia is like a flower سافىيە گۈلگە ئوخشايدۇ. These fixed grammatical formats help to judge the metaphor of the text from the syntactic level. We summarized the Uyghur metaphorical language rules are as follows.

- **Type 1:** unmarked metaphorical expression
  Description: The source and the target vocabulary are directly connected.
  Language rules: (1) pearl+teeth=white teeth; (2) add elements after adjectives. That is to use source+target to represent the target to modify the source.
- **Type 2:** marked metaphorical expression
  Description: Express the similarity between the target and the source through additional components.
  Language rules: For example, target+سىممە+source. چە،چىللىك means similar numbers. چىللىپ means similar behavior.

Moreover, Uyghur metaphors can be divided into three types, namely anal-ogy, metonymy, and simile. The language rules for each type of metaphor are as follows.

- **Type 1:** analogy
  Description: Express the current things through the characteristics of other things.
  Language rules: Generally divided into anthropomorphic and simulant (plants, animals, natural objects, etc.). For example, شىپاڭلاتماق "a dog wags its tail" to express pitiful prayers.
- **Type 2:** metonymy
  Description: Use target instead of source function.
  Language rules: There are many kinds of Uyghur metonymy, including humans, animal organs, plants, and celestial bodies. For example, سايىماخۇن "bedside table" expresses fear of his wife.
- **Type 3:** simile
  Description: Use the symbolic meaning of things to express metaphors.
  Language rules: Uyghur simile has strong language characteristics. For example, use قوز lamb to mean a good child.

Moreover, Uyghur metaphors have strong language characteristics affected by the language environment. For instance, the "cat" in animals means "greedy" instead of "docile and cute". "Rabbit" stands for "cowardly and pitiful" rather than "tamed and agile." To take another example: The "pumpkin" and "gourd" in plants are metaphors for "fool." "soaked tea" means "failure", etc. Uyghur metaphorical emotional expression is closely related to language environment. For example, we usually use monkeys to express cleverness, but in Uyghur lan-guage it is a symbol of cunning. Based on all the above, we can conclude that metaphor detection tasks in different languages have huge differences in meth-ods. Therefore, it is imperative to establish a framework for Uyghur metaphor detection.

3   Our Proposed Model

In this section, we first introduce the research ideas of the paper. Then, in order to describe the flow of the method in detail, we will explain each function of the model from bottom to top.

3.1   Basic Idea

In this paper, we regard Uyghur metaphor detection as a sequence tagging task. Considering the characteristics of metaphor expression containing rich emotions, we model emotional information into the network to consider its impact on metaphor detection performance. In order to further enrich the feature expres-sion ability, we propose to use POS and position to construct multi-attention representations of words.

3.2 Model Structure

Word embedding and emotional embedding are used as model input, and BiL-STM is utilized to generate semantic representation. Moreover, we propose multi-attention to model the interaction between words and context. Our model structure is shown in Fig. 1.

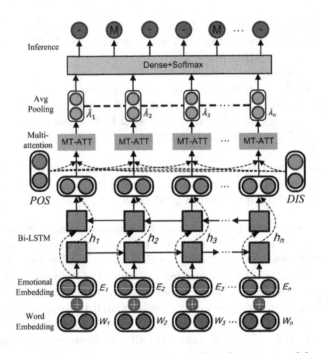

Fig. 1. The architecture of the metaphor detection model.

Emotional Embedding Layer: In order to obtain the emotional embedding, we construct the emotional dictionary from the word level. The steps to construct an emotional embedding are as follows.

First, we extract the word w_i in the sentence and annotate its emotional information as emo_i. In order to reduce the complexity of the task, we use three emotional polarities (positive, negative, neutral) to annotate each word.

Second, we represent each emotional information as $[w_i, emo_i]$ and map it to a multi-dimensional continuous value vector $V_{emo}=[w_i, emo_i] \in \mathbf{R}^L$, where L is the dimension of the emotional information vector, and V_{emo} is the emotional embedding of the w_i.

In order to accurately represent each word, we use GloVe (g) and ELMo (e) respectively as the common representation of word vectors, and combine emotional embedding as the input of the model. The final model input can be obtained by the following equation.

$$V_{input} = V_{emo} \oplus [V_g, V_e] \tag{1}$$

where V_{emo} represents emotional embedding, $[V_g, V_e]$ represents word embedding matrix, and $\oplus$ is splicing operation.

BiLSTM Layer: BiLSTM generates a hidden state based on the current input and the previous state. Therefore, entering the current word will affect the following words. We use BiLSTM on the input V_{input} to summarize the both direction information of the words and generate the hidden vector sequence $[h_1, h_2, h_3 ... h_t ... h_n]$, thereby obtaining the representation of the sentence. h_t can be expressed as follows.

$$h_t = f_{BiLSTM}(V_{emo}, V_g, V_e), \overrightarrow{h}_{t-1}, \overleftarrow{h}_{t+1} \tag{2}$$

Multi-attention Layer: A single word embedding cannot reveal enough word semantic information. For example, incorrect annotations in the dictionary or words that do not exist in the dictionary will affect the performance of the metaphor detection model. In order to reduce the impact of this problem, we propose to use multi-attention to focus on two aspects of semantic information from part of speech and position.

Since metaphors usually have specific POS tags, it is easier to identify metaphors by adding POS information. In addition, POS tags can ideally solve the impact of out-of-vocabulary words on the performance of the experiment. We therefore propose POS attention. Specifically, we combine the word w_i and the POS tag POS_i to generate a POS representation of the word. Then combine w_i and POS_i into $WPOS_i$ in a splicing manner, and map $WPOS_i$ to a multidimensional vector V_{WPOS_i}. Finally, a POS representation dictionary $POSDic = V_{WPOS_1}, V_{WPOS_2} ... V_{WPOS_L}$ of length L is generated. For a single sentence, we find the POS representation of each word in the dictionary as input for attention. The calculation method is shown in the following equation.

$$a_i = innerproduct(V_{WPOS_i}, POSDic) \tag{3}$$

$$a_i^c = \frac{exp(a_i)}{\sum_{j=1}^n exp(a_j)} \tag{4}$$

In addition, a_i^c can also be obtained by the following equation.

$$a_i^c = \beta \times \frac{exp(a_i)}{\sum_{j=1}^n exp(a_j)} \tag{5}$$

where β is an adjustable parameter to control the importance of different POS vectors.

The position between each word hides important semantic information. We generally believe that position is closely related to the semantic connection of words. In order to indicate the relative position of each word, we use the matrix L to record the absolute distance between the current word and other words in the sentence. Then calculate the value of α.

$$\alpha = 1 - \frac{L_i + 1}{n + 1} \tag{6}$$

Then, map the distance value in the matrix L to a multi-dimensional vector, namely $L_i \in R^k$, and then calculate the input matrix.

$$input_i^L = \frac{L_i + x_i}{2} \tag{7}$$

where x_i and input are the word vector and position attention input of the $i - th$ word. Finally, we combine POS attention and position attention as a multi-attention representation.

Avg Pooling Layer: In order to reduce parameters, prevent overfitting, and improve the generalization ability of the model, we use average pooling after multi-attention. The calculation method is as follows.

$$A = [c_1, c_2, c_3...c_n] \tag{8}$$

$$A = [\tilde{c}_1, \tilde{c}_2, \tilde{c}_3...\tilde{c}_n,] \tag{9}$$

where A represents the output of the multi-attention, and $\tilde{A}$ represents the output of the average pooling.

Inference Layer: Our inference layer consists of a dense and softmax. Each sentence passes through the dense layer, and finally uses $softmax$ to predict the probability distribution.

$$A = \widehat{y}_i = softmax(W \cdot dense_i + b) \tag{10}$$

$$softmax_i = \frac{exp(y_i)}{\sum_{j=1}^{N} exp(y_j)} \tag{11}$$

where $\widehat{y}_i$ is the prediction probability, W is the final optimal weight of the dense layer after model training process. $dense_i$ is an output of dense layer and b is a bias term.

We regard Uyghur metaphor detection as a sequence tagging task, and the loss function is formulated as follows.

$$L(\hat{y}, y) = -\sum_{s \in S}\sum_{i=1}^{N} w_{y_i} y_i log(\hat{y}_i) \tag{12}$$

where y_i is the ground-truth label of i_{th} word, $\hat{y}_i$ is the predicted label, and w_{y_i} is the loss weight of the metaphor label y_i.

4   Experiment

The proposed multi-attention model is applied to the task of Uyghur metaphor detection. Specifically, we reveal the richer semantic information of words from POS and position respectively, which makes the semantic expression of words and its context more complete. Besides, due to the close relationship between metaphorical expression and emotion, we propose that emotion embedding allows the model to learn emotion consistency information during the training process.

4.1 Dataset

The current researches on metaphor detection mostly focus on languages that have a complete corpus such as English. Due to the grammatical complexity and metaphor diversity of Uyghur language, metaphor detection in Uyghur has not yet been developed.

Therefore, we conducted data collection and annotation for the task of Uyghur metaphor detection. Specifically, we manually annotated 5605 Uyghur metaphorical sentences under the guidance of Uyghur language experts. For a single sentence, we annotated the metaphor of each word. The corpus is randomly divided into training set, validation set and test set. The results are shown in Table 1.

Table 1. Splitting of training set, validation set and test set, where "MW" represents metaphorical words, and "Non-MW" represents non-metaphorical words.

Sets	Samples	MW	Non-MW
Training set	4605	7951	33281
Validation set	500	806	4266
Test set	500	715	4609
Total	5605	9472	42156

Our corpus annotation members consist of 9 people, 7 of whom are familiar with Uyghur. In order to ensure the reliability of the Uyghur metaphor corpus, a sentence is annotated by multiple members. For ambiguous annotations, the final result is determined by Uyghur language experts.

4.2 Experimental Details

In our experiment, GloVe [30] and ELMo are used to initialize word vectors. The word vectors are pre-trained on Uyghur text with a size of about 78M, and the dimensions are 200 and 1024 respectively. Moreover, the uniform distribution $U(--0.1, 0.1)$ is employed to initialize all out-of-vocabulary words. The dimension of BiLSTM hidden states is 200, and batch size is 64.

4.3 Result Analysis

We set up three sets of experiments to verify the effectiveness of the multi-attention and emotional embedding. First, we add POS and position attention in turn to illustrate the impact on model performance. Then, we compared the performance of our model with and without emotional embedding. Finally, we conducted ablation experiments to verify the effectiveness of multi-feature input and multi-attention mechanisms.

Influence of Additional Features. Multi-attention can focus on semantic information from multiple levels. In order to illustrate the effectiveness of multi-attention, we assemble POS attention and position attention respectively under the same experimental setting.

Table 2. The influence of additional features.

Models	Precision	Recall	F-score
None	57.8	65.1	61.2
+POS	60.5	66.2	63.2
+Position	58.4	65.7	61.8
+POS+Position	**62.8**	**67.2**	**64.9**

Table 2 shows the performance of the model under different additional features. The experimental results show that both POS attention and position attention help to improve the performance of Uyghur metaphor detection. Especially the performance of the model with POS attention is significantly improved by 2%. Since each word in the corpus has a specific POS tag, which contains useful information for identifying metaphors. Therefore, the model can identify metaphors more easily after adding POS attention.

After adding position attention, the performance of the model is also slightly improved, which shows that the position information between words is related to metaphors. However, the improvement in model performance is not ideal. This is because the corpus contains short sentences. E.g., (using underlines for metaphors) I'm hunting. (ID: S106-4).

In all words, the model combining POS attention and position attention achieves the best performance. This shows that the multi-attention mechanism can reveal the semantic information of words from multiple aspects and promote more complete feature expression.

Influence of Emotional Embedding. Metaphors promote rich emotional expression, so metaphors are closely related to emotional information. In order to better illustrate the effectiveness of the proposed emotional embedding, we conducted experiments under different experimental settings. More specifically, we compared the performance of models with and without emotional embedding under different iterations. For each iteration, we use the same parameter settings.

Figure 2 shows the experimental results of our model with or without emotional embedding. It can be seen from the experimental results that our model achieves the best performance with emotional embedding. This is because metaphorical expressions are mostly to express richer and stronger emotions. With the help of emotional embedding, our model can learn emotional consistency information from words and its context emotional level respectively. The probability of words being classified as metaphors will increase when words

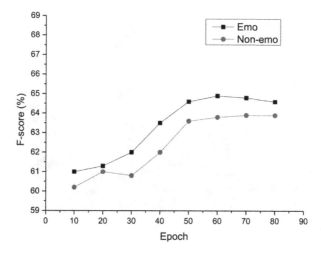

Fig. 2. Experiment results of different models, where "Emo" represents the model with emotional embedding and "Non-emo" presents the model without emotional embedding.

conflict with the emotional polarity of the context. In addition, observing the experimental results, we can see that the F-score stops increasing as the epoch is greater than 60, we therefore set the epoch to 60.

Ablation Experiment. Multi-feature input and multi-attention mechanisms are used to represent the richer semantic features of words, which play a key role in the improvement of classification performance. In order to better illustrate the effectiveness of each module in the model, we conducted a set of experiments under different module assemblies. In particular, we remove the multi-feature input (-MTinput) and multi-attention mechanisms (-MTatt) respectively. For each time, we report the performance of the model on the test set. For all these experiments, we kept the rest of the model unchanged.

It can be seen from the experimental results in Fig. 3 that the performance of metaphor detection is greatly reduced when the multi-feature input is removed. This is due to insufficient semantic expression of words, which leads to the model not being able to fully learn the semantics and context of words. This also clarifies that the Uyghur metaphor detection task is more dependent on the semantics of words and their contextual representations. The performance of the model that removes the multi-attention mechanism is also reduced. This result is because the multi-attention mechanism is a method based on multi-feature input. With the help of multi-feature input, the multi-attention mechanism enables the model to learn the semantics of words and the importance of each word from multiple levels during the training process. Therefore, multi-feature input and multi-attention mechanisms complement each other to achieve better semantic representation, thereby achieving more ideal classification performance.

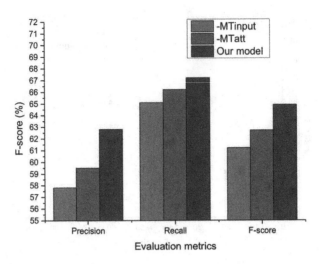

Fig. 3. Experimental results of the model under different module assembly.

5 Conclusion

This paper proposed a Uyghur metaphor detection model that combines multi-attention and emotional embedding. We used POS attention and position attention to construct multi-attention representations of words, which allows the model to reveal text semantics from multiple aspects. In addition, due to the correlation between metaphor and emotional expression, we construct an emotionally related dictionary as the source of emotional embedding. Emotional embedding is utilized to learn the emotional consistency features of words and its context. The experimental results verify the effectiveness of the model for Uyghur metaphor detection.

Acknowledgments. This work was supported by the National Natural Science Foundation of China (61962057), the Key Program of the National Natural Science Foundation of China (U2003208), and the Major Science and Technology Projects in the Autonomous Region (2020A03004-4).

References

1. Cameron, L.: Metaphor in Educational Discourse. A&C Black, London (2003)
2. Steen, G.: A Method for Linguistic Metaphor Identification: From MIP to MIPVU. John Benjamins Publishing, Amsterdam (2010)
3. Shutova, E., Teufel, S.: Metaphor Corpus Annotated for Source-Target Domain Mappings. In: LREC, pp. 2–2 (2010)
4. Lakoff, G., Johnson, M.: Metaphors We Live by. University of Chicago Press, Chicago (2008)
5. Rentoumi, V., Vouros, G.A., Karkaletsis, V., Moser, A.: Investigating metaphorical language in sentiment analysis: a sense-to-sentiment perspective. ACM Trans. Speech Lang. Proces. (TSLP) 9(3), 1–31 (2012)

6. Mohler, M., Bracewell, D., Tomlinson, M., Hinote, D.: Semantic signatures for example-based linguistic metaphor detection. In: The 1th Workshop on Metaphor in NLP, pp. 27–35 (2013)

7. Dodge, E.K., Hong, J., Stickles, E.: MetaNet: deep semantic automatic metaphor analysis. In: The 3th Workshop on Metaphor in NLP, pp. 40–49 (2015)

8. Gao, G., Choi, E., Choi, Y., Zettlemoyer, L.: Neural metaphor detection in context. In: The 2018 Conference on Empirical Methods in Natural Language Processing, pp. 607–613 (2018)

9. Stowe, K., Moeller, S., Michaelis, L., Palmer, M.: Linguistic analysis improves neural metaphor detection. In: CoNLL, pp. 362–371 (2019)

10. Le, D., Thai, M., Nguyen, T.: Multi-task learning for metaphor detection with graph convolutional neural networks and word sense disambiguation. In: AAAI, pp. 8139–8146 (2020)

11. Klebanov, B.B., Leong, C.W., Gutierrez, E.D., Shutova, E., Flor, M.: Semantic classifications for detection of verb metaphors. In: ACL, pp. 101–106 (2016)

12. Birke, J., Sarkar, A.: A clustering approach for nearly unsupervised recognition of nonliteral language. In: EACL, pp. 329–336 (2006)

13. Zhang, D., Lin, H., Yang, L., Zhang, S., Xu, B.: Construction of a Chinese corpus for the analysis of the emotionality of metaphorical expressions. In: ACL, pp. 144–150 (2018)

14. Badryzlova, Y., Lyashevskaya, O.: Metaphor shifts in constructions: the Russian metaphor corpus. In: AAAI, pp. 127–130 (2017)

15. Alkhatib, M., Shaalan, K.: Paraphrasing Arabic metaphor with neural machine translation. Procedia Comput. Sci. **142**, 308–314 (2018)

16. Shutova, E.: Metaphor identification as interpretation. In: The 2th Joint Conference on Lexical and Computational Semantics, pp. 276–285 (2013)

17. Dunn, J., Beltran de Heredia, J., Burke, M., Gandy, L., Kanareykin, S., Kapah, O., Argamon, S.: Language-independent ensemble approaches to metaphor identification. pp. 276–285 (2014)

18. Tsvetkov, Y., Mukomel, E., Gershman, A.: Cross-lingual metaphor detection using common semantic features. In: The 1th Workshop on Metaphor in NLP, pp. 45–51 (2013)

19. Shutova, E., Kiela, D., Maillard, J.: Black holes and white rabbits: metaphor identification with visual features. In: The 2016 Conference of the North American Chapter of the Association for Computational Linguistics, pp. 160–170 (2016)

20. Mao, R., Lin, C., Guerin, F.: Word embedding and wordnet based metaphor identification and interpretation. In: The 56th Annual Meeting of the Association for Computational Linguistics, pp. 1222–1231 (2018)

21. Do Dinh, E.L., Gurevych, I.: Token-level metaphor detection using neural networks. In: The 4th Workshop on Metaphor in NLP, pp. 28–33 (2016)

22. Swarnkar, K., Singh, A.K.: Di-LSTM contrast: A deep neural network for metaphor detection. In: The Workshop on Figurative Language Processing, pp. 115–120 (2018)

23. Pramanick, M., Gupta, A., Mitra, P.: An lSTM-CRF based approach to token-level metaphor detection. In: The Workshop on Figurative Language Processing, pp. 67–75 (2018)

24. Wu, C., Wu, F., Chen, Y., Wu, S., Yuan, Z., Huang, Y.: Neural metaphor detecting with CNN-LSTM model. In: The Workshop on Figurative Language Processing, pp. 110–114 (2018)

25. Mao, R., Lin, C., Guerin, F.: End-to-end sequential metaphor identification inspired by linguistic theories. In: The 57th Annual Meeting of the Association for Computational Linguistics, pp. 3888–114 (2019)
26. Group, P. : MIP: A method for identifying metaphorically used words in discourse. Metaph. symb. **22**(1), 1–39 (2007)
27. Wilks, Y.: A preferential, pattern-seeking, semantics for natural language inference. In: Words and Intelligence, pp. 83–102 (2007)
28. Wilks, Y.: MIP: making preferences more active. Artif. Intell. **11**(3), 197–223 (1978)
29. Pennington, J., Socher, R., Manning, C.D.: Glove: Global vectors for word representation. In: The 2014 conference on empirical methods in natural language processing, pp. 1532–1543 (2014)
30. Peters, M., et al. : Deep Contextualized word representations. In: The 2018 Conference of the North American Chapter of the Association for Computational Linguistics: Human Language Technologies, pp. 2227–2237 (2018)

Incorporating Translation Quality Estimation into Chinese-Korean Neural Machine Translation

Feiyu Li, Yahui Zhao[✉], Feiyang Yang, and Rongyi Cui

Department of Computer Science and Technology, Yanbian University,
977 Gongyuan Road, 133002 Yanji, China
{2019010431,yhzhao,2018010454,cuirongyi}@ybu.edu.cn

Abstract. Exposure bias and poor translation diversity are two common problems in neural machine translation (NMT), which are caused by the general of the teacher forcing strategy for training in the NMT models. Moreover, the NMT models usually require the large-scale and high-quality parallel corpus. However, Korean is a low resource language, and there is no large-scale parallel corpus between Chinese and Korean, which is a challenging for the researchers. Therefore, we propose a method which is to incorporate translation quality estimation into the translation process and adopt reinforcement learning. The evaluation mechanism is used to guide the training of the model, so that the prediction cannot converge completely to the ground truth word. When the model predicts a sequence different from the ground truth word, the evaluation mechanism can give an appropriate evaluation and reward to the model. In addition, we alleviated the lack of Korean corpus resources by adding training data. In our experiment, we introduce a monolingual corpus of a certain scale to construct pseudo-parallel data. At the same time, we also preprocessed the Korean corpus with different granularities to overcome the data sparsity. Experimental results show that our work is superior to the baselines in Chinese-Korean and Korean-Chinese translation tasks, which fully certificates the effectiveness of our method.

Keywords: Machine translation · Chinese-Korean machine translation · Reinforcement learning · Quality estimation

1 Introduction

With the rapid development of deep learning, neural machine translation (NMT) has attracted much attention in recent years. This is because it has superior performance and does not require much manual intervention [1,3]. Korean is the official language of the Korean ethnic group in China, and is also used in the Korean Peninsula, the United States, the Russian Far East and other areas where Koreans congregate. Transnational and transregional is a key feature of Korean. As the Korean ethnic group is one of the 24 ethnic minorities in China that

S. Li et al. (Eds.): CCL 2021, LNAI 12869, pp. 45–57, 2021.
https://doi.org/10.1007/978-3-030-84186-7_4

have their own language [20], the research on Chinese-Korean neural machine translation plays an important role in promoting the language work of ethnic minorities and strengthening the communication and unity of ethnic groups.

At present, most NMT models adopt the teacher forcing strategy in training [17]. Teacher forcing is to minimize the difference between the source sentence and the reference and force the predicted translation to be infinitely close to the reference. First of all, as there is usually no reference for sentence prediction, exposure bias will be brought, which may affect the performance and robustness of the model [9]. Secondly, as there are a large number of synonyms and similar expressions in the language, the translation model cannot always generate the ground truth word even with teacher forcing in training. Using this strategy will greatly curb the diversity of translation and make many reasonable translations in an unreachable state [21]. In addition, domestic research on Chinese-Korean machine translation started late, the foundation is poor, the lack of large-scale parallel corpus. Therefore, there are many challenges to improving the performance of Chinese-Korean machine translation under the condition of low resources.

We attempt to introduce a sentence-level evaluation mechanism to guide the training of the model, so that the prediction can not converge completely to the ground truth word, to alleviate exposure bias and poor translation diversity. The evaluation mechanism is reference-free evaluation, which is called quality estimation (QE) in conference on machine translation. The main idea is that the evaluation model can estimate the quality of the unseen translation without reference after supervised training. The instruct mechanism is guided by the reinforcement learning (RL) of policy optimization, which enables the model to optimize the target sequence at the sentence level. In order to alleviate the acknowledged instability and variance difficulty fitting problems of reinforcement learning, we trained MLE and RL together and referred to the baseline reward method proposed by Weaver L. [16]. Moreover, previous studies have directly used BLEU value as a reward [18,19], which would lead to serious bias in the model and exacerbate the problem of poor translation diversity. Therefore, we propose a reward function based on QE score. Meanwhile, monolingual corpus and Korean text preprocessing with different granularity are used in the training process overcome data sparsity and improve the quality of machine translation for low-resource languages.

2   Related Work

Machine Translation Quality Estimation: Machine translation quality estimation is different from the evaluation indexes of machine translation such as BLEU [8], TER [12], METEOR [7], etc. It can automatically give the quality prediction of machine-generated translation without relying on any reference translation. The most commonly used quality score is the human translation edit rate (HTER). QuEst is a model proposed by Specia L. [13], which is used for quality estimation tasks. As the baseline model of machine translation quality estimation task, the model consists of feature extraction module and machine learning module. In order to solve the problem of machine translation quality estimation, Kim H. [5] firstly applied the machine translation model to the quality estimation task and proposed a translation quality estimation model based on RNN. Fan K. [2] replace the machine translation model based on RNN with Transformer model on the basis of Predictor-Estimator, and proposed a bilingual expert model, which improved the performance and interpretability of the model evaluation.

Reinforcement Learning-based NMT: There is a lot of research showing that the advantages of reinforcement learning in sequence generation tasks. Ranzato M. [9] proposed a novel sequence level training algorithm that directly optimizes the metric used at the test time. Wu L. [18]conducted a systematic study on how to train better NMT models using reinforcement learning with several large-scale translations tasks. Keneshloo Y. [4] considered seq2seq problems from the RL point of view and provide a formulation combining the power of RL methods in decision-making with seq2seq models that enable remembering long-term memories.

3   Methodology

In this section, we describe the construction and training of a Chinese-Korean neural machine translation model that incorporates translation quality assessment. We introduce the model architecture, sentence-level translation quality estimation methods, reward function design and training of the whole model.

3.1   Model Overview

To alleviate exposure bias and poor translation diversity, we propose a Chinese-Korean machine translation model that incorporates translation quality estimation. The model introduces an evaluation mechanism at the sentence level to guide the model prediction not to converge completely to the ground truth word. The specific framework structure of the model is shown in Fig. 1, which mainly includes two modules: machine translation and machine translation quality estimation. The translation module adopts the encoder-decoder architecture, and the framework is consistent with Transformer. The evaluation module adopts sentence-level machine translation quality estimation model Bilingual Expert. The training adopts reinforcement learning.

In the process of machine translation, NMT system, as an agent of reinforcement learning, obtains the environmental state information at the current moment through continuous interaction with the environment. The environmental state information is the source sentence x under the time step t and the above $P(y_t|x, \hat{y}_{<t})$ of the generated target sentence. Where $\hat{y}_{<t}$ represents the target sentence predicted by the model before the time step t. According to the state of the current environment, the agent decides to choose the next selected word, obtains the reward value of the word selection operation in the current state

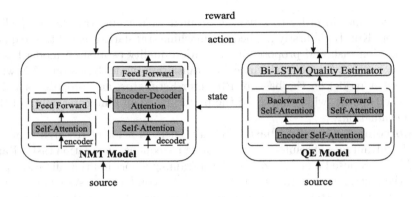

Fig. 1. The architecture of the Chinese-Korean machine translation module

and enters the next state, and finally finds the optimal strategy of translation through reinforcement learning.

As shown in Fig. 2, the machine translation task is described as: Training a machine translation model M_θ with parameter θ under the condition that the Chinese-Korean parallel corpus is given. The machine translation model M_θ translates the given source sentence sequence $\mathbf{x} = (x_1, x_2, ..., x_n)$ into a target sentence sequence $\mathbf{y} = (y_1, y_2, ..., y_m)$, where n, m are sequence lengths of source and target sentences respectively. In the time step t, the state S_t defines the target sentence $\mathbf{y} = (y_1, y_2, ..., y_m)$ generated by the machine translation model M_θ under the current time step, and the action a_t is defined as the selection of the next word y_{t+1} in the current environment. Training a machine translation quality estimation model Q_φ with parameter φ under the condition that the condition of the translation data and HTER scores is given. After supervised training, the quality estimation model Q_φ acts as a generator of reward functions to give quality scores to the unseen translations, and the machine translation model M_θ is guided to interact with the environment to produce the next word y_{t+1}.

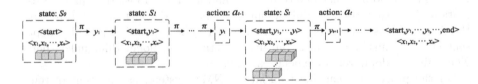

Fig. 2. The schematic diagram of the decision process in translation

3.2 Generate Rewards Through Sentence-Level Quality Estimation

An excellent translation usually includes multidimensional evaluation, such as fidelity and faithfulness, so it is difficult to abstract machine translation tasks into simple optimization problems. Therefore, instead of manually setting a single rule as the source of the reward function, we use the output of the machine translation quality estimation model as part of the reward. The model can generate a more comprehensive score through a relatively complex network structure, which is more relevant to human evaluation and more tolerant to the diversity of translation.

The model Q_φ in this paper uses the same network structure as Bilingual Expert. The model includes a word prediction module based on bidirectional Transformer and a regression prediction model based on Bi-LSTM. Bidirectional Transformer includes three parts encoder self attention, e forward and backward self-attentions and token reconstruction. It acquires hidden state features h by pre-training on large-scale parallel corpus. The encoder part corresponds to $q(h|x, y)$, and the decoder part corresponds to $p(y|h)$. The calculation formula is as follows:

$$q(h|x, y) = \prod_t q(\overrightarrow{h_t}|x, y_{<t})q(\overleftarrow{h_t}|x, y_{<t}) \tag{1}$$

$$p(y|h) = \prod_t p(y_t|\overrightarrow{h_t}, \overleftarrow{h_t}) \tag{2}$$

The hidden state $h = (h_1, ..., h_m)$ is a combination of the forward and backward hidden state, which captures the deep translation features of the sentence. The final features are as follows:

$$f = \text{Concat}\left(\overrightarrow{h}_t, \overleftarrow{h}_t, e_{t-1}, e_{t+1}, f^{mm}\right) \tag{3}$$

where, e_{t-1}, e_{t+1} is the embedding concatenation of two neighbor tokens, and f^{mm} is the mis-matching features. Finally, the features are input to Bi-LSTM for training to get the predicted HTER score:

$$\text{HTER}' = \text{sigmoid}\left(w^T [\text{Bi} - \text{LSTM}(f)]\right) \tag{4}$$

The loss function in the training process is:

$$\text{argmin}||\text{HTER} - \text{HTER}'||_2^2 \tag{5}$$

The scalar value obtained in Eq. (4) is the evaluation of the generated translation by the machine translation quality estimation module. Compared with BLEU, it has deeper translation characteristics. Therefore, our method uses this score to guide the machine translation module and can achieve the effect of the prediction can not converge completely to the ground truth word.

3.3   Reward Computation

It is critical to set up appropriate rewards for RL training. In previous researches on NMT, it is assumed that the effective predictive value of each word item in the generated target sentence is unique, that is to say, there is a fixed reference for each sentence. Therefore, both the minimum-risk training method [10] and the reinforcement learning method [4,19] use the BLEU score of similarity between the generated sentence and the reference as the training target. However, in natural language, the same source sentence fragment can correspond to multiple reasonable translations, so the reward based on BLEU cannot give a reasonable reward or punishment for words other than the target language. As a result, most reasonable translations are denied, which greatly limits the improvement of translation effect by reinforcement learning and exacerbates the problem of poor diversity of machine translations. Thus we set the reward as:

$$R\left(\hat{y}_t\right) = \alpha Score_{BLEU}\left(\hat{y}_t\right) + \frac{1-\alpha}{Score_{QE}\left(\hat{y}_t\right)+1} \tag{6}$$

where, $Score_{BLEU}\left(\hat{y}_t\right)$ is the normalized BLEU between the generated translation and the ground truth, and $Score_{QE}\left(\hat{y}_t\right)$ is the normalized QE evaluation score of the generated translation. The super parameter α is used to balance the weight between BLEU and QE scores, so as to avoid the problem that the QE score may aggravate the instability of training after introducing it. In this way, the training can be converged quickly and the diversity of translation can be fully considered.

In the machine translation task, the agent needs to take dozens of actions to generate a complete target sentence, but after generating a complete sequence, only one terminal reward can be obtained, and sequence-level reward cannot distinguish the contribution of each word item to the total reward. Therefore, there is a problem of reward sparsity during the training, which will lead to slow convergence speed of the model or even failure to learn. Reward shaping can alleviate this problem. In reward shaping, the instant reward at each decoding step t is imposed, and the rewards correspond to word levels. The rewards are set as follows:

$$r_t\left(\hat{y}_t\right) = R\left(\hat{y}_t\right) - R\left(\hat{y}_{t-1}\right) \tag{7}$$

During the training, an accumulative reward is calculated as the current sequence reward after each sampling action is completed, and the reward difference between two continuous time steps is the word level reward. In this way, the model can get an instant reward for the current time step after each action, thus alleviating the problem of reward sparse.

$$R(\hat{y}_t) = \sum_{t=1}^{T} r_t\left(\hat{y}_t\right) \tag{8}$$

Experiments have shown that using reward shaping does not change the optimal strategy. Since the reward of the whole sequence is the sum of the reward value of each word item level, which is consistent with the reward of the sequence level, the total reward of the sequence will not be affected.

3.4   The Training of Reinforcement Learning

The idea of reinforcement learning is that the agent selects an action to execute according to the current environment, and then the environment shifts with a certain probability and gives a reward to the agent, and the agent repeats the above process for the purpose of maximizing the reward. Specifically, in the translation task, the NMT model is regarded as the agent making decisions, and the random strategy $\pi(a_t|\mathbf{s_t};\Theta)$ is adopted to select candidate words from the word list as an action. During the training, the agent learns better translation through the reward given by the environment after the target sentence is generated by the decoder.

$$\pi(a_t|\mathbf{s_t};\Theta) = \sigma(\mathbf{W} * \mathbf{s}_t + \mathbf{b}) \tag{9}$$

where $\pi(a_t|\mathbf{s_t};\Theta)$ is the probability of choosing an action, and σ is the sigmoid function, $\Theta = \{\mathbf{W}, \mathbf{b}\}$ is a parameter of the policy network. During the training, the action from all the conditional aim-listed probability $p(y_t|x, y_{<t})$ of the given source sentence and the word selected below, and the goal is to pursue the maximum expected reward, i.e.:

$$a_t^* = \mathrm{argmax}_a \pi(a|\mathbf{s_t};\Theta) \tag{10}$$

When the complete target sentence is generated, the quality estimation score of the sentence is used as the label information to calculate the reward. Then the Policy Gradient method [11] of reinforcement learning algorithm is used to maximize the expected revenue, as shown follows:

$$
\begin{aligned}
J(\Theta) &= \sum_{i=1}^{N} E_{\hat{y}\sim p(\hat{y}|x^i)} R(\hat{y}) \\
&= \sum_{i=1}^{N} \sum_{\hat{y}\in Y} p(\hat{y} \mid x^i) R(\hat{y})
\end{aligned}
\tag{11}
$$

where Y is the space composed of candidate translated sentences and $R(\hat{y})$ is the sentence-level reward of the translation. Because the state at time step $t+1$ is completely determined by the state at time step t, the probabilities $p(\mathbf{s_1})$ and $p(\mathbf{S_{t+1}}|\mathbf{S_t}, a_t)$ are equal to 1. The gradient update strategy as shown follows:

$$\nabla_\Theta J(\Theta) = -\frac{1}{N} \sum_{n=1}^{N} \sum_{t=1}^{L} (R_L - b) \nabla_\Theta \log \pi_\Theta(a_t|\mathbf{s_t}) \tag{12}$$

where N is the number of turns, $b \approx E[R_L]$.

The action space of reinforcement learning-based machine translation is considerable and discrete. Its size is the capacity of the entire word list. So we use beam search to sample the actions. It reduces the computational cost and increases the probability of high quality translation results in the decoding stage. The principle of beam search is shown in Fig. 3.

In order to stabilize the process of reinforcement training and alleviate the large variance that may be brought by reinforcement learning, we combined MLE training goals with RL goal. The specific step is to retain the cross-entropy loss function of traditional machine translation in the loss function, and then combine

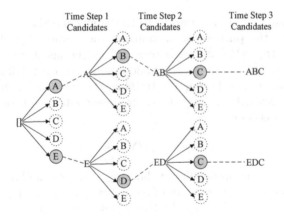

Fig. 3. The schematic diagram of beam search

it with the reinforcement learning training objective linearly. The loss function after mixing is shown below:

$$L_{\text{combine}} = \gamma \times L_{\text{mle}} + (1 - \gamma) L_{\text{rl}} \tag{13}$$

where, L_{combine} is the binding loss function, L_{mle} is the cross entropy loss function, L_{rl} is the reinforcement learning reward, and γ is the super parameter controlling the weight between L_{mle} and L_{rl}. Different γ will affect the performance of the final translation results.

4   Experiments

4.1   Datasets

The data set used in the experiments comes from the corpus constructed by the laboratory to undertake the "China-Korea Science and Technology Information Processing Comprehensive Platform" project [14]. The original corpus includes more than 30,000 documents and more than 160,000 parallel sentence pairs, covering 13 fields such as biotechnology, Marine environment and aerospace. To alleviate the problem about data sparsity, we also used additional monolingual corpus in the experiment. The detailed data information obtained after preprocessing according to the task in this paper is shown in Table 1. The HTER score for quality estimation tasks is automatically calculated by the TERCOM tool.

4.2   Preprocessing

Large-scale corpus word embedding can provide sufficient priori information for the model, accelerate the convergence rate of the model, and effectively improve the effect of downstream tasks. However, as a low-resource language, Korean lacks a large corpus, so there will be a large number of low-frequency words in

Table 1. Description of data in Chinese-Korean machine translation

Category	Languages	Size(sentence)
Parallel corpora		163,449
Monolingual corpora	Chinese	115,000
	Korean	115,000
QE corpora	Chinese	163,449
	Korean	163,449

the corpus, which will lead to low quality of word vectors. To solve this problem, we use more flexible Korean language granularity for word embedding to alleviate the data sparsity problem.

Korean is a phonetic alphabet. From a phonetic point of view, the Korean language consists of phonemes that form syllables according to rules, and syllables that form sentences. Since the number of phonemes and syllables is relatively fixed (67 phonemes and 11172 syllables), the scale of dictionary construction using such granularity is very small, and the existence of low-frequency words can be significantly reduced compared with other granularities. And from a semantic point of view, word may have clearer morphological and linguistic features. Therefore, we use phoneme, syllable and word to preprocess the corpus of the Korean text. Phonemes are obtained by the open source phoneme decomposition tool hgtk, syllables are obtained by reading characters directly, and word is obtained by the word segmentation tool Kkam.

4.3 Setting

Our translation module is implemented on an encoder-decoder framework based on self-attention, and the Transformer system adopts the same model configuration as described by [15], which is implemented using Tensor2Tensor, an open source tool built on Google Brain. We set dropout to 0.1, and the word vector dimension to 512. The MLE training gradient optimization algorithm uses the Adam [6] algorithm and learning rate decay. In the feature extraction part of our machine translation quality estimation module, the number of encoder and decoder layer is 2, the number of hidden units of feedforward sublayer is 1024, and the number of head of attention is 4. In our quality estimation part, the network structure is single-layer Bi-LSTM, the hidden layer unit is set as 512, the gradient optimization algorithm uses Adam, and the learning rate is set as 0.001. During reinforcement learning training, the MLE model was used for parameter initialization, the learning rate was set as 0.0001, and the beam search width was set as 6.

4.4 Main Results and Analysis

In order to test the translation performance of the model, we conducted Chinese-Korean machine translation experiments under the same hardware and corpus environment and calculated the BLEU and QE of the test set respectively. The results are shown in Table 2.

Table 2. The score of translation performance

Model	zh-ko		ko-zh	
	BLEU ↑	QE ↓	BLEU ↑	QE ↓
LSTM+attention	11.52	83.95	12.61	83.20
Transformer	15.82	30.91	20.20	27.41
Our Method	**21.67**	**25.58**	**23.46**	**25.16**

As can be seen from Table 2, our method can exceed the baseline model for both the Chinese-Korean and Korean-Chinese translation tasks. Compared with LSTM+attention, BLEU increases by 10.15 and QE score decreases by 58.37 in Chinese-Korean direction, and BLEU increases by 10.85 and QE score by 58.04 in Korean-Chinese direction. Compared with Transformer, BLEU increases by 5.85 and QE score decreases by 5.33 in Chinese-Korean direction, and BLEU increases by 3.26 and QE score by 2.25 in Korean-Chinese direction. Therefore, the introduction of evaluation module effectively improves the performance of Chinese-Korean machine translation.

4.5 Performance Verification About QE

To ensure the rationality and effectiveness of our method, we verify the performance of the machine translation quality estimation module. Pearson's Correlation Coefficient, Mean Average Error (MAE) and Root Mean Squared Error (RMSE) used in WMT competition were used as verification indexes. Pearson's Correlation Coefficient is used to measure the correlation between the predicted value and the ground truth. The higher the positive correlation, the better the performance of QE module. MAE and RMSE represent the mean and square root of the absolute error between the predicted value and the true value respectively, the smaller the value, the better. The baseline system adopts the open source system QuEst ++ [13], which is the official baseline system of WMT2013-2019. The specific experimental results are shown in Table 3.

Table 3. Verification results for QE module performance

Model	Bilingual Expert	QuEst++
Pearson's ↑	**0.476**	0.397
MAE ↓	**0.118**	0.136
RMSE ↓	**0.166**	0.173

As can be seen from the experimental results in Table 3, compared with the baseline system of QE task, the Bilingual Expert used in the experiment has a better performance improvement, with Pearson correlation coefficient increased by 0.079, MAE decreased by 0.018, and RMSE decreased by 0.007. Our results have a high correlation with manual evaluation, thus proving the effectiveness of using the machine translation quality assessment module in this experiment. In conclusion, it is reasonable to use the machine translation quality estimation module to optimize the translation module.

4.6　Example of Translation Results

Examples of translation results for different models are shown in Table 4.

Table 4. Examples of translation results

Model	Sentence
Source	스마트폰을 이용해서 정밀한 길이 측정이 가능해요.
LSTM+attention	智能手机以测量的。
Transformer	通过使用智能手机的方式准确测量。
Our Method	使用智能手机精确测量长度成为可能。
Source	이미지가 저장이 안되고 보기만 가능해요.
LSTM+attention	图像查看因为他不能保存。
Transformer	图像只可以查看并不能保存。
Our Method	图像不能保存只能查看。
Source	而其余的英雄们回到了他们的日常生活。
LSTM+attention	내 가 다음 날 들은 일상 생활 로 돌아 갔어 .
Transformer	그리고 나머지 영웅 들은 일상 생활 로 돌아 갔어 .
Our Method	그리고 나머지 영웅 들은 일상 생활로 돌아 갔어.
Source	这些都与织物在螺杆或机头中的流动有关。
LSTM+attention	스트레스 없이 모두 외부 에 의해 바꿀 수 있어.
Transformer	이들은 나사 또는 머리의 직물 흐름과 관련이 있습니다.
Our Method	이러한 것들은 스크류나 헤드 내에서 생지의 유동과 관계가 있어.

As can be seen from the translation examples in Table 4, the translation obtained by our method is more accurate in both directions, the fluency and fidelity of the translation are in line with the target language specification, and the quality of the translation is significantly better than that of the other baseline models, which proves that our method can effectively improve the performance of the Chinese-Korean neural machine translation model.

5　Conclusion

In order to alleviate the exposure bias and poor translation diversity problems caused by teacher forcing in machine translation, we propose a Chinese-Korean neural machine translation model that incorporates machine translation quality estimation. The model introduces an evaluation mechanism at the sentence level to guide the training of the model, so that the prediction cannot converge completely to the ground truth word. The evaluation mechanism is reference-free evaluation. The instruct mechanism is guided by the reinforcement learning. The experimental results clearly show that our approach can effectively improve the performance of Chinese-Korean neural machine translation.

Acknowledgements. This research work has been funded by the National Language Commission Scientific Research Project (YB135-76), the Yanbian University Foreign Language and Literature First-Class Subject Construction Project (18YLPY13).

References

1. Bahdanau, D., Cho, K.H., Bengio, Y.: Neural machine translation by jointly learning to align and translate. In: 3rd International Conference on Learning Representations, ICLR 2015 - Conference Track Proceedings, pp. 1–15 (2015)
2. Fan, K., Wang, J, Li, B.: Bilingual expert1 can find translation errors. In: Proceedings of the AAAI Conference on Artificial Intelligence, pp. 6367–6374 (2019)
3. Junczys-Dowmunt, M., Dwojak, T., Hoang, H.: Is neural machine translation ready for deployment? In: A Case Study on 30 Translation Directions (2016)
4. Keneshloo, Y., Shi, T., Ramakrishnan, N.: Deep reinforcement learning for sequence-to-sequence models, pp. 2469–2489 (2020)
5. Kim, H., Lee, J.-H., Na, S.H.: Predictor-estimator using multilevel task learning with stack propagation for neural quality estimation. In: Proceedings of the 2nd Conference on Machine Translation, pp. 562–568 (2017)
6. Kingma, D.P., Ba, J.: Adam: a method for stochastic optimization. In: 3rd International Conference on Learning Representations (ICLR), pp. 1–15 (2015)
7. Lavie, A., Agarwal, A.: Meteor: an automatic metric for MT evaluation with high levels of correlation with human judgments. In: Proceedings of the Second Workshop on Statistical Machine Translation, pp. 228–231 (2007)
8. Papineni, K., Roukos, S., Zhu, W.-T.: Bleu: a method for automatic evaluation of machine translation. In: Proceedings of the 40th Annual Meeting of the Association for Computational Linguistics, 2002, pp. 311–318 (2002)
9. Ranzato, M., Chopra, S., Auli, M.: Sequence level training with recurrent neural networks. In: 4th International Conference on Learning Representations (ICLR 2016), pp. 1–16 (2016)
10. Shiqi, S., Yong, C., He, Z.: Minimum risk training for neural machine translation. In: Proceedings of the 54th Annual Meeting of the Association for Computational Linguistics. Stroudsburg, pp. 1683–1692. Association for Computational Linguistics. (2016)
11. Silver, D., Lever, G., Hees, N.: Deterministic policy gradient algorithms. In: 31st International Conference on Machine Learning (ICML), pp. 605–619 (2014)
12. Snover, M., Dorr, B., Schwartz, R.: A study of translation edit rate with targeted human annotation. In: Proceedings of the 7th Conference of the Association for Machine Translation of the Americas: Visions for the Future of Machine Translation (AMTA), pp. 223–231 (2006)
13. Specia, L., Shah, K., de Souza, J.: Quest++a translation quality estimation framework. In: Proceedings of the 51st ACL: System Demonstrations, pp. 79–84 (2013)

14. Mingjie, T., Yahui, Z., Cui, R.: Identifying word translations in scientific literature based on labeled bilingual topic model and co-occurrence features. In: Proceedings of Chinese Computational Linguistics and Natural Language Processing Based on Naturally Annotated Big Data, pp. 76–87 (2018)

15. Vaswani, A., Shazeer, N., Parmar, N.: Attention is all you need. In: The proceedings of Advances in Neural Information Processing Systems 30: Annual Conference on Neural Information Processing Systems. pp. 5998–6008 (2017)

16. Weaver, L., Tao, N.: The optimal reward baseline for gradient-based reinforcement learning. In: Proceedings of the Seventeenth Conference on Uncertainty in Artificial Intelligence, pp. 538–545 (1999)

17. Williams, R.J., Zipser, D.: A learning algorithm for continually running fully recurrent neural networks. Neural Computation, **1**, 270–280 (1989)

18. Wu, L., Tian, F., Qin, T.: A study of reinforcement learning for neural machine translation. In: Proceedings of the 2018 Conference on Empirical Methods in Natural Language Processing (EMNLP), pp. 3612–3621 (2018)

19. Zhen, Y., Wei, C., Wang, F.: Improving neural machine translate on with conditional sequence generative adversarial nets p. arXiv preprint arXiv:1703.04887 (2017)

20. Yongshou, J.: Current situation and future research direction of Chinese-Korean translation theory. In: Korean Language in China, pp. 66–73 (2020)

21. Zhang, W., Feng, Y., Meng, F.: Bridging the gap between training and inference for neural machine translation. In: 57th Annual Meeting of the Association for Computational Linguistics, Proceedings of the Conference, ACL. arXiv preprint arXiv:1906.02448 (2019)

Social Computing and Sentiment Analysis

Emotion Classification of COVID-19 Chinese Microblogs Based on the Emotion Category Description

Xianwei Guo[1,2], Hua Lai[1,2], Yan Xiang[1,2(✉)], Zhengtao Yu[1,2], and Yuxin Huang[1,2]

[1] Faculty of Information Engineering and Automation,
Kunming University of Science and Technology, Kunming, China
[2] Yunnan Key Laboratory of Artificial Intelligence,
Kunming University of Science and Technology, Kunming, China

Abstract. Emotion classification of COVID-19 Chinese microblogs helps analyze the public opinion triggered by COVID-19. Existing methods only consider the features of the microblog itself, without combining the semantics of emotion categories for modeling. Emotion classification of microblogs is a process of reading the content of microblogs and combining the semantics of emotion categories to understand whether it contains a certain emotion. Inspired by this, we propose an emotion classification model based on the emotion category description for COVID-19 Chinese microblogs. Firstly, we expand all emotion categories into formalized category descriptions. Secondly, based on the idea of question answering, we construct a question for each microblog in the form of 'What is the emotion expressed in the text X?' and regard all category descriptions as candidate answers. Finally, we construct a question-and-answer pair and use it as the input of the BERT model to complete emotion classification. By integrating rich contextual and category semantics, the model can better understand the emotion of microblogs. Experiments on the COVID-19 Chinese microblog dataset show that our approach outperforms many existing emotion classification methods, including the BERT baseline.

Keywords: COVID-19 Chinese microblogs · Emotion classification · Emotion category description · BERT

1 Introduction

The COVID-19 pandemic is spreading all over the world, and fighting the pandemic is a protracted battle. It is also an important battle to analyze the COVID-19 related data continuously generated on social media and quickly grasp the public opinion that the pandemic may trigger. Public opinion caused by the pandemic will have an important impact on the decision-making of the government and relevant departments. Automatic emotion classification of COVID-19

© Springer Nature Switzerland AG 2021
S. Li et al. (Eds.): CCL 2021, LNAI 12869, pp. 61–76, 2021.
https://doi.org/10.1007/978-3-030-84186-7_5

related data on social media is helpful to assess the risk of public opinion. Common social media platforms at home and abroad include Weibo, Facebook, Twitter, and so on, which are important ways for netizens to express their opinions and emotions [1].

As one of the largest social media platforms in China, Sina Weibo has generated massive amounts of COVID-19 microblog data. Through analysis, we found that the text of the COVID-19 Chinese microblog is short, lacks context, and has nonstandard expressions. For example, the two posts '武汉加油，中国必赢 *(Come on Wuhan, China will win)*' and '#新型冠状病毒# 比恐慌更可怕的是怠慢。—《看见》（讲述　非典时所写到）'. Generally, existing emotion classification methods only utilize the features of the microblog itself for modeling and do not consider the semantics of emotion categories well. Therefore, they cannot analyze the emotion of nonstandard COVID-19 Chinese microblog text well. To solve this problem, we present an emotion classification method based on the emotion category description. Based on the idea of question answering, the semantic information of categories is fused to help the model understand the emotion of microblogs. In the next paragraph, we will analyze the impact of category semantic information on the emotion classification of COVID-19 Chinese microblogs.

Emotion classification requires identifying specific sentiments in the text, such as happiness, anger, sadness, and fear [2,3]. Traditional supervised emotion classification models generally transform the categories into digital labels, as the supervised signal to guide the learning process of the model. For example, traditional models use '1' for happiness emotion and '2' for anger emotion. Normally, the digital label will be represented as a one-hot vector, and be used to calculate the training loss. Then the backpropagation algorithm is used to minimize the objective function to train the model. Traditional models do not adequately consider the semantic information of emotion categories, which means that they do not better learn the meaning of emotion categories and cannot accurately classify microblogs into corresponding emotion categories. When judging the emotions expressed in a text, humans usually have some prior knowledge. For example, the microblog "这人什么心态！能不能对自己和他人都负点责，丈夫确诊隔离了，妻子还往外跑，还坐火车，什么心态'. When humans are judging the emotion of this microblog, the embodiment of prior knowledge is that they know the specific meaning of anger emotion, that is, someone is agitated because of extreme dissatisfaction, so it is easy to correctly judge the emotion of this microblog. However, the models do not have prior knowledge, so they are not ideal for microblog emotion learning. In other words, if the models can grasp the prior knowledge that humans have, they will better understand the emotion of the text.

By asking what the emotion of a certain microblog expresses and then giving an answer, this process of judging the emotions of the text is similar to the question-answering task. Inspired by this, we introduce the idea of question answering into the emotion classification task of COVID-19 Chinese microblogs. The main contributions are summarized as follows:

1) We propose an emotion classification model of COVID-19 Chinese microblogs based on the emotion category description. Firstly, all emotion categories of microblogs to be classified are expanded into formalized category descriptions, as a candidate answer set. Secondly, we construct a question for each microblog in the form of 'What is the emotion expressed in the text X?'. Then, the question and all category descriptions are constructed into a question-and-answer pair as the input of the pre-trained BERT model. Finally, by fusing rich contextual and category semantic information, the model completes the emotion classification of COVID-19 Chinese microblogs.

2) We present three emotion category description strategies, which consider words, extended words and emotion definitions to describe three different granularity of category information, respectively.

3) Experimental results show that our approach outperforms many existing emotion classification methods on the COVID-19 Chinese microblog dataset.

2   Related Work

Emotion classification of COVID-19 Chinese microblogs is essentially a sentiment classification task. Recently, with the rise of deep learning, existing sentiment classification studies are usually based on deep learning methods. Neural network models, such as recurrent neural networks (RNNs), convolutional neural networks (CNNs), and Transformers, have been proven effective in many sentiment classification tasks.

Tang et al. [4] first used CNN or LSTM to encode a single sentence and then used gated RNN to encode the internal relations and semantic connections between sentences. Finally, they obtained the representation of the document to complete sentiment classification. Wang et al. [5] proposed a context-aware bidirectional LSTM model, which used forward and backward LSTMs to jointly encode the context information of the text. It has achieved good results in the emotion classification of Chinese microblogs. Kim [6] proposed for the first time to use convolutional neural networks to extract text sequence features for sentence-level text classification and completed sentiment classification on movie review datasets. Since then, a series of sentiment classification methods based on CNN have been produced [7,8]. Johnson et al. [9] proposed a word-level DPCNN, which extracted long-distance text dependencies by continuously deepening the network. They performed sentiment classification on review datasets such as Amazon and achieved the best results at the time. He et al. [10] completed the enhancement of emotional semantics by mapping the commonly used emoji vector representation and the word vector representation of the text to the same emotional space. Then they used a multi-channel convolutional neural network to classify the emotion of Chinese microblogs. In the above research, although the RNN-based model can effectively process serialized text, it also has the problem of sequence dependence and cannot be calculated in parallel. Although the CNN-based model can be processed in parallel, its ability to capture long-distance features is weak due to its mechanism of extracting text features through sliding convolution windows. Besides, CNN-based models generally use a pooling layer to integrate text features, but this will lose the location information of the text, which is another serious problem. To improve these problems, Vaswani et al. [11] proposed the Transformer model, which completely took the Self-Attention mechanism as the basic structure of the model, and abandoned the loop structure of RNN and the convolution structure of CNN. At the same time, Transformer not only has all the advantages of RNN and CNN but also solves the problem of sequence dependence of RNN and the problem of CNN's weak ability to capture long-distance features. Devlin et al. [12] proposed the BERT model based on Transformer, which opened the prelude to the development of pre-training language models and refreshed the records of a series of NLP tasks including sentiment classification tasks. Since then, a series of Transformer-based models have been proposed, which can be referred to as Transformers [13,14].

To sum up, the current methods of sentiment classification mainly focus on neural network models such as RNNs, CNNs, and Transformers. These methods only perform modeling based on the text semantics and fail to utilize the semantic information of classification categories. Studies have shown that the semantic information of categories is effective for classification problems. For example, Rios and Kavuluru [15] integrated the semantics of the category into the model in the form of word embedding, which improved the performance of the text classification task. Chai et al. [16] guided the learning process of the model by using all the category descriptions as questions and the classified text as the answer, thereby enhancing the performance of the text classification task. Their method requires the model to ask about the emotion of the text N times, where N is the number of categories. This process seems not easy to understand, that is, when humans judge the category of a text, they usually understand the semantics of its category, and then combine their knowledge to make judgments.

Different from [16], we construct a question for the input microblog text, and then use all category descriptions as candidate answer set to classify the emotion of the text by a question answering (QA) based method. By constructing a question-and-answer pair to combine each microblog with category descriptions, the model can focus on both the category information related to the microblog and the microblog information related to categories. We also introduce the attention mechanism to focus on the important information in the candidate answer set. Emotion classification requires not only understanding the semantics of the text, but also the emotions contained in the text. Therefore, how to accurately abstract and integrate the semantic information of emotion categories to help the model better understand the emotions of the text is an important issue that we focus on.

3 Methods

In this section, we present our emotion classification method for COVID-19 Chinese microblogs, which contains two parts: (1) the definition and strategy of emotion category description (Sect. 3.1), and (2) the emotion classification fine-tuning based on a question answering (QA) method (Sect. 3.2).

3.1   Definition and Strategy of Emotion Category Description

The definition of emotion category description is to extend emotion categories into formalized descriptions according to a certain strategy. We use three strategies to construct descriptions.

Keyword-Based Category Description. We use six keywords as the description of six emotion categories of happiness, anger, sadness, fear, surprise, and neutral. The construction examples are shown in Table 1.

Table 1. Construction examples of keyword-based category description.

Emotion category	Description(Chinese/English)
happiness	快乐 (happiness)
anger	愤怒 (anger)
sadness	悲伤 (sadness)
fear	恐惧 (fear)
surprise	惊奇 (surprise)
neutral	中性 (neutral)

Keyword Expansion-Based Category Description According to the affective lexicon ontology of Dalian University of Technology [17], the synonyms corresponding to the five Chinese category keywords of '快乐 (happiness)', '愤怒 (anger)', '悲伤 (sadness)', '恐惧 (fear)', and '惊奇 (surprise)' are searched, and they are used as emotion category descriptions together. For the neutral category, the Chinese keywords '中性 (neutral)' and '无情绪 (no emotion)' are spliced together as the emotion category description. There are two versions of the keyword expansion-based category description, which are shown in Table 2 and Table 3.

Table 2. Construction examples of keyword expansion-based category description (Version 1).

Emotion category	Description(Chinese/English)
happiness	快乐、高兴 (happiness, joy)
anger	愤怒、生气 (anger, angry)
sadness	悲伤、悲痛 (sadness, grief)
fear	恐惧、害怕 (fear, afraid)
surprise	惊奇、奇怪 (surprise, strange)
neutral	中性、无情绪 (neutral, no emotion)

Table 3. Construction examples of keyword expansion-based category description (Version 2).

Emotion category	Description(Chinese/English)
happiness	快乐、高兴、幸福、开心 (happiness, joy, blessed, happy)
anger	愤怒、生气、气愤、恼火 (anger, angry, indignant, annoyed)
sadness	悲伤、悲痛、失望、内疚 (sadness, grief, disappointed, guilty)
fear	恐惧、害怕、慌张、害羞 (fear, afraid, panic, shy)
surprise	惊奇、奇怪、惊讶、吃惊 (surprise, strange, flummox, amazed)
neutral	中性、无情绪 (neutral, no emotion)

Emotion Definition-Based Category Description. We determined the specific definition of each emotion category through Baidu Encyclopedia and then adapted it to the category descriptions. The construction examples are shown in Table 4.

Table 4. Construction examples of emotion definition-based category description.

Emotion category	Description(Chinese/English)
happiness	快乐，表示感到高兴或满意的一种状态。 (Happiness, means that someone is glad or satisfied.)
anger	愤怒，表示因极度不满而情绪激动。 (Anger, means that someone is agitated because of extreme dissatisfaction.)
sadness	悲伤，表示因心情不好而伤感。 (Sadness, means that someone is sad because of a bad mood.)
fear	恐惧，表示因陷入某种危险情境而害怕。 (Fear, means that someone is afraid because of being caught in a dangerous situation.)
surprise	惊奇，表示感到惊讶奇怪。 (Surprise, means that someone is surprised and strange.)
neutral	中性，表示不包含任何情绪。 (Neutral, means that it does not contain any emotions.)

3.2 Emotion Classification Fine-Tuning Based on a Question Answering (QA) Method

We use the Chinese pre-trained BERT model (BERT-Base, Chinese) released by Google as the basic model. There are two input forms of pre-trained BERT to fine-tune downstream classification tasks: one is the single sentence input, and the other is the sentence pair input. We adopt the second form, by constructing a question for the microblog and using the category descriptions of all emotion categories as a candidate answer set to construct a question-and-answer pair as the input of the pre-trained BERT model. The structure of our model is shown in Fig. 1.

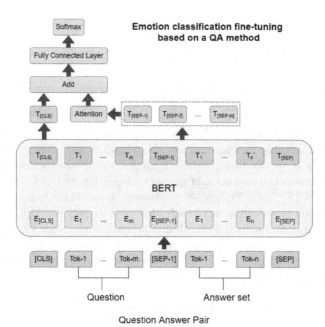

Fig. 1. Structure of emotion classification model based on the emotion category description

First, we introduce the method of constructing the question-and-answer pair. Given a microblog and all emotion categories $\{Y_c|X\} = \{Y_c|x_1, x_2, \ldots, x_n\}, c = 1, 2, \ldots, N$. Y_c represents a category of emotions, and $X = \{x_1, x_2, \ldots, x_n\}$ represents a microblog. Based on the idea of question answering, a microblog X is used to construct a question to ask the model what is the emotion expressed in the text X, and all emotion category descriptions are used as a set of candidate answers. Then, $\{Y_c|X\} = \{Y_c|x_1, x_2, \ldots, x_n\}, c = 1, 2, \ldots, N$ can be represented as a question-and-answer pair: $\{Y_c|X\} = $ '$[CLS]$What is the emotion expressed in the text $X?[SEP-1]$Category description of $Y_1[SEP-2]$Category description of $Y_2[SEP-N]$Category description of $Y_N[SEP]$'. Among them, '$[CLS]$' represents a special classification token, and the hidden state of '$[CLS]$' can be used to represent the semantics of text for classification tasks. '$[SEP-1]$', '$[SEP-2]$', etc. represent the separator tokens, which are used to separate each category description in the answer set. An construction example of the question-and-answer pair is shown in Fig. 2.

The basic structure of the BERT is the Transformer. We omit the specific description of the model and instead focus on how to use a pre-trained BERT model to fine-tune downstream emotion classification task based on category description. In order to fine-tune, we first initialize the BERT model with pre-trained parameters, and then input the constructed question-and-answer pairs into the model, as shown in Fig. 1. In the process of training, the model will

Fig. 2. An construction example of the question-and-answer pair

be continuously fine-tuned according to the input labeled data and adjusted to the final model suitable for the emotion classification task. After the question-and-answer pair is encoded by the BERT model, we obtain the hidden state of the special tokens as the contextual representations, denoted as $h_{[CLS]} \in R^{768 \times 1}$ and $h_{[SEP-n]} = \{h_{[SEP-1]}, h_{[SEP-2]}, \ldots, h_{[SEP-N]}\} \in R^{768 \times 1}$. $h_{[SEP-n]}$ is the contextual hidden representation of each answer (category description). For the current question of 'What is the emotion expressed in the text X?', the contextual hidden representation of the category description corresponding to the real label should be more important. Therefore, we used the attention mechanism to process $h_{[SEP-n]}$, which is given as follows:

$$a_n = h_{[SEP-n]}^T q \tag{1}$$

$$\alpha_n = \frac{exp(a_n)}{\sum_{t=1}^{N} exp(a_t)} \tag{2}$$

$$h_{[SEP]}^{att} = \sum_{n=1}^{N} \alpha_n h_{[SEP-n]} \tag{3}$$

where $h_{[SEP-n]}^T$ is the transpose of $h_{[SEP-n]}$, $q \in R^{768 \times 1}$ is the randomly initialized attention query vector, and α_n is the attention distribution.

Then we calculate the fused semantic representation by the formula $h_{add} = h_{[CLS]} + h_{[SEP]}^{att}$ and input it into a fully connected layer to obtain the emotion category score vector $s \in R^{N \times 1}$. Furthermore, we use the Softmax function to normalize s to obtain the conditional probability distribution $P_i(s)$. The formulas are as follows:

$$s = W_1 h_{add} + b_1 \tag{4}$$

$$P_i(s) = \frac{exp(s_i)}{\sum_{j=1}^{N} exp(s_j)} \tag{5}$$

where $W_1 \in R^{N \times 768}$ is the weight matrix, $b_1 \in R^{N \times 1}$ is the bias vector, and N is the number of emotion categories. The cross-entropy loss function is used to train and update the parameters of the model through a backpropagation algorithm, the formula is as follows:

$$loss = -\sum_{x \in T} \sum_{i=1}^{N} P_i^t(x) log_2(P_i^p(x)) \tag{6}$$

where T is the training set, x is one of the samples in the training set. $P_i^t(x)$ is the ground truth probability distribution of the emotion category of x, and $P_i^p(x)$ is the predicted probability distribution of the emotion category of x.

We propose several models based on the above three strategies, named BERT-KCD, BERT-KECD, and BERT-EDCD. Among them, BERT-KCD (BERT with Keyword-based Category Description) represents the integration of keyword-based category descriptions into the BERT model. BERT-KECD (BERT with Keyword Expansion-based Category Description) represents the integration of keyword expansion-based category descriptions into the BERT model. There are two versions of BERT-KECD, corresponding to the two types of extended keywords in Table 2 and Table 3, named BERT-KECD-v1 and BERT-KECD-v2, respectively. BERT-EDCD (BERT with Emotion Definition-based Category Description) represents the integration of emotion definition-based category descriptions into the BERT model.

4 Experiments

4.1 Experimental Dataset

The experimental dataset comes from 'The Evaluation of Weibo Emotion Classification Technology, SMP2020-EWECT[1]' on 'The Ninth China National Conference on Social Media Processing'. Each microblog is manually labeled with one of six categories: happiness, anger, sadness, fear, surprise, and neutral. Table 5 shows the statistical information of the experimental dataset.

Table 5. Statistical information of the COVID-19 microblog dataset.

Emotion category	Training set	Validation set	Testing set
Happiness	4423	923	1540
Anger	1322	314	463
Sadness	649	165	219
Fear	555	75	190
Surprise	197	47	68
Neutral	1460	476	520
Total	8606	2000	3000

4.2 Baseline Models

We compared the model with seven other baseline models, the baseline models are as follows:

---

[1] https://smp2020ewect.github.io/.

MNB (Multinomial Naïve Bayes) [18]: It achieves excellent performance in many sentiment classification tasks. The smoothing factor-alpha of MNB is set to 1.0.

SVM (Support Vector Machines) [19]: It is widely used in sentiment classification tasks and has achieved excellent results. The regularization constant C of SVM is set to 1.0, and the kernel function is linear.

BLSTM (Bidirectional Long Short-Term Memory) [20]: The model extracts context-related text features for sentiment classification through bidirectional LSTM. It uses a single-layer bidirectional LSTM network with 256 hidden layer units.

CNN (Convolutional Neural Networks) [6]: The classic convolutional neural network proposed by Kim, which uses CNN to extract deep semantic features for text sentiment classification. The convolution kernel sizes of the model are 3, 4, and 5. There are 100 convolution kernels of each size.

DPCNN (Deep Pyramid Convolutional Neural Networks) [9]: The model performs deep convolution operations at the word level and extracts long-distance text features for sentiment classification. It has achieved the best results at the time on multiple review datasets such as Amazon. The hyper-parameters are consistent with [9].

HAN (Hierarchical Attention Networks [21]: The model extracts word-level and sentence-level features through hierarchical bidirectional GRU and Attention mechanisms, and obtain the semantic representation of the entire text for sentiment classification. The best results are obtained in many sentiment classification tasks, and the parameter settings of the model are consistent with [21].

BERT (Bidirectional Encoder Representations from Transformers) [12]: The Chinese pre-trained BERT model (BERT-Base, Chinese)[2] released by Google, which refreshed a series of NLP task records including sentiment classification tasks. Our models are based on this model to fine-tune the emotion classification task.

4.3   Implementation Details

All experimental codes are based on Python 3.6.5 and Tensorflow 1.15.0 and run on the Linux CUDA platform. For baseline models, the learning rate of BLSTM, CNN, DPCNN, and HAN is 0.001, and the batch size is 64. We use the pre-trained word vectors[3] disclosed in [22] for neural network models. The dimension of each word vector is 300. All neural network models use Adam optimizer. Furthermore, the hyper-parameter settings of the BERT series model are shown in Table 6.

We counted the sequence length of the COVID-19 microblog data, the sequence length of the question, and the sequence length of each emotion category description strategy, respectively. After that, the max sequence length of the BERT, BERT-KCD, BERT-KECD-v1, BERT-KECD-v2, and BERT-EDCD models are taken as 128, 160, 180, 210, and 240, respectively.

---

[2] https://github.com/google-research/bert.
[3] https://github.com/Embedding/Chinese-Word-Vectors.

Table 6. Hyper-parameters setting.

Parameters	Values
Max sequence length	128/160/180/210/240
Batch size	32
Dropout	0.1
Learning rate	2e−5

4.4   Experimental Results

We use Precision, Recall, F1, Macro_Precision, Macro_Recall, Macro_F1, and Micro_F1 as the evaluation metrics.

Emotion Classification of COVID-19 Chinese Microblogs. To verify the effectiveness of our model, we compare it with some existing mainstream emotion classification models. Among them, BERT-KECD-v2 is our model. The experimental results are shown in Table 7.

Table 7. Experimental results of COVID-19 microblog emotion classification.

Models	Macro_Precision	Macro_Recall	Macro_F1	Micro_F1
MNB	48.20%	42.68%	43.40%	68.53%
SVM	51.83%	47.09%	48.66%	68.47%
BLSTM	52.93%	49.22%	49.53%	72.97%
CNN	67.25%	55.48%	57.23%	74.97%
DPCNN	66.27%	53.68%	55.20%	74.67%
HAN	64.98%	56.84%	56.43%	75.20%
BERT	67.28%	64.98%	65.63%	79.17%
BERT-KECD-v2	70.74%	65.77%	67.73%	79.83%

Table 7 shows that compared with other deep learning models, the traditional machine learning models such as MNB and SVM have poor performance. The Micro_F1 of MNB and SVM are only 68.53% and 68.47%, respectively. It can be seen that the Micro_F1 of BLSTM, CNN, DPCNN, HAN, and BERT are 72.97%, 74.97%, 74.67%, 75.20%, and 79.17%, respectively. The performance of these deep learning-based models is significantly better than MNB and SVM.

Besides, the Micro_F1 of BERT-KECD-v2 is 79.83%, which is significantly better than the above five deep learning models. Compared with BLSTM, the Macro_Precision, Macro_Recall, Macro_F1, and Micro_F1 of BERT-KECD-v2 have increased by 17.81%, 16.55%, 18.20%, and 6.86%, respectively. Compared with DPCNN, the four metrics of BERT-KECD-v2 have increased by 4.47%,

12.09%, 12.53%, and 5.16%, respectively. Compared with BERT, the four metrics of BERT-KECD-v2 have increased by 3.46%, 0.79%, 2.10%, and 0.66%, respectively. The experimental results prove the effectiveness of our model and show the advantages of our model in the COVID-19 microblog emotion classification.

Validation of Category Description Strategy. We compare the five models of BERT, BERT-KCD, BERT-KECD-v1, BERT-KECD-v2, and BERT-EDCD to verify the effectiveness of the proposed category description strategies. The experimental results are shown in Table 8.

Table 8. Validation results of category description strategy.

Models	Macro_F1	Micro_F1
BERT	65.63%	79.17%
BERT-KCD	67.03%	79.43%
BERT-KECD-v1	67.51%	79.93%
BERT-KECD-v2	67.73%	79.83%
BERT-EDCD	66.82%	79.50%

Table 8 shows that the three different category description strategies could improve the performance of the BERT model. Compared with the BERT model, the Macro_F1 and Micro_F1 of BERT-KCD increased by 1.40% and 0.26%, respectively. The Macro_F1 and Micro_F1 of BERT-KECD-v1 increased by 1.88% and 0.76%, respectively. The Macro_F1 and Micro_F1 of BERT-KECD-v2 increased by 2.10% and 0.66%, respectively. In our analysis, this is because the two keyword-based category descriptions represent part of the semantic information of the category, which can help the model understand the emotion of the text. Moreover, the richer the keywords, the more obvious the performance improvement of the model.

Besides, compared with BERT, the Macro_F1 and Micro_F1 of BERT-EDCD increased by 1.19% and 0.33%, respectively. In our hypothesis, because the BERT-EDCD based on the emotion definition description carries richer category information, it should perform best. However, the experimental results show that the keyword-based models surprisingly achieve better results than BERT-EDCD. In our analysis, there are two possible reasons why the BERT-EDCD did not work as expected. One is that the definition of emotion is not precise enough. As a comparison, keywords may more intuitively reflect the semantics of the emotion category and are easily accessible. The other is that the structure of BERT is not conducive to handling long sequences. The max sequence length of BERT-EDCD is 240, which is the maximum length among all BERT series models.

Overall, the proposed models based on the category description have improved performance compared to the basic model BERT. The experimental results show that the three description strategies proposed in this paper are effective for the COVID-19 microblog emotion classification task.

Effective Verification of Our Model on Each Emotion Category. To prove the effectiveness of our model on each emotion category, we compare the Precision, Recall, and F1 of the BERT and BERT-KECD-v2 models. The experimental results are shown in Figs. 3, 4, and 5.

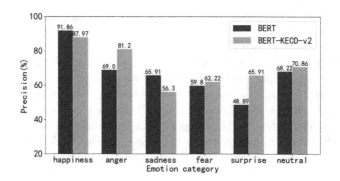

Fig. 3. Comparison results of Precision on each emotion category

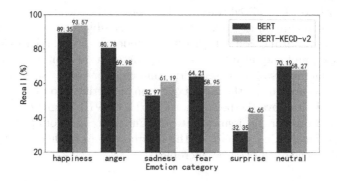

Fig. 4. Comparison results of Recall on each emotion category

From Figs. 3, 4, and 5, it can be seen that BERT-KECD-v2 has overall better performance than the basic model BERT. Compared with the BERT model, the Precision of the anger and surprise categories of BERT-KECD-v2 increased the most significantly, with an increase of 12.20% and 17.02%, respectively. In addition, BERT-KECD-v2 and BERT perform poorly on the three categories

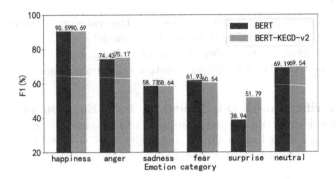

Fig. 5. Comparison results of F1 on each emotion category

of sadness, fear, and surprise. In our analysis, it can be attributed to the small number of training samples in these three categories, which are only 649, 555, and 197, respectively. In contrast, there are 4423 training samples in the happiness category, which enables the model to be fully trained and thus has the best classification performance in this category. The experimental results show that our model effectively improves the performance of the COVID-19 microblog emotion classification.

5   Conclusion

This paper proposes an emotion classification method based on the emotion category description for COVID-19 Chinese microblog data. By extending the emotion category into a formalized category description, the semantic information of the category is integrated to guide the model to classify emotions. Experimental results show that our method can effectively model the nonstandard COVID-19 microblog text, and the introduced category description semantic information helps the model understand the semantics and emotions of the irregular text. It also proves that introducing the idea of question answering into the BERT model can significantly improve the performance of the COVID-19 microblog emotion classification. Moreover, the issue of category imbalance in emotion classification is a challenge for existing studies. In the future, we will further investigate the category imbalance of COVID-19 microblogs.

Acknowledgement. This work was supported by the General Projects of Basic Research in Yunnan Province [NO. 202001AT070047, NO. 202001AT070046] and the National Key Research and Development Plan [No. 2018YFC0830105].

References

1. Wang, X., Xing, Y., Zhang, L., Li, S.: Research on the development trend of network public opinion at home and abroad in the social media environment. Inf. Documentation Serv. **4**, 6–14 (2017)
2. Wen, S., Wan, X.: Emotion classification in microblog texts using class sequential rules. In: Proceedings of the 28th Association for the Advance of Artificial Intelligence (AAAI 2014), pp. 187–193 (2014)
3. Valentina, S., Pearl, P.: Dystemo: distant supervision method for multicategory emotion recognition in tweets. ACM Trans. Intell. Syst. Technol. **8**(1), 1–22 (2016)
4. Tang, D., Qin, B., Liu., T.: Document modeling with gated recurrent neural network for sentiment classification. In: Proceedings of the 2015 conference on Empirical Methods on Natural Language Processing (EMNLP 2015), pp. 1422–1432 (2015)
5. Wang, Y., Feng, S., Wang, D., Zhang, Y., Yu, G.: Context-aware Chinese microblog sentiment classification with bidirectional LSTM. In: Proceedings of 18th Asia-Pacific Web Conference (APWeb 2016), pp. 594–606 (2016)
6. Kim, Y.: Convolutional neural networks for sentence classification. In: Proceedings of the 2014 conference on Empirical Methods on Natural Language Processing (EMNLP 2014), pp. 1746–1751 (2014)
7. Zhang, X., Zhao, J., LeCun, Y.: Character-level convolutional networks for text classification. In: Proceedings of Advances in Neural Information Processing Systems 28 (NIPS 2015), pp. 649–657 (2015)
8. Conneau, A., Schwenk, H., LeCun, Y., Barrault, L.:Very deep convolutional networks for text classification. In: Proceedings of the 15th Conference of the European Chapter of the Association for Computational Linguistics (EACL 2017), pp. 1107–1116 (2017.)
9. Johnson, R., Zhang, T.: Deep pyramid convolutional neural networks for text categorization. In: Proceedings of the 55th Annual Meeting of the Association for Computational Linguistics (ACL 2017), pp. 562–570 (2017)
10. He, Y., Sun, S., Niu, F., Li, F.: A deep learning model enhanced with emotion semantics for microblog sentiment analysis. Chin. J. Comput. **40**(4), 773–790 (2017)
11. Vaswani, A., et al.: Attention is all you need. In: Proceedings of Advances in Neural Information Processing Systems 30 (NIPS 2017), pp. 5998–6008 (2017)
12. Devlin, J., Chang., M,. Lee, K., Toutanova, K.: BERT: pre-training of deep bidirectional transformers for language understanding. In: Proceedings of the 2019 Conference of the North American Chapter of the Association for Computational Linguistics: Human Language Technologies (NAACL-HLT 2019), pp. 4171–4186 (2019)
13. Lan, Z., Chen, M., Goodman, S., Gimpel, K.: ALBERT: a lite BERT for self-supervised learning of language representations. In: Proceedings of the 8th International Conference on Learning Representations (ICLR 2020) (2020)
14. Clark, K., Luong, M., Le, Q., Manning, C.: ELECTRA: pre-training text encoders as discriminators rather than generators. In: Proceedings of the 8th International Conference on Learning Representations (ICLR 2020) (2020)
15. Rios, A., Kavuluru, R.: Few-shot and zero-shot multi-label learning for structured label spaces. In: Proceedings of the 2018 Conference on Empirical Methods in Natural Language Processing (EMNLP 2018), pp. 3132–3142 (2018)

16. Chai, D., Wu, W., Han, Q., Wu, F., Li, J.: Description based text classification with reinforcement learning. In: Proceedings of the 37th International Conference on Machine Learning (ICML 2020) (2020)
17. Xu, L., Lin, H., Pan, Y., Ren, H., Chen, J.: Constructing the affective lexicon ontology. J. China Soc. Sci. Tech. Inf. **27**(2), 180–185 (2008)
18. Bermingham, A., Smeaton, A.: Classifying sentiment in microblogs: Is brevity an advantage?. In: Proceedings of the 19th ACM International Conference on Information and Knowledge Management (CIKM 2010), pp. 1833–1836 (2010)
19. Pang, B., Lee, L., Vaithyanathan, S.: Thumbs up? Sentiment classification using machine learning techniques. In: Proceedings of the 2002 Conference on Empirical Methods on Natural Language Processing (EMNLP 2002), pp. 79–86 (2002)
20. Graves, A., Jaitly, N., Mohamed, A.: Hybrid speech recognition with deep bidirectional LSTM. In: Proceedings of the 2013 IEEE Automatic Speech Recognition and Understanding Workshop (ASRU 2013), pp. 273–278 (2013)
21. Yang, Z., Yang, D., Dyer, C., He, X.: Hierarchical attention networks for document classification. In: Proceedings of the 15th Annual Conference of the North American Chapter of the Association for Computational Linguistics: Human Language Technologies (NAACL-HLT 2016), pp. 1480–1489 (2016)
22. Li, S., Zhao, Z., Hu, R., Li, W.: Analogical reasoning on Chinese morphological and semantic relations. In: Proceedings of the 56th Annual Meeting of the Association for Computational Linguistics (ACL 2018), pp. 138–143 (2018)

Multi-level Emotion Cause Analysis by Multi-head Attention Based Multi-task Learning

Xiangju Li, Shi Feng, Yifei Zhang, and Daling Wang[✉]

School of Computer Science and Engineering, Northeastern University,
No. 195, Chuangxin Road, Hunnan District, Shenyang 110207, China
1610543@stu.neu.edu.cn, {fengshi,zhangyifei,wangdaling}@cse.neu.edu.cn

Abstract. Emotion cause analysis (ECA) aims to identify the potential causes behind certain emotions in text. Lots of ECA models have been designed to extract the emotion cause at the clause level. However, in many scenarios, only extracting the cause clause is ambiguous. To ease the problem, in this paper, we introduce multi-level emotion cause analysis, which focuses on identifying emotion cause clause (ECC) and emotion cause keywords (ECK) simultaneously. ECK is a more challenging task since it not only requires capturing the specific understanding of the role of each word in the clause but also the relation between each word and emotion expression. We observe that ECK task can incorporate the contextual information from the ECC task, while ECC task can be improved by learning the correlation between emotion cause keywords and emotion from the ECK task. To fulfill the goal of joint learning, we propose a multi-head attention based multi-task learning method which utilizes a series of mechanisms including shared and private feature extractor, multi-head attention, emotion attention and label embedding to capture features and correlations between the two tasks. Experimental results show that the proposed method consistently outperforms the state-of-the-art methods on a benchmark emotion cause dataset.

Keywords: Emotion cause analysis · Emotion cause clause · Emotion cause keywords · Multi-task learning

1 Introduction

Emotion cause analysis (ECA), a new field in emotion analysis, attempts to comprehend a given text, and then extracts potential causes that lead to emotion expressions in the text. There has been an increasing interest in the research community on ECA more recently since it is widely used in many scenarios. For example, restaurants are eager to find out why people like or dislike their

Supported by National Key R&D Program of China (2018YFB1004700), National Natural Science Foundation of China (61772122, 61872074).

S. Li et al. (Eds.): CCL 2021, LNAI 12869, pp. 77–93, 2021.
https://doi.org/10.1007/978-3-030-84186-7_6

food or services from users' comments or reviews. Similarly, instead of gauging public opinions towards policies or political issues just using frequency counts, governments would like to further know the triggering factors of certain attitudes expressed online.

ECA is a challenging emotion analysis task since it requires a comprehensive understanding of natural languages and the ability to do further inference. Restricted by the lack of annotated corpora, early studies used rule-based methods [1,17] and crowd-sourcing methods [24] to tackle this task. Until recently, Gui et al. [11] released a reasonable ECA corpus, based on which they developed the first deep learning model for the task [10], and by following that, various other ECA approaches were proposed and achieved superior results [5,7,20,21,27].

Example 1. [x_1] : Entertainment reporter Jucy interviewed that LAM Raymond and Xinyue Zhang to get married in 2020. [x_2] : Also, [x_3] : she interviewed a piece of explosive news that **the wedding ceremony of Tina Tang will be held.** [x_4] : The Tina Tang's fans were very happy and congratulated her[1]. *(Original text*: [x_1] : 娱乐记者Jucy采访到林峰和张馨月在2020年结婚. [x_2] : 同时, [x_3] : 她采访到一条唐艺昕将要举办婚礼的爆炸性新闻. [x_4] : 唐艺昕的粉丝非常开心并且庆祝她.*)*

Most of the existing studies identify which clause contains the emotion cause. Example 1 shows a piece of text from Sina Weibo, in which the emotion word is "*happy*" and the exact emotion cause of "*happy*" is "*the wedding ceremony of Tina Tang will be held*". We call all the words in the exact emotion cause as the emotion cause keywords and the clause which contains the emotion word as the emotion clause. For instance, in this example, the emotion clause, emotion cause clause and the emotion cause keywords are clause [x_4], clause [x_3] and { "*the*", "*wedding*", "*ceremony*", "*of*", "*Tina*", "*Tang*", "*will*", "*be*", "*held*"}, respectively. With the existing methods, the emotion cause clause [x_3] is expected to be extracted because the cause of "*happy*" is "*the wedding ceremony of Tina Tang will be held*" that is a part of [x_3].

However, only identifying which clause contains the emotion cause is flawed and ambiguous. In Example 1, the content "*she interviewed a piece of explosive news*" in [x_3] is not the cause of "Tina Tang's fans happiness. If [x_4] becomes "*The reporter felt very happy and immediately won the boss's praise*", the content "*she interviewed a piece of explosive news*" is the cause and "*the wedding ceremony of Tina Tang will be held*" is not the cause in [x_3]. Therefore, with only an emotion cause clause extracted, it is common that one cannot exactly tell the real stimulus of a given emotion.

Extracting the exact emotion cause is very challenging. It needs not only deep text understanding including the role of each word in the emotion expression, but also requires specific semantic inference based on what is understood. Meanwhile, it is difficult to precisely determine the boundary of the cause segment, which

---

[1] Each instance in the ECA corpus contains presumably a unique emotion and at least one emotion cause clause. A clause is typically a text segment separated by punctuation marks (e.g., ',', '.', '?', '!', etc.) in the given document.

differs from the traditional Question Answering (QA) task for why questions. Because the emotion clause expressions in ECA triggering the cause finding are typically much more diverse and ambiguous, and the real cause to be extracted is generally much finer-grained. We argue that rather than only locating the coarse-grained emotion cause clause or precisely finding the exact cause segment(s), it would be more practical to adopt a hybrid extraction strategy considering clause level and word level to help us get the emotion cause.

In this paper, we attempt to extract the Emotion Cause Clause (ECC) and Emotion Cause Keywords (ECK) simultaneously. Given an emotion event, the goal of ECC task is to identify which clause contains the stimulants of emotion. ECK is a finer-grained emotion cause analysis task, which aims to identify which word(s) in the clause contribute to stimulate the emotion expression. Basically, ECK is more difficult to identify than ECC but more light-weighted than the exact emotion cause identification. The ECK task requires not only capturing the relationship between the words and emotion expression but also understanding the role of each word in the clause. However, it does not only need to identify the complete and precise cause content but also the keywords that help us better understand the emotion cause from the clause. For example, we can find that the specific cause of *"happy"* emotion in Example 1 can be better conveyed if both the emotion cause clause $[x_3]$ and emotion cause keywords, e.g., *"wedding"*, *"ceremony"*, *"Tina"* and *"Tang"* are identified.

To this end, we propose a Multi-head Attention based Multi-task learning network for Multi-level Emotion Cause Analysis (MamMeca). In the MamMeca, both ECC and ECK tasks make use of the semantic information of the text and the emotion expression to infer the cause of the emotion, for which the ECK and ECC mutually enhance each other in the unified framework. The proposed model consists of a shared feature extractor and a private feature extractor, where multi-head attention and label embedding mechanisms are designed to facilitate capturing the relationship between the two tasks. The contribution of our paper is three-fold:

- We present a multi-level ECA problem, based on the hypothesis that ECC and ECK tasks together can help us better identify the specific emotion cause and both tasks can benefit each other by mutual enhancement. To the best of our knowledge, this work is the first attempt to incorporate the two sub-tasks into a unified framework for ECA.
- We propose an extensible and effective multi-head attention based multi-task neural network for multi-level ECA. The model utilizes a shared private feature extractor to get effective representations of the keywords and clause. Meanwhile, multi-head attention and label embedding mechanisms are designed to further capture the inter-task correlations.
- Our results on a dominating benchmark dataset validate the feasibility and effectiveness of our proposed MamMeca model.

2   Related Work

Various learning methods have been applied to emotion cause analysis, which are mainly categorized as rule-based models, feature-driven models and feature-learning models.

Rule-Based Models. Lee et al. [17] first gave the formal definition of emotion cause analysis task and constructed a small-scale corpus from the Academia Sinica Balanced Chinese Corpus. Based on the corpus, Lee et al. [18] developed a rule-based system for emotion cause detection based on various linguistic rules. Some studies then extended rule-based approaches to informal texts such as Gao et al. [9]. Li et al. [19] also constructed an automatic rule-based system to detect the cause event of emotional post on Chinese microblog posts.

Feature-Driven Models. Chen et al. [1] developed two sets of linguistic features based on linguistic cues and a multi-label approach, and utilized SVM to detect emotion causes. Similarly, Gui et al. [12] extended the linguistic rules as features and used SVM model for emotion cause extraction. More recently, Gui et al. [11] released a Chinese emotion cause corpus based on public city news, which has inspired a large-scale ECA research campaign. Meanwhile, they presented a multi-kernel SVM approach for emotion cause extraction. Xu et al. [28] used LambdaMART algorithm incorporating both emotion-independent features and emotion-dependent features to identify emotion cause clause. The above models have achieved highly competitive results for ECA task, but the models heavily depend on the design of effective features.

Feature-Learning Models. Inspired by deep learning, Gui et al. [10] utilized the deep memory network model to capture the relationship between the clause and the emotion word, and then identified the emotion cause clause. Yu et al. [31] presented a hierarchical network-based clause selection framework for ECA, which considered three levels (word-phrase-clause) of information. Li et al. [21] proposed a co-attention mechanism to capture the relationship between the emotion expression and the candidate clause, and then extracted the emotion cause clause. Li et al. [20] took advantage of clues provided by the context of the emotion word and proposed a multi-attention-based neural network to identify which clause contained emotion cause. Ding et al. [5] proposed a neural network architecture to incorporate the relative position of the clause and the prediction label of previous clauses information for emotion cause clause extraction. Xia et al. [27] proposed a hierarchical network architecture based on RNN and Transformer to capture the different levels features for emotion cause clause identification. Fan et al. [7] designed a regularized hierarchical neural network (RHNN) which utilized the discourse context information and the relative position information for emotion cause clause extraction. Hu et al. [14] proposed a graph convolutional network to fuse the semantics and structural information, which automatically learned how to selectively attend the relevant clauses useful for emotion cause analysis. Recently, Xia et al. [26] proposed a new task: emotion-cause pair extraction, which aims to extract all potential pairs of emo-

Table 1. An example of illustrating the ECC task.

Clause	Content	y^c
x_1	Entertainment ... get married in 2020. (娱乐...在2020年结婚.)	0
x_2	Also, (同时，)	0
x_3	she interviewed ... held. (她采访到...新闻.)	1
x_4	The Tina Tang's fans ... congratulated her. (唐艺昕的粉丝...庆祝她.)	0

tion clause and corresponding cause clause in a text. Following this, many deep learning models [6, 8, 14, 25, 30] were designed for this task.

Discussion. Most of the previous studies attempt to extract which clause contains the emotion causes for a given emotion cause event. It is not enough to identify which clause contains the emotion cause in many application scenario, and Example 1 has illustrated this situation clearly. Only Gui et al. [10] utilized the emotion cause keyword to identify which clause contains the emotion cause, however, they still extract the emotion cause at clause level. That is, the clause is identified as emotion cause clause if it contains the emotion cause keyword in their model. Different from the previous studies, we propose to extract both the emotion cause clause and the indicative emotion cause keywords in one shot which is the first of such effort.

3 Methodology

3.1 Task Definition

Given a document d, which is a passage about an emotion cause event, it contains an emotion expression and the cause of the emotion. The document usually consists of multiple clauses $\{x_1, x_2, \cdots, x_m\}$, and each $x_i = \{w_{i1}, w_{i2}, \cdots, w_{in_i}\}$ is a clause where w_{ij} is the j-th word of x_i. Each document is assumed to have *a unique emotion* and at least one corresponding **emotion cause clause**. Let $x^e = \{w_1^e, \cdots, w_{l_e}^e, \cdots, w_{n_e}^e\}$ be the **emotion clause** containing the concerned emotion word $w_{l_e}^e$ which is the l_e-th word of x^e. In our work, both ECC and ECK tasks are seen as a binary classification problem. The expected labels of the clause or word obtained by the model is either 1 (yes) or 0 (no).

ECC Task. The goal of ECC task is to identify which clause stimulates the emotion expression. Then, the task can be formulated as

$$p_i^{y^c} = f_{ECC}(x_i, x^e) \tag{1}$$

where the function f_{ECC} identifies whether the clause x_i stimulates the emotion expressed in the emotion clause x^e, and $p_i^{y^c}$ is the predicted probability of x_i ($y^c = 1$ if x_i stimulates the emotion expressed in the x^e, or $y^c = 0$ otherwise). Table 1 illustrates ECC task clearly.

Table 2. An example of illustrating the ECK task. (Entertainment: 娱乐; reporter: 记者; the wedding ceremony: 婚礼; her: 她)

w	Entertainment	reporter	...	that	the	wedding	ceremony	...	her	.
y^w	0	0	...	0	1	1	1	...	0	0

ECK Task. ECK task aims to identify which word participates to stimulate the emotion $w_{l_e}^e$, which is formulated as

$$p_{ij}^{y^w} = f_{ECK}\left(w_{ij}, x_i, x^e\right) \tag{2}$$

where x_i and x^e are the i-th clause and the emotion clause of document, respectively. w_{ij} is the j-th word of x_i. The function f_{ECK} outputs the probability that the word w_{ij} stimulates the emotion expression or not, $p_{ij}^{y^w}$ is the predicted probability for $w_{ij} \in x_i$, and $y^w \in \{1,0\}$. To illustrate this definition, we show the labels of words in Example 1 in Table 2.

3.2   Model Description

In this section, we introduce our proposed MamMeca model that will learn task-shared feature (Sect. 3.2) and the task-private feature (Sect. 3.2). The architecture of MamMeca is given in Fig. 1, which mainly consists of three components: (1) task-shared feature extracting layer; (2) task-private feature extracting layer; and (3) classification layer. The task-shared feature extracting layer aims to capture the common features of the ECC and ECK tasks, which mainly contains two parts: shared Bi-GRU and emotion attention mechanism. After this layer, we can obtain the emotion weighted word representations, which will be further fed into the private feature extracting layer. Task-private feature extracting layer mainly contains three parts: private Bi-GRU, multi-head attention mechanism, and label embedding mechanism. Private Bi-GRUs are used for ECC task and ECK tasks to get the word level and clause level representations respectively. The emotion cause keywords must appeared in emotion cause clause which can be seen the definitions of two tasks in Sect. 3.1. Hence, the labeling embedding and multi-head attention mechanisms are designed to enhance the performance of the ECC task and ECK task by using the predicted word labels in ECK task and the clause presentation obtained in ECC task. The classification layer aims to get the class distribution of the clauses and words for ECC and ECK tasks respectively.

Task-Shared Feature Extracting Layer. This layer extracts common features shared between the two tasks, which contains two parts: (1) shared Bi-GRU encoder; (2) emotion attention mechanism.

Shared Bi-GRU Encoder. Bi-directional gated recurrent units (Bi-GRU) leverages gates to control the information flow from previous and future words,

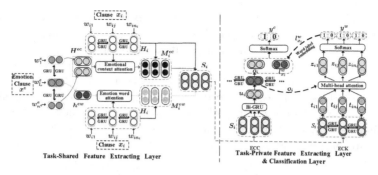

Fig. 1. The architecture of the MamMeca model. The model contains three main parts: Task-Shared feature Extracting Layer, Task-Private Feature Extracting Layer and Classification Layer. Task-Shared Feature Extracting Layer contains shared Bi-GRU encoder and emotion attention mechanism. This layer aims to capture the shared features for the ECC and ECK tasks. Task-Private Feature Extracting Layer includes private Bi-GRU encoder for specific task extraction, multi-head attention mechanism for enhancing the word representation by using the clause representation obtained by ECC task, Label embedding mechanism for enhancing the clause representation by utilizing the word label obtained in ECK task. Classification Layer is able to get the class distribution of the words and clauses, respectively.

which can better capture long term dependencies than basic RNNs, and are often chosen in practice [2]. Thus, we adopt Bi-GRU to incorporate information from both the forward and the backward directions of input sequence. In this work, we first map each word into a low dimensional embedding space and then feed the whole document into a Bi-GRU word encoder to extract word sequence features.

$$\overrightarrow{h}_{ij} = \overrightarrow{GRU}(\overline{w}_{ij}), \quad \overleftarrow{h}_{ij} = \overleftarrow{GRU}(\overline{w}_{ij}), \quad j \in \{1, \cdots, n_i\} \tag{3}$$

where $\overline{w}_{ij} \in \mathbb{R}^{d_w}$ is the embedding vector for the word w_{ij} in clause x_i at time step j and n_i is the length of clause x_i. The j-th word representation in the clause x_i can be expressed as $h_{ij} = [\overrightarrow{h}_{ij} \oplus \overleftarrow{h}_{ij}]$, where $\oplus$ denotes concatenation, $h_{ij} \in \mathbb{R}^{2d_h}$, and d_h is the size of Bi-GRU hidden vector. Therefore, we can obtain the representation matrix $H_i = [h_{i1}; h_{i2}; \cdots; h_{in_i}]$ $(H_i \in \mathbb{R}^{n_i \times 2d_h})$ of clause x_i. Symmetrically, we can obtain the emotion word $(w_{l_e}^e)$ representation vector $h^{ew} \in \mathbb{R}^{2d_h}$ and the emotion context word (w_i^e) representation $h_i^{ec} \in \mathbb{R}^{2d_h}$ $(i \in \{1, \ldots, l_e - 1, l_e + 1, \ldots, n_e\})$.

Emotion Attention Mechanism. The relationship between the candidate cause clause and the emotion clause plays an important role in emotion cause identification, which has been verified in [21]. We introduce an emotion attention mechanism to extract such words that are important to the emotion expression of the clause and aggregate the representation of these informative words to construct the clause vector. Specifically, we differentiate emotion word and emotion context which usually express different types of information. The emotion word "*happy*" in Example 1 aims to convey the emotion polarity directly while the emotion context "*The Tina Tang's fans were very - and congratulated her*" provides the related event information about the emotion, such as "*Tina Tang's fans congratulated her*" (dubbed as *emotion event*). These two types information play different roles in emotion cause identification. Hence, we get separate clause representations based on emotion word attention and emotion context attention.

(1) *Emotion word attention.* Emotion word attention is applied over the words embedding to allow the model to focus on words that contribute highly to the emotion category expression of the clause:

$$m_{ij}^{ew} = \alpha_{ij} * h_{ij}; \qquad \alpha_{ij} = \frac{\exp(h_{ij}^\top h^{ew})}{\sum_{j'=1}^{n_i} \exp(h_{ij'}^\top h^{ew})} \qquad (4)$$

where h^{ew} is the emotion word vector obtained by Bi-GRU encoder, α_{ij} is the attention weight indicating the importance of word w_{ij}, and m_{ij}^{ew} is the emotion word attention-based representation of w_{ij}. We then obtain the emotion weighted representation of x_i as $M_i^{ew} = [m_{i1}^{ew}; \ldots ; m_{in_i}^{ew}]$ where $M_i^{ew} \in \mathbb{R}^{n_i \times 2d_h}$.

(2) *Emotion context attention.* Emotion context attention allows the model to focus on words that contribute to the emotion event of the clause. The relation matrix between the clause x_i and the emotion context is constructed as $A = (H_i W_1) * (H^{ec} W_2)^\top$, where $H^{ec} = [h_1^{ec}; \cdots ; h_{l_e-1}^{ec}; h_{l_e+1}^{ec}; \ldots ; h_{n_e}^{ec}]$, $H^{ec} \in \mathbb{R}^{(n_e-1) \times 2d_h}$, $W_1, W_2 \in \mathbb{R}^{2d_h \times 2d_h}$ are trainable parameters. Each element a_{jk} ($j \in \{1, \ldots, n_i\}$, $k \in \{1, \ldots, l_e - 1, l_e + 1, \ldots, n^e\}$) of A represents the relationship between the j-th word of clause x_i and the k-th word of emotion context of x^e. The importance of the j-th word of x_i to the emotion event expression can be obtained as follows:

$$\beta_{ij} = \frac{\exp(\theta_{ij})}{\sum_{j'=1}^{n_i} \exp(\theta_{ij'})}; \qquad \theta_{ij} = max(a_{j1}, a_{j2}, \ldots, a_{jn_i}) \qquad (5)$$

θ_{ij} represents the most influential values for the emotion context obtained by x_i. Then we can obtain the new representation of x_i considering the emotion context as: $M_i^{ec} = [m_{i1}^{ec}; \ldots ; m_{in_i}^{ec}]$ where $m_{ij}^{ec} = \beta_{ij} * h_{ij}$, $M^{ec} \in \mathbb{R}^{n_i \times 2d_h}$.

Finally, the high-level representation of the clause x_i can be obtained by combining the original clause representation, the emotion word attention weighted clause representation and the emotion context attention weighted clause representation:

$$S_i = Relu((M_i^{ew} \oplus M_i^{ec}) * W_3) \oplus H_i \qquad (6)$$

where $W_3 \in \mathbb{R}^{4d_h \times 2d_h}$ is the trainable parameter.

Task-Private Feature Extracting Layer. This layer extracts private features that are specific to each task being updated exclusively, which contains three parts: (1) private Bi-GRU encoder; (2) multi-head attention mechanism; (3) label embedding mechanism.

Private Bi-GRU Encoder. For the ECC task, two private Bi-GRUs are utilized, one applied at word level and the other at clause level.

To capture the task-specific information, a private Bi-GRU is used at word level to get the representation of x_i as $u_i = [\overrightarrow{GRU}(s_{in_i}) \oplus \overleftarrow{GRU}(s_{i1})]$ $i \in \{1, 2, \ldots, m\}$, where s_{i1} and s_{in_i} are the first and the n_i-th word vectors of S_i (see Eq. (6)). The semantic expression of a clause is usually impacted by its context. Hence, we utilize another Bi-GRU applied at clause level to model the latent relation among different clauses on top of u_i. The clause-level representation of x_i can be obtained as $o_i = [\overrightarrow{GRU}(u_i) \oplus \overleftarrow{GRU}(u_i)]$, where $o_i \in \mathbb{R}^{2d_h}$.

For the ECK task, we utilize a single Bi-GRU to obtain the specific word representation for each word w_{ij} as $t_{ij} = [\overrightarrow{GRU}(s_{ij}) \oplus \overleftarrow{GRU}(s_{ij})]$ $j \in \{1, 2, \ldots, n_i\}$, where $s_{ij} \in \mathbb{R}^{2d_h}$ is the word vector of w_{ij} in S_i.

Multi-head Attention Mechanism. ECC and ECK tasks are closely related as the emotion cause keywords must appear in emotion cause clause. Our core idea is to utilize the cause clause representation generated by the ECC task to enhance the learning of cause keyword representation in the ECK task. We exploit multi-head attention mechanism to capture word correlation in each clause, based on which the high-level word representation is obtained for further classification.

Let τ denote the number of heads in the multi-head attention. We first linearly project the queries, keys and values by using different linear projections: $q_{ij} = t'_{ij}W^q$, $k_{ij} = t'_{ij}W^k$, $v_{ij} = t_{ij}W^v$. Where $t'_{ij} = t_{ij} \oplus o_i$ and $t'_{ij} \in \mathbb{R}^{4d_h}$, $W^q \in \mathbb{R}^{d_k \times d_k}$, $W^k \in \mathbb{R}^{d_k \times d_k}$ and $W^v \in \mathbb{R}^{d_k/2 \times d_k}$ are trainable parameters, and $d_k = 2d_h/\tau$. Then the attention value of the j-th word to the k-th word of clause x_i can be computed below:

$$\eta_{jk} = \frac{\exp(q_{ij} * k_{ik}^\top)}{\Sigma_{k'=1}^{n_i} \exp(q_{ij} * k_{ik'}^\top)} \tag{7}$$

The final representation of the j-th word is obtained by fusing the attention weighted vector and the query (q_{ij}): $z'_{ij} = \eta_{jk}v_{ik} + q_{ij}$, where z'_{ij} is the word representation taking into account word correlations in the clause.

Label Embedding Mechanism. The emotion cause keywords can provide important signals for locating the emotion cause clause. Therefore, we can enhance the ECC representation learning using the cause keyword labels obtained by the ECK task.

Let $l_{y^w} \in R^{d_w}$ be the embedding vector of keyword label y^w. Note that the clause, which contains emotion cause keywords, is the emotion cause clause. Therefore, the keyword label in the clause x_i also plays an important role in emotion cause clause identification. Let $\{y_{i1}^w, y_{i2}^w, \cdots, y_{in_i}^w\}$ represent the keywords labels predicted by ECK task (see Sect. 3.2). Then, the predicted keywords label embedding vector of x_i can be presented as: $l_{x_i}^w = [l_{y_{i1}^w} \oplus l_{y_{i2}^w} \oplus \cdots \oplus l_{y_{in_i}^w}] * W_l$. Finally, we obtain the new clause vector by concatenating the label embedding vector and the original clause representation vector as $o'_i = [o_i \oplus l_{x_i}^w]$.

Classification Layer. In the classification layer, the class distribution of a keyword w is computed using softmax as $p_{ij}^{y^w} = softmax(W_w z_{ij} + b_w)$, where z_{ij} is the combination of τ representation vectors (z'_{ij}), and W_w and b_w are learnable parameters. Similarly, the class distribution of clause x_i is computed as $p_i^{y^c} = softmax(W_c o'_i + b_c)$, where W_c and b_c are training parameters.

3.3   Training and Parameter Learning

Given a document d, the loss functions of ECC task and ECK task can be defined as follows:

$$\mathcal{L}_{ECC} = -\sum_{x_i \in d} \mathcal{G}(x_i) \log(p_i^{\mathcal{G}(x_i)}) \qquad \mathcal{L}_{ECK} = -\sum_{x_i \in d} \sum_{w_{ij} \in x_i} \mathcal{Y}(w_{ij}) \log(p_{ij}^{\mathcal{Y}(w_{ij})})$$

(8)

where $\mathcal{G}(x_i)$ and $\mathcal{Y}(w_{ij})$ denote the ground-truth label of x_i and w_{ij}, respectively, and $p_i^{\mathcal{G}(x_i)}$ and $p_{ij}^{\mathcal{Y}(w_{ij})}$ are the corresponding class probability predicted. The final loss function of the proposed model is given as:

$$\mathcal{L} = \lambda_1 \mathcal{L}_{ECC} + \lambda_2 \mathcal{L}_{ECK}$$

(9)

where λ_1 and λ_2 are hyper-parameters.

In the training phrase, we use Adam [16] to optimize the final loss function. After learning the parameters, we feed the test instance into the model and take the label with the highest probability as the predicted category.

4   Experiments and Results

4.1   Dataset and Settings

Dataset. Our experiments are conducted on a Chinese emotion cause analysis dataset publicly available and widely used for ECA evaluation which was collected from Sina News[2] by Gui et al. [11]. The dataset is manually annotated with the clause labels and keyword labels which contains 2,105 documents, 11,799 clauses and 2,167 emotion cause clauses. Most of the documents contain one emotion cause clause. Each clause is word segmented by Jieba[3] and the average number of words in the clause is 7.

Experimental Settings. We follow the settings of previous works to split the datasets for train/test [10,27]. We apply fine-tuning for the word vectors, which can help us improve the performance. The word vectors are initialized by word embeddings that are pre-trained on the emotion cause dataset with CBOW [22], where the dimension is 100. The trainable model parameters are given initial values by sampling from uniform distribution $\mathcal{U}(-0.01, +0.01)$. The learning rate is initialized as 0.001. Dropout [13] is taken to prevent overfitting, and the

---

[2] http://hlt.hitsz.edu.cn/?page%20id=694.
[3] https://github.com/fxsjy/jieba.

dropout rate is 0.5. The size of Bi-GRU hidden states d_h is set as 50. λ_1 and λ_2 are set as 1.0 and 0.75, respectively. Both the batch size and epochs are set to 20. The metrics of both tasks we use in evaluation include precision (P), recall (R) and F1 score ($F1$), which are the most commonly used evaluation metrics for emotion cause analysis [10, 27].

4.2 Comparison of Different Methods

For the ECC task we compare our proposed model with the following three groups models. (1) **Group I (Rule-based and knowledge-based models):** *RB* extracts the emotion cause by utilizing two sets of linguistic rules proposed by Lee et al. [17]. *KB* is a knowledge-based method [24] that uses the Chinese Emotion Cognition Lexicon [29] as the common-sense knowledge base. (2) **Group II (Feature-driven models):** *SVM (RB+KB), SVM (Word2vec)* and *SVM(n-grams)* use linguistic rules [17] plus Emotion Cognition Lexicon [29], Word2vec embeddings [23], and n-grams as features, respectively, to train a SVM classifier. *SVM (MK)* uses the multi-kernel SVMs based on structured representation of events to extract emotion cause [11]. *LambdaMART* utilizes LambdaMART algorithm incorporating emotion independent and dependent features to identify emotion cause [28]. (3) **Group III (Feature-learning models):** *ConvMS-Memnet* is a convolutional multiple-slot deep memory network for the ECC task [10]. *CANN* [21] and *MANN* [20] takes advantage of the emotion context information and designed different attention model to capture the relationship between the emotion clause and clause for ECC task. *PAE-DGL* is a reordered prediction model, which incorporates relative position information and dynamic global label for emotion cause extraction [5]. *RTHN* is a transformer hierarchical network which utilizes RNN to encode multiple words in each clause and transforms to learn the correlation between multiple clauses in a document [27]. *RHNN* is a regularized hierarchical neural network [7]. *FSS-GCN* is a graph convolutional networks with fusion of semantic and structure for emotion cause clause identification [14].

Among these methods, only **RB** and **ConvMS-Memnet** are able to identify emotion cause keywords. To test the performance on ECK task, we compare the proposed model with the rule-based model (**RB**), feature-driven model (**SVM**), and Feature-learning models (**ConvMS-Memnet, Bi-GRU, Bi-LSTM**). Furthermore, we compare the proposed model with question answering which is relevant to the ECA problem. In our experiment, we adopt **BERT** ($BERT_{BASE}$ version[4]) [4], a pre-trained bidirectional Transformer-based language model which achieves a good performance on various public question answering datasets recently [3, 15].

---

[4] https://storage.googleapis.com/bert_models/2018_11_03/chinese_L\discretionary-12_H\discretionary-768_A\discretionary-12.zip.

Results and Analysis. Table 3 shows the results of our proposed MamMeca model and baselines on ECC task. We can observe that: (1) MamMeca outperforms state-of-art baselines for ECC task on all the evaluation metrics, which clearly confirms the effectiveness of joint identification of emotion cause clause and keywords with our multi-task learning framework. (2) The $F1$ value obtained by MamMeca model outperforms the strongest baseline RHNN by 3.1%, which verifies the effectiveness of incorporating the label embedding and emotion attention mechanisms. (3) MamMeca outperforms the BERT-based QA model, which further verifies advantage of our proposed model. This is because standard QA task assumes that the question is a complete question expression while in our case the emotion clause is most likely incomplete or ambiguous rendering a more challenging problem. MamMeca can better deal with it since the complex relationship between the emotion clause and the cause clause can be captured with the joint learning.

Table 3. Results on ECC task. The results with superscript ◇ are reported in Gui et al. [10], and the rest are reprinted from the corresponding publications.

Compared with Group I and Group II				Compared with Group III			
Method	P	R	$F1$	**Method**	P	R	$F1$
RB◇	0.675	0.429	0.524	ConvMS-Memnet◇	0.708	0.689	0.696
KB◇	0.267	0.713	0.389	CANN	0.772	0.689	0.727
RB+KB◇	0.544	0.531	0.537	MANN	0.784	0.759	0.771
SVM (RB+KB)◇	0.592	0.531	0.560	PAE-DGL	0.762	0.691	0.742
SVM (n-grams)◇	0.420	0.4375	0.429	RTHN	0.770	0.766	0.768
SVM (Word2vec)◇	0.430	0.423	0.414	RHNN	0.811	0.773	0.791
SVM (MK)◇	0.659	0.693	0.675	FSS-GCN	0.786	0.757	0.771
LambdaMART	0.772	0.750	0.761	BERT	0.782	0.757	0.769
MamMeca	**0.849**	**0.798**	**0.822**	MamMeca	**0.849**	**0.798**	**0.822**

Table 4. Results on ECK task.

Method	P	R	$F1$
RB	0.228	0.643	0.337
SVM (Word2vec)	0.024	0.006	0.010
Bi-LSTM	0.150	0.332	0.207
Bi-GRU	0.149	0.311	0.202
ConvMS-Memnet	0.625	0.614	0.620
BERT	0.710	0.749	0.729
MamMeca	**0.714**	**0.774**	**0.742**

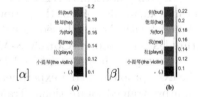

Fig. 2. Visualization of attention. Darker color represents lower attention weight.

Table 4 shows the results of the emotion cause keyword extraction. From this table, we find that our MamMeca model outperforms all the baselines including the state-of-the-art model ConvMS-Memnet [10] and the strong QA model

BERT. It gains improvement more than 12% in $F1$ compared to ConvMS-Memnet, which indicates that the proposed model's strong ability to capture the relationships between the emotion expression and the candidate cause words expressions. BERT achieves a good performance on many QA datasets, however performs worse than MamMeca on the ECK task as well. It further confirms that the QA models is not a better choice for tackling the ECA problem. In general, the emotion cause extraction is concerned about the cause of the given emotion expression instead of the relevance or similarity between the question and text.

4.3 Ablation Study

To understand the effect of different components, we compare several sub-networks of our model.

Full is the full MamMeca model. We use **Full-X** to represent the model without component **X**, where **X** can be ECK, ECC, EA, MA and LE corresponding to ECK private parameters, ECC private parameters, Emotion Attention, Multi-head Attention, and Labeling Embedding mechanisms, respectively.

The performance of above models are shown in Tables 5 and 6. As expected, the results in F1-score of the sub-networks all drop. This clearly demonstrates the usefulness of these components. Both Full-ECK and Full-ECC are worse which confirms that joint training of two tasks is helpful for learning the effective features. On the one hand, the word label predicted by ECK task is able to provide the important emotion cause signal which help inferring that whether the clause is the emotion cause clause. For example, if there are some words are predicted as emotion cause keywords, the model will increase the probability of the current clause being predicted as an emotion cause clause. On the other hand, the clause representation obtained by ECC task is able to give a positive impact for emotion cause keyword prediction. That is, if the current clause is predicted as emotion cause clause, the words in this clause more likely be the emotion cause keywords. Full gains 1.6% improvement in $F1$ over Full-EA, which indicates that the emotion attention can provide important information for emotion cause keywords extraction. In Table 5, when removing the word label embedding mechanism, the $F1$ score of Full-LE decreases 2.9%, which indicates the word label embedding from ECK task is conducive to ECC task. Also, Full gains 10.5% improvement in $F1$ over Full-MA indicating that the ECC task can enhance the performance of the ECK task by multi-head attention mechanism in Table 6. We also find that Full-ECK outperforms the strong baseline RHNN, which maybe due to the case that considering the emotion word and context differently is effective.

Table 5. Ablation test results of ECC task.

Model	P	R	$F1$
Full	**0.849**	**0.798**	**0.822**
Full-ECK	0.807	0.786	0.796
Full-EA	0.818	0.821	0.819
Full-LE	0.830	0.761	0.793
Full-MA	0.816	0.779	0.796

Table 6. Ablation test results of ECK task.

Model	P	R	$F1$
Full	**0.714**	**0.774**	**0.742**
Full-ECC	0.662	0.690	0.674
Full-EA	0.689	0.771	0.726
Full-LE	0.696	0.745	0.718
Full-MA	0.621	0.655	0.637

4.4   Case Study

To show how emotion attention and self-attention mechanisms work, we visualize the attention weights α_{ij} (in Eq. (4)) and β_{ij} (in Eq. (5)) with heatmap. Example 2 illustrates the detail with a training example.

Example 2. [x_1] : 后士凤心中充满感激。[x_2] : 她说: [x_3] : 虽然我们并不熟悉, [x_4] : 但他却为我拉小提琴, [x_5] : 我十分开心。(**In English:** [x_1] : Shifeng Hou's heart is full of gratitude. [x_2] : She said: [x_3] : we are not familiar, [x_4] : but he **plays the violin for me**, [x_5] : and I'm very happy.)

Figure 2(a) and (b) represent the attention distribution of emotion word and emotion context to the each word of x_4. In Fig. 2(a), "*but*", and "." have low attention score as they are indeed irrelevant with respect to the emotion cause expression. Figure 2(b) shows that the words "*for*" and "*me*" in clause x_4 are paid more attention by the emotion context, which means that the emotion cause has a close relation with these two words. From Fig. 2(a) and (b), we can easily find the words "*me*", "*plays*", "*the violin*" in the clause x_4 have higher attention weights than "*but*" and punctuation ".", implying that the words, which help express the cause, are more important and thus captured by the emotion attention mechanism. These again verify the effectiveness of our proposed emotion attention mechanism on emotion cause analysis.

4.5   Error Analysis

We notice that for some passages which have the long distance between the emotion word and the cause, our model may have a difficulty in detecting the correct emotion cause keywords. We show an example to illustrate this situation (see Example 3). From the example, we can find the emotion cause of the emotion "angry" is "the old lady who was helped up ran to the front of the bus and sat down on the ground". However, the emotion cause keywords obtained by our model is "Seeing this scene". It is a challenging task to properly model the words which have long-distance with the emotional expression. In the feature, we will explore different network architecture with consideration of the various relationship between the words and emotion expression.

Example 3. 没想到，徐连林刚准备发动汽车离开车站，那位**被扶起的老太以迅雷不及掩耳之势跑到了公交车前一屁股坐在了地上**. 站在公交车前部的乘客都将这一幕看得一清二楚，看到这一幕，车上的乘客立刻炸开了锅，激烈争论起来。其中一部分乘客很气愤，一边数落徐连林"不该多事"一边给他"上课"："我叫你们别去管这事吧。"

In English: Unexpectedly, when Lianlin Xu was just about to activate the car and leave the station, **the old lady who was helped up ran to the front of the bus and sat down on the ground.** The passengers standing in the front of the bus saw the scene clearly. Seeing this scene, the passengers on the bus immediately burst into a boiling pot and argued fiercely. Some of the passengers were very <u>angry</u>. They accused Lianlin Xu of "not being too busy" while giving him a lesson: "I told you not to take care of this."

5 Conclusions

In this paper, we study the multi-task learning approach to identify emotion cause at clause level and word level simultaneously. We propose an effective multi-head attention based multi-task learning network, which utilizes shared-private feature extractor, multi-head attention mechanism and label embedding mechanism to enable two tasks to interact with each other for better learning the task-oriented representations. Results on benchmark dataset for ECA task demonstrate that our model can effectively extract multi-level emotion causes, and outperform the strong QA-based system and other strong ECA baselines by large margins. In the future, we plan to focus on extracting the specific cause(s) in a more accurate granularity for improving emotion cause analysis.

References

1. Chen, Y., Lee, S.Y.M., Li, S., Huang, C.R.: Emotion cause detection with linguistic constructions. In: Proceedings of the 23rd International Conference on Computational Linguistics, pp. 179–187 (2010)
2. Cho, K., et al.: Learning phrase representations using RNN encoder-decoder for statistical machine translation. In: Moschitti, A., Pang, B., Daelemans, W. (eds.) EMNLP 2014, pp. 1724–1734 (2014)
3. Cui, Y., et al.: A span-extraction dataset for Chinese machine reading comprehension. arXiv preprint arXiv:1810.07366 (2018)
4. Devlin, J., Chang, M., Lee, K., Toutanova, K.: BERT: pre-training of deep bidirectional transformers for language understanding. In: NAACL-HLT 2019, pp. 4171–4186 (2019)
5. Ding, Z., He, H., Zhang, M., Xia, R.: From independent prediction to re-ordered prediction: integrating relative position and global label information to emotion cause identification. In: The Thirty-Third AAAI Conference on Artificial Intelligence, AAAI 2019 (2019)
6. Ding, Z., Xia, R., Yu, J.: ECPE-2D: emotion-cause pair extraction based on joint two-dimensional representation, interaction and prediction. In: Jurafsky, D., Chai, J., Schluter, N., Tetreault, J.R. (eds.) ACL, pp. 3161–3170 (2020)

7. Fan, C., et al.: A knowledge regularized hierarchical approach for emotion cause analysis. In: EMNLP-IJCNLP 2019, pp. 5618–5628 (2019)

8. Fan, C., Yuan, C., Du, J., Gui, L., Yang, M., Xu, R.: Transition-based directed graph construction for emotion-cause pair extraction. In: Jurafsky, D., Chai, J., Schluter, N., Tetreault, J.R. (eds.) ACL 2020, pp. 3707–3717 (2020)

9. Gao, K., Xu, H., Wang, J.: Emotion cause detection for Chinese micro-blogs based on ECOCC model. In: Pacific-Asia Conference on Knowledge Discovery and Data Mining, pp. 3–14 (2015)

10. Gui, L., Hu, J., He, Y., Xu, R., Lu, Q., Du, J.: A question answering approach to emotion cause extraction. In: Proceedings of the 2017 Conference on Empirical Methods in Natural Language Processing, pp. 1953–1602 (2017)

11. Gui, L., Wu, D., Xu, R., Lu, Q., Zhou, Y.: Event-driven emotion cause extraction with corpus construction. In: EMNLP, pp. 1639–1649 (2016)

12. Gui, L., Yuan, L., Xu, R., Liu, B., Lu, Q., Zhou, Y.: Emotion cause detection with linguistic construction in Chinese Weibo text. In: Natural Language Processing and Chinese Computing, pp. 457–464 (2014)

13. Hinton, G.E., Srivastava, N., Krizhevsky, A., Sutskever, I., Salakhutdinov, R.R.: Improving neural networks by preventing co-adaptation of feature detectors. arXiv preprint arXiv:1207.0580 (2012)

14. Hu, G., Lu, G., Zhao, Y.: FSS-GCN: a graph convolutional networks with fusion of semantic and structure for emotion cause analysis. Knowl. Based Syst. **212**, 106584 (2021)

15. Hu, M., Peng, Y., Huang, Z., Li, D.: A multi-type multi-span network for reading comprehension that requires discrete reasoning. In: Inui, K., Jiang, J., Ng, V., Wan, X. (eds.) EMNLP-IJCNLP 2019, pp. 1596–1606 (2019)

16. Kingma, D.P., Ba, J.: Adam: a method for stochastic optimization. In: International Conference on Learning Representations, pp. 1–15 (2015)

17. Lee, S.Y.M., Chen, Y., Huang, C.R.: A text-driven rule-based system for emotion cause detection. In: Proceedings of the NAACL HLT 2010 Workshop on Computational Approaches to Analysis and Generation of Emotion in Text, pp. 45–53 (2010)

18. Lee, S.Y.M., Ying, C., Huang, C.R., Li, S.: Detecting emotion causes with a linguistic rule-based approach. Comput. Intell. **29**(3), 390–416 (2013)

19. Li, W., Hua, X.: Text-based emotion classification using emotion cause extraction. Expert Syst. Appl. **41**(4), 1742–1749 (2014)

20. Li, X., Feng, S., Wang, D., Zhang, Y.: Context-aware emotion cause analysis with multi-attention-based neural network. Knowl.-Based Syst. **174**, 205–218 (2019)

21. Li, X., Song, K., Feng, S., Wang, D., Zhang, Y.: A co-attention neural network model for emotion cause analysis with emotional context awareness. In: Proceedings of the 2018 Conference on Empirical Methods in Natural Language Processing, pp. 4752–4757 (2018)

22. Mikolov, T., Chen, K., Corrado, G., Dean, J.: Efficient estimation of word representations in vector space. In: Bengio, Y., LeCun, Y. (eds.) ICLR 2013 (2013)

23. Mikolov, T., Sutskever, I., Chen, K., Corrado, G.S., Dean, J.: Distributed representations of words and phrases and their compositionality. In: Advances in Neural Information Processing Systems, pp. 3111–3119 (2013)

24. Russo, I., Caselli, T., Rubino, F., Boldrini, E., Martínez-Barco, P.: EMOCause: an easy-adaptable approach to emotion cause contexts. In: Proceedings of the 2nd Workshop on Computational Approaches to Subjectivity and Sentiment Analysis, pp. 153–160 (2011)

25. Tang, H., Ji, D., Zhou, Q.: Joint multi-level attentional model for emotion detection and emotion-cause pair extraction. Neurocomputing **409**, 329–340 (2020)
26. Xia, R., Ding, Z.: Emotion-cause pair extraction: a new task to emotion analysis in texts. In: Korhonen, A., Traum, D.R., Màrquez, L. (eds.) ACL 2019, pp. 1003–1012 (2019)
27. Xia, R., Zhang, M., Ding, Z.: RTHN: A RNN-transformer hierarchical network for emotion cause extraction. In: Proceedings of the Twenty-Eighth International Joint Conference on Artificial Intelligence, IJCAI 2019 (2019)
28. Xu, B., Lin, H., Lin, Y., Diao, Y., Yang, L., Xu, K.: Extracting emotion causes using learning to rank methods from an information retrieval perspective. IEEE Access **7**, 15573–15583 (2019)
29. Xu, R., et al.: A new emotion dictionary based on the distinguish of emotion expression and emotion cognition. J. Chinese Inf. Process. **27**(6), 82–90 (2013)
30. Yu, J., Liu, W., He, Y., Zhang, C.: A mutually auxiliary multitask model with self-distillation for emotion-cause pair extraction. IEEE Access **9**, 26811–26821 (2021)
31. Yu, X., Rong, W., Zhang, Z., Ouyang, Y., Xiong, Z.: Multiple level hierarchical network-based clause selection for emotion cause extraction. IEEE Access **7**, 9071–9079 (2019)

Text Generation and Summarization

Using Query Expansion in Manifold Ranking for Query-Oriented Multi-document Summarization

Quanye Jia, Rui Liu[✉], and Jianying Lin

State Key Laboratory of Software Development Environment, Beihang University,
Beijing 100191, People's Republic of China
{jiaquanye,lr,jianying.lin}@buaa.edu.cn

Abstract. Manifold ranking has been successfully applied in query-oriented multi-document summarization. It not only makes use of the relationships among the sentences, but also the relationships between the given query and the sentences. However, the information of original query is often insufficient. So we present a query expansion method, which is combined in the manifold ranking to resolve this problem. Our method not only utilizes the information of the query term itself and the knowledge base WordNet to expand it by synonyms, but also uses the information of the document set itself to expand the query in various ways (mean expansion, variance expansion and TextRank expansion). Compared with the previous query expansion methods, our method combines multiple query expansion methods to better represent query information, and at the same time, it makes a useful attempt on manifold ranking. In addition, we use the degree of word overlap and the proximity between words to calculate the similarity between sentences. We performed experiments on the datasets of DUC 2006 and DUC2007, and the evaluation results show that the proposed query expansion method can significantly improve the system performance and make our system comparable to the state-of-the-art systems.

Keywords: Multi-document summarization · WordNet · Query expansion · Manifold ranking

1 Introduction

Query-focused multi-document summarization is to create a summary from a set of documents that answers the information requirements expressed in the query. Compared to generic summarization, query-focused summarization requires the summary biased to a specific query besides the general requirement for a summary. In contrast to the task of question answering (QA) that mainly focuses on simple factoid questions and results in precise answers such as person, location or date, etc., in the case of query-focused summarization, the queries are mostly real-world complex questions and the information provided by the queries is

© Springer Nature Switzerland AG 2021
S. Li et al. (Eds.): CCL 2021, LNAI 12869, pp. 97–111, 2021.
https://doi.org/10.1007/978-3-030-84186-7_7

insufficient. So how to understand the query and expand the query has a certain impact on the quality of the query-oriented summary.

Currently most query oriented abstract systems are based on general summarization systems. By incorporating features that are related to the given query (e.g., the relevance of a sentence to the query), a generic summarization system can be adapted to a query-focused one. Due to the limit of information that a query can express, some systems also expand the query using some external resources such as WordNet, by which the synonyms of the query words can be obtained as expansion words [30]. However, such approaches to query expansion are restricted in query itself and external resources, because they cannot be applied to words not in WordNet such as named entities which frequently occur in queries, and much context related information cannot be captured by only synonyms. Lin [14] uses graph-based sentence ranking and sentence-to-word relationships to implement query expansion, which can improve the effect of graph ranking. However, the paper uses a single query expansion method and only adopts TextRank instead of manifold ranking as graph ranking algorithm. So we improved the query expansion method based on the above problems.

The main contributions of this work are: (1) We used multiple query expansion methods to expand the query, including the query itself (using Wordnet synonyms) and the expanded query based on the document set (mean expansion, variance expansion and TextRank expansion). (2) We improved the calculation of sentence similarity matrix in the manifold ranking process, using sentence TF-ISF similarity, word overlap similarity, and sentence proximity information. Our experiments on the dataset DUC 2006 and DUC 2007 show that the summarization results with query expansion are much better than that without query expansion, achieving the state-of-the-art performance.

2   Related Work

Graph-based ranking algorithms have been successfully used in text summarization. [8] proposed LexRank for generic text summarization. They construct a connected similarity graph where nodes represent sentences and edges represent cosine similarities between sentences. A random walk is applied on the graph until converging to a stationary distribution, by which the sentences can be ranked. Topic-sensitive LexRank [16] has been applied to the task of query-focused summarization where the relevance of a sentence to the query is taken into account when performing random walk. [26] proposed a hypergraph based vertex-reinforced random walk framework for multi-document summarization. [24] applied a manifold ranking algorithm to query-focused summarization which can simultaneously make full use of both the relationships among all the sentences in the documents and the relationships between the given query and the sentences.

Regarding manifold ranking algorithms, subsequent researchers have made a lot of improvements. [21] introduced the matrix factorization model on the basis of Xiaojun Wan's model, and constructed the sentence relational matrix by using the cosine value of the decomposed matrix. [4] used the syntactic and shallow semantic kernels to calculate the correlation between sentences based on the model of Xiaojun Wan. On the basis of Xiaojun Wan's model, [13] improves the relationship matrix by mixing the cosine similarity of decomposed matrix and TF-ISF cosine similarity in a certain proportion. At the same time, this paper uses lifelong learning topic model to enhance the co-occurrence effect of matrix factorization words. Finally, this paper adds statistical features, mix various scoring methods together, and then extract summaries with the strategy of de-redundancy. [2] not only considers the internal relevance propagation within the sentence set (or within the theme cluster set), but also considers the mutual reinforcement between the ranking of the sentence set and the ranking of the cluster set. These models mainly focus on the calculation of sentence similarity and sentence ranking, but pay little attention to queries.

Query understanding is a research hotspot in the field of search engines. [1] proposed the use of a corpus to expand the original query. [9] proposed a Wikipedia-based user query intent understanding. [17] considered the distribution of expansions in documents, statistical associations with query terms, and semantic associations. [25] and [18] combined query expansion with word meaning elimination to enhance the validity of the query. To the best of our knowledge, few people have done the research on the effect of query understanding and query expansion on manifold ranking. Therefore, based on the above research, we use multiple query expansion methods to improve the effect of model based on manifold ranking.

3   Method

In this section, this paper will introduce the model architecture, as Fig. 1 shows.

For the preprocessing step, this paper performed sentence segmentation, word segmentation, word stemming and TF-ISF calculation for each sentence in the documents and the query. This paper utilized titles and narratives as queries. When computing the sentence similarities, stop words, emails, digital characters, punctuation marks and single characters were removed, and only the content words (i.e., noun, verb, adjective and adverb) were used after the morphological analysis based on treetagger.

For the extraction process of automatic text summarization, this paper selects the method used by [24].

Manifold ranking, query expansion and sentence similarity calculation and will be introduced in detail in following sections.

3.1   Manifold Ranking

Manifold ranking [28,29] is a universal graph ranking algorithm and it is initially used to rank data points along their underlying manifold structure. The prior assumption of manifold-ranking is: (1) sentences on the same structure are likely to have the same ranking scores; (2) sentences similar to the query

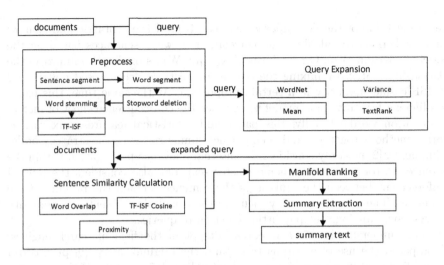

Fig. 1. Model architecture.

sentence have higher ranking scores. The process of the manifold ranking algorithm can be understood as follows: construct a similarity matrix W between all sentences, initialize a vector $y = [y_0, y_1, \cdots, y_n]^T$, in which $y_0 = 1$ because x_0 is the query sentence and $y_i = 0 (1 \leq i \leq n)$ for all the sentences in the documents. Then the ranking algorithm transmits the scores of query sentences to the scores of sentences similar to query points through the similarity between sentences, and then the scores of each sentence are propagated to the scores of other similar sentences, iterating many times until the scores of each sentence remain unchanged. The algorithm for manifold ranking is as Formula 1 shows:

$$L(f) = \alpha_{mr} * \sum_{i,j=1}^{N} W_{ij} \left| \frac{1}{\sqrt{D_{ii}}} f_i - \frac{1}{\sqrt{D_{jj}}} f_j \right|^2 + (1 - \alpha_{mr}) * \sum_{i=1}^{N} |f_i - y_i|^2$$

$$\text{s.t.} D_{ii} = \text{sum}(W_{i*}), Y = [y_1\, y_2\, \cdots\, y_N]^T = [1\, 0 \cdots 0]^T \tag{1}$$

Where $F = [f_1\quad f_2\quad \cdots\quad f_n]^T$ is the sentence score vector for the solution, and Y is the vector indicating the query sentence. This paper assumes that when A_i is the query sentence or the query sentence $y_i = 1$, and otherwise is 0, W_{ij} indicates the relationship between the i-th sentence and the j-th sentence. There are many ways to calculate the relationship between sentences, such as Manhattan distance, Euler distance, inner product calculation method, cosine similarity, etc., or a combination of various methods. D_{ij} represents the sum of the elements of the i-th row in the W matrix, and α_{mr} represents the proportion of the similarity between the sentences in the manifold ranking.

3.2   Query Expansion

In the query-oriented multi-document summarization, each topic has one or two sentences as the query, but the information of these sentences is often insufficient. Therefore, the query sentence should be expanded. There are two general methods: one is to expand according to the content of documents. The other is to expand based on the query itself, which usually depends on the understanding of the query. For each topic, we extract query sentences and sentences of multiple documents to form a sentence-word matrix $A^{(0)} = \left[\, T^{(0)} \quad X \, \right]^{\mathrm{T}}$ of N * M, and then use the above two methods to expand the query sentence, the formula for expanding query sentences is as Formula 2 shows:

$$T^{(1)} = \text{sim\_word}(T^{(0)}) + \theta_m \text{mean}(X) + \theta_v \text{variance}(X) + \theta_r \text{TextRank}(A^{(0)}) \tag{2}$$

Where A is a sentence-word matrix, the word weight is calculated using the TF-ISF (as [14] use) formula, T is query sentences, X is all document sentences, N is the number of the sentences, M is the number of all various content words in the documents, $\text{sim\_word}(T^{(0)})$ is the expansion of the word meaning of the query sentence, $\text{mean}(X)$ is the mean expansion of the query sentence, $\text{variance}(X)$ is the variance expansion of the query sentence, $\text{TextRank}(A^{(0)})$ is the TextRank expansion of the query sentence.

Query Expansion by WordNet. WordNet is a large English word sense database [7]. It not only explains the meaning of words, but also links the meaning of words together. In WordNet, each of synset (semantics) represents a basic lexical concept. A word can have more than one synonym, and a synonym can correspond to more than one word. Synonymous phrases are connected by semantic relations such as hypernymy-hyponymy, synonym-antonym, meronymy-holonyms and implication, forming a network structure with synonymous phrases as nodes and semantic relations as edges. Therefore, semantic similarity can be calculated by the length of the edge between two semantic nodes.

Semantic similarity calculation is based on [19], which mainly uses the path length between concepts. The shorter the path between two nodes is, the higher the semantic similarity between the two concepts is. In this paper, the similarity is calculated using the following formula 3:

$$\text{sim}(syn_1, syn_2) = \frac{a}{a + d} \tag{3}$$

Where $syn1$, $syn2$ refers to the semantic node of a word, d represents the number of upper edges of the shortest path between the semantic nodes $syn1$ and $syn2$ in the semantic graph, and a represents an adjustable parameter.

For most applications, there is no data with word meaning labels, so algorithms are needed to provide us with similarities between words rather than between meanings or concepts. For any algorithm based on a semantic dictionary, according to Resnik [20], we can approximate the similarity by using the maximum similarity between two word semantic items. Therefore, based on word meaning similarity, we can define word similarity as following formula 4:

$$sim(w_1, w_2) = \max_{\substack{syn_1 \in senses(w_1) \\ syn_2 \in senses(w_2)}} sim(syn_1, syn_2) \tag{4}$$

Where $w1$, $w2$ represent the target word, $syn1$, $syn2$ refer to the sense of the word. For a sentence-word matrix $A^{(0)}$ of $N * M$, the similarity of all words in the document can be obtained first, and then the word similarity matrix of $M * M$ can be obtained. According to the matrix, each word of the query sentence can be expanded. The expanded formula is as Formula 5 shows.

$$t_i^{(1)} = \max_j(t_j^{(0)} \cdot sim(w_j, w_i)) \tag{5}$$

$t_i^{(0)}$ represents the i-th word feature before expansion, $t_i^{(1)}$ represents the expanded i-th word feature, and w_i represents the i-th word item, then $T^{(1)} = sim\_word(T^{(0)}) = \begin{bmatrix} t_1^{(1)} & t_2^{(1)} & \cdots & t_M^{(1)} \end{bmatrix}$ is the query sentence after the word meaning expansion. By expanding the meaning of the query sentence, it can make the query sentence reflect the features of the similar words in the original sentence, and solve the problem that the feature of the query sentence is not salient because the original query sentence has too few words.

Word Similarity Filtering. According to the above word similarity calculation method, we can extract all words from the dataset to calculate the word similarity matrix. However, most of the similarities between words have no practical significance, so they need to be filtered. There are two kinds of filtering methods in this paper: vertical filtering and horizontal filtering.

Vertical filtering is based on the path length between words. For example, computer and football are obviously not similar, but because they can be connected by superordinate entities, even if the shortest path between them is very long, they will still get smaller values after similarity calculation. This degree of similarity has little significance for feature calculation, so smaller values can be discarded. In this way, not only computational meaningful features can be expanded, but also sparse matrix storage can be used to reduce the waste of storage space. Therefore, this paper will use vertical filtering parameter as L to filter the similarities which the shortest path is greater than L.

Horizontal filtering is based on the number of similarities between a word and other words. When a word is similar to most words, it shows that the feature of the word itself is not salient, so it needs to be filtered. For example, the verbs "use" and "do" are similar to almost all verbs. If these verbs are expanded, noise will be introduced into the textual features of the query sentences, which will not highlight the topic feature of the query sentences. Therefore, this paper will use horizontal filtering parameter as C to retain the words which are similar to no more than C other words.

Query Expansion by Mean and Variance. Due to the large amount of information contained in the text, the average TF-ISF value and the variance TF-ISF value of each word in the document can be obtained to expand the query sentence, as shown in the following Formula 6 and 7 show.

$$\text{mean}(X) = \begin{bmatrix} \bar{X}_{*1} & \bar{X}_{*2} & \cdots & \bar{X}_{*M} \end{bmatrix}$$
$$= \begin{bmatrix} \dfrac{1}{N-1} \sum_{i=1}^{N-1} X_{i1} & \dfrac{1}{N-1} \sum_{i=1}^{N-1} X_{i2} & \cdots & \dfrac{1}{N-1} \sum_{i=1}^{N-1} X_{iM} \end{bmatrix} \quad (6)$$

$$\text{variance}(X) = \begin{bmatrix} \dfrac{1}{N-2} \sum_{i=1}^{N-1} (X_{i1} - \bar{X}_{*1})^2 & \cdots & \dfrac{1}{N-2} \sum_{i=1}^{N-1} (X_{iM} - \bar{X}_{*M})^2 \end{bmatrix} \quad (7)$$

N represents the number of all sentences, X_{ij} represents the i-th document sentence and the j-th word item feature, $\bar{X}_{*j}$ represents the average of the j-th word item feature in the documents.

Query Expansion by TextRank. In this section, we select from the document set both informative and query relevant words based on TextRank (Sect. 3.2 in [14]) results, add them into the original query and use the updated query to perform manifold ranking again. Our query expansion approach goes as follows:

1. Normalize X by $S = D^{-1}X$ to make the sum of each row equal to 1, where D is the diagonal matrix with (i, i)-element equal to the sum of the ith row of X.

2. Calculate vector y by $y = S^T p^*$, where p^* is the vector of sentence ranking scores derived in the last step in the TextRank algorithm described in Lin et al. [14], and y represents the word scores.

3. Rank all the words based on their scores in y and select the top c words as query expansions. c is a parameter representing the number of expansion words, which is set in the experiments.

4. Add the top c words into the query sentence T, the formula is as follows:

$$TextRank(A^{(0)}) = \begin{bmatrix} t'_1, & t'_2, & \cdots, & t'_n \end{bmatrix} \quad (8)$$

$$t'_i = \begin{cases} 1, \text{when the words corresponding to t are in the top c words} \\ 0, \text{when the words corresponding to t are not in the top c words} \end{cases} \quad (9)$$

Where t'_i represents the i-th word feature after expansion. This algorithm is to make use of both the sentence importance and sentence-to-word relationships to select the expansion words. By this step, salient words occurring in the important sentences are more likely to be selected, and because the higher ranked sentences are biased towards the query, the words selected in this way are also biased towards the query.

3.3 Sentence Similarity Calculation

The W matrix is used to measure the similarity between two sentences, which is critical in manifold ranking. If the calculation method of W is closer to the similarity between sentences, the quality of the summary is higher. This paper mainly uses TF-ISF cosine similarity, word overlap and proximity similarity to calculate. The TF-ISF cosine similarity and word overlap use the word frequency characteristics of the sentence itself. Word overlap only considers the number of words co-occurring in two sentences, regardless of the occurrence of words in other sentences, which is beneficial to increase the number of different words in the summary. The adjacency similarity mainly considers the relationship between the relative positions of two sentences, so that the meaning of the sentence in the context is reflected. The three are organically combined, as shown in the following Formula 10:

$$W_{ij} = \alpha_A \cdot \cos(A_{i*}^{(1)}, A_{j*}^{(1)}) + \alpha_{overlap} \cdot SS_{ij} + \alpha_{peer} \cdot P_{ij} \tag{10}$$

Where $\cos(A_{i*}^{(1)}, A_{j*}^{(1)})$ represents the cosine similarity of the TF-ISF of the sentence $A_{i*}^{(1)}$ and $A_{j*}^{(1)}$. SS_{ij} indicates the degree of word overlap of the sentence $A_{i*}^{(0)}$ and $A_{j*}^{(0)}$, that is, the ratio of the number of words appearing together in two sentences to the minimum word length of two sentences.

For the position information of the sentence, this paper mainly considers the context of the sentence. P_{ij} is used to represent the adjacency matrix of the sentence. One topic has multiple documents, and one document has multiple sentences. These sentences constitute the context and the sentences are in a specific document. There is a specific context, each sentence must have a certain relationship with the adjacent sentence, so the adjacency matrix can be used to quantify the relationship between the sentences. The closer the sentences are, the higher the similarity score is. The representation is as shown in the following Formula 11:

$$P_{ij} = \begin{cases} 0.1^{|i-j|}, \text{i and j are in the same document} \\ 0, \text{ i and j are not in the same document} \end{cases} \tag{11}$$

In addition, when solving the cosine similarity matrix of TF-ISF between sentences, this paper uses the expanded query sentence as the query sentence. When calculating the sentence word overlap matrix, the original query sentence is used as the query sentence.

4   Experimental Setup

4.1 Datasets and Evaluation

The summary data sets used in this experiment are DUC2006 [5] and DUC2007 [6]. DUC 2006 and DUC 2007 are query-oriented multi-document summary datasets, each data set contains multiple topics, each topic consists of multiple related documents, and each topic provides a title and a narrative as a query. For each topic, we take the first 250 words as the model summary from the results and compare it with the expert summary by using ROUGE [12] toolkit. We report F1 scores of ROUGE-1, ROUGE-2, ROUGE-W and ROUGE-SU4 metrics.

4.2 Parameter Settings

The parameters of the proposed methods are determined according to the overall effect of the model on DUC data set. The parameter a is 1, L is 4, C is 5000, the mean parameter θ_m is 1, the variance parameter θ_v is 1, r_t is 0.4, the α_{mr} is 1, the cosine similarity parameter α_A is 0.9, the word coverage parameter $\alpha_{overlap}$ is 0.1, and the adjacency similarity parameter α_{peer} is 0.4. ω as a de-redundancy parameter, takes 8 (refer to [24]). In the TextRank expansion, the c is 100 and d is 0.6 (refer to [14]).

5 Results

5.1 Comparison with Query Expansion Methods

In this experiment, our aim is to examine the efficiency of the combination of these query expansion methods. The results are reported in Table 1.

From the report, we can figure out that the performance of the method using expansion of query is better than that using the original query. In other words, the results of ori+MEAN+VAR+TextRank are better than one that combines a subset of query expansion methods.

5.2 Comparison with Related Methods

In this experiment, we will compare the model performance with the other existing well-known methods. We compare QE-WMVT-Mani (Manifold ranking with word similarity, mean, variance and TextRank query expansion) with other methods: 1) Random (For each topic, randomly extracts a certain number of sentences as a summary); 2) Lead (Under each topic, select a certain number of sentences in the document of the most recent time as a summary); 3) MV-CNN [27]; 4) AttSum [3]; 5) VAEs-A [11]; 6) QODE [15]; 7) C-Attention [10]; 8) MultiMR [23]; 9) JMFMR [22]; 10) JTMMR [13]; 11) JLTMMR [13]; 12) JLTMMR + SF [13]; 13) RDRP_AP [2]; 14) HERF [26]; 15) TextRank [14]; 16) QE-T+TextRank [14].

For MV-CNN, QODE, AttSum, VAEs-A, C-Attention, MultiMR, JMFMR, JTMMR, JLTMMR, JLTMMR+SF, RDRP_AP and HERF models, this paper chooses results from their corresponding papers. For the QE-T+TextRank method, because the original paper does not give the F value, this paper re-experiments it, and the preprocessing process is the same as the original paper. r_t is set to 0.4, and other parameters are the same as the original paper.

Table 1. ROUGE evaluation results of expanded query sentences and original query sentences in DUC2006 and DUC2007.

Dataset	Extension method	Rouge-1	Rouge-2	Rouge-W	Rouge-SU4
DUC2006	ori	0.40331	0.08809	0.13960	0.14637
	ori+MEAN	0.40695	0.09057	0.14099	0.14827
	ori+VAR	0.40604	0.09013	0.14060	0.14835
	ori+SIM_WORD	0.41019	0.08785	0.14084	0.14691
	ori+TextRank	0.41658	0.09174	0.14274	**0.15102**
	ori+SIM_WORD+MEAN	0.41129	0.08895	0.14078	0.14762
	ori+SIM_WORD+VAR	0.41112	0.08879	0.14113	0.14762
	ori+MEAN+VAR +SIM_WORD	0.41060	0.08934	0.14090	0.14723
	ori+MEAN+VAR +SIM_WORD+TextRank	**0.41674**	**0.09202**	**0.14279**	0.15071
DUC2007	ori	0.42010	0.10372	0.14559	0.16051
	ori+MEAN	0.42612	0.10638	0.14773	0.16329
	ori+VAR	0.42149	0.10444	0.14616	0.16102
	ori+SIM_WORD	0.42565	0.10195	0.14564	0.15991
	ori+TextRank	0.43557	0.11174	0.15088	0.16668
	ori+SIM_WORD+MEAN	0.42907	0.10447	0.14706	0.16163
	ori+SIM_WORD+VAR	0.42699	0.10184	0.14604	0.16016
	ori+MEAN+VAR +SIM_WORD	0.43042	0.10574	0.14768	0.16295
	ori+MEAN+VAR +SIM_WORD+TextRank	**0.43982**	**0.11185**	**0.15159**	**0.16870**

Table 2. Rouge results of multi-document summarization on DUC 2006.

Dataset	System	Rouge-1	Rouge-2	Rouge-W	Rouge-SU4
DUC2006	Random	0.35352	0.05381	0.12135	0.111
	Lead	0.3396	0.05412	0.11748	0.10961
	MV-CNN	0.3865	0.0791	–	0.1409
	QODE	0.4015	0.0928	–	0.1479
	AttSum	0.409	**0.094**	–	–
	VAEs-A	0.396	0.089	–	0.143
	C-Attention	0.393	0.087	–	0.141
	MultiMR	0.40306	0.08508	0.13997	–
	JMFMR	0.41244	0.0887	–	0.14585
	JTMMR	0.40014	0.08160	0.13652	0.14062
	JLTMMR	0.4043	0.08126	0.13774	0.14032
	JLTMMR+SF	0.41045	0.08926	0.14214	0.14798
	RDRP_AP	0.39615	0.08975	–	0.13905
	TextRank	0.39056	0.0871	0.13536	0.14299
	QE-T+TextRank	0.40115	0.08985	0.13804	0.14619
	QE-WMVT+Mani	**0.41674**	**0.09202**	**0.14279**	**0.15071**

Table 3. Rouge results of multi-document summarization on DUC2007.

Dataset	System	Rouge-1	Rouge-2	Rouge-W	Rouge-SU4
DUC2007	Random	0.36896	0.06654	0.125	0.12312
	Lead	0.35461	0.06639	0.1219	0.11926
	MV-CNN	0.4092	0.0911	–	0.1534
	QODE	0.4295	**0.1163**	–	0.1685
	AttSum	0.4392	0.1155	–	–
	VAEs-A	0.421	0.11	–	0.164
	C-Attention	0.423	0.107	–	0.161
	MultiMR	0.42041	0.10302	0.14595	–
	JTMMR	0.42717	0.10181	0.1458	0.15761
	JLTMMR	0.43349	0.10375	0.14786	0.16002
	JLTMMR+SF	0.43734	0.10439	0.1496	0.1625
	RDRP_AP	0.43775	0.11563	–	**0.16904**
	HERF	0.42444	0.11211	–	0.16307
	TextRank	0.4033	0.10196	0.14063	0.15463
	QE-T+TextRank	0.41554	0.10597	0.14379	0.16074
	QE-WMVT+Mani	**0.43982**	**0.11185**	**0.15159**	**0.16870**

Table 2 and 3 show the ROUGE scores of related methods. The bolded results highlight the best results in the set of experiments. From the results, we have the following observations:

1. Our proposed method (QE-WMVT+Mani) outperforms other related methods generally. Random and Lead sentences show the poorest ROUGE scores.

2. The effect of deep learning method is not very prominent in the query-oriented multi-document summarization, only slightly prominent in the score of ROUGE-2, and significantly lower in the score of ROUGE-1 than our proposed method. This shows that the method of deep learning needs to be improved.

3. Manifold ranking models based on query sentence expansion (QE-WMVT + Mani) have improved compared with the manifold ranking model MultiMR, JMFMR, JTMMR, JLTMMR, JLTMMR+SF and RDRP_AP. This shows that the manifold ranking based on the expansion of the query sentence is generally better than the manifold ranking based on the original query sentence.

4. TextRank with TextRank expansion (QE-T+TextRank) can improve the effect of TextRank, but the overall performance is not as good as our method. This is because the TextRank method of the original query sentence is significantly worse than manifold ranking of the original query sentence (see result in Sect. 5.1) and the query expansion method has limited ability to improve the model.

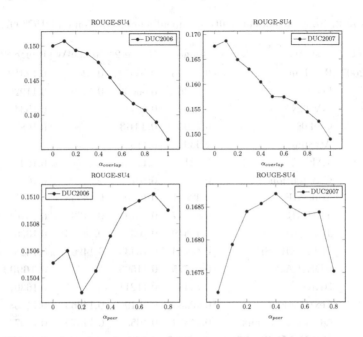

Fig. 2. ROUGE-SU4 scores vs. $\alpha_{overlap}$ ($\alpha_{overlap} + \alpha_A = 1$) and α_{peer}.

5.3   Influence of Sentence Similarity Parameter Tuning

In our method, $\alpha_{overlap}$ is used to tune the trade-off between the TF-ISF and word overlap feature. α_{peer} is utilized to add the context feature. We carry out experiments with different $\alpha_{overlap}$ and α_{peer} to see their influence.

Figure 2 presents the ROUGE-SU4 evaluation results of our method on DUC2006 and DUC2007 respectively for different values of $\alpha_{overlap}$ and α_{peer}. In general, with an increase in $\alpha_{overlap}$ and α_{peer}, the performance first increases, reaches its peak value and then is degraded. For DUC2006 and DUC2007, a value of $\alpha_{overlap}$ around 0.1 can attain the best ROUGE-SU4 result. This indicates that TF-ISF is a critical feature in sentence similarity, but word overlap feature can improve the effect of the model. For the DUC2006 dataset, a value of α_{peer} around 0.7 can attain the best ROUGE-SU4 result. For DUC2007, $\alpha_{peer} = 0.4$ produces peak values of ROUGE scores. Although the best value of α_{peer} differ for DUC2006 and DUC2007, they can slightly improve the model results. This indicates that there are positional connections between sentences.

5.4   Query Relevance Performance

We also perform the qualitative analysis to exam the query relevance performance of our model. We randomly choose some queries in the test datasets and calculate the relevance scores of sentences from our model. We then extract the top ranked sentences and check whether they are able to meet the query need.

Examples for query and model summaries are shown in Table 4. We also give the sentences from our model without query expansion for comparison.

With manual inspection, we find that most query-focused sentences in our model with query expansion can answer the query to a large extent. All these aspects are mentioned in reference summaries. The sentences from our model without query expansion, however, are usually short and simply repeat the key words in the query. The advantage of query expansion is apparent in query relevance ranking.

Table 4. Sentences recognized to focus on the query.

Query: What has been the argument in favor of a line item veto? How has it been used? How have US courts, especially the Supreme Court, ruled on its constitutionality?
Our model without query expansion Summary:
Supreme Court to Rule on Line Item Veto
Court rules Line Item Veto unconstitutional
Court strikes down president's Line Item Veto power
Supreme Court Rules Line Item Veto Law Unconstitutional
Clinton Vetoes Congress Rejection of His Line Item Veto
QE-WMVT+Mani Summary: Supreme Court to Rule on Line Item Veto
In a 6-3 decision the high court ruled the Line Item Veto Act violated the constitution's separation of powers between Congress which approves legislation and the president who either signs it into law or vetoes it
Court strikes down president's Line Item Veto power
Clinton is the first president to have line item veto authority
In a 6-3 ruling the court said the 1996 Line Item Veto Act unconstitutionally allows the president by altering a bill after its passage to create a law that was not voted on by either house of Congress

Different from single document summary, multi-document summary has more information, and the central idea of each document under each topic is different. Therefore, the amount of valid information, such as ROUGE, is more suitable for evaluating the effect of multi-document summarization. Although our model is based on extraction method, it is simple and efficient, and the amount of valid information can be comparable to the state-of-the-art models.

6   Conclusion

In this paper, we propose a query-oriented multi-document summarization model based on query expansion and manifold ranking. We use WordNet, mean, variance, TextRank to expand query. We also use the cosine values of TF-ISF, word overlap values and the relative position of sentences to calculate the relation between sentences. We applied our model to DUC2006 and DUC2007, and found that our model is better than one that combines a subset of query expansion methods and achieves competitive performance with the state-of-the-art results.

References

1. Abdelali, A., Cowie, J., Soliman, H.S.: Improving query precision using semantic expansion. Inf. Process. Manage. **43**(3), 705–716 (2007)
2. Cai, X., Li, W.: Mutually reinforced manifold-ranking based relevance propagation model for query-focused multi-document summarization. IEEE Trans. Audio Speech Lang. Process. **20**(5), 1597–1607 (2012)
3. Cao, Z., Li, W., Li, S., Wei, F.: AttSum: joint learning of focusing and summarization with neural attention (2016)
4. Chali, Y., Hasan, S.A., Imam, K.: Using syntactic and shallow semantic kernels to improve multi-modality manifold-ranking for topic-focused multi-document summarization. In: IJCNLP (2011)
5. Dang, H.T.: Overview of DUC 2006. In: In Document Understanding Conference (2006)
6. Dang, H.T.: Overview of DUC 2007. In: In Document Understanding Conference (2007)
7. Fellbaum, C., Miller, G.: WordNet: An Electronic Lexical Database. MIT Press, Cambridge (1998)
8. G., E., Radev, D.: LexRank: graph-based lexical centrality as salience in text summarization. J. Artif. Intell. **22**(1), 457–479 (2004)
9. Hu, J., Wang, G., Lochovsky, F., Sun, J.T., Chen, Z.: Understanding user's query intent with Wikipedia. In: WWW Madrid! Track Search, pp. 471–480 (2009)
10. Li, P., Lam, W., Bing, L., Guo, W., Li, H.: Cascaded attention based unsupervised information distillation for compressive summarization, pp. 2081–2090, January 2017
11. Li, P., Wang, Z., Lam, W., Ren, Z., Bing, L.: Salience estimation via variational auto-encoders for multi-document summarization. In: AAAI 2017, pp. 3497–3503 (2017)
12. Lin, C.Y., Hovy, E.: Automatic evaluation of summaries using n-gram co-occurrence statistics. In: Proceedings of the 2003 Conference of the North American Chapter of the Association for Computational Linguistics on Human Language Technology, vol. 1, pp. 71–78 (2003)
13. Lin, J., Liu, R., Jia, Q.: Joint lifelong topic model and manifold ranking for document summarization (2019)
14. Lin, Z., Wu, L., Huang, X.: Using query expansion in graph-based approach for query-focused multi-document summarization. Inf. Process. Manage. **45**(1), 35–41 (2009)
15. Liu, Y., Zhong, S., Li, W.: Query-oriented multi-document summarization via unsupervised deep learning. Expert Syst. Appl. **2** (2015)
16. Otterbacher, J., Erkan, G., Radev, D.R.: Using random walks for question-focused sentence retrieval, pp. 915–922 (2005)
17. Pal, D., Mitra, M., Datta, K.: Improving query expansion using wordnet. J. Am. Soc. Inf. Sci. **65**(12), 2469–2478 (2014)
18. Pinto, F.J., Martinez, A.F., Perez-Sanjulian, C.F.: Joining automatic query expansion based on thesaurus and word sense disambiguation using WordNet (2009)

19. Rada, R., Mili, H., Bicknell, E., Blettner, M.: Development and application of a metric on semantic nets. IEEE Trans. Syst. Man Cybernet. **19**(1), 17–30 (1989)
20. Resnik, P.: Using information content to evaluate semantic similarity in a taxonomy (1995)
21. Tan, J., Wan, X., Xiao, J.: Joint matrix factorization and manifold-ranking for topic-focused multi-document summarization. In: International ACM SIGIR Conference on Research and Development in Information Retrieval (2015)
22. Tan, J., Wan, X., Xiao, J.: Joint matrix factorization and manifold-ranking for topic-focused multi-document summarization. In: Proceedings of the 38th International ACM SIGIR Conference on Research and Development in Information Retrieval, pp. 987–990. ACM (2015)
23. Wan, X., Xiao, J.: Graph-based multi-modality learning for topic-focused multi-document summarization, pp. 1586–1591, January 2009
24. Wan, X., Yang, J., Xiao, J.: Manifold-ranking based topic-focused multi-document summarization. In: International Joint Conference on Artifical Intelligence (2007)
25. Wang, R., Kong, F.: Semantic query expansion based on unsupervised word sense disambiguation. J. China Soc. Sci. Tech. Inf. (2011)
26. Xiong, S., Ji, D.: Query-focused multi-document summarization using hypergraph-based ranking. Inf. Process. Manage. **52**(4), 670–681 (2016)
27. Zhang, Y., Er, M.J., Zhao, R., Pratama, M.: Multiview convolutional neural networks for multidocument extractive summarization. IEEE Trans. Cybern. **47**, 3230–3242 (2016)
28. Zhou, D., Bousquet, O., Lal, T.N., Weston, J., Olkopf, B.S.: Learning with local and global consistency. Adv. Neural Inf. Process. Syst. **16**(3) (2004)
29. Zhou, D., Weston, J., Gretton, A., Bousquet, O., Schölkopf, B.: Ranking on data manifolds. In: Advances in Neural Information Processing Systems, pp. 169–176 (2004)
30. Zhou, L., Lin, C.Y., Hovy, E.: A BE-based multi-document summarizer with query interpretation. In: Proceedings of Document Understanding Conference, Vancouver, BC, Canada. Citeseer (2005)

Jointly Learning Salience and Redundancy by Adaptive Sentence Reranking for Extractive Summarization

Ximing Zhang and Ruifang Liu[✉]

Beijing University of Posts and Telecommunications, Beijing 100876, China
{ximingzhang,lrf}@bupt.edu.cn

Abstract. Extractive text summarization seeks to extract indicative sentences from a source document and assemble them to form a summary. Selecting salient but not redundant sentences has always been the main challenge. Unlike the previous two-stage strategies, this paper presents a unified end-to-end model, learning to rerank the sentences by modeling salience and redundancy simultaneously. Through this ranking mechanism, our method can improve the quality of the overall candidate summary by giving higher scores to sentences that can bring more novel information. We first design a summary-level measure to evaluate the cumulating gain of each candidate summaries. Then we propose an adaptive training objective to rerank the sentences aiming at obtaining a summary with a high summary-level score. The experimental results and evaluation show that our method outperforms the strong baselines on three datasets and further boosts the quality of candidate summaries, which intensely indicate the effectiveness of the proposed framework.

Keywords: Extractive text summarization · Salience · Redundancy

1 Introduction

Extractive summarization aims to create a summary by identifying and concatenating the most important sentences in a document [8,15,18,26,30]. To choose an appropriate sentence from the document, we need to consider two aspects: *salience*, which represents how much information the sentence carries; and *redundancy*, which represents how much information in the sentence is already included in the previously selected sentences [21]. The former focuses on *sentence-level* importance, while the latter considers the relationship between sentences at the *summary-level*.

The main challenge of extractive summarization is how to combine salience and redundancy simultaneously. Most previous methods [6,11] only consider salience at the sentence-level, where they usually model sentence-selecting as a sequence labeling task. Several approaches for modeling redundancy between selected sentences are generally classified into two types: heuristics-based and

© Springer Nature Switzerland AG 2021
S. Li et al. (Eds.): CCL 2021, LNAI 12869, pp. 112–126, 2021.
https://doi.org/10.1007/978-3-030-84186-7_8

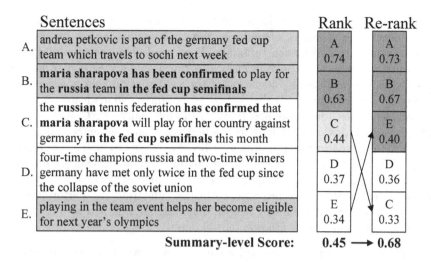

Fig. 1. A summary sample whose sentences are scored by the baseline model(labeled as Rank) and our model(labeled as Re-rank) respectively. Both models only select sentences with top-3 scores as candidate summaries. Sentences on blue background constitute the gold summary. Phrases painted in the same color indicate N-gram overlap.

model-based approaches [22]. The former such as Trigram Blocking [15] is not adaptive since they usually apply the same rule to all the documents, which results in limited effects on a few specific datasets. The latter depends heavily on feature engineering or a neural post-processing module to model redundancy. [3] extracts Ngram-matching and semantic-matching features to indicate the redundancy of a candidate sentence. [28] proposes a two-step pipeline that first scores salience, then learns to balance salience and redundancy as a text-matching task. The critical drawbacks are error propagation and high computation cost. Compared to these models, we aim to propose an efficient and unified end-to-end model to jointly learn salience and redundancy without extra post-processing modules.

In this paper, we present a unified end-to-end ranking-based method for extractive summarization. Unlike previous methods, we train a single unified model that is both salience-aware and redundancy-aware to extract high-quality sentences at the summary level. The principle idea is that the best summary should consist of candidate sentences that can make the largest cumulating metric gain. As shown in Fig. 1, only modeling salience when scoring sentences leads to give a high score to a salient but redundant sentence. Due to the redundancy between the sentences labeled as B and C in Fig. 1, the readers can not get the most overall information from the candidate summary consisting of the top 3 sentences. Our method aims to give higher scores to the sentences containing novel but important information by globally considering its contribution to the summary-level metric gain, and generates a candidate summary with a high summary-level score. Specifically, similar to the intuition in [1,9], we first define a new summarization evaluation measure - Normalized Cumulating Gain based on the Overlap of N-gram (NCGON) between candidate sentences and gold summary, which can better evaluate the overall quality of candidate summaries by considering the distances of multiple N-gram matching scores between the candidate summary and golden summary. Then we design a novel redundancy-aware ranking loss - Adaptive Summary Ranking Loss (AdpSR-Loss) modified by NCGON. We penalize the deviation of the predicted ranking position probabilities of sentence pairs from the desired probabilities, which leads our model to rerank sentences adaptively according to the difference of NCGON between candidate summaries, and finally find the best candidate summary end-to-end.

We conduct experiments on a range of benchmark datasets. The experimental results demonstrate that the proposed ranking framework achieves improvements over previous strong methods on a large range of benchmark datasets. Comprehensive analyses are provided to further illustrate the performance of our method on different summary length, matching N-gram and so on.

Our contributions are as follows:

1. We propose a unified end-to-end summary-level extractive summarization model, jointly learning salience and redundancy of candidate sentences without an extra post-processing stage.
2. We consider extractive summarization as a sentence ranking task and present a new objective AdpSR-Loss based on the summary-level measure NCGON.
3. Without using extra models to reduce redundancy, we outperform the strong baseline methods by a large margin. Experimental results and analysis show the effectiveness of our approach.

2   Related Work

Neural networks have achieved great success in the task of text summarization. There are two main lines of research: abstractive and extractive. The abstractive paradigm [20,23] focuses on generating a summary word-by-word after encoding the full document. The extractive approach [6] directly selects sentences from the document to assemble into a summary. Recent research on extractive summarization spans a large range of approaches. These Extractive summarization models often use a dedicated sentence selection step aiming to address redundancy after sentence scoring step which deals with salience.

2.1   Salience Learning

With the development of neural networks, great progress has been made in extractive document summarization. Most of them focus on the encoder-decoder framework and use recurrent neural networks [8] or Transformer [27] encoders for the sentence scoring [25]. These architectures are widely used and also extended with reinforcement learning [30]. More recently, summarization methods based on BERT [7] have been shown to achieve state-of-the-art performance [15,26,29] for extractive summarization. The development of the above sentence representation and scoring models do help to achieve improvements on selecting salient sentences.

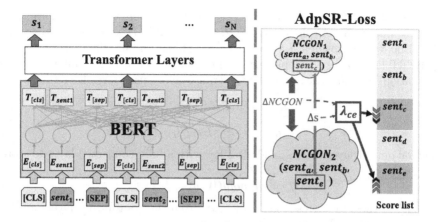

Fig. 2. Overview of the proposed model. The left part is the basic sentence extraction architecture based on BERT and the right part represents the AdpSR-Loss mechanism. Δs represents the difference between the scores of $sent_c$ and $sent_e$ predicted by model in the left part.

2.2 Redundancy Learning

There are relatively fewer methods that study sentence selection to avoid redundancy. In the non-neural approaches, Maximal Marginal Relevance [5] based methods select the content that has the maximal score and is minimally redundant with the previously constructed partial summary. Integer Linear Programming based methods [17] formulate sentence selection as an optimizing problem under the summary length constraint. Trigram blocking [15] filter out sentences that have trigram overlap with previously extracted sentences. In the neural approaches, [30] propose to jointly learn to score and select sentences with a sequence generation model. [3] proposed redundancy-aware models by modeling salience and redundancy using neural sequence models. [28] proposes a two-step pipeline that first scores salience, then learns to balance salience and redundancy as a text-matching task.

Compared to these methods, our method aims to propose an efficient one-stage method to jointly learn salience and redundancy without extra redundancy-aware models, and have a good generalization on a large range of benchmark datasets.

3 Method

In this section, we first introduce the overall architecture including the sentence scoring model and our training mechanism in Sect. 3.1. Then we introduce our designed reranking training objective in Sect. 3.2.

3.1 Overall Architecture

Sentence Scoring Model. Given a single document consisting of sentences $[sent_1, sent_2, \ldots, sent_m]$, where $sent_i$ is the i-th sentence in the document, our task is to extract a certain number of sentences to represent the main information of source document. As shown in Fig. 2, using BERT [7] as a sentence encoder BERTEnc, we add token $[CLS]$ before each sentence and use the vector from the top BERT layer h_i as the representation of $sent_i$. To learn more inter-sentence information, several transformer layers TransEnc are used after BERT:

$$h_i = BERTEnc(sent_i) \tag{1}$$

$$h_i^L = TransEnc(h_i) \tag{2}$$

The final output layer is a sigmoid classifier with a fully-connected layer to score each sentence:

$$\hat{y}_i = sigmoid(W_o h_i^L + b_o) \tag{3}$$

where h_i^L is the i-th representation of $sent_i$ from the transformer layers and $\hat{y}_i$ is the extraction probability of each sentence.

Training Mechanism. We train the model with two different losses – binary cross entropy loss and a new adaptive ranking loss (AdpSR-Loss, Fig. 2) successively. The binary-cross entropy loss aims at scoring each sentences. The novel ranking loss is designed to rerank the sentences and obtain the better sentence combination. Using this strategy, the summary-level gain of top k sentences can be efficiently improved end-to-end without introducing a second-stage model.

3.2   Extraction Summarization as Ranking

NCGON: A Summary-Level Evaluation Measure. We refer to $D = \{sent_n | n \in (1, N)\}$ as a single document consisting of n sentences and $C = \{sent_k | sent_k \in D\}$ as a candidate summary consisting of k sentences extracted from D. A sequence of labels $\{y_n | n \in (1, N)\}$ $(y_n \in (0, 1))$ for $\{sent_n\}$ are given as ground truth by a greedy algorithm similar to [18] by maximizing the ROUGE score against the gold summary C^* written by human. Let $\{s_n | n \in (1, N)\}$ represent the scores of sentences $\{sent_n\}$ predicted by model. Traditional method simply ranks $\{sent_n\}$ according to $\{s_n\}$ in descending order and selects top k sentence as candidate summary C^t, and the remaining sentences are abandoned directly by model, which does not consider the overall gain for obtaining summaries.

To better evaluate the gain of sentence choosing at a summary-level, we consider the overlap between candidate and gold summary as cumulating gain (CG). As Fig. 1 shows, the CG of C^t may not be the highest due to the redundancy, although the score of each $sent_i$ is top k highest individually. We use function as followed to measure the overlap of N-grams (N = 2,3,4) between the extracted summary C^t and the gold summary C^*:

$$R_n(C^*, C^t) = Rouge\text{-}N(C^*, C^t) \tag{4}$$

Consequently, the cumulating gain based on the overlap of N-grams (CGON) as follows:

$$CGON = \sum_{n=2,3,4} R_n(C^*, C^t) \tag{5}$$

Swapping the position of $sent_c$ and $sent_e$ in Fig. 1 makes CGON improve by minimizing the overlap between top k sentences. Therefore, using CGON as the measure, we can better find the summary-level candidate containing most over-all information rather than only using cross entropy loss which tends to select sentences with higher individual gain. To fairly quantify the CGON of differ-ent candidates, we normalize it by $CGON_{max}$ as our final evaluation measure NCGON:

$$NCGON = \frac{CGON}{CGON_{max}} \tag{6}$$

where $CGON_{max}$ is the CGON of the ground truth candidate consisting of sentences whose labels are equal to 1.

AdpSR-Loss: Adaptive Summary Ranking Loss. Inspired by the Learning to Rank (LTR) structure [4,9], we model extractive summarization as a pair-wise ranking problem and aim at reranking the sentence list to find the best candidate summary consisting of sentences with top k scores. Based on the method in 2.1, we get the original score list using cross entropy loss. Considering one pair of sentences $\{sent_i, sent_j\}$, where $i \in (1, k)$ and $j \in (k+1, N)$. Let $U_i > U_j$ denote that $sent_i$ is ranked higher than $sent_j$. $\{s_i, s_j\}$ is mapped to a learned probability P_{ij} as followed, which indicates the probability of ranking position between $sent_i$ and $sent_j$ predicted by model.

$$P_{ij} = P(U_i > U_j) = \frac{1}{1 + e^{-(s_i - s_j)}} \tag{7}$$

We apply the cross entropy loss, which penalizes the deviation of the model output probabilities P_{ij} from the desired probabilities $\overline{P}_{ij}$: let $\overline{P}_{ij} \in \{0, 1\}$ be the ground truth of the ranking position of $\{sent_i, sent_j\}$. Then the cost is:

$$\begin{aligned} L_{ij} &= -\overline{P}_{ij} log P_{ij} - (1 - \overline{P}_{ij}) log(1 - P_{ij}) \\ &= log(1 + e^{-(s_i - s_j)}) \end{aligned} \tag{8}$$

To better optimize the evaluation measure NCGON through L_{ij}, we modify L_{ij} by simply multiplying ΔNCGON, which represents the size of the change in NCGON given by swapping the rank positions of $sent_i$ and $sent_j$ (while leav-ing the rank positions of all other sentences unchanged). Specific to extractive summarization, there is no need to calculate all pairs of sentences in D, we only focus on $\{sent_i, sent_j | i \in (1, k), j \in (k+1, N)\}$ due to the fact that only the overlap of N-grams between top k sentences in the ranking list and gold sum-mary make sense. Finally, the Adaptive Summary Ranking Loss (AdpSR-Loss) can be written as:

$$L = \sum_{i,j} L_{ij} = \sum_{i,j} log(1 + e^{-(s_i - s_j)}) |\Delta NCGON| \tag{9}$$

Our experiments have shown that such a ranking loss actually optimizes NCGON directly, which leads our model to extract a better candidate summary.

Understanding How AdpSR-Loss Works. To explain AdpSR-Loss commonly, we define the gradient of the cost L_{ij} as $|\lambda_{ij}|$, which we can easily get through derivation:

$$\lambda_{ij} = -\frac{1}{1 + e^{(s_i - s_j)}}|\Delta NCGON| \tag{10}$$

$|\lambda_{ij}|$ can be interpreted as a force: if $sent_j$ is more salient than $sent_i$, which means choosing $sent_j$ could obtain higher cumulating gain than $sent_i$, then $sent_j$ will get a push upwards of size $|\lambda_{ij}|$. By multiplying $\Delta NCGON$, the gradient is endowed with practical physical meaning: sentences that can bring more relative gain will be given greater power to improve their ranking position adaptively.

4   Experiments

4.1   Datasets

We conduct empirical studies on three benchmark single-document summarization datasets, CNN/DailyMail [10], Xsum [19] and WikiHow [13] as followed.

CNN/DailyMail is a widely used summarization dataset for single-document summarization, which contains news articles and associated highlights as summaries.

XSum is a one-sentence summary dataset to answer the question "What is the article about?". We use the splits of Narayan et al. (2018a) for training, validation, and testing.

WikiHow is a new large-scale dataset using the online WikiHow knowledge base.

Table 1 shows the full statistics of three datasets. All the sentences are split with the Stanford CoreNLP toolkit [16] and pre-processed following [15]. We tokenize sentences into subword tokens, and truncate documents to 512 tokens.

Table 1. Datasets overview. The data in Doc. and Sum. indicates the average length of documents and summaries in the test set respectively. # Ext denotes the number of sentences that should extract in different datasets.

Datasets	Source	# Pairs			# Tokens		# Ext
		Train	Valid	Test	Doc.	Sum.	
CNN/DM	News	287,084	13,367	11,489	766.1	58.2	3
XSum	News	203,028	11,273	11,332	430.2	23.3	2
WikiHow	KB	168,126	6,000	6,000	580.8	62.6	4

4.2   Implementation Details

Our baseline BertSum [15] using BERT [7] as a sentence encoder, the vectors from the top BERT layer of token [CLS] before each sentence are used as the representation of each sentence. To learn more inter-sentence information, several transformer layers are used after BERT and a sigmoid classifier is stacked to score each sentence. In order to avoid interference of other factors, we reimplement the model BERTSUM in our training environment according to the default parameters of [15] using the base version of BERT[1], and compare our method on this baseline fairly. All the models are trained on 2GPUs (GTX 1080 Ti) with gradient accumulation per two steps. We use Adam optimizer [12] with $\beta_1 = 0.9, \beta_2 = 0.999$ and adopt the learning rate schedule as [24] with a warming-up strategy. We train the model with cross entropy loss for 50,000 steps to obtain the original score for each sentence and AdpSR-Loss afterward with max learning rate $2e^{-5}$ for 3000 steps to obtain the final scores.

4.3 Evaluation Metric

We adopt ROUGE [14] for evaluation metric, which is the standard evaluation metric for summarization. We report results in terms of unigram and bigram overlap (ROUGE-1 and ROUGE-2) as a means of assessing informativeness, and the longest common subsequence (ROUGE-L) as a means of assessing fluency.

Table 2. Results on CNN/DM test set. Results with ∗ mark are taken from the corresponding papers.

Model	R-1	R-2	R-L
LEAD	40.43	17.62	36.67
ORACLE	52.59	31.23	48.87
BanditSum* [8]	41.50	18.70	37.60
NeuSum* [30]	41.59	19.01	37.98
HiBert* [26]	42.37	19.95	38.83
BertExt* [2]	42.29	19.38	38.63
BertSum[15]	42.54	19.86	39.00
BertSum + Tri-Blocking	42.86(+0.32)	19.87(+0.01)	39.29(+0.29)
BertSum + Reranking (Ours)	**42.94(+0.40)**	**20.04(+0.18)**	**39.31(+0.31)**

4.4 Experimental Results

Table 2 summarizes the results of CNN/DM dataset using ROUGE-F1 evaluation. The first block in the table includes the results of an extractive ORACLE system as an upper bound and a LEAD-3 baseline (which simply selects the first three sentences in a document). The second block summarizes the strong extractive summarization baselines on CNN/DM. The third block shows our proposed method results on R-1, R-2 and R-L compared to BERTSUM and BERTSUM without Tri-blocking. Compared with BERTSUM without Tri-blocking,

---

[1] https://github.com/huggingface/pytorch-pretrained-BERT.

Table 3. Results on test sets of WikiHow and XSum. *Num* indicates how many sentences are extracted as a summary.

Model	R-1	R-2	R-L
XSum (Num = 2)			
LEAD	16.30	1.60	11.95
ORACLE	29.79	8.81	22.66
BERTSUM[15]	22.83	4.38	16.96
BERTSUM + Tri-Blocking	22.72(-0.11)	4.18(-0.20)	17.21(+0.25)
BERTSUM + Reranking (Ours)	**23.32(+0.49)**	**4.51(+0.13)**	**17.71(+0.75)**
WikiHow (Num = 4)			
LEAD	24.97	5.83	23.24
ORACLE	35.59	12.98	32.68
BERTSUM[15]	30.08	8.39	28.00
BERTSUM + Tri-Blocking	30.00(-0.08)	8.25(-0.14)	27.95(-0.05)
BERTSUM + Reranking (Ours)	**31.64(+1.56)**	**8.49(+0.10)**	**29.31(+1.31)**

our method achieves 0.40/0.18/0.31 improvements on R-1, R-2 and R-L. Also, we outperforms BERTSUM with Tri-blocking, which is the most commonly used and effective method to remove redundancy on CNN/DM.

Table 3 presents results on the XSum and WikiHow dataset. Our model achieves 0.49/0.13/0.75 improvements on R-1, R-2, and R-L in the Xsum dataset and 1.56/0.10/1.31 improvements on R-1, R-2, and R-L in the WikiHow dataset. Notably, using Tri-Blocking on these two datasets leads to a decrease in performance. Compared with using Tri-Blocking for redundancy removal, our method has improvements on all the three datasets, which illustrates that the reranking mechanism has better generalization ability on summary-level sentence extracting systems.

In addition, we find the scores improve especially on XSum and WikiHow by a large margin. And the baseline model tends to choose longer sentences than our model. Through calculation, we can get the average sentence length (19.5/11.65/15.6) of the three datasets (CNNDM/XSum/ WikiHow), which indicates our summary-level model is more powerful than the sentence-level framework, especially when the gold summaries consist of shorter sentences.

5   Qualitative Analysis

To further analyze the main results in Sect. 4.4, we carry out detailed evaluation and analysis. We first give an ablation study of our method in Sect. 5.1, and then analyze the performance of our method on different summary length in Sect. 5.2. Considering that the main metric ROUGE is based on the N-gram overlap, we conduct a fine-grained analysis in Sect. 5.3. Finally, we implement a human evaluation and case study in Sect. 5.4 and 5.5.

5.1 Ablation Study

To study the effectiveness of each component of the AdpSR-Loss, we conduct several ablation experiments on the CNN/DM dataset. "w/o R-N" denotes that we remove the Rouge-N gain and only use the sum of remaining gain to weight the AdpSR-loss. As the results shown in Table 4, we could find R-2 is an essential element in NCGON to increase the outcome of our strategy, comparing to R-3 and R-4 that we guessed in advance. However, increasing the weight of R-2 in NCGON can not bring better results.

Table 4. Results of removing different components of our ranking loss on the CNN/DM dataset.

Model	R-1	R-2	R-L
OUR METHOD	**42.94**	**20.04**	**39.31**
w/o R-2	42.69	19.96	39.13
w/o R-3	42.80	20.00	39.23
w/o R-4	42.83	20.00	39.21
w/o (R-3+R-4)	42.70	19.98	39.21
w/o (R-2+R-4)	42.65	19.96	39.09

5.2 Effect of Summary Length

To analyze the performance of our method on different summary length, we divide the test set of CNN/DM into 5 intervals based on the length of gold summaries (X-axis in Fig. 3). We evaluate the performance of our method and the baseline BERTSUM in various parts, and the improvements of the sum of scores on R-1, R-2, R-L are drawn as bars (left y-axis ΔR). As shown in Fig. 3, the ROUGE increases more significantly on documents with short summaries, which means our model can efficiently extract short but salient sentences instead of long sentences prone to redundancy. This further proves that the joint consideration of salience and redundancy during model training is effective on obtaining extractive summaries, especially on summaries consisting of short sentences.

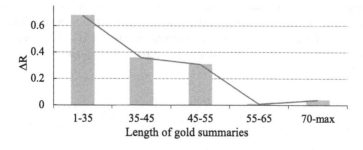

Fig. 3. Datasets splitting experiment. The X-axis indicates the length of gold summaries, and the Y-axis represents the ROUGE improvement of OUR MODEL over BERTSUM on this subset.

5.3   Analysis of N-Gram Frequency

To further analyze the difference between system summary and the oracle, we count the n-gram frequency in the source document of the matching n-gram and unmatching n-gram between system summary and oracle in Table 5 on CNN/DM. Here F-match means the n-gram frequency of the matching n-grams and F-unmatch means the n-gram frequency of the unmatching n-grams.

Table 5. Results of the n-gram frequency of BertSum on the CNN/DM and WikiHow test set.

Datasets	F-match	F-unmatch
BERTSUM	5.53	1.68
BERTSUM + Tri-blocking	5.52	1.68
BERTSUM + Reranking	5.50	1.69

Compared with matching n-grams, the average n-gram frequency of unmatching n-grams is much lower, and the frequency is almost the same in all the comparing models (the frequency of unmatching n-grams on the three models are similar to equal), which shows that finding sentences containing perl but important n-grams with lower n-gram frequency is an important aspect of improving the summary quality.

5.4   Human Evaluation

As we know, although the ROUGE metric has long been regarded as a classical evaluation metric in summarization, it does not always reflect the quality of salience and redundancy. Hence, we conduct human evaluation to further analyze. We follow the human evaluation method proposed in [30], which is widely used in the extractive summarization task. We randomly sample 200 documents and ask five volunteers to evaluate the summaries of the two model outputs. They rank the output summaries as 1 (best) or 2 (worst) regarding informativeness, redundancy and overall quality, and they evaluate the summaries by the fair and anonymous ranking method. Table 6 shows the human evaluation results. Our method performs better than the baseline model BERTSUM, especially in Rdnd, which demonstrates our model is more redundancy-aware.

Table 6. Average ranks of BERTSUM and our method on CNN/DM in terms of informativeness (Info), redundancy (Rdnd) and overall quality by human (lower is better).

Model	Info	Rdnd	Overall
BERTSUM	1.56	1.64	1.59
BERTSUM + Reranking	**1.44**	**1.36**	**1.41**

Article #1 (label: 0, 2, 15)	Candidate #1
john carver says his newcastle players have a point to prove to themselves at liverpool on monday night . the magpies are in danger of being sucked into what had previously seemed an unlikely relegation battle given their alarming run of form . amid accusations of the team ' playing with their flip-flops on ', united have lost four on the spin - scoring just once - and carver admits he does not know where their next point is coming from . it is unlikely to arrive at anfield , a venue at which they last won in 1994 . they were beaten 1-0 by north-east rivals sunderland seven days ago -- a performance carver labelled 'em barrassing ' -- and there was a showdown meeting at the club 's training ground this week in which the head coach let his feelings be known . carver , though , is hoping that personal pride will kick in when they travel to Merseyside ... i have to accept what people say . as long as it is done the right way and in a constructive manner , i can accept that , ' he said . newcastle were beaten in last weekend 's tyne/wear derby by jermain defoe 's stunning volley. carver has seen his newcastle team lose four in a row and fall into relegation danger ...	**BERTSUM (candidate: 0, 1, 7)** • john carver says his newcastle players have a point to prove to themselves at liverpool on monday night . • john carver says his newcastle players have a point to prove when they face liverpool on monday. • the magpies are in danger of being sucked into what had previously seemed an unlikely relegation battle given their alarming run of form . **OURS (candidate: 0, 1, 15)** • john carver says his newcastle players have a point to prove to themselves at liverpool on monday night . • the magpies are in danger of being sucked into what had previously seemed an unlikely relegation battle given their alarming run of form • newcastle were beaten in last weekend 's tyne/wear derby by jermain defoe 's stunning volley.
Article #2 (label: 3, 4, 9)	**Candidate #2**
two hours before the miami open semifinal , novak djokovic practiced his returns in an empty stadium ...' but i managed to get a lot of serves back . that was one of the keys in the match , making him play and getting into the rally and making him work extra . novak djokovic beat john isner in straight sets to reach the miami open on friday night . the no 1- seeded djokovic closed to within one win of his fifth key biscayne title and will face andy murray . the no 1-seeded djokovic closed to within one win of his fifth key biscayne title . his opponent sunday will be two-time champion andy murray... djokovic 's biggest hole while serving was a love- 30 deficit late in the first set . he responded with consecutive aces and escaped . djokovic is aiming to win his fifth title in miami and will take on scotsman murray in sunday 's final . djokovic 's first break gave him a 2-1 edge in the second set , and that margin grew to 5-1 . he finished with just eight unforced to 31 by isner , who lost 70 percent of his second- serve points ...	**BERTSUM (candidate: 3, 5, 6)** • the no 1-seeded djokovic closed to within one win of his fifth key biscayne title and will face andy murray. • novak djokovic beat john isner in straight sets to reach the finalof the miami open on friday night. • his opponent sunday will be two-time champion andy murray , who defeated tomas berdych 6-4 , 6-4 . **OURS (candidate: 3, 4, 9)** • novak djokovic beat john isner in straight sets to reasch the finalof the miami open on friday night . • the no 1-seeded djokovic closed to within one win of his fifth key biscayne title and will face andy murray. • djokovic is aiming to win his fifth title in miami and will take on scotsman murray in sunday 's final.

Fig. 4. Example output articles, candidate summaries from BERTSUM and our method from CNN/DM dataset. Sentences painted in green color are the golden sentences. Phrases painted with the grey highlight indicate N-gram overlaps.

5.5 Case Study

We investigate two examples of extracted output in Fig. 4. As the first case illustrates, comparing to BERTSUM, our reranking method effectively avoids the n-gram overlap between the chosen sentences, and chooses more correct sentences, which means our method could find the sentence combination with less redundancy and more salient information. Also, we also find that during evaluating BERTSUM and our method, both of these two systems tend to choose sentences with similar positions. As the cases show, BERTSUM chooses the sentences at position 0, 1 in the first case and position 5, 6 in the second case.

Besides, BERTSUM also tends to choose sentences that are closer to the beginning of the article. Comparing to BERTSUM, our reranking method alleviates this problem to a certain extent, but the trend still exists, which may also have a correlation with the characteristics of the CNN/DM dataset itself.

6   Conclusions

In this paper, we propose a novel end-to-end summary-level method jointly learning salience and redundancy in the extractive summarization task. Experiments on three benchmark datasets confirm the effectiveness of our one-stage mechanism based on the adaptive ranking objective. Moreover, we find that our method delivers more benefits to short-sentence summaries. We believe the power of this ranking-based summarization framework has not been fully exploited especially on the design of cumulating gain and fine-grained learning, and we hope to provide new guidance for future summarization work.

Acknowledgements. We would like to thank the CCL reviewers for their valuable comments and Keqing He, Pengyao Yi and Rui Pan for their generous help and discussion.

References

1. Ai, Q., Bi, K., Guo, J., Croft, W.B.: Learning a deep listwise context model for ranking refinement. In: The 41st International ACM SIGIR Conference on Research and Development in Information Retrieval, pp. 135–144 (2018)
2. Bae, S., Kim, T., Kim, J., Lee, S.: Summary level training of sentence rewriting for abstractive summarization. In: Proceedings of the 2nd Workshop on New Frontiers in Summarization, Hong Kong, China, pp. 10–20. Association for Computational Linguistics, November 2019. https://doi.org/10.18653/v1/D19-5402, https://www.aclweb.org/anthology/D19-5402
3. Bi, K., Jha, R., Croft, W., Elikyilmaz, A.: AREDSUM: adaptive redundancy-aware iterative sentence ranking for extractive document summarization. arXiv abs/2004.06176. arXiv:2004.06176 (2020)
4. Burges, C., et al.: Learning to rank using gradient descent. In: Proceedings of the 22nd International Conference on Machine Learning, pp. 89–96 (2005)
5. Carbonell, J.: The use of MMR, diversity-based reranking for reordering documents and producing summaries. In: Proceedings of the 21st Annual International ACM SIGIR Conference on Research and Development in Information Retrieval, vol. 1998, pp. 335–336 (1998)
6. Cheng, J., Lapata, M.: Neural summarization by extracting sentences and words. In: Proceedings of the 54th Annual Meeting of the Association for Computational Linguistics (Volume 1: Long Papers), Berlin, Germany, pp. 484–494. Association for Computational Linguistics, August 2016. https://doi.org/10.18653/v1/P16-1046, https://www.aclweb.org/anthology/P16-1046
7. Devlin, J., Chang,M.-W., Lee, K., Toutanova, K.: BERT: pre-training of deep bidirectional transformers for language understanding. arXiv preprint arXiv:1810.04805 (2018)

8. Dong, Y., Shen, Y., Crawford, E., van Hoof, H., Cheung, J.C.K.: Banditsum: extractive summarization as a contextual bandit. In: Proceedings of the 2018 Conference on Empirical Methods in Natural Language Processing, pp. 3739–3748 (2018)
9. Donmez, P., Svore, K.M., Burges, C.J.C.: On the local optimality of lambdarank. In: Proceedings of the 32nd International ACM SIGIR Conference on Research and Development in Information Retrieval, pp. 460–467 (2009)
10. Hermann, K., et al.: Teaching machines to read and comprehend. In: NIPS (2015)
11. Kedzie, C., McKeown, K., Daumé, H.: Content selection in deep learning models of summarization. In: EMNLP (2018)
12. Kingma, D.P., Ba, A.: Adam: A method for stochastic optimization. CoRR abs/1412.6980 (2015)
13. Koupaee, M., Wang, W.Y.: Wikihow: a large scale text summarization dataset. arXiv abs/1810.09305. arXiv:1810.09305 (2018)
14. Lin, C.-Y.: Rouge: a package for automatic evaluation of summaries. In: Text Summarization Branches Out (2004)
15. Liu, Y., Lapata, M.: Text summarization with pretrained encoders. In: EMNLP/IJCNLP (2019)
16. Manning, C.D., Surdeanu, M., Bauer, J., Finkel, J.R., Bethard, S., McClosky, D.: The stanford CoreNLP natural language processing toolkit. In: Proceedings of 52nd Annual Meeting of the Association for Computational Linguistics: System Demonstrations, pp. 55–60 (2014)
17. McDonald, R.: A study of global inference algorithms in multi-document summarization. In: Amati, G., Carpineto, C., Romano, G. (eds.) ECIR 2007. LNCS, vol. 4425, pp. 557–564. Springer, Heidelberg (2007). https://doi.org/10.1007/978-3-540-71496-5_51
18. Nallapati, R., Zhai, F., Zhou, B.: Summarunner: a recurrent neural network based sequence model for extractive summarization of documents (2016)
19. Narayan, S., Cohen, S.B., Lapata, M.: Don't give me the details, just the summary! topic-aware convolutional neural networks for extreme summarization. arXiv abs/1808.08745. arXiv:1808.08745 (2018)
20. Narayan, S., Cohen, S.B., Lapata, M.: Don't give me the details, just the summary! topic-aware convolutional neural networks for extreme summarization. In: Proceedings of the 2018 Conference on Empirical Methods in Natural Language Processing, Brussels, Belgium. Association for Computational Linguistics, October–November 2018
21. Peyrard, M.: A simple theoretical model of importance for summarization. In: Proceedings of the 57th Annual Meeting of the Association for Computational Linguistics, pp. 1059–1073 (2019)
22. Ren, P., Wei, F., Chen, Z., Ma, J., Zhou, M.: A redundancy-aware sentence regression framework for extractive summarization. In: COLING, pp. 33–43 (2016)
23. Abigail See, Peter J. Liu, and Christopher D. Manning. Get to the point: Summarization with pointer-generator networks. In: Proceedings of the 55th Annual Meeting of the Association for Computational Linguistics (Volume 1: Long Papers), Vancouver, Canada, pp. 1073–1083. Association for Computational Linguistics, July 2017. https://doi.org/10.18653/v1/P17-1099, https://www.aclweb.org/anthology/P17-1099
24. Vaswani, A., et al.: Attention is all you need. arXiv abs/1706.03762. arXiv:1706.03762 (2017)

25. Wang, D., Liu, P., Zheng, Y., Qiu, X., Huang, X.-J.: Heterogeneous graph neural networks for extractive document summarization. In: Proceedings of the 58th Annual Meeting of the Association for Computational Linguistics, pp. 6209–6219 (2020)
26. Zhang, X., Wei, F., Zhou, M.: Hibert: document level pre-training of hierarchical bidirectional transformers for document summarization. In: Proceedings of the 57th Annual Meeting of the Association for Computational Linguistics, pp. 5059–5069 (2019)
27. Zhong, M., Wang, D., Liu, P., Qiu, X., Huang, X.: A closer look at data bias in neural extractive summarization models (2019)
28. Zhong, M., Liu, P., Chen, Y., Wang, D., Qiu, X., Huang, X.: Extractive summarization as text matching. arXiv abs/2004.08795. arXiv:2004.08795 (2020)
29. Zhong, M., Liu, P., Wang, D., Qiu, X., Huang, X.-J.: Searching for effective neural extractive summarization: What works and what's next. In: Proceedings of the 57th Annual Meeting of the Association for Computational Linguistics, pp. 1049–1058 (2019)
30. Zhou, Q., Yang, N., Wei, F., Huang, S., Zhou, M., Zhao, T.: Neural document summarization by jointly learning to score and select sentences. In: Proceedings of the 56th Annual Meeting of the Association for Computational Linguistics (Volume 1: Long Papers) (2018)

Incorporating Commonsense Knowledge into Abstractive Dialogue Summarization via Heterogeneous Graph Networks

Xiachong Feng[1]([✉]), Xiaocheng Feng[1,2], and Bing Qin[1,2]

[1] Harbin Institute of Technology, Shenzhen, China
{xiachongfeng,xcfeng,bqin}@ir.hit.edu.cn
[2] Peng Cheng Laboratory, Shenzhen, China

Abstract. Abstractive dialogue summarization is the task of capturing the highlights of a dialogue and rewriting them into a concise version. In this paper, we present a novel multi-speaker dialogue summarizer to demonstrate how large-scale commonsense knowledge can facilitate dialogue understanding and summary generation. In detail, we consider utterance and commonsense knowledge as two different types of data and design a Dialogue Heterogeneous Graph Network (D-HGN) for modeling both information. Meanwhile, we also add speakers as heterogeneous nodes to facilitate information flow. Experimental results on the SAM-Sum dataset show that our model can outperform various methods. We also conduct zero-shot setting experiments on the Argumentative Dialogue Summary Corpus, the results show that our model can better generalized to the new domain.

Keywords: Dialogue summarization · Commonsense knowledge · Graph neural networks.

1 Introduction

Automatic summarization is a fundamental task in Natural Language Processing, which aims to condense the original input into a shorter version covering salient information and has been continuously studied for decades [12,22]. Recently, online multi-speaker dialogue/meeting has become one of the most important ways for people to communicate with each other in their daily works. Especially due to the spread of COVID-19 worldwide, people are more dependent on online communication. In this paper, we focus on dialogue summarization, which can help people quickly grasp the core content of the dialogue without reviewing the complex dialogue context.

Recent works that incorporate additional commonsense knowledge in the dialogue generation [32] and dialogue context representation learning [29] show that even though neural models have strong learning capabilities, explicit knowledge can still improve response generation quality. It is because that a dialog system can understand conversations better and thus respond more properly if it

© Springer Nature Switzerland AG 2021
S. Li et al. (Eds.): CCL 2021, LNAI 12869, pp. 127–142, 2021.
https://doi.org/10.1007/978-3-030-84186-7_9

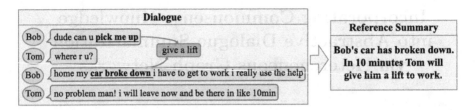

Fig. 1. An example of dialogue-summary pair. Green for speakers, blue for utterances, and pink for commonsense knowledge. In order to generate "give a lift" in the reference summary, the summarization model needs to understand the commonsense knowledge behind "pick up" and "car broke down".

can access and make full use of large-scale commonsense knowledge. However, current dialogue summarization systems [3,6,13,15,33] ignore the exploration of commonsense knowledge, which may limit the performance. In this work, we examine the benefit of incorporating commonsense knowledge in the dialogue summarization task and also address the question of how best to incorporate this information. Figure 1 shows a positive example to illustrate the effectiveness of commonsense knowledge in the dialogue summarization task. Bob asks Tom for help because his car has broken down. On the one hand, by introducing commonsense knowledge according to the *pick up* and *car broke down*, we can know that Bob expects Tom to *give him a lift*. On the other hand, commonsense knowledge can serve as a bridge between non-adjacent utterances that can help the model better understanding the dialogue.

In this paper, we follow the previous setting [32] and also use ConceptNet [26] as a large-scale commonsense knowledge base, while the difference is that we regard knowledge and text(utterance) as heterogeneous data in a real multi-speaker dialogue. We propose a model named **D**ialogue **H**eterogeneous **G**raph **N**etwork (D-HGN) for incorporating commonsense knowledge by constructing the graph including both utterance and knowledge nodes. Besides, our heterogeneous graph also contains speaker nodes at the same time, which has been proved to be a useful feature in dialogue modeling. In particular, we equip our heterogeneous graph network with two additional designed modules. One is called message fusion, which is specially designed for utterance nodes to better aggregate information from both speakers and knowledge. The other one is called node embedding, which can help utterance nodes to be aware of position information. Compared to homogeneous graph network in related works [6,13,15,33], we claim that the heterogeneous graph network can effectively fuse information and contain rich semantics in nodes and links, and thus more accurately encode the dialogue representation.

We conduct experiments on the SAMSum corpus [7], which is a large-scale chat summarization corpus. We analyze the effectiveness of integration of knowledge and heterogeneity modeling. The human evaluation also shows that our approach can generate more abstractive and correct summaries. To evaluate whether commonsense knowledge can help our model better generalize to the new domain,

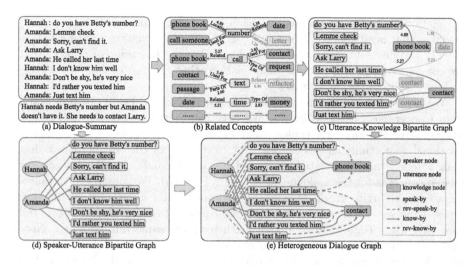

Fig. 2. Illustration of heterogeneous dialogue graph construction process.

we also perform zero-shot setting experiments on the Argumentative Dialogue Summary Corpus [19], which is a debate summarization corpus. In the end, we give a brief summary of our contributions: (1) We are the first to incorporate commonsense knowledge into dialogue summarization task. (2) We propose a D-HGN model to encode the dialogue by viewing utterances, knowledge and speakers as heterogeneous data. (3) Our model can outperform various methods.

2 Heterogeneous Dialogue Graph Construction

In this section, we describe the graph notation and the graph construction process, which consists of three steps, including (1) utterance-knowledge bipartite graph construction, (2) speaker-utterance bipartite graph construction and (3) heterogeneous dialogue graph construction.

2.1 Graph Notation

Our heterogeneous dialogue graph (HDG) is defined as a directed graph $G = (\mathcal{V}, \mathcal{E}, \mathcal{A}, \mathcal{R})$, where each node $v \in \mathcal{V}$ and each edge $e \in \mathcal{E}$. Different types of nodes and edges are associated with their type mapping functions $\tau(v) : \mathcal{V} \to \mathcal{A}$ and $\phi(e) : \mathcal{E} \to \mathcal{R}$.

2.2 Utterance-Knowledge Bipartite Graph Construction

Current dialogue summarization corpus has no knowledge annotations. To ground each dialogue to commonsense knowledge, we make use of ConceptNet [26] to incorporate knowledge. ConceptNet is a semantic network that contains

34 relations in total and represents each knowledge tuple by $R = (h, r, t, w)$ meaning that head concept h and tail concept t have a relation r with a weight of w. It contains not only world facts such as *"Paris is the capital of France"* that are constantly true, but also informal relations that are part of daily knowledge such as *"Call is used for Contact"*.

We use each word in the utterance as a query to retrieve a one-hop graph from ConceptNet, as done by [9]. We only consider nouns, verbs, adjectives, and adverbs. We filter out tuples where (1) r is in a pre-defined list of useless relations[1] (e.g. "number" is antonym of "letter"), (2) the weight of r is less than 1 (e.g. "text" is related to "refactor", weight: 0.9). Finally, we can get related concepts for the dialogue, as shown in Fig. 2(b). We construct utterance-knowledge bipartite graph by viewing utterances and knowledge as different types of nodes. As shown in Fig. 2(c), we connect two utterances to one tail concept t using edge *know-by* if they both have the same tail concept t. Note that two utterances may connect to multiple tail concepts, we choose the one with the highest average weight of relations (e.g. "phone book" is better than "date"). If there are multiple identical knowledge nodes, we also combine them to a single one (e.g. two "contact" nodes are combined into one node).

2.3 Speaker-Utterance Bipartite Graph Construction

Given multiple speakers and corresponding utterances in a dialogue, we construct the speaker-utterance bipartite graph by viewing speakers and utterances as different types of nodes. As shown in Fig. 2(d), we construct *speak-by* edges from speakers to utterances based on who said the utterances.

2.4 Heterogeneous Dialogue Graph Construction

We combine the utterance-knowledge bipartite graph and the speaker-utterance bipartite graph as our heterogeneous dialogue graph, as shown in Fig. 2(e). Additionally, we add a reverse edge *rev-know-by* and *rev-speak-by* to facilitate information flow over the graph. Finally, there are three types of nodes, where $\mathcal{A}$ becomes *speaker*, *utterance*, and *knowledge* and four types of edges, where $\mathcal{R}$ becomes *speak-by*, *know-by*, *rev-speak-by* and *rev-know-by*.

3 Dialogue Heterogeneous Graph Network

In this section, we describe the details of our dialogue heterogeneous graph network (D-HGN), including three components: node encoder, graph encoder and pointer decoder. The model is shown in Fig. 3.

---

[1] We pre-define the useless relation list, including Antonym, EtymologicallyDerivedFrom, NotHasProperty, DistinctFrom, NotCapableOf, EtymologicallyRelatedTo and NotDesires.

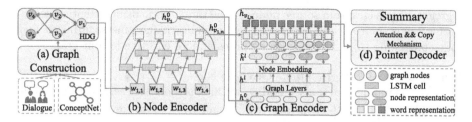

Fig. 3. Illustration of our D-HGN model. (a) Graph construction receives a dialogue and ConceptNet and outputs a heterogeneous dialogue graph (HDG). (b) Node encoder receives a sequence of words for a node and produces initial node and word representations. (c) Graph encoder first conducts graph operations for initial node representations. Then a node embedding module is added after graph layers to make nodes to be aware of position information. Finally, the initial word representations and corresponding updated node representations are concatenated as final word representations. (d) Pointer decoder can either generate summary words from the vocabulary or copy from the input words.

3.1 Node Encoder

The role of node encoder is to give each node $v_i \in \mathcal{V}$ an initial representation $h_{v_i}^0$, where v_i consists of $|v_i|$ words $[w_{i,1}, w_{i,2}, ... w_{i,|v_i|}]$. Note that speaker and knowledge may have multiple words. We employ a Bi-LSTM as the node encoder that encodes input node forwardly and backwardly to generate two sequences of hidden states $\left(\overrightarrow{h_1}, \overrightarrow{h_2}, \ldots, \overrightarrow{h_{|v_i|}}\right)$ and $\left(\overleftarrow{h_1}, \overleftarrow{h_2}, \ldots, \overleftarrow{h_{|v_i|}}\right)$.

$$
\begin{aligned}
\overrightarrow{h_n} &= \text{LSTM}_f\left(x_n, \overrightarrow{h_{n-1}}\right) \\
\overleftarrow{h_n} &= \text{LSTM}_b\left(x_n, \overleftarrow{h_{n+1}}\right)
\end{aligned}
\tag{1}
$$

x_n denotes the embedding of $w_{i,n}$. The forward and backward hidden states are concatenated as the initial node representation $h_{v_i}^0 = [\overrightarrow{h_{|v_i|}}; \overleftarrow{h_1}]$ and initial word representation $h_{v_i,n}^0 = [\overrightarrow{h_n}; \overleftarrow{h_n}]$. $h_{v_i}^0$ will be passed to the graph encoder to learn high-level representations. $h_{v_i,n}^0$ will be concatenated with updated node representations to get final word representations.

3.2 Graph Encoder

Graph encoder is used to digest the structural information and get updated node representations. We employ Heterogeneous Graph Transformer [10] as our graph encoder, which models heterogeneity by type-dependent parameters and can be easily applied to our graph. It includes: (a) heterogeneous mutual attention, which calculates attention scores $\text{Attn}(s, e, t)$ between source nodes and the target node. (b) heterogeneous message passing, which prepares the message vector $\text{Msg}(s, e, t)$ for each source node and (c) target-specific aggregation, which aggregates messages from source nodes to the target node using attention scores as the weight. Specifically, we design two modules named message fusion and node embedding to make the learning process more effective for our graph.

Heterogeneous Mutual Attention. Given an edge $e = (s, t)$ with their node and edge type mapping functions τ and ϕ, we first project source and target node representations from $(l\text{-}1)$-th layer $h_s^{(l-1)}$ and $h_t^{(l-1)}$ into key vector $k_s^{(l)}$ and query vector $q_t^{(l)}$ with type-dependent linear projection.

$$k_s^{(l)} = \text{K\_Linear}_{\tau(s)}^{(l)}\left(h_s^{(l-1)}\right)$$
$$q_t^{(l)} = \text{Q\_Linear}_{\tau(t)}^{(l)}\left(h_t^{(l-1)}\right) \tag{2}$$

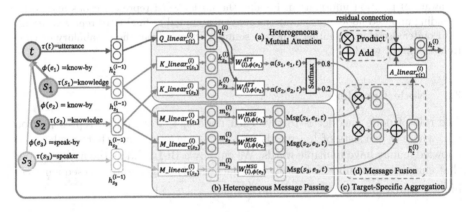

Fig. 4. Illustration of one graph layer. Given a target node of utterance type and source nodes of knowledge and speaker type. Firstly, we use (a) heterogeneous mutual attention to calculate the attention scores by type-dependent linear projection. Secondly, we use (b) heterogeneous message passing to prepare the message vector for each source node. Thirdly, we use (c) target-specific aggregation to aggregate messages to the target node. Specifically, we propose a message fusion module that uses attention scores as the weight to average the knowledge vectors and add speaker information additionally.

Next, to integrate edge type information, we calculate unnormalized score $\alpha(s, e, t)$ between t and s by adding a edge-based matrix $W_{(l),\phi(e)}^{ATT}$. Finally, for each target node t, we conduct Softmax for all $s \in N(t)$ to get the final normalized attention scores $\text{Attn}^{(l)}(s, e, t)$, where $N(t)$ denotes neighbors of target node t. Note that if target node is of utterance type and source node is of speaker type, we do not calculate the attention score between these two types of nodes. See more detail at *message fusion* module. The process is shown in Fig. 4(a).

$$\alpha(s, e, t) = \left(k_s^{(l)} W_{(l),\phi(e)}^{ATT} q_t^{(l)^\top}\right)$$
$$\text{Attn}^{(l)}(s, e, t) = \underset{\forall s \in N(t)}{\text{Softmax}}\left(\alpha(s, e, t)\right) \tag{3}$$

Heterogeneous Message Passing. We first project source node representation $h_s^{(l-1)}$ into the vector $m_s^{(l)} = \text{M\_Linear}_{\tau(s)}^{(l)}\left(h_s^{(l-1)}\right)$ with type-dependent linear projection and then followed by a edge-based matrix $W_{(l),\phi(e)}^{MSG}$ to get the message vector. The process is shown in Fig. 4(b).

$$\text{Msg}^{(l)}(s,e,t) = m_s^{(l)} W_{(l),\phi(e)}^{MSG} \tag{4}$$

Target-Specific Aggregation. We divide this process into two cases based on the type of target node: (1) $\tau(t) \neq utterance$, (2) $\tau(t) = utterance$. For the first case, We use attention vector as the weight to average messages: $\widetilde{h}_t^{(l)} = \oplus_{\forall s \in N(t)}\left(\text{Attn}^{(l)}(s,e,t) \otimes \text{Msg}^{(l)}(s,e,t)\right)$. For the second case, we design a Message Fusion module to aggregate messages to utterance node more effectively. After getting aggregated message vector $\widetilde{h}_t^{(l)}$, we maps it back to $\tau(t)$-type distribution with a linear projection followed by residual connection to get the updated representation $h_t^{(l)}$, as shown in Fig. 4(c).

$$h_t^{(l)} = \text{A\_Linear}_{\tau(t)}^{(l)}\left(\text{Sigmoid}\left(\widetilde{h}_t^{(l)}\right)\right) + h_t^{(l-1)} \tag{5}$$

Message Fusion. Dialogue summaries often describe *"who did what"*, thus speaker information is required for utterances. However, if target node of utterance type aggregates messages from source nodes of knowledge and speaker type, it will prefer more to the speaker node while giving up using knowledge nodes, since attention is a normalized distribution. Therefore, in our message fusion module, we use attention weights for knowledge nodes to average corresponding messages and add speaker information additionally. The process is shown in Fig. 4(d).

$$s_k = (\forall s \in N(t) \wedge \tau(s) = knowledge), s_s = (\forall s \in N(t) \wedge \tau(s) = speaker)$$

$$\widetilde{h}_t^{(l)} = \bigoplus_{s \in s_k}\left(\text{Attn}^{(l)}(s,e,t) \otimes \text{Msg}^{(l)}(s,e,t)\right) + \text{Msg}^{(l)}(s_s,e,t) \tag{6}$$

Node Embedding. In this section, a module named Node Embedding is designed to make utterance nodes to be aware of position information in source dialogue. This is because original heterogeneous graph cannot directly model the chronological order between utterances, while an ideal dialogue summary needs to refer to the order of corresponding dialogue utterances. In detail, for speaker and knowledge nodes, we fix their position to 0. For each utterance node v_i, it associates with a position p_{v_i}, which is the ranking of utterances in the original dialogue. As shown in Fig. 3(c), we add position information for each node: $\hat{h}_{v_i}^{(l)} = h_{v_i}^{(l)} + W^{pos}[p_{v_i}]$, where W^{pos} denotes a learnable node embedding matrix. After getting the output representation $\hat{h}^{(l)}$ for each node, we concatenate updated node representation $\hat{h}_{v_i}^{(l)}$ and corresponding initial word representations $h_{v_i,n}^0$ followed by a linear projection F_Linear to get final word representations $h_{v_i,n}$.

$$h_{v_i,n} = \text{F\_Linear}([\hat{h}_{v_i}^{(l)}; h_{v_i,n}^0]) \tag{7}$$

3.3 Pointer Decoder

We employ a LSTM with attention and copy mechanism to generate summaries. At each decoding time step t, the LSTM reads the previous word embedding x_{t-1} and previous context vector c_{t-1} as inputs to compute the new hidden state $s_t = \text{LSTM}\,(x_{t-1}, c_{t-1}, s_{t-1})$. We use the average of all word representations s_0 in the graph to initialize the decoder.

$$s_0 = \text{Average}(\sum_{v_i \in G} \sum_{n \in [1, |v_i|]} h_{v_i, n}) \tag{8}$$

The context vector c_t is computed as in [1], which is then used to calculate generation probability p_{gen} and the final probability distribution $P(w)$, as done by [24].

3.4 Training

For each heterogeneous dialogue graph G that is paired with a ground truth summary $Y^* = [y_1^*, y_2^*, ..., y_{|Y^*|}^*]$, we minimize the negative log-likelihood of the target words sequence.

$$L = - \sum_{t=1}^{|Y^*|} \log p\,(y_t^* | y_1^* \cdots y_{t-1}^*, G) \tag{9}$$

4 Experiments

Dataset Following the latest works [6,7], we conduct experiments on two different settings. Firstly, we train and evaluate our model on the SAMSum corpus [7], which contains dialogues around chit-chats topics. Secondly, we train using SAMSum corpus and use the Argumentative Dialogue Summary Corpus (ADSC) [19] as the test set to perform zero-shot setting experiments. Each dialogue in ADSC dataset owns 5 different summaries and is mainly around debate topics. Table 1 shows the knowledge related statistics of two datasets.

Table 1. Knowledge related statistics on SAMSum and ADSC datasets. # is the number of dialogues. Coverage represents the percentage of dialogues with at least one knowledge node. Average Know represents the average number of knowledge nodes per dialogue.

Dataset	Split	#	Coverage	Average Know
SAMSum	Train	14732	94.43%	19.60
	Valid	818	95.72%	18.23
	Test	819	93.89%	19.77
ADSC	Full	45	100%	6.50

Implementation Details. The word embedding size is set to 100 and initialized with the pre-trained GloVe vector. The dimension of node encoder and pointer decoder is set to 300. The dimension of graph encoder is set to 200. The graph layer number is set to 1. Dropout is set to 0.5. We use Adam with the learning rate of 0.001 and use gradient clipping with a maximum gradient norm of 2. In the test process, beam size is set to 10, minimum decoded length is 19^2.

Evaluation Metrics. We employ the standard F_1 scores for ROUGE-1, ROUGE-2, and ROUGE-L metrics [14] to measure summary qualities. These three metrics evaluate the accuracy on unigrams, bigrams, and longest common subsequence between the groundtruth and the generated summary [3].

Baseline Models. We compare our model with several baselines.

- **LONGEST-3** chooses the longest three utterances as the summary.
- **TextRank** [18] is a graph-based extractive method.
- **SummaRunner** [20] extract utterances based on a hierarchical RNN model.
- **Transformer** [27] is a Seq2Seq model that utilizes self-attention operations.
- **PGN** [24] is a Seq2Seq model equipped with copy mechanism.
- **HRED** [25] is a hierarchical Seq2Seq model.
- **Abs RL** [4] is a pipeline model that first selects salient utterances based on a extractive model then produces the summary based on a abstractive model using diversity beam search. The extractive model is trained using utterance-level extraction labels. The overall model is jointly trained using reinforcement learning.
- **Abs RL Enhance** [7] is based on **Abs RL**, which appends all speakers after each utterance, because the original model may select utterances of a single speaker that will lead to no other speaker information.
- **D-GAT**, **D-GCN** and **D-RGCN** are variants of our model that replace heterogeneous graph layers with homogeneous graph layers, including GAT [28], GCN [11] and RGCN [23][4].

4.1   Automatic Evaluation

Table 2 shows the results on SAMSum corpus. The D-HGN stands for our full model, which outperforms various baselines. Compared with HRED that uses no additional auxiliary information such as commensence knowledge or utterance-level extraction labels, D-RGCN that uses commensence knowledge can achieve 0.97% improvement on ROUGE-1, 0.94% on ROUGE-2, 1.28% on ROUGE-L, which shows the effectiveness of knowledge integration. Compared with homogeneous networks like D-RGCN, D-HGN that based on heterogeneous graph networks can achieve 0.67% improvement on ROUGE-1, 1.00% on ROUGE-2, 0.63% on ROUGE-L, which verifies the effectiveness of heterogeneity modeling.

---

[2] Our codes are available at: https://github.com/xcfcode/DHGN.

[3] https://pypi.org/project/pyrouge/.

[4] Note that D-GAT also use message fusion module to update representations for utterance nodes.

Table 2. Test set results on the SAMSum Dataset, where "R-1" is short for "ROUGE-1", "R-2" for "ROUGE-2", "R-L" for "ROUGE-L". "Know.", "Heter.", "Utter." and "RL" indicate whether knowledge, heterogeneity modeling, utterance-level extraction labels and reinforcement learning are used or not.

Type	Model	Know.	Heter.	Utter.	RL	R-1	R-2	R-L
Extractive	LONGEST-3	✗	✗	✗	✗	32.46	10.27	29.92
	TextRank	✗	✗	✗	✗	29.27	8.02	28.78
	SummaRunner	✗	✗	✗	✗	33.76	10.28	28.69
Abstractive	Transformer	✗	✗	✗	✗	36.62	11.18	33.06
	PGN	✗	✗	✗	✗	40.08	15.28	36.63
	HRED	✗	✗	✗	✗	40.39	16.13	37.65
Pipeline	Abs RL	✗	✗	✓	✓	40.96	17.18	39.05
	AbsRL Enhance	✗	✗	✓	✓	41.95	18.06	39.23
Ours	D-GCN	✓	✗	✗	✗	41.33	16.98	38.70
	D-GAT	✓	✗	✗	✗	41.08	16.89	38.61
	D-RGCN	✓	✗	✗	✗	41.36	17.07	38.93
	D-HGN	✓	✓	✗	✗	**42.03**	**18.07**	**39.56**

4.2   Human Evaluation

We conduct human evaluation to verify the quality of the generated summaries, including abstractiveness (contains higher-level conceptual words), informativeness (covers adequate information) and correctness (associates right names with actions). We randomly sample 50 dialogues with corresponding generated summaries to conduct the evaluation. We hired five graduates to perform human evaluation. For each metric, the score ranges from 1 (worst) to 5 (best). The results are shown in Table 3.

Table 3. Human evaluation results.

Model	Abstractiveness	Informativeness	Correctness
PGN	2.70	2.68	2.49
AbsRL enhance	2.94	3.23	2.43
D-HGN	**3.26**	**3.25**	**2.92**
W/o *knowledge*	3.09	3.16	2.80
W/o *speaker*	3.23	3.21	2.60

Our model achieves higher scores. Compared with D-HGN, D-HGN(w/o knowledge) gets a lower score in abstractiveness, which indicates knowledge incorporation can help our model express deeper meanings. D-HGN(w/o speaker) performs worse than D-HGN in correctness, which shows effectiveness of heterogeneity modeling by viewing speakers as heterogeneous data. AbsRL Enhance performs worst in correctness, which may due to the utterances extraction will break the coherence of dialogue contexts.

4.3 Ablation Study

We conduct two types of ablation studies to verify the effectiveness of different types of nodes and two modules we propose. As shown in Table 4, without knowledge integration(w/o knowledge), the model suffers the performance drop, which shows incorporating knowledge can help our model better modeling the dialogue context. For speaker nodes, directly remove them in the graph will lead to no speaker in the final summary. Instead, we append the speakers in front of utterances(w/o speaker). The results show that modeling speakers as heterogeneous data will do good the final summary generation process.

Table 4. Ablation study for two modules

Model	ROUGE-1	ROUGE-2	ROUGE-L
D-HGN	**42.03**	**18.07**	**39.56**
W/o *message fusion*	41.29	17.09	38.74
w/o *node embedding*	41.99	17.85	38.89

As shown in Table 5, we remove the message fusion module(w/o message fusion), the results show that it is worth to design specific message fusion method according to different types of nodes. Besides, without taking position information into account(w/o node embedding), our model will lose some performance.

Table 5. Ablation study for different types of nodes

Model	ROUGE-1	ROUGE-2	ROUGE-L
D-HGN	**42.03**	**18.07**	**39.56**
w/o *knowledge*	41.52	17.38	38.76
w/o *speaker*	41.06	17.17	38.92

4.4 Zero-Shot Setting

To verify whether knowledge can help our model better generalize to the new domain, we directly test models on the ADSC Corpus. The results are shown in Table 6. The homogeneous model D-GAT that uses knowledge can get better results than other baselines. The D-HGN gets the best score. We contribute this to the fact that knowledge can help our models better understand the dialogue in the new domain.

Table 6. ROUGE F_1 results on the argumentative dialogue summary corpus.

Model	ROUGE-1	ROUGE-2	ROUGE-L
PGN	28.69	4.77	22.39
AbsRL Enhance	30.00	4.87	22.27
D-GAT	32.90	5.46	22.47
D-HGN	**33.55**	**5.68**	**22.75**

4.5 Visualization

To examine whether our D-HGN can learn easily distinguishable representations, we extract node representations from the last graph layer for the SAMSum test set. We apply t-SNE [17] to these vectors. The results are shown in Fig. 5. We find that our model can generate more discrete and easily distinguishable representations. Besides, D-GAT also tends to separate representations of different types of nodes, which indicates explicitly heterogeneity modeling is a more reasonable approach.

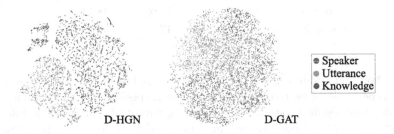

Fig. 5. Visualization of node representations generated by the last graph layer of D-HGN and D-GAT.

4.6 Case Study

Figure 6 shows summaries generated by different models and the visualization of knowledge-to-utterance attention weights learned by our D-HGN model, the darker the color, the higher the weights. Our model incorporates two knowledge nodes, one is *birthday party* according to "bday party", "happy" and "cake", the other one is *some people* according to "Tom" and "boyfriend". We can see that our D-HGN model pays more attention to *birthday party* rather than *some people*. On the one hand, incorporating *birthday party* helps our model generate a more formal summary (using birthday rather than bday). On the other hand, *birthday party* connects non-adjacent utterances around the birthday topic, which helps our model generate a more informative and detailed summary (including cake).

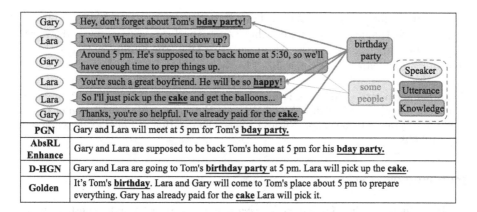

PGN	Gary and Lara will meet at 5 pm for Tom's **bday party.**
AbsRL Enhance	Gary and Lara are supposed to be back Tom's home at 5 pm for his **bday party.**
D-HGN	Gary and Lara are going to Tom's **birthday party** at 5 pm. Lara will pick up the **cake**.
Golden	It's Tom's **birthday**. Lara and Gary will come to Tom's place about 5 pm to prepare everything. Gary has already paid for the **cake** Lara will pick it.

Fig. 6. Example summaries generated by different models for one dialogue.

5   Related Work

Previous works used feature engineering [30], template-based [21] and graph-based [2] methods for extractive dialogue summarization. Although extractive methods are widely used, the results tend to be incoherent and poorly readable. Therefore, current works mainly focus on abstractive methods, which can produce more readable and fluency summaries. They tend to incorporate additional auxiliary information to help better modeling the dialogue. [8] incorporated dialogue acts to model the interactive status of the meeting. [15] tackled the problem of customer service summarization, which first produced a sequence of pre-defined keywords then generated the summary. [16] generated summaries for nurse-patient conversation by incorporating topic information. [6] first removed useless utterances by utilizing discourse labels and then generated summaries. [13] combined vision and textual features in a unified hierarchical attention framework to generate meeting summaries. [33] employed a hierarchical transformer framework and incorporated part-of-speech and entity information for meeting summarization. [3] incorporated topic and stage information to model the dialogue. [31] used topic words to alleviate the factual inconsistency problem. [5] used dialogue discourse to model the interaction between utterances. In this paper, we facilitate dialogue summarization task by incorporating commonsense knowledge and further model utterances, commonsense knowledge and speakers as heterogeneous data.

6   Conclusion

In this paper, we improve abstractive dialogue summarization by incorporating commonsense knowledge. We first construct a heterogeneous dialogue graph by introducing knowledge from a large-scale commonsense knowledge base. Then we present a Dialogue Heterogeneous Graph Network (D-HGN) for this task by viewing utterances, knowledge and speakers in the graph as heterogeneous nodes. We additionally design two modules named message fusion and node embedding to facilitate information flow. Experiments on the SAMSum dataset show the effectiveness of our model that can outperform various methods. Zero-shot setting experiments on the Argumentative Dialogue Summary Corpus show that our model can better generalized to the new domain.

Acknowledgments. This work is supported by the National Key R&D Program of China via grant 2020AAA0106502 and National Natural Science Foundation of China (NSFC) via grant 61906053 and Natural Science Foundation of Heilongjiang via grant YQ2019F008.

References

1. Bahdanau, D., Cho, K., Bengio, Y.: Neural machine translation by jointly learning to align and translate. In: Bengio, Y., LeCun, Y. (eds.) 3rd International Conference on Learning Representations (ICLR 2015), May 7–9 2015, San Diego, Conference Track Proceedings (2015)
2. Bui, T., Frampton, M., Dowding, J., Peters, S.: Extracting decisions from multi-party dialogue using directed graphical models and semantic similarity. In: Proceedings of the SIGDIAL 2009 Conference, pp. 235–243. Association for Computational Linguistics, London (2009)
3. Chen, J., Yang, D.: Multi-view sequence-to-sequence models with conversational structure for abstractive dialogue summarization. In: Proceedings of the 2020 Conference on Empirical Methods in Natural Language Processing (EMNLP), pp. 4106–4118. Association for Computational Linguistics (2020). https://doi.org/10.18653/v1/2020.emnlp-main.336
4. Chen, Y.C., Bansal, M.: Fast abstractive summarization with reinforce-selected sentence rewriting. In: Proceedings of the 56th Annual Meeting of the Association for Computational Linguistics (Volume 1: Long Papers), pp. 675–686. Association for Computational Linguistics, Melbourne, Australia (2018). https://doi.org/10.18653/v1/P18-1063
5. Feng, X., Feng, X., Qin, B., Geng, X., Liu, T.: Dialogue discourse-aware graph convolutional networks for abstractive meeting summarization. arXiv preprint arXiv:2012.03502 (2020)
6. Ganesh, P., Dingliwal, S.: Abstractive summarization of spoken and written conversation. arXiv preprint arXiv:1902.01615 (2019)
7. Gliwa, B., Mochol, I., Biesek, M., Wawer, A.: SAMSum corpus: a human-annotated dialogue dataset for abstractive summarization. In: Proceedings of the 2nd Workshop on New Frontiers in Summarization, pp. 70–79. Association for Computational Linguistics, Hong Kong (2019). https://doi.org/10.18653/v1/D19-5409
8. Goo, C.W., Chen, Y.N.: Abstractive dialogue summarization with sentence-gated modeling optimized by dialogue acts. In: 2018 IEEE Spoken Language Technology Workshop (SLT), pp. 735–742. IEEE (2018)
9. Guan, J., Wang, Y., Huang, M.: Story ending generation with incremental encoding and commonsense knowledge. In: The Thirty-Third AAAI Conference on Artificial Intelligence, AAAI 2019, The Thirty-First Innovative Applications of Artificial Intelligence Conference (IAAI 2019), The Ninth AAAI Symposium on Educational Advances in Artificial Intelligence (EAAI 2019), Honolulu, Hawaii, USA, January 27–February 1, 2019, pp. 6473–6480. AAAI Press (2019). https://doi.org/10.1609/aaai.v33i01.33016473

10. Hu, Z., Dong, Y., Wang, K., Sun, Y.: Heterogeneous graph transformer. In: Huang, Y., King, I., Liu, T., van Steen, M. (eds.) WWW '20: The Web Conference 2020, Taipei, Taiwan, April 20–24 2020, pp. 2704–2710. ACM/IW3C2 (2020). https://doi.org/10.1145/3366423.3380027

11. Kipf, T.N., Welling, M.: Semi-supervised classification with graph convolutional networks. In: 5th International Conference on Learning Representations (ICLR 2017), April 24–26 2017, Conference Track Proceedings. OpenReview.net, Toulon, France (2017)

12. Chuang, W.T., Yang, J.: Text summarization by sentence segment extraction using machine learning algorithms. In: Terano, T., Liu, H., Chen, A.L.P. (eds.) PAKDD 2000. LNCS (LNAI), vol. 1805, pp. 454–457. Springer, Heidelberg (2000). https://doi.org/10.1007/3-540-45571-X_52

13. Li, M., Zhang, L., Ji, H., Radke, R.J.: Keep meeting summaries on topic: abstractive multi-modal meeting summarization. In: Proceedings of the 57th Annual Meeting of the Association for Computational Linguistics. pp. 2190–2196. Association for Computational Linguistics, Florence, Italy (2019). https://doi.org/10.18653/v1/P19-1210

14. Lin, C.Y.: ROUGE: a package for automatic evaluation of summaries. In: Text Summarization Branches Out, pp. 74–81. Association for Computational Linguistics, Barcelona (2004)

15. Liu, C., Wang, P., Xu, J., Li, Z., Ye, J.: Automatic dialogue summary generation for customer service. In: Teredesai, A., Kumar, V., Li, Y., Rosales, R., Terzi, E., Karypis, G. (eds.) Proceedings of the 25th ACM SIGKDD International Conference on Knowledge Discovery & Data Mining, KDD 2019, August 4–8 2019. pp. 1957–1965. ACM, Anchorage, AK, USA (2019). https://doi.org/10.1145/3292500.3330683

16. Liu, Z., Ng, A., Lee, S., Aw, A.T., Chen, N.F.: Topic-aware pointer-generator networks for summarizing spoken conversations. arXiv preprint arXiv:1910.01335 (2019)

17. van der Maaten, L.: Accelerating t-SNE using tree-based algorithms. J. Mach. Learn. Res. **15**, 3221–3245 (2014)

18. Mihalcea, R., Tarau, P.: TextRank: Bringing order into text. In: Proceedings of the 2004 Conference on Empirical Methods in Natural Language Processing, pp. 404–411. Association for Computational Linguistics, Barcelona (2004)

19. Misra, A., Anand, P., Fox Tree, J.E., Walker, M.: Using summarization to discover argument facets in online idealogical dialog. In: Proceedings of the 2015 Conference of the North American Chapter of the Association for Computational Linguistics: Human Language Technologies, pp. 430–440. Association for Computational Linguistics, Denver, Colorado (2015). https://doi.org/10.3115/v1/N15-1046

20. Nallapati, R., Zhai, F., Zhou, B.: Summarunner: a recurrent neural network based sequence model for extractive summarization of documents. In: Singh, S.P., Markovitch, S. (eds.) Proceedings of the Thirty-First AAAI Conference on Artificial Intelligence, February 4–9 2017, pp. 3075–3081. AAAI Press, San Francisco (2017)

21. Oya, T., Mehdad, Y., Carenini, G., Ng, R.: A template-based abstractive meeting summarization: Leveraging summary and source text relationships. In: Proceedings of the 8th International Natural Language Generation Conference (INLG), pp. 45–53. Association for Computational Linguistics, Philadelphia (2014). https://doi.org/10.3115/v1/W14-4407

22. Paice, C.D.: Constructing literature abstracts by computer: techniques and prospects. Inf. Proces, Manag. **26**(1), 171–186 (1990)

23. Schlichtkrull, M., Kipf, T.N., Bloem, P., van den Berg, R., Titov, I., Welling, M.: Modeling relational data with graph convolutional networks. In: Gangemi, A., Navigli, R., Vidal, M.-E., Hitzler, P., Troncy, R., Hollink, L., Tordai, A., Alam, M. (eds.) ESWC 2018. LNCS, vol. 10843, pp. 593–607. Springer, Cham (2018). https://doi.org/10.1007/978-3-319-93417-4_38

24. See, A., Liu, P.J., Manning, C.D.: Get to the point: summarization with pointer-generator networks. In: Proceedings of the 55th Annual Meeting of the Association for Computational Linguistics (Volume 1: Long Papers). pp. 1073–1083. Association for Computational Linguistics, Vancouver, Canada (2017). https://doi.org/10.18653/v1/P17-1099

25. Serban, I.V., Sordoni, A., Bengio, Y., Courville, A.C., Pineau, J.: Building end-to-end dialogue systems using generative hierarchical neural network models. In: Schuurmans, D., Wellman, M.P. (eds.) Proceedings of the Thirtieth AAAI Conference on Artificial Intelligence, February 12–17 2016, pp. 3776–3784. AAAI Press, Phoenix (2016)

26. Speer, R., Havasi, C.: Representing general relational knowledge in ConceptNet 5. In: Proceedings of the Eighth International Conference on Language Resources and Evaluation (LREC 2012), pp. 3679–3686. European Language Resources Association (ELRA), Istanbul, Turkey (2012)

27. Vaswani, A., et al.: Attention is all you need. In: Guyon, I., et al. (eds.) Advances in Neural Information Processing Systems 30: Annual Conference on Neural Information Processing Systems 2017, December 4–9 2017, pp. 5998–6008. Long Beach (2017)

28. Velickovic, P., Cucurull, G., Casanova, A., Romero, A., Liò, P., Bengio, Y.: Graph attention networks. In: 6th International Conference on Learning Representations (ICLR 2018). April 30 - May 3 2018, Conference Track Proceedings. OpenReview.net, Vancouver, BC, Canada (2018)

29. Wang, T., Zhang, Y., Liu, X., Sun, C., Zhang, Q.: Masking orchestration: multi-task pretraining for multi-role dialogue representation learning (2020)

30. Xie, S., Liu, Y., Lin, H.: Evaluating the effectiveness of features and sampling in extractive meeting summarization. In: 2008 IEEE Spoken Language Technology Workshop. IEEE (2008)

31. Zhao, L., Xu, W., Guo, J.: Improving abstractive dialogue summarization with graph structures and topic words. In: Proceedings of the 28th International Conference on Computational Linguistics. pp. 437–449. International Committee on Computational Linguistics, Barcelona, Spain (Online) (2020). https://doi.org/10.18653/v1/2020.coling-main.39

32. Zhou, H., Young, T., Huang, M., Zhao, H., Xu, J., Zhu, X.: Commonsense knowledge aware conversation generation with graph attention. In: Lang, J. (ed.) Proceedings of the Twenty-Seventh International Joint Conference on Artificial Intelligence (IJCAI 2018) July 13–19 2018, pp. 4623–4629. Stockholm, Sweden, ijcai.org (2018). https://doi.org/10.24963/ijcai.2018/643

33. Zhu, C., Xu, R., Zeng, M., Huang, X.: A hierarchical network for abstractive meeting summarization with cross-domain pretraining. In: Proceedings of the 2020 Conference on Empirical Methods in Natural Language Processing: Findings. pp. 194–203 (2020)

Information Retrieval, Dialogue and Question Answering

Enhancing Question Generation with Commonsense Knowledge

Xin Jia[1], Hao Wang[2], Dawei Yin[2], and Yunfang Wu[1(✉)]

[1] MOE Key Lab of Computational Linguistics, School of EECS, Peking University,
Beijing, China
{jemmryx,wuyf}@pku.edu.cn
[2] Baidu Inc., Beijing, China
way_wh@yeah.net, yindawei@acm.org

Abstract. Question generation (QG) is to generate natural and grammatical questions that can be answered by a specific answer for a given context. Previous sequence-to-sequence models suffer from a problem that asking high-quality questions requires commonsense knowledge as backgrounds, which in most cases can not be learned directly from training data, resulting in unsatisfactory questions deprived of knowledge. In this paper, we propose a multi-task learning framework to introduce commonsense knowledge into question generation process. We first retrieve relevant commonsense knowledge triples from mature databases and select triples with the conversion information from source context to question. Based on these informative knowledge triples, we design two auxiliary tasks to incorporate commonsense knowledge into the main QG model, where one task is Concept Relation Classification and the other is Tail Concept Generation. Experimental results on SQuAD show that our proposed methods are able to noticeably improve the QG performance on both automatic and human evaluation metrics, demonstrating that incorporating external commonsense knowledge with multi-task learning can help the model generate human-like and high-quality questions.

Keywords: Question generation · Commonsense knowledge · Multi-task learning

1 Introduction

Question Generation (QG) has become an essential task for NLP, which aims to generate grammatical and fluent questions for a given context and answer. QG can create question-answer pairs as data augmentation for Question Answering (QA) [9,32,38]. Moreover, it is also useful in education [13,14] and business applications [24], such as creating materials for language beginners, helping build chatbots, etc.

Existing question generation methods can be roughly grouped into two categories. First, rule-based methods utilize handcrafted paradigms to perform

© Springer Nature Switzerland AG 2021
S. Li et al. (Eds.): CCL 2021, LNAI 12869, pp. 145–160, 2021.
https://doi.org/10.1007/978-3-030-84186-7_10

Table 1. A real example in the training set of SQuAD, which demonstrates the vital effect of commonsense knowledge on QG.

Context Passage:

The European Parliament and the Council of the European Union have powers of amendment and veto during the legislative process.

Reference question:

Which governing bodies have legislative veto power?

Generated question by the baseline model:

What has the powers of amendment and veto during the legislative process?

Extracted commonsense knowledge:

('council', 'RelatedTo', 'governing')

('parliament', 'Hypernymy', 'legislative bodies')

---

declarative-to-interrogative sentence transformations [6,12], but these methods often consume lots of efforts from domain experts and usually cover limited areas. Second, neural network-based methods typically model the question generation task in a fully data-driven manner [8,33,41], which has made much progress in recent years.

One key issue, however, is still up in the air: human beings often ask questions with commonsense knowledge that may exist in their brain but not appear in the given context. Take the instance in Table 1 as an example. To generate the human-like high-quality question, one must have the corresponding commonsense knowledge that the "European Parliament" and "Council of the European Union" are of "governing bodies". Lacking such commonsense knowledge results in unsatisfactory questions that simply copy some words from the source context. This directs us to introduce commonsense backgrounds to bridge the knowledge gap between the given contexts and generated questions.

Actually, previous NLP works have investigated the structured commonsense knowledge to help text generation, such as story generation [11,37] and response generation [40]. They model commonsense knowledge from external databases as additional context through attention mechanism [1,40]. However, simply employing these methods may not perform well in question generation task. Moreover, there remain two open issues in modeling external knowledge for text generation: 1) existing methods [1,40] usually utilize all extracted knowledge triples indiscriminately, which neglects the fact that some triples may not provide useful information, and introduce noises. 2) existing methods simply plug the knowledge triples into encoders, which may not fully leverage the information of these knowledge triples.

We here propose more sophisticated modeling on knowledge triples to help question generation: relevant triples, which cover the knowledge gap between source contexts and generated questions, are selected in the very beginning; furthermore, we not only utilize knowledge triples as additional inputs, but also design auxiliary tasks to help the QG model deeply absorb commonsense knowledge. To the best of our knowledge, we are the first to incorporate structured commonsense knowledge into question generation via a multi-task learning framework.

Specifically, we first retrieve all context-relevant knowledge triples from ConceptNet [30] and WordNet [23], and keep the triples where the head concept appears in the context and the tail concept appears in the reference question, i.e., ("council", "RelatedTo", "governing") and ("parliament", "Hypernymy", "legislative bodies") in Table 1.

Then, we design a multi-task learning framework that combines the main QG task and two triple-based auxiliary tasks: Concept Relation Classification and Tail Concept Generation to benefit the question generation process. The two auxiliary tasks can provide useful knowledge information and optimize the parameters of the main QG model.

We conduct extensive experiments on SQuAD dataset, and our proposed model outperforms strong baselines and achieves comparable state-of-the-art performance, demonstrating that incorporating commonsense knowledge with multi-task learning is able to improve the performance of question generation. We will release our data and code for future research.

2 Related Work

Traditionally, QG is tackled by rule-based methods [6,13,18] that rely heavily on extensive hand-crafted rules. Different from these, neural network-based methods are completely data-driven and trainable in an end-to-end fashion [8,15–17,25, 29,39,41–43]. For better representing the input context, the answer position and token lexical features (e.g. NER, POS and word case) are treated as supplements for the neural encoder [29,41]. Pointer or copy mechanisms [10,28,39] are also utilized to overcome the OOV problem in question generation process.

In order to optimize the parameters of QG model, recent works adopt the multi-task learning framework with different auxiliary tasks. Zhou [42] use language modeling as a low-level task to provide coherent representations of the input context for the high-level QG task. To improve the accuracy of the start-up word generation, Zhou [43] treat question type prediction as an auxiliary task and use the predicted word to initialize the decoding process. Jia [15] acquire built-in paraphrase knowledge through back-translation, and introduce paraphrase knowledge into QG process. Different from these works, this paper introduces external commonsense knowledge into QG through multi-task learning framework.

In addition to unstructured knowledge as Jia [15] used, many works in text generation utilize structured knowledge from mature database like ConceptNet [30]. Yang [37] employ external commonsense knowledge through dynamic memory mechanism to generate more diverse essays. Guan [11] apply structured commonsense knowledge through multi-source attention to facilitate story comprehension and generate coherent endings. Zhou [40] incorporate commonsense knowledge graphs through graph attention to create more appropriate and informative responses. Instead of combining knowledge triples into encoding process, our model creatively incorporates commonsense knowledge into QG procedure

Table 2. The statistics of retrieved knowledge triples from ConceptNet and WordNet for the training set of SQuAD.

	SQuAD	Knowledge-equipped	Pure
ConceptNet	–	12432	–
WordNet	–	41049	–
Whole	75722	46455 (61.3%)	29267 (38.7%)

with two new auxiliary tasks. In this way, the commonsense knowledge can be effectively absorbed by question generation.

3   Knowledge Extraction

To incorporate structured commonsense knowledge into question generation, the most basic step is to extract proper knowledge triples for each training sample. We will describe the details of the knowledge extraction in this section.

In order to obtain more commonsense knowledge, we extract structured knowledge triples from two commonly used databases: ConceptNet and Word-Net. ConceptNet is a semantic network composed of triples $(\mathbf{h}, \mathbf{r}, \mathbf{t})$ denoting that the head concept $\mathbf{h}$ has a relation $\mathbf{r}$ with tail concept $\mathbf{t}$. WordNet is a lexical database organized in accordance with psycholinguistic theories, where lexicalized concepts are organized by semantic relations (synonymy, hyponymy, etc.). For each sample in the training set of SQuAD, we use each non-stop word in the context passage as a query to retrieve corresponding triples from both ConceptNet and WordNet.

In the process of question generation, only those triples that provide essential knowledge for source contexts and target questions are useful rather than all retrieved triples. Therefore, we design a rule to filter triples: keep only those triples where the head concept $\mathbf{h}$ appears in the context passage and the tail concept $\mathbf{t}$ appears in the question (in the case of reversed, we swap the head and tail concepts), since these triples can directly provide conversion information between the input context and the output question. For example, in Table 1, we can extract triples like ("council", "RelatedTo", "governing"), ("council", "RelatedTo", "city") and ("council", "Synonymy", "assembly") for the word "council", while we only maintain ("council", "RelatedTo", "governing") since it directly provides the needed information for generating the right question.

From ConceptNet, there are 12,432 training samples that have extracted knowledge triples, and from WordNet there are 41,049 training samples successfully extract knowledge triples. For each training sample, we merge these triples from ConceptNet and WordNet and then remove repeated ones. Finally we obtain 46,455 knowledge-equipped samples and each sample has 1.7 corresponding triples on average, as clearly shown in Table 2.

Accordingly, we divide the origin SQuAD training set into two parts: **commonsense knowledge-equipped samples:** (context passage, answer, question, knowledge triples) and **pure samples**: (context passage, answer, question). We will explain how to use these two parts of data in the following sections.

As shown in Table 3, among these extracted knowledge triples, there are mainly 6 types of relations, where "Synonymy" and "RelatedTo" contribute to the largest proportion, with 41% and 38% respectively.

Table 3. Relation types of retrieved commonsense knowledge triples. We use "Others" to uniformly denote other relations whose proportion is less than 1%.

Type	Proportion	Type	Proportion
Synonymy	41%	RelatedTo	38%
IsA	6%	Hypernymy	6%
Hyponymy	3%	Others	6%

4   Model Description

In this section, we will describe our proposed question generation model, as is illustrated in Fig. 1. Based on the extracted knowledge triples which can provide commonsense transition information, we incorporate this knowledge into question generation via multi-task learning by employing two triple-based auxiliary tasks.

4.1   Multi-task Learning Framework

For the commonsense knowledge triple $(\mathbf{h}, \mathbf{r}, \mathbf{t})$, where the head concept $\mathbf{h}$ appears in the context passage and tail concept $\mathbf{t}$ appears in the question, it directly provides the conversion information needed for QG. To help the main QG model have a deeper understanding of this commonsense transition, we design two auxiliary tasks: Relation Classification (RC) and Tail Concept Generation (TG). We describe our main QG model and two auxiliary tasks, as well as the unified model in the following parts.

Main Task: QG Baseline Model. Given a context passage p and a specific answer a, QG targets to generate a grammatical question that can be answered by a based on the content of p. We perform sequence-to-sequence generation and adopt the model proposed by Zhang [38] as our main QG model.

First, we employ a two-layer bi-directional LSTMs as the encoder, which takes feature-enriched embedding e_i as input and outputs a list of hidden representations H:

$$H_i = [\overrightarrow{h_i}; \overleftarrow{h_i}] \tag{1}$$

$$\overrightarrow{h_i} = \overrightarrow{LSTM}([e_i; \overrightarrow{h_{i-1}}]) \tag{2}$$

$$\overleftarrow{h_i} = \overleftarrow{LSTM}([e_i; \overleftarrow{h_{i-1}}]) \tag{3}$$

$$e_i = [w_i; a_i; n_i; p_i] \tag{4}$$

where w_i, a_i, n_i, p_i respectively represents the embedding of words, answer position (BIO), Name Entity (NER) and Part-of-Speech (POS). For word embedding, we follow the settings of Zhang [38] and use ELMo [27] or BERT [5] to obtain contextualized word representations.

To aggregate long-term dependencies within the context passage, we add a gated self-attention mechanism to the encoder outputs H for $\hat{H}$:

$$\hat{h_i^p} = g_i * f_i^p + (1 - g_i) * h_i^p \tag{5}$$

We obtain self-attention context vector f_i^p through self-matching mechanism on H and then use a learnable gate g_i to balance how much f_i^p and h_i^p will contribute to the output $\hat{H}$.

The decoder we used is a two-layer uni-directional LSTM. At each decoding step t, the decoder state s_t is updated dynamically by an attention mechanism on $\hat{H}$:

$$s_{t+1} = LSTM([y_t, \tilde{s}_t]) \tag{6}$$
$$\tilde{s}_t = tanh(W^e[c_t; s_t]) \tag{7}$$
$$c_t = \hat{H}\alpha_t, \alpha_t = softmax(\hat{H}^T W^h s_t) \tag{8}$$

For each target word y_t, its probability generated from vocabulary is computed by a maxout neural network and softmax function:

$$\hat{u}_t = tanh(W^d[c_t; s_t]) \tag{9}$$
$$u_t = [max\{\hat{u}_{t,2k-1}, \hat{u}_{t,2k}\}]_k \tag{10}$$
$$P_{vocab} = softmax(W^o u_t) \tag{11}$$

Besides, the pointer mechanism will also be applied to calculate the probability of copying a word from the source context. Finally, the probability distribution is a combination of these two modes with a gate p_g:

$$P(y_t|y_{<t}) = p_g P_{vocab} + (1 - p_g)P_{copy} \tag{12}$$

The training objective is to minimize the negative log likelihood of the target sequence $\mathbf{q}$:

$$\mathcal{L}_q = -\frac{1}{T_q} \sum_{t=1}^{T_q} log(P(y_t = \mathbf{q}_t)) \tag{13}$$

Auxiliary Task-1: Relation Classification. This task is designed to predict the correct relationship between the head concept $\mathbf{h}$ and tail concept $\mathbf{t}$. We use a two-layer bi-directional LSTM to encode the $(\mathbf{h}, \mathbf{t})$ pair and obtain the hidden representation R. Then we conduct co-attention mechanism between R and the context passage representation $\hat{H}$ to get the co-dependent context $\hat{R}$:

$$\hat{R} = [\hat{H}; RA^H]A^R \tag{14}$$

$$R = LSTM([\mathbf{h}; \mathbf{t}]) \tag{15}$$

$$A^H = softmax((R^T\hat{H})^T) \tag{16}$$

$$A^R = softmax(R^T\hat{H}) \tag{17}$$

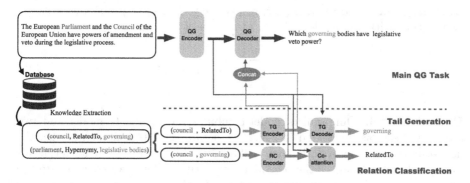

Fig. 1. The illustration of our proposed QG framework. First, we conduct knowledge extraction. In the training stage, the two triple-based auxiliary tasks provide commonsense knowledge for the main QG model and the QG model also serves as a context for TG and RC tasks.

Based on $\hat{R}$, we use a feed-forward layer f and softmax function to predict the class of relationship:

$$y_r = softmax(f(\hat{R})) \tag{18}$$

$$\mathcal{L}_r = -\sum_r(\hat{y}_r \log(y_r)) \tag{19}$$

where $\mathcal{L}_r$ is the loss function, and $\hat{y}_r$ is the one-hot label of the relationship class as listed in Table 3.

Auxiliary Task-2: Tail Concept Generation. Correspondingly, given the head concept $\mathbf{h}$ and relationship $\mathbf{r}$, generating a proper tail concept $\mathbf{t}$ also needs a deep understanding of the commonsense knowledge between them, so we design a second auxiliary task: Tail Concept Generation. Specially, we adopt another two-layer bi-directional LSTM to encode $(\mathbf{h}, \mathbf{r})$:

$$T = LSTM([\mathbf{h}; \mathbf{r}]) \tag{20}$$

In the decoding process, we use a uni-directional LSTM to generate tail concept words sequentially based on the head-relation pair. Additionally, QG passage can serve as a background so we add the context representation of QG passage $\hat{H}$ (computed in Eq. 5) into tail concept decoding:

$$s_{j+1} = LSTM([y_j, \tilde{s}_j]) \tag{21}$$

$$\tilde{s}_j = tanh(W^t[c_j; k_j; s_j]) \tag{22}$$

$$c_j = \hat{H}\alpha_j, \alpha_j = softmax(\hat{H}^T W^h s_j) \tag{23}$$

$$k_j = T\gamma_j, \gamma_j = softmax(T^T W^k s_j) \tag{24}$$

where y_j refers to the j-th tail concept word, c_j and k_j represent the context of QG passage and head-relation pair, respectively. The word probability calculation is same with QG model. The loss function of tail concept generation is:

$$\mathcal{L}_t = -\frac{1}{T_t} \sum_{j=1}^{T_t} log(P(y_j = \mathbf{t}_j)) \tag{25}$$

Unified Model. Our unified model combines the main QG model and two auxiliary tasks, as illustrated in Fig. 1. In detail, the QG context passage, the head-tail pair and head-relation pair will firstly be encoded by QG encoder, head-tail encoder and head-relation encoder, respectively. Then we concatenate the head-relation and head-tail encoder outputs to obtain the complete knowledge triple representation:

$$K = concat([T; R]) \tag{26}$$

$$k_t = K\beta_t, \beta_t = softmax(K^T W^h s_t) \tag{27}$$

We use this representation as additional commonsense context for QG decoding process. Therefore, the Eq. 7 can be rewritten as:

$$\tilde{s}_t = tanh(W^e[c_t; k_t; s_t]) \tag{28}$$

The overall training objective is the combination of the main QG task and two auxiliary tasks:

$$\mathcal{L} = \mathcal{L}_q + \mathcal{L}_r + \mathcal{L}_t \tag{29}$$

As we mentioned above, each knowledge-equipped training sample has 1.7 corresponding triples on average. Since our proposed auxiliary tasks only take one knowledge triple as input at a time, we need to choose one triple among several extracted triples. As shown in Table 3, the types of "synonymy" and "relatedto" have the largest proportions in all extracted triples. We test the effects of prioritizing "synonymy" triples and prioritizing "relatedto" triples. According to experiment results, priority use of "synonymy" is slightly better than the priority use of "relatedto" triples. Based on this, we prefer to use "synonymy" triples in our models.

4.2 Iterative Training Framework

Actually, only 61.3% percent of the training samples have extracted knowledge triples, and our unified model can only be trained on these knowledge-equipped samples. In order to make full use of the remaining pure training samples, we adopt an iterative training framework (ITF) that alternately utilizes knowledge-triple-equipped training data and pure training data.

Our unified model is composed of three parts: QG model, RC model and TG model, where RC and TG models serve as auxiliary tasks to provide commonsense knowledge for QG. During iterative training, the unified model is firstly trained based on knowledge-triple-equipped training data for N steps. Then we switch to the pure training data for another N steps and only update the QG model's parameters and leave the other parameters related to using knowledge triples frozen. In this way, the pure training samples can also contribute to the unified model.

5 Experimental Settings

5.1 Dataset and Metrics

As the most commonly used QG dataset, SQuAD is composed of (passage, answer, question) samples. Follow the setting of Zhang [38], we split the accessible parts of SQuAD into a training set (75,722 samples), a development set (10,570 samples) and a test set (11,877 samples).

We evaluate the performance of our models using BLEU [26], ROUGE-L [20] and METEOR [4].

5.2 Baseline Models

We compare our method with the following previous works on SQuAD.

- **Rule-based methods:** PCFG-Trans [13], Syn-QG [6]
- **Pre-trained models:** ACS-QG [21], UNILM [34], ERNIE-GEN [35], UNILMv2 [2], ProphetNet [36]
- **Seq2Seq models:** NQG++ [41], M2S+cp [29], A-P-Hybrid [31], s2s-a-ct-mp-gsa [39], Q-type [43], sent-Relation [19], Capture Great Context [22], NQG-RL-GS [3], QPP&QAP [38], Paraphrase-QG [15]

Besides, in order to present the different performance between our multi-task learning framework of using knowledge triples and previous methods, we also implement the graph attention method [40], which is proposed to facilitate conversation generation task, as a comparing model. This method constructs the retrieved triples into a graph and attentively reads the knowledge triples within each graph through a dynamic graph attention mechanism.

Table 4. Experimental results of our unified model comparing with previous works

Categories	Models	BLEU-1	BLEU-2	BLEU-3	BLEU-4	ROUGH-L	METEOR
Rule-based	PCFG-Trans	28.77	17.81	12.64	9.47	31.68	18.97
	Syn-QG	**45.55**	**30.24**	**23.84**	**18.72**	–	–
Pre-trained	ACS-QG	**52.30**	**36.70**	**28.00**	22.05	53.25	25.11
	UNILM	–	–	–	23.75	52.04	25.61
	ERNIE-GEN	–	–	–	25.57	53.31	26.89
	UNILMv2	–	–	–	26.30	53.19	27.09
	ProphetNet	–	–	–	**26.72**	**53.79**	**27.64**
Seq2Seq	NQG++	42.36	26.33	18.46	13.51	41.60	18.18
	M2S+cp	–	–	–	13.91	42.72	18.77
	A-P-Hybrid	43.02	28.14	20.51	15.64	–	–
	s2s-a-ct-mp-gsa	44.51	29.07	21.06	15.82	44.24	19.67
	Q-type	43.11	29.13	21.39	16.31	–	–
	Sent-Relation	44.40	29.48	21.54	16.37	44.73	20.68
	Paraphrase-QG	44.32	29.88	22.28	17.21	–	20.96
	Capture Great Context	**46.60**	**31.94**	23.44	17.76	45.89	21.56
	NQG-RL-GS	–	–	–	17.94	46.02	21.76
	QPP&QAP	–	–	–	18.65	**46.76**	**22.91**
	QG baseline model (with ELMo)	44.99	30.03	22.05	16.70	45.15	21.11
	+ graph attention	45.34	30.18	22.29	17.14	45.04	20.72
	Our Unified model (EMLo)	45.44	30.64	22.63	17.31	46.02	21.58
	QG baseline model (with BERT)	46.12	31.41	23.51	18.19	46.41	21.69
	Our Unified model (BERT)	46.36	31.74	**23.91**	**18.65**	46.65	21.84

5.3   Implementation Details

Following the settings of Zhang [38], we tokenize and obtain POS/NER features by Stanford Corenlp[1]. The QG encoder, TG encoder, RC encoder, QG decoder, and TG decoder are all 2-layer LSTMs with the hidden size of 600. We set the probability of dropout to 0.3 for each layer. We use beam search with size of 10 for decoding. In the Iterative Training Framework (ITF), we set N to 3000 and each training mode will be iteratively trained 3 times. In order to reduce the volatility of the training process, we averaged the 5 models closest to the best performing checkpoint on the development set.

The pre-trained ELMo [27] word embedding is character-level and we keep it fixed during training. For the BERT [5] version model, we utilize BERT embeddings as the replacement of ELMo. WordPiece tokenizer is applied to tokenize each word and POS/NER tags are also extended to its corresponding word pieces. During inference, we map the word-piece outputs to normal words through post-processing.

6   Results

6.1   Main Results

The main experimental results are shown in Table 4. For fair comparisons, we divide previous works into three categories: rule-based methods, pre-trained

---

[1] http://stanfordnlp.github.io/CoreNLP/.

language modeling-based methods and Seq2Seq methods, where rule-based methods and pre-trained models only serve as references and different Seq2Seq models are our comparing methods.

Our unified model (with BERT) outperforms all but one previous Seq2Seq models and obtains comparable results (18.65 BLEU-4 score) with the best QPP&QAP method [38]. QPP&QAP method relies on two pre-trained models: question paraphrasing classification and question answering models to provide rewards for policy gradient, and is further equipped with reinforcement learning mechanism, which is more complicated than ours.

Compared with the best performance of rule-based method Syn-QG, our evaluation score of BLEU-4 is very close while our other scores are obviously better.

Although the performance of our model is still far from the methods that are based on pre-training language modeling, our method is much simpler and provides an effective way to use diverse knowledge databases to supplement knowledge triples into QG. Our idea of introducing knowledge into the generation process is completely different from pre-trained models. Actually, pre-training turns out to be a useful strategy for helping models with a large number of parameters based on large-scale unsupervised training corpus [2,7,21,35,36]. However, it is extremely time-consuming and computationally expensive.

Compared with these, our method is lightweight and only takes several related knowledge triples as additional inputs instead of huge amounts of data.

Besides, directly using top K extracted triples[2] as additional inputs through graph attention [40] has a slight improvement over the QG baseline model (17.14 vs. 16.70 on BLEU-4). Compared with this traditional method, our proposed multi-task learning framework has better performance, achieving a 0.61 increase over the QG baseline model (with ELMo). Meanwhile, after applying our proposed framework, we also achieve a 0.46 BLEU-4 improvement over the much stronger QG baseline model (with BERT), which demonstrates that our proposed methods can indeed improve the performance of question generation.

Table 5. Ablation studies of our unified model (with ELMo) trained on **knowledge-equipped data**.

Model	BLEU-1	BLEU-2	BLEU-3	BLEU-4	ROUGH-L	METEOR
QG baseline model	44.47	29.28	21.23	15.90	44.30	20.52
Unified model (w.o. ITF)	44.91	29.96	**21.96**	**16.67**	**45.48**	**21.28**
- TG	44.30	29.47	21.60	16.32	45.10	20.70
- RC	**46.21**	**30.37**	21.95	16.38	44.04	21.10
- TG-RC	45.04	29.71	21.57	16.15	44.84	20.77

---

[2] We set K to 3*m in our experiments, where m represents the number of content words in each paragraph.

6.2 Ablation Study

We perform model variants and conduct ablation tests for better understanding the effect of different components of our model, and the results are shown in Table 5.

Through our Iterative Training Framework (ITF), we can simultaneously utilize the knowledge-equipped training data and pure training data. In order to better displaying the improvement brought by the introduction of external commonsense knowledge alone, we also conduct experiments by training the

Table 6. Two real cases in the test set of SQuAD. We bolded the answer and highlight the parts of the passage and the question that require commonsense knowledge conversion. The baseline model refers to QG baseline model (with BERT) and Unified model refers to Unified model (BERT) with ITF.

Context Passage-1:

Gateway National Recreation Area contains over 26,000 acres (**10,521.83** ha) in total , most of it surrounded by New York City , including the Jamaica Bay Wildlife Refuge in Brooklyn and Queens , over 9,000 acres (36 km2) of salt marsh , islands , and water , including most of Jamaica Bay . Also ......

Answer:

10,521.83

Reference:

how large is the gateway national recreation area in hectares ?

Baseline:

how many acres contains gateway national recreation area ?

Unified:

how many hectares does the gateway national recreation area have?

Context Passage-2:

Australia : The event was held in Canberra , Australian Capital Territory on April 24 , and covered around 16 km of Canberra 's central areas , from Reconciliation Place to Commonwealth Park . Upon its arrival in Canberra , the Olympic flame was presented by Chinese officials to local Aboriginal elder **Agnes Shea** , of the Ngunnawal people . She , in turn ......

Answer:

Agnes Shea

Reference:

what is the name of the aboriginal elder who received the torch from chinese officials ?

Baseline:

who was the olympic flame presented to ?

Unified:

who received the torch from chinese officials ?

model only on the **knowledge-equipped training data**. In this case, as shown in Table 5, the QG baseline model (with ELMo) only obtains a 15.90 BLEU-4 score. After applying our proposed components, our unified model boosts the BLEU-4 score to 16.67. That means, with the help of extracted commonsense knowledge and our multi-task learning framework, 61.3% training data (as listed in Table 2) yields very close results to the performance of training on the whole SQuAD dataset (16.70 in Table 4).

To confirm the effect of each component we proposed, we conduct ablation experiments over the unified model based on the **knowledge-equipped training data**. Without the Tail Generation auxiliary task, the performance of our unified model has a drop of 0.35. Besides, the unified model has a performance degradation of 0.29 if removing the Relation Classification task. In the case of removing two auxiliary tasks at the same time (using knowledge triple as an additional input for attention mechanism), the effect of the model will drop by 0.52. These experimental results verify the effectiveness of each auxiliary task and the combination of them.

6.3 Analysis of Auxiliary Tasks

In addition to the main QG task, we also evaluate the performance of two triple-based auxiliary tasks. The Relation Classification task has six categories and its accuracy reaches 66%, which has a 25% increase over the most-frequent-category baseline (41%). For RC task, the relationship between the head and tail concept is generally unique.

On the contrary, Tail Concept Generation is a much difficult task because given head concept and relation, the tail concept is not unique. We use BLEU-1 as evaluate metric and obtain a 6.23 score on this task.

6.4 Human Evaluation

For a text generation task, the automatic metrics like BLEU, ROUGH, and METEOR have limitations to evaluate the quality of generated questions. Therefore, we conduct human evaluation to compare the performance of Unified model (BERT) and QG baseline model (with BERT). We randomly select 100 samples and ask three annotators to score the generated questions of two models, according to: **Relevancy**: whether the question is relevant to the context passage; **Fluency**: whether the question is grammatical and fluent; **Answerability**: whether the question can be answered by the given answer. The rating score is set to [0, 2]. The evaluation results are shown in Table 7. Our unified model receives higher scores on all three metrics. Moreover, the high Spearman correlation coefficients guarantee the validity of our human evaluation results.

Table 7. Human evaluation results.

Model	Relevancy	Fluency	Answerability
baseline/unified	1.39/1.41	1.74/1.80	1.44/1.47
Spearman	0.65	0.80	0.73

6.5 Case Study

To clearly display the output questions, two real cases in the test set of SQuAD are shown in Table 6. Generating the right questions needs commonsense knowledge that cannot be directly obtained from the given passage, such as "ha" is the short form of "hectares" in case-1 and "flame" is synonymy with "torch" in case-2. For the baseline model, lacking such commonsense knowledge results in unsatisfactory questions which only copy some words from the passage. Compared with the baseline model, our unified model can better deal with these cases with the help of external commonsense knowledge.

7 Conclusion

In this paper, we propose a new multi-task learning framework to introduce commonsense knowledge into QG. We first extract relevant structured knowledge triples from external databases, ConceptNet and WordNet. Based on these knowledge triples, we design two auxiliary tasks to help the main QG model deeply absorb the commonsense knowledge. Both the automatic and human evaluation results verify the effectiveness of our proposed methods. In the future, we will explore new ways to use multiple knowledge triples simultaneously in the multi-task learning framework. Besides, we may also apply our framework in other text generation tasks, such as conversation generation and story generation.

Acknowledgments. This work is supported by the National Natural Science Foundation of China (62076008, 61773026) and the KeyProject of Natural Science Foundation of China (61936012).

References

1. Bai, G., He, S., Liu, K., Zhao, J.: Variational attention for commonsense knowledge aware conversation generation. In: Tang, J., Kan, M.-Y., Zhao, D., Li, S., Zan, H. (eds.) NLPCC 2019. LNCS (LNAI), vol. 11838, pp. 3–15. Springer, Cham (2019). https://doi.org/10.1007/978-3-030-32233-5_1
2. Bao, H., et al.: UniLMv2: pseudo-masked language models for unified language model pre-training. ArXiv arXiv:2002.12804 (2020)
3. Chen, Y., Wu, L., Zaki, M.J.: Natural question generation with reinforcement learning based graph-to-sequence model. ArXiv arXiv:1910.08832 (2019)

4. Denkowski, M.J., Lavie, A.: Meteor universal: language specific translation evaluation for any target language. In: WMT@ACL (2014)
5. Devlin, J., Chang, M.W., Lee, K., Toutanova, K.: BERT: pre-training of deep bidirectional transformers for language understanding. In: NAACL-HLT (2019)
6. Dhole, K.D., Manning, C.D.: Syn-QG: syntactic and shallow semantic rules for question generation. In: ACL (2020)
7. Dong, L., et al.: Unified language model pre-training for natural language understanding and generation. In: NeurIPS (2019)
8. Du, X., Shao, J., Cardie, C.: Learning to ask: neural question generation for reading comprehension. In: ACL (2017)
9. Duan, N., Tang, D., Chen, P., Zhou, M.: Question generation for question answering. In: EMNLP (2017)
10. Gu, J., Lu, Z., Li, H., Li, V.O.K.: Incorporating copying mechanism in sequence-to-sequence learning. ArXiv arXiv:1603.06393 (2016)
11. Guan, J., Wang, Y., Huang, M.: Story ending generation with incremental encoding and commonsense knowledge. In: AAAI (2019)
12. Heilman, M., Smith, N.A.: Question generation via overgenerating transformations and ranking (2009)
13. Heilman, M., Smith, N.A.: Good question! Statistical ranking for question generation. In: HLT-NAACL (2010)
14. Jia, X., Zhou, W., Sun, X., Wu, Y.: EQG-RACE: examination-type question generation. ArXiv arXiv:2012.06106 (2020)
15. Jia, X., Zhou, W., Sun, X., Wu, Y.: How to ask good questions? Try to leverage paraphrases. In: ACL (2020)
16. Kim, Y., Lee, H., Shin, J., Jung, K.: Improving neural question generation using answer separation. In: AAAI (2018)
17. Ko, W.J., Chen, T.Y., Huang, Y., Durrett, G., Li, J.J.: Inquisitive question generation for high level text comprehension. In: EMNLP (2020)
18. Labutov, I., Basu, S., Vanderwende, L.: Deep questions without deep understanding. In: ACL (2015)
19. Li, J., Gao, Y., Bing, L., King, I., Lyu, M.R.: Improving question generation with to the point context. ArXiv arXiv:1910.06036 (2019)
20. Lin, C.Y.: ROUGE: a package for automatic evaluation of summaries. In: ACL 2004 (2004)
21. Liu, B., Wei, H., Niu, D., Chen, H., He, Y.: Asking questions the human way: scalable question-answer generation from text corpus. In: Proceedings of the Web Conference 2020 (2020)
22. Luu, A.T., Shah, D.J., Barzilay, R.: Capturing greater context for question generation. In: AAAI (2020)
23. Miller, G.: WordNet: a lexical database for English. Commun. ACM **38**, 39–41 (1995)
24. Mostafazadeh, N., Misra, I., Devlin, J., Mitchell, M., He, X., Vanderwende, L.: Generating natural questions about an image. In: Proceedings of the 54th Annual Meeting of the Association for Computational Linguistics (Volume 1: Long Papers), pp. 1802–1813. Association for Computational Linguistics, Berlin, August 2016. https://doi.org/10.18653/v1/P16-1170. https://www.aclweb.org/anthology/P16-1170
25. Nema, P., Mohankumar, A.K., Khapra, M.M., Srinivasan, B.V., Ravindran, B.: Let's ask again: refine network for automatic question generation. ArXiv arXiv:1909.05355 (2019)

26. Papineni, K., Roukos, S., Ward, T., Zhu, W.J.: BLEU: a method for automatic evaluation of machine translation. In: Proceedings of the 40th Annual Meeting of the Association for Computational Linguistics, pp. 311–318. Association for Computational Linguistics, Philadelphia, July 2002. https://doi.org/10.3115/1073083. 1073135. https://www.aclweb.org/anthology/P02-1040

27. Peters, M.E., et al.: Deep contextualized word representations. ArXiv arXiv:1802.05365 (2018)

28. See, A., Liu, P.J., Manning, C.D.: Get to the point: summarization with pointer-generator networks. In: ACL (2017)

29. Song, L., Wang, Z., Hamza, W., Zhang, Y., Gildea, D.: Leveraging context information for natural question generation. In: NAACL-HLT (2018)

30. Speer, R., Chin, J., Havasi, C.: ConceptNet 5.5: an open multilingual graph of general knowledge. ArXiv arXiv:1612.03975 (2017)

31. Sun, X., Liu, J., Lyu, Y., He, W., Ma, Y., Wang, S.: Answer-focused and position-aware neural question generation. In: EMNLP (2018)

32. Tang, D., Duan, N., Qin, T., Zhou, M.: Question answering and question generation as dual tasks. ArXiv arXiv:1706.02027 (2017)

33. Wang, S., et al.: PathQG: neural question generation from facts. In: EMNLP (2020)

34. Wang, W., Wei, F., Dong, L., Bao, H., Yang, N., Zhou, M.: MiniLM: deep self-attention distillation for task-agnostic compression of pre-trained transformers. ArXiv arXiv:2002.10957 (2020)

35. Xiao, D., et al.: ERNIE-GEN: an enhanced multi-flow pre-training and fine-tuning framework for natural language generation. ArXiv arXiv:2001.11314 (2020)

36. Yan, Y., et al.: ProphetNet: predicting future n-gram for sequence-to-sequence pre-training. ArXiv arXiv:2001.04063 (2020)

37. Yang, P., Li, L., Luo, F., Liu, T., Sun, X.: Enhancing topic-to-essay generation with external commonsense knowledge. In: ACL (2019)

38. Zhang, S., Bansal, M.: Addressing semantic drift in question generation for semi-supervised question answering. ArXiv arXiv:1909.06356 (2019)

39. Zhao, Y., Ni, X., Ding, Y., Ke, Q.: Paragraph-level neural question generation with maxout pointer and gated self-attention networks. In: EMNLP (2018)

40. Zhou, H., Young, T., Huang, M., Zhao, H., Xu, J., Zhu, X.: Commonsense knowledge aware conversation generation with graph attention. In: IJCAI (2018)

41. Zhou, Q., Yang, N., Wei, F., Tan, C., Bao, H., Zhou, M.: Neural question generation from text: a preliminary study. In: Huang, X., Jiang, J., Zhao, D., Feng, Y., Hong, Y. (eds.) NLPCC 2017. LNCS (LNAI), vol. 10619, pp. 662–671. Springer, Cham (2018). https://doi.org/10.1007/978-3-319-73618-1_56

42. Zhou, W., Zhang, M., Wu, Y.: Multi-task learning with language modeling for question generation. ArXiv arXiv:1908.11813 (2019)

43. Zhou, W., Zhang, M., Wu, Y.: Question-type driven question generation. In: Proceedings of the 2019 Conference on Empirical Methods in Natural Language Processing and the 9th International Joint Conference on Natural Language Processing (EMNLP-IJCNLP), pp. 6032–6037. Association for Computational Linguistics, Hong Kong, November 2019. https://doi.org/10.18653/v1/D19-1622. https://www.aclweb.org/anthology/D19-1622

Topic Knowledge Acquisition and Utilization for Machine Reading Comprehension in Social Media Domain

Zhixing Tian[1,2](✉), Yuanzhe Zhang[1], Kang Liu[1,2], and Jun Zhao[1,2]

[1] National Laboratory of Pattern Recognition, Institute of Automation, Chinese Academy of Sciences, Beijing 100190, China
{zhixing.tian,yzzhang,kliu,jzhao}@nlpr.ia.ac.cn
[2] University of Chinese Academy of Sciences, Beijing 100049, China

Abstract. In this paper, we focus on machine reading comprehension in social media. In this domain, one normally posts a message on the assumption that the readers have specific background knowledge. Therefore, those messages are usually short and lacking in background information, which is different from the text in the other domain. Thus, it is difficult for a machine to understand the messages comprehensively. Fortunately, a key nature of social media is clustering. A group of people tend to express their opinion or report news around one topic. Having realized this, we propose a novel method that utilizes the topic knowledge implied by the clustered messages to aid in the comprehension of those short messages. The experiments on TweetQA datasets demonstrate the effectiveness of our method.

Keywords: Machine Reading Comprehension · Social media · Topic knowledge

1 Introduction

As an increasing number of people share and obtain information from social media, social media is now becoming an important real-time information source. The unprecedented volume, variety of user-generated content, and the user interaction network constitute new opportunities for understanding social behavior and building socially intelligent systems. It is important and challenging to teach a machine to automatically understand the content presented in social media.

Although considerable progress has been made in the task of Machine Reading Comprehension (MRC), most of the previous works only focus on the comprehension of the other domains, such as news [1,5], story [11,16] and Wikipedia [17,27], and there are very few works addressing the problem of social media MRC. Table 1 shows an example of social media comprehension. Different from the other domains, in social media domain, one normally posts a message on the assumption that the readers have specific background knowledge.

© Springer Nature Switzerland AG 2021
S. Li et al. (Eds.): CCL 2021, LNAI 12869, pp. 161–176, 2021.
https://doi.org/10.1007/978-3-030-84186-7_11

Those messages are generally short and contain limited contextual information as shown in the table. Thus, it is difficult for a machine to understand them thoroughly based only on the text itself. In this example, if only look at the message, without background knowledge, a machine reader would be puzzled about its topic and could not answer the question.

Table 1. An example of machine reading comprehension in social media domain.

Message: *in case you missed it, here's your first look at jamie bell's the thing: # fantasticfour empire magazine (@ empiremagazine) april 9, 2015*
Question: *what movie is being shared here?* **Answer:** *fantastic four*

To obtain background knowledge for a machine reader, one feasible method is to introduce external knowledge from the knowledge base like ConceptNet [10] and WordNet [4], as the previous works do in other domains [9,22,24]. Unfortunately, due to the nature of informality and diversity of the messages in social media, some key phrases of the short messages cannot be found in those pre-constructed knowledge base. For example, in Table 1, the token *fantasticfour*, which indicates the topic of the message, cannot be found in ConceptNet or WordNet.

By studying social media messages, we find that a significant nature of them is clustering. That is to say, on a social media platform, a group of people tend to express their opinion or report news around one topic. Specifically, those topic-relevant messages is commonly clustered by the hashtag, which is marked with "#" symbol (e.g., #fantasticfour in Table 1) and ubiquitous in social media domain. Thus, given a social media message, we can find a group of relevant messages based on the hashtag. As shown in Table 2, there are a series of topic-relevant messages clustered by the hashtag "#fantasticfour". Through those messages, we would know the topic is a science fiction film, from Marvel, about some superheroes and so on. Those hashtag-clustered messages tend to share a common topic and can be considered as a knowledge source of the topic of the given message. To this end, we propose a novel method, which obtains and utilizes the topic knowledge from the hashtag-clustered messages, to address the problem of lack of background knowledge in the task of social media comprehension.

Given a message and a question, we extract the hashtag from the message and retrieve the relevant messages based on the hashtag. Subsequently, we refine topic knowledge from the retrieved messages. Moreover, we construct a neural network, dubbed as Topic Knowledge Reader (TKR). The refined knowledge will be fused into the TKR model and contribute to the process of reading comprehension and question answering. We conduct experiments on the TweetQA dataset [26]. The result shows the effectiveness of our method.

To summarize, the major contributions of this paper are as follows:

– In the task of machine reading comprehension, we investigate the problem of lack of background knowledge in social media domain. We propose to utilize the nature of clustering of social media to obtain the knowledge from the other relevant messages.

Table 2. An example of hashtag-clustered messages in social media.

Message0: *#FantasticFour was way better than I thought it would be! It's def a science fiction film tho this ain't for the mindless action crowd.*
Message1: *GREAT SCIENCE FICTION FILM! #fantasticFour*
Message2: *Move over, Captain Marvel! #FantasticFour*
Message3: *#2020WillBeTheYearFor the #FantasticFour to unite with all of these cool characters in @Marvel Ultimate Alliance 3!*
Message4: *#FantasticFour Remains the Worst Superhero Film of the Decade*
Message5: *@MarkHarrisNYC delivers a postmortem of #FantasticFour and the state of superhero films*

- We propose a particular knowledge acquisition approach, which retrieves and refines topic knowledge from those relevant messages clustered by the hashtag which exists generally in the social media messages.
- We build a machine reading comprehension model, TKR, to utilize the refined knowledge in a targeted manner and conduct experiments on the public dataset, which demonstrates the effectiveness of our method.

2   Related Work

Social Media NLP: Over the past few years, social media has revolutionized the way we communicate. Massive amount of information in form of text is continuously generated by the users, which creates enormous challenges for NLP community to analyze and understand those text automatically. In recent years, several NLP techniques and datasets for processing social media text have been proposed. Dos Santos and Gatti [3] use a deep convolutional neural network that exploits from character-level to sentence-level information to perform sentiment analysis of short texts. Vo and Zhang [21] splits the context and employs distributed word representations and neural pooling functions to extract features from tweets. Zhou and Chen [29] propose a graphical model, named location-time constrained topic (LTT), to capture the content, time, and location of social messages. Singh et al. [18] develop an event classification and location prediction system which uses the Markov model for location inference. Qian et al. [13] jointly discover subevents from microblogs of multiple media types–user, text, and image, and design a multimedia event summarization process.

Machine Reading Comprehension: Due to the fast development of deep learning techniques and large-scale datasets, Machine Reading Comprehension (MRC) has gained increasingly wide attention over the past few years. Richardson et al. [15] build the multiple choice dataset MCTest, and this dataset encourages the early research of machine reading comprehension, and a strand of MRC models [11,16] are inspired by the dataset. Hermann et al. [5] propose a cloze test dataset CNN & Daily Mail, which is large-scale and more suitable than MCTest for deep learning methods. Based on this dataset, Hermann et al. [5] propose an attention-based LSTM [6] model named Attentive Reader. Moreover, Rajpurkar et al. [14] release the span extraction dataset, SQuAD, which has become the most popular MRC dataset over recent years. This dataset enlightens a lot of classical MRC model, like BiDAF [17] and R-Net [25]. In addition, the multi-hop MRC dataset HotpotQA [27] has gained recent wide attention. This dataset addresses the problem of multiple clues based question answering.

3   Method

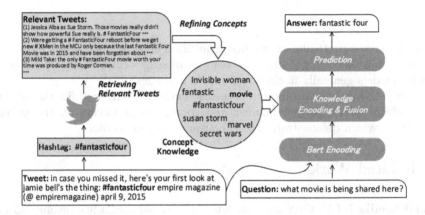

Fig. 1. The framework of our proposed method. The left half is the process of obtaining topic knowledge. The right half is the reading comprehension model, Topic Knowledge Reader (TKR)

Figure 1 shows the framework of our method. Note that it is an example of the tweets, but the method is universal for the messages from other social media platforms. Given a tweet and a question, we obtain the answer by the following steps: First, we extract the hashtag from the tweet, meanwhile, we encode the tweet and the question by the BERT encoder. Second, we retrieve relevant tweets that contain the same hashtag. Third, we refine the topic knowledge from the retrieved tweets. Next, the knowledge is encoded and fused with the BERT representation. Finally, the model predicts the answer based on the knowledge aware representation of the tweet and question.

3.1   Knowledge Acquisition

We regard the set of tweets clustered by the hashtag as the resource of knowledge and obtain topic knowledge from them. We first retrieve relevant tweets, then from those tweets, we gather common concepts. Meanwhile, we maintain a hashtag pool to score each concept. Finally, we refine the concepts by select the top-k scored ones as the topic knowledge.

Retrieving Relevant Tweets. Given a tweet text T, we first extract the hashtag H. Next, we use H as the query and retrieve the relevant tweets. So we have a set S consisting of tweets that contain the same hashtag. The tweets in S tend to share the same topic with the given tweet T, and the information of them is helpful to understand T comprehensively. Next, we remove the non-English tweets from S, and then delete the non-normal strings in the text of S such as the URL which starts with "*http*" and the reference of the picture which starts with "*pic*".

Exceptionally, for those tweets that contain no hashtag, we utilize a hashtag extractor to extract hashtag words from the tweet. The extractor is composed of a BERT encoder and a span pointer. We input the tweet, T, to the BERT model and obtain the representation $P = \{p_0, p_1, ..., p_n\}$, where p_i is the i-th word of the tweet. Then we extract the hashtag from the tweet by a pointer:

$$Start_i = \frac{exp(w_0^T p_i)}{\sum_j exp(w_0^T p_j)} \qquad End_i = \frac{exp(w_1^T p_i)}{\sum_j exp(w_1^T p_j)} \tag{1}$$

where w_0 and w_1 are trainable vectors. The pointer labels the probability of each word as the start and the end of the hashtag, respectively. We calculate the score of a span by multiplying those two probabilities and take the span with the max score as the hashtag. Figure 2 shows an example. We train the extractor model on a hashtag extraction dataset proposed by Zhang et al. (2016). [28]. Evaluated on the test set of the dataset, our extractor achieves 85.1% accuracy.

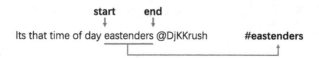

Fig. 2. An example of extracting hashtag for those tweets without hashtag

Gathering Relevant Concepts. Having retrieved the tweets with the same topic, we gather the fine-grained knowledge, i.e., concepts that connect to the topic. We tokenize every tweet text in S and obtain a set of tokens. Then we segment each token to get the concept. Due to the nature of informality, some tokens from the tweets could contain multiple words, like the hashtag "#secret-wars", thus we conduct a segmentation on each token. After that, we obtain a set C consisting of concepts (e.g., "movie", "marvel", and "secret wars").

Maintaining Hashtag Pool. To further refine the concepts, we maintain a hashtag pool. First of all, a large scale of recent tweets is collected as the original corpus. Based on the corpus, we collect the hashtags, then find the relevant tweets and obtain the concept set C for each hashtag follow above-mentioned process. Those hashtags and their all relevant concepts are added to the empty hashtag pool as the initialization. When a new tweet, T, is given during application, we update the hashtag pool by adding the hashtag of T and the relevant concepts, C, to the pool.

Refining Topic Knowledge. Given tweet T, by above-mentioned steps, we have concepts C, then we apply Term Frequency-Inverse Document Frequency (TF-IDF) to score each concept. The score for $concept_i$ in C is calculated by:

$$score_i = \frac{n_i}{N} log \frac{|P|}{|p_i| + 1} \tag{2}$$

where n_i is the frequency of $concept_i$ in C, N is the total count of concepts in C, P denotes the hashtag pool. Thus, $|P|$ is the total number of the hashtags in the hashtag pool, and $|p_i|$ is the number of hashtags, whose relevant concepts contain the $concept_i$, in the hashtag pool. Finally, the top-k scored concepts are selected as the topic knowledge K of the tweet T.

3.2 Topic Knowledge Reader

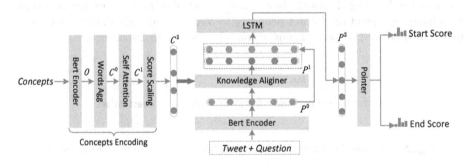

Fig. 3. The detail architecture of Topic Knowledge Reader (TKR).

As shown in Fig. 3, we propose a reading comprehension model, named Topic Knowledge Reader (TKR), to fuse the refined concepts and then answer the question. The inputs of the model are

- the given tweet $T = \{t_0, t_1, ..., t_{n-1}\} \in \mathbb{R}^n$, where n is the number of words in the tweet, t_i is the i-th word in T.
- the question $Q = \{q_0, q_1, ..., q_{m-1}\} \in \mathbb{R}^m$, where m is the number of words in the question, q_i is the i-th word in Q.

- the concept knowledge $K = \{k_{00}, k_{01}, ..., k_{ij}, ..., k_{(l-1)x}\} \in \mathbb{R}^y$, where y is the number of words of all concepts, k_{ij} refers to the j-th word of i-th concept.
- the concept score $S = \{s_0, s_1, ..., s_{l-1}\} \in \mathbb{R}^l$.

The output of the model is the predicted answer.

Encoding Tweet and Question. We first concatenate the question Q and the tweet T. The combination passage is

$$D = \{[CLS], t_0, t_1, ..., t_{n-1}, [SEP], q_0, q_1, ...q_{m-1}, [SEP]\} \tag{3}$$

where we add the special word "[CLS]" and "[SEP]", which follows the process from Devlin et al. [2]. Then, we employ BERT [2] to encode the tweet and the question together, thus we have the question-aware representation of the passage:

$$P^0 = BERT(D) \in \mathbb{R}^{(m+n+3) \times h} \tag{4}$$

Encoding Concepts. We encode the concepts, before the step of fusion. To obtain the original representation, we apply BERT encoder for the concepts as well. Analogously, we add the special word "[CLS]" to the single sequence of knowledge words,

$$K = \{[CLS], k_{00}, k_{01}, ..., k_{ij}, ..., k_{(l-1)x}\} \tag{5}$$

and then the pre-trained model, BERT, is applied on encoding the concepts:

$$O = BERT(K) \in \mathbb{R}^{(y+1) \times h} \tag{6}$$

Words Aggregation: As there are multiple words in some concepts, we aggregate the words of each concept by mean pooling, then obtain the one-vector representation, c_i^0, for each concept:

$$c_i^0 = \frac{1}{N} \sum_{j \in [0,N)} c_{ij}^0 \qquad C^0 = \{c_0^0, c_1^0, ..., c_i^0, ..., c_{l-1}^0\} \in \mathbb{R}^{l \times h} \tag{7}$$

Self Attention: Though no sequential relation exists among those concepts, they are still interrelated. Thus, we use the self-attention mechanism to perform a non-sequence context encoding on the concepts:

$$c_i^1 = \sum_j \alpha_{ij} c_j^0 \qquad \alpha_{ij} = \frac{exp(\sigma(W_q c_i^0) \cdot \sigma(W_k c_j^0))}{\sum_{j'} exp(\sigma(W_q c_i^0) \cdot \sigma(W_k c_{j'}^0))} \tag{8}$$

where σ is the activation function, $W_q \in \mathbb{R}^{h \times h}$ and $W_k \in \mathbb{R}^{h \times h}$ are trainable matrixes. Thus we have self-aligned concepts $C^1 = \{c_0^1, c_1^1, ..., c_{l-1}^1\} \in \mathbb{R}^{l \times h}$.

Score Scaling: We then scale the concepts by the score $S \in \mathbb{R}^l$ assigned in the step of knowledge refining:

$$C^2 = SC^1 \in \mathbb{R}^{l \times h} \tag{9}$$

C^2 denotes the final representation of the Concepts.

Topic Knowledge Fusion. The Concepts are fused into the passage by:

$$p_i^1 = \sum_j \beta_{ij} c_j^2 \qquad \beta_{ij} = \frac{exp(\sigma(W_p p_i^0) \cdot \sigma(W_c c_j^2))}{\sum_{j'} exp(\sigma(W_p p_i^0) \cdot \sigma(W_c c_{j'}^2))} \qquad (10)$$

where σ is the activation function, $W_p \in \mathbb{R}^{h \times h}$ and $W_c \in \mathbb{R}^{h \times h}$ are trainable matrixes. Thus, we obtain the concepts-aware passage representation $P^1 = \{p_0^1, p_1^1, ..., p_{m+n+2}^1\} \in \mathbb{R}^{(m+n+3) \times h}$. A bidirectional LSTM is applied to conduct an additional sequential context encoding and aggregate the original question-aware passage representation P^0 and the concepts-aware passage representation P^1.

$$P^2 = BiLSTM([P^0; P^1]) \in \mathbb{R}^{(m+n+3) \times h} \qquad (11)$$

Prediction. We employ two Linear layers to point the start position and the end position of the answer in the passage, respectively, and then normalize the prediction scores:

$$\tilde{Start}_i = \frac{exp(w_s^T p_i^2)}{\sum_j exp(w_s^T p_j^2)} \qquad \tilde{End}_i = \frac{exp(w_e^T p_i^2)}{\sum_j exp(w_e^T p_j^2)} \qquad (12)$$

$w_s \in \mathbb{R}^h$ and $w_e \in \mathbb{R}^h$ are trainable weight vectors. We utilize Negative Log Likelihood (NLL) as the loss function during training. Moreover, during the evaluation, we obtain the score of each span of the tweet by multiplying its start score and the end score and then select the text span with the max score as the answer.

4   Experiment

4.1   TweetQA Dataset

We conduct experiments on the recently released social media MRC dataset, TweetQA. Each instance of the dataset is a triple consisting of a tweet text, a human proposed question, and a list of human-annotated answers. The dataset is composed of 10692 training triples, 1086 development triples, and 1979 test triples. It is the first large-scale MRC dataset over social media data.

4.2   Implement Detail

Preprocess: As we employ BERT [2] to encode the text, we tokenize the text by the default tokenizer of BERT. Since the answer spans are not labeled in the train set, we annotate the approximate answer span in each tweet by selecting the span that achieves the best F1 score.

Knowledge Acquisition: To simulate the real-world scenario where a social media MRC system works, we regard the train set as the original corpus for the initialization of the Hashtag Pool. During evaluating, we update the Hashtag Pool by the hashtag and the relevant concepts, from the development set and the test set. Based on the experimental analysis, we select top-8 scored concepts for each hashtag at the step of refining the knowledge.

Training: We select the instances that contain the span whose F1 score no less than 0.6 to train the model in the way of weakly supervised. As a result, 8238 instances are used during training. We employ Adam optimizer to train the model. The learning rate is set to 3×10^{-5}, the model is fine-tuned for 3 epochs, and the dropout rate of BERT is set to 0.1. The BERT model we choose is the pre-trained *bert-base* [2] model, distinguished from the *bert-large* model. The hidden size of BERT is 768.

Evaluation: As the answer in TweetQA is not always a span of given tweet, following Xiong et al. [26], we use the metrics for natural language generation to evaluate the models, namely BLEU-1, Meteor, and Rouge-L. The answers of the test set are not released, so we submit our prediction to the official evaluating platform of TweetQA[1] and receive the response of the performance results.

4.3   Baselines

- **Query Matching:** a simple IR baseline [7], which is adapted to the TweetQA Task by Xiong et al. [26].
- **BiDAF:** a popular neural baseline [17] of Machine Reading Comprehension, which extract answers from the original tweet text.
- **Gerative QA:** a RNN-based generative model [19]. The model employs both copy and coverage mechanisms during the process of generating.
- **BERT Extraction:** a recently proposed pre-trained model [2]. Following [2], we construct a BERT based answer extraction model by inputting the representation of passage, P^0, obtained from Eq. 4 directly to the prediction layer formulated by Eq. 12.
- **BERT Generation:** Because part of the answers of TweetQA are not a span of the tweet text, we build a BERT based generative model. We use BERT as the encoder same with Eq. 4. Following [20], we employ a pointer generator, which selects words from both the tweet and the vocabulary, to decode the answer. The generative model is trained on all instances of the train set.
- **Knowledge Concat:** We also introduce another simpler method to fuse the topic knowledge, named "Knowledge Concat". The model directly concatenates topic knowledge (i.e., the selected concepts) with the sequence of tweet and question before BERT encoding and finally, same with TKR, conduct a span prediction.
- **KAR:** Knowledge Aided Reader (KAR) [23] is a recently proposed MRC model, which utilizes the knowledge from WordNet. The model conducts mutual attention and self-attention based on the connections among the words of the question and the passage. The connections are built based on the

---

[1] https://tweetqa.github.io/.

knowledge from WordNet. For a fair comparison, we change the model by utilizing BERT as the basic encoder instead of the original embedding layers composed of Glove [12], CNN [8], and LSTM.

4.4 Main Results

Table 3. The results on TweetQA dataset. Extract-UB denotes the upper bound of extractive methods.

Model	Dev set			Test set		
	BLEU-1	Meteor	Rouge-L	BLEU-1	Meteor	Rouge-L
Human	–	63.7	70.9	70.0	66.7	73.5
Extract-UB	–	68.8	74.3	75.1	69.8	75.6
Query matching	–	12.0	17.0	11.2	12.1	17.4
BiDAF	–	31.6	38.9	34.9	31.4	38.6
Gerative QA	–	32.1	39.5	36.1	31.8	39.0
BERT generation	48.5	42.0	51.8	49.1	42.1	52.3
BERT extraction	61.0	58.4	64.2	63.2	60.9	65.8
Knowledge concat	65.5	61.6	67.9	66.9	63.4	69.1
KAR	66.9	63.2	68.9	67.7	64.1	69.8
TKR	**68.7**	**64.7**	**70.6**	**69.0**	**65.6**	**71.2**

As shown in Table 3, our model, TKR, surpasses the recently proposed Knowledge Aided Reader (KAR) and achieves competitive performance. From our point of view, due to the limitation of the knowledge from the pre-constructed knowledge base (WordNet), KAR suffers from the sparsity problem of knowledge extraction for the diverse and informal expressions in social media domain. Besides, TKR outperforms all of the other baselines significantly, especially the BERT based model, BERT Extraction. The model, BERT Extraction, is exactly the rest architecture of TKR when we ablate topic knowledge from TKR. Thus, the comparison between TKR and BERT Extraction can directly demonstrate the advantages of our methods to acquire and utilize topic knowledge for social media comprehension.

Moreover, Knowledge Concat performs better than BERT Extraction, which also validates the effectiveness of the knowledge. Besides, comparing Knowledge Concat with TKR, we find that TKR performs better. This is because TKR can integrate our refined knowledge to the MRC model in a more targeted manner.

4.5 Different Number of Concepts

To further verify the effectiveness of the topic knowledge, we study the relationship between the number of employed concepts and the performance of TKR. We choose top-k concepts during the step of refining, where k changes from 2

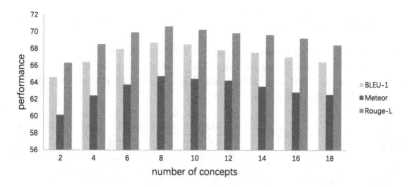

Fig. 4. The performance of TKR with different number (k) of concepts on the dev set of TweetQA.

to 18, and then train and evaluate TKR at different settings of k. As shown in Fig. 4, by increasing the number, k, from 2 to 18, the performance of TKR first rises rapidly until k reaches 8 and then drop down slowly. The gain of performance from $k = 2$ to $k = 8$ proves our topic knowledge effective. The loss of performance from $k = 8$ to $k = 18$ is caused by the noise introduced by the concepts with low scores.

Table 4. Sampled cases which show the effect of the different numbers of concepts. The concepts in blue are the *Top-5* scored ones, and those in black are the *13–18th* scored concepts.

Tweet: *Vets saw a fetus on Mei Xiang's ultrasound today. Paws crossed 4 viable pregnancy #PandaStory National Zoo (@NationalZoo)*
Hashtag: *PandaStory*
Concepts: *panda life, national zoo, panda, awesome ip, Mei Xiang, ..., youtube, conservation, giant, yuan meng panda, panda fans on instagram*
Question0: *where is mei xiang located?*
Answer0: *at the national zoo*
Question1: *what is the name of the panda?*
Answer1: *mei xiang*

To probe the effect of employing different numbers of concepts more intuitively, we sample and analyze some cases from the development set. Table 4 shows one of them, where *Question0* and *Question1* are two questions proposed based on the same tweet as shown in the table. In this example, we find that the *top-5* concepts describe the topic comprehensively, build a semantic connection between some key concepts in the tweet including *panda, Mei Xiang,* and *National Zoo*, and finally contribute to answering the questions. On the contrary, the *13–18th* scored concepts tend to deviate from the topic, and as noisy-like information, they even damage the Reading Comprehension model, TKR, when introduced into the model.

4.6   Ablation Study

Table 5. Ablation study on the development set. -*score scale* denotes TKR without the module of score scaling. -*self attn* denotes TKR without self attention. -*word agg* denotes TKR without word aggregation. - *LSTM* denotes TKR that use a dense layer instead of LSTM for the knowledge fusion

Model	BLEU-1	ΔBLEU-1	Meteor	ΔMeteor	Rouge-L	ΔRouge-L
TKR	68.7	–	64.7	–	70.6	–
- score scale	68.0	−0.7	64.3	−0.4	70.0	−0.6
- self attn	67.9	−0.8	64.0	−0.7	69.7	−0.9
- word agg	68.1	−0.6	64.1	−0.6	70.1	−0.5
- LSTM	68.4	−0.3	64.5	−0.2	70.3	−0.3

To study the effect of some key modules of TKR, we conduct ablation experiments on the development set. As shown in Table 5, all of the three knowledge encoding module, including *score scale*, *self attention* and *word agg*, contribute to the overall performance. The results demonstrate that those modules. which are designed for the topic knowledge in a targeted manner, indeed help the model to encode the knowledge and further to absorb it. Furthermore, the performance of - *LSTM* is slightly behind the original TKR, which proves that the sequential information captured by the additional context encoding is beneficial for the comprehension.

4.7   Extractive vs. Generative

Table 6. Sampled cases that show the difference between the generative model and the extractive one.

Tweet: *thank you to @ marvel for sending gifts for our patients. they were thrilled to receive them! ...*
Question: ru*what were patients of Seattle children's thrilled to receive?*
Answer: *gifts from marvel*
Generative prediction: *gifts of marvel*
Extractive prediction: *gifts*
Tweet: *our prayers are with the students, educators & families at independence high school & all the first responders on the scene. # patriotpride ...*
Question: *at which school were first responders on the scene for?*
Answer: *independence high school*
Generative prediction: *the school*
Extractive prediction: *independence high school*

As shown in Table 3, compared with BERT Generation, the extractive models including BERT Extraction and TKR achieve better performance, though there is no identically matching substring in the tweet for part of answers. Table 6 shows two sampled cases which tell the difference between the extractive model and the generative one. As shown in the table, the generative model performs better in some cases where the answer is supposed to be synthesized based on the question and tweet. On the contrary, the generative model lag behind the extractive one, when the answer is an uninterrupted snippet of the tweet. However, as studying more cases, we find that even in many cases, where the answer need to synthesize, the generative model fails to provide a qualified answer. We consider that much more data is needed to train a qualified generative MRC model.

4.8 Weakly Supervised Training

To train the extractive model, TKR, we annotate the answer span in the tweets by the F1 score. We train the model to locate the annotated span. As the annotated span may not be the true answer, it is a process of weakly supervised training. As shown in Table 7, we study the relationship between the span score of training data and the performance which is evaluated on the development set. As shown in the table, by reducing the threshold of span score, increasing training data is involved, meanwhile the performance first rises until *span score = 0.6* and then drop down. That is to say, in the process of introducing different amount of the weakly supervised training data, *span score = 0.6* is the point where the difference between the benefit from the positive example and the damage from the noise is maximized.

Table 7. The performance on development set with different scale of train data. *SpanScore* $\geq$ *i* denotes that the model is trained by the instances containing the span whose F1 score is no less than *i*. *data* and *proportion* refer to the scale and the proportion of the selected training data.

Span score	Data	Proportion	BLEU-1	Meteor	Rouge-L
≥ 1	6899	64%	67.0	63.5	69.1
≥ 0.8	7442	69%	67.9	64.3	69.2
≥ 0.6	8238	77%	**68.7**	**64.7**	**70.6**
≥ 0.4	8978	83%	68.2	64.2	70.3
≥ 0.2	9283	86%	67.3	63.3	69.3
> 0	9314	87%	67.1	63.0	69.2

5   Conclusion

In this paper, we focus on machine reading comprehension in social media domain. We propose a novel method to address the problem of lacking in background knowledge in this task. Utilizing the nature of clustering of social media, we retrieve and refine topic knowledge from the relevant messages, and then integrate the knowledge into an MRC model, TKR. Experimental results show that our proposed method outperforms the recently proposed models and the BERT-based baselines, which proves the method effective overall. By introducing different amount of topic knowledge, we demonstrate the effectiveness of our refined knowledge. Moreover, the ablation study further validates the contribution of the key modules of TKR for utilizing the knowledge.

Acknowledgements. This work is supported by the National Natural Science Foundation of China (No. 61976211, No. 61922085, No. 61906196). This work is also supported by the Key Research Program of the Chinese Academy of Sciences (Grant NO. ZDBS-SSW-JSC006), the Open Project of Beijing Key Laboratory of Mental Disorders (2019JSJB06).

References

1. Chen, D., Bolton, J., Manning, C.D.: A thorough examination of the CNN/daily mail reading comprehension task. In: Proceedings of the 54th Annual Meeting of the Association for Computational Linguistics (Volume 1: Long Papers), pp. 2358–2367. Association for Computational Linguistics, Berlin, August 2016. https://doi.org/10.18653/v1/P16-1223. https://www.aclweb.org/anthology/P16-1223
2. Devlin, J., Chang, M.W., Lee, K., Toutanova, K.: BERT: pre-training of deep bidirectional transformers for language understanding. arXiv preprint arXiv:1810.04805 (2018)
3. Dos Santos, C., Gatti, M.: Deep convolutional neural networks for sentiment analysis of short texts. In: Proceedings of COLING 2014, The 25th International Conference on Computational Linguistics: Technical Papers, pp. 69–78 (2014)
4. Fellbaum, C. (ed.): WordNet: An Electronic Lexical Database. MIT Press, Cambridge (1998)
5. Hermann, K.M., et al.: Teaching machines to read and comprehend. In: Advances in Neural Information Processing Systems, pp. 1693–1701 (2015)
6. Hochreiter, S., Schmidhuber, J.: Long short-term memory. Neural Comput. **9**, 1735–1780 (1997)
7. Kočiský, T., et al.: The narrativeQA reading comprehension challenge. Trans. Assoc. Comput. Linguist. **6**, 317–328 (2018)
8. Krizhevsky, A., Sutskever, I., Hinton, G.E.: ImageNet classification with deep convolutional neural networks. In: NIPS (2012)
9. Lin, H., Sun, L., Han, X.: Reasoning with heterogeneous knowledge for commonsense machine comprehension. In: Proceedings of the 2017 Conference on Empirical Methods in Natural Language Processing, Copenhagen, Denmark, pp. 2032–2043, September 2017. https://doi.org/10.18653/v1/D17-1216. https://www.aclweb.org/anthology/D17-1216
10. Liu, H., Singh, P.: ConceptNet - a practical commonsense reasoning tool-kit. BT Technol. J. **22**(4), 211–226 (2004). https://doi.org/10.1023/B:BTTJ.0000047600.45421.6d
11. Narasimhan, K., Barzilay, R.: Machine comprehension with discourse relations. In: Proceedings of the 53rd Annual Meeting of the Association for Computational Linguistics and the 7th International Joint Conference on Natural Language Processing (Volume 1: Long Papers), pp. 1253–1262 (2015)

12. Pennington, J., Socher, R., Manning, C.D.: GloVe: global vectors for word representation. In: EMNLP (2014)
13. Qian, X., Li, M., Ren, Y., Jiang, S.: Social media based event summarization by user-text-image co-clustering. Knowl.-Based Syst. **164**, 107–121 (2019)
14. Rajpurkar, P., Zhang, J., Lopyrev, K., Liang, P.: SQuAD: 100,000+ questions for machine comprehension of text. arXiv preprint arXiv:1606.05250 (2016)
15. Richardson, M., Burges, C.J., Renshaw, E.: MCTest: a challenge dataset for the open-domain machine comprehension of text. In: Proceedings of the 2013 Conference on Empirical Methods in Natural Language Processing, pp. 193–203. Association for Computational Linguistics, Seattle, October 2013. https://www.aclweb.org/anthology/D13-1020
16. Sachan, M., Dubey, K., Xing, E., Richardson, M.: Learning answer-entailing structures for machine comprehension. In: Proceedings of the 53rd Annual Meeting of the Association for Computational Linguistics and the 7th International Joint Conference on Natural Language Processing (Volume 1: Long Papers), pp. 239–249 (2015)
17. Seo, M., Kembhavi, A., Farhadi, A., Hajishirzi, H.: Bidirectional attention flow for machine comprehension. arXiv preprint arXiv:1611.01603 (2016)
18. Singh, J.P., Dwivedi, Y.K., Rana, N.P., Kumar, A., Kapoor, K.K.: Event classification and location prediction from tweets during disasters. Ann. Oper. Res. **283**, 737–757 (2019). https://doi.org/10.1007/s10479-017-2522-3
19. Song, L., Wang, Z., Hamza, W.: A unified query-based generative model for question generation and question answering (2017)
20. Tay, Y., et al.: Simple and effective curriculum pointer-generator networks for reading comprehension over long narratives. In: Proceedings of the 57th Annual Meeting of the Association for Computational Linguistics, Florence, Italy, pp. 4922–4931, July 2019. https://www.aclweb.org/anthology/P19-1486
21. Vo, D.T., Zhang, Y.: Target-dependent twitter sentiment classification with rich automatic features. In: Twenty-Fourth International Joint Conference on Artificial Intelligence (2015)
22. Wang, C., Jiang, H.: Explicit utilization of general knowledge in machine reading comprehension. In: Proceedings of the 57th Annual Meeting of the Association for Computational Linguistics, Florence, Italy, pp. 2263–2272, July 2019. https://www.aclweb.org/anthology/P19-1219
23. Wang, C., Jiang, H.: Explicit utilization of general knowledge in machine reading comprehension. In: ACL (2019)
24. Wang, L., Sun, M., Zhao, W., Shen, K., Liu, J.: Yuanfudao at SemEval-2018 Task 11: three-way attention and relational knowledge for commonsense machine comprehension. arXiv preprint arXiv:1803.00191 (2018)
25. Wang, W., Yang, N., Wei, F., Chang, B., Zhou, M.: Gated self-matching networks for reading comprehension and question answering. In: Proceedings of the 55th Annual Meeting of the Association for Computational Linguistics (Volume 1: Long Papers), pp. 189–198. Association for Computational Linguistics, Vancouver, July 2017. https://doi.org/10.18653/v1/P17-1018. https://www.aclweb.org/anthology/P17-1018
26. Xiong, W., et al.: TWEETQA: a social media focused question answering dataset. In: Proceedings of the 57th Annual Meeting of the Association for Computational Linguistics, pp. 5020–5031. Association for Computational Linguistics, Florence, July 2019. https://www.aclweb.org/anthology/P19-1496

27. Yang, Z., et al.: HotpotQA: a dataset for diverse, explainable multi-hop question answering. In: Proceedings of the 2018 Conference on Empirical Methods in Natural Language Processing, pp. 2369–2380. Association for Computational Linguistics, Brussels, October–November 2018. https://doi.org/10.18653/v1/D18-1259. https://www.aclweb.org/anthology/D18-1259
28. Zhang, Q., Wang, Y., Gong, Y., Huang, X.: Keyphrase extraction using deep recurrent neural networks on Twitter. In: Proceedings of the 2016 Conference on Empirical Methods in Natural Language Processing, pp. 836–845. Association for Computational Linguistics, Austin, November 2016. https://doi.org/10.18653/v1/D16-1080. https://www.aclweb.org/anthology/D16-1080
29. Zhou, X., Chen, L.: Event detection over Twitter social media streams. VLDB J.-Int. J. Very Large Data Bases **23**(3), 381–400 (2014)

Category-Based Strategy-Driven Question Generator for Visual Dialogue

Yanan Shi, Yanxin Tan, Fangxiang Feng, Chunping Zheng,
and Xiaojie Wang[✉]

Beijing University of Posts and Telecommunications, Beijing, China
{apilis,fxfeng,zhengchunping,xjwang}@bupt.edu.cn

Abstract. GuessWhat?! is a task-oriented visual dialogue task which has two players, a guesser and an oracle. Guesser aims to locate the object supposed by oracle by asking several Yes/No questions which are answered by oracle. How to ask proper questions is crucial to achieve the final goal of the whole task. Previous methods generally use an word-level generator, which is hard to grasp the dialogue-level questioning strategy. They often generate repeated or useless questions. This paper proposes a sentence-level category-based strategy-driven question generator (CSQG) to explicitly provide a category based questioning strategy for the generator. First we encode the image and the dialogue history to decide the category of the next question to be generated. Then the question is generated with the helps of category-based dialogue strategy as well as encoding of both the image and dialogue history. The evaluation on large-scale visual dialogue dataset GuessWhat?! shows that our method can help guesser achieve 51.71% success rate which is the state-of-the-art on the supervised training methods.

Keywords: Visual dialogue · Question generation · Question category

1 Introduction

Goal-oriented Visual Dialogue is one of the multi-modal task which has gained increasing attentions. It can usually be explained as a task which using a couple of visual related turns of communication to reach a specific goal, such as Guess-Which [2], VisDial [5], Multimodal dialogs (MMD) [17], GuessWhat?! [28] and so-on.

GuessWhat?! [28] is a two-player game which can be named guesser and oracle. Given an image, the oracle choose an object in the image without telling the guesser. Guesser reads the image and asks several Yes/No questions, at the meanwhile Oracle uses 'Yes', 'No' and 'N/A' to answer them. After a fixed turns of questioning, Guesser should guess the chosen object in image and one game finishes. As is mentioned above, there are two sub tasks: one is ask question (QGen), and the other is Guess. Figure 1 gives a sample in GuessWhat?! dataset.

To reach the goal of the entire game, the agent for task QGen should be able to make use of information from both the dialogue history and the image. de

© Springer Nature Switzerland AG 2021
S. Li et al. (Eds.): CCL 2021, LNAI 12869, pp. 177–192, 2021.
https://doi.org/10.1007/978-3-030-84186-7_12

Vries et al. [28] use an encoder-decoder model to generate questions, which is the first QGen model. After that, some researchers manage to improve the generation quality by attention mechanism. Zhuang et al. [32] proposed a parallel attention (PLAN) network. Deng et al. [6] proposed accumulated attention mechanism (A-ATT). Xu et al. [29] focused on boosting the agent to generate highly efficient questions and obtains reliable visual attentions and proposed Answer-Driven Visual State Estimator (ADVSE). On the other hand, Strub et al. [24] introduced reinforcement learning (RL). Pang et al. [14] proposed visual dialogue state tracking (VDST) to represent visual dialogue state and update with the change of the distribution on objects.

When humans play the game, they often have some strategies which aims to minimize the scope of the object and reaches the goal as quickly as possible. For example, In visual dialogue dataset GuessWhat?!, people might first ask the type of the object, once they determine the right type of the object, they then ask the attributes of the object and so on. In fact, we can find there are more than 80% of records ask questions about an object at the first turn, however only 4.03% ask questions about color at the same place. The sample in dataset GuessWhat?! [28] is shown in Fig. 1.

However, most of previous works didn't explicitly focus on the question generating strategy. de Vries et al. [28] encodes the image and dialogue history into a fixed length representation vector and generates the response. Deng et al. [6] and Zhuang et al. [32] improve the model by adding strong attention mechanisms but still lack of improvement on generation side. In the absence of specialized strategy modules, such simple encoding way is often difficult to fit the question generation switching strategy [28]. Xu et al. [29] used an answer-based attention transfer method. They used the transferring of attention to shift the focus point of the problem, but still lacks of explicit influence on question generation.

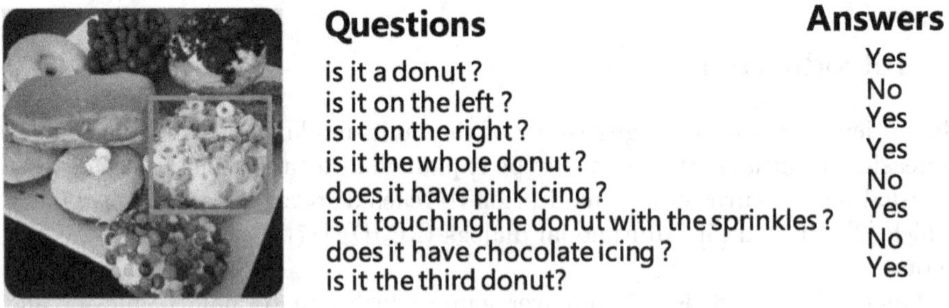

Questions	Answers
is it a donut?	Yes
is it on the left?	No
is it on the right?	Yes
is it the whole donut?	Yes
does it have pink icing?	No
is it touching the donut with the sprinkles?	Yes
does it have chocolate icing?	No
is it the third donut?	Yes

Fig. 1. This is a sample of visual dialogue in dataset GuessWhat?!. From the question sequence, we can see that human beings use the explicit category strategy to help locate the target.

We proposed a category-based strategy-driven question generator (CSQG) to solve the above issues. We design a module of category prediction before response generation. We first encode the image and historical information using faster-RCNN [16] and multi-layer GRU [4] respectively. Then we use text-image-based attention and combine the representation into context vectors. After that, we predict the probability distribution of the category of the question to be generated basing on the context vectors and a prior-probability of the category of next question, which is an explicit strategy on categories of questions learned from training data. Experimental results on GuessWhat?! [28] show that our model effectively learns the information of category and achieve state-of-the-art performance on supervising learning domain.

In summery, our contributions are mainly as follows.

1. We propose a category-based strategy-driven question generator to explicitly introduce the strategy for question generation at dialogue level. An explicit prior-probability is used to control the category transfer together with encoding of the dialogue history and the image.
2. Experiments on large Visual Dialogue dataset GuessWhat?! show that our improved model achieves state-of-the-art performance on supervising learning setting.

In the rest of this paper, we firstly gives a general introduction to the relate work in Sect. 2. Section 3 introduces the design of category information and how to label it on each question. Section 3 describes our CSQG model in detail. Section 4 shows the experiment results on large visual dialogue dateset and at last we gives our conclusion in Sect. 5.

2   Related Work

Task-oriented visual dialogue is a new and critical research direction for dialogue systems. It needs not only the ability of traditional text-based dialogue [3,19,26], but also image representation and multi-modal fusion capabilities [16,22]. As one of the widely used dataset, GuessWhat?! [28] has three sub-tasks, QGen, Oracle and Guesser. QGen is extensively studied due to its key role in the entire game [14,28,29].

There are mainly two lines of researches on QGen task. One line is the improvement on model structure. de Vries et al. [28] proposed the first model for the task as a baseline. It uses the fc8-layer of VGG-16 model [22] to provide image embedding and a hierarchical RNN-based model to encode the dialogue history. The generator is performed by RNN-based model which concatenates image and context representation and generates responses. In order to have more clear image representation, ResNet [11] is introduced by the work from Shekhar et al. [21] which extracts the image representation by the second to last layer of ResNet152. Due to the lack representation capability of static embedding, object detection models are involved. Xu et al. [29] and Pang et al. [14] introduced Faster-RCNN [16] and receive a better performance. Many of other researches focus on the improvement of linguistic information encoding. Other than the baseline method, some attention-based models are proposed to get dynamic visual and linguistic embedding. PLAN network [32] is proposed to provide dynamic visual information at each round by computing the attention on different regions. Accumulated Attention [6] consists of three kinds of attention which are query, image and objects attention. Answer-Driven Visual State Estimator [29] uses an answer-based attention transfer method to exploit different answers in different ways to update the visual attention at each step.

The other line is on model learning. Strub et al. [24] first used reinforce learning on this task. Temperature Policy Gradient [31] is proposed to make balance of exploration and exploitation while selecting words. Visual dialogue state tracking (VDST) [14] is proposed to update the change of the distribution on objects.

Our work is basically on the first line. Inspired by human cognitive psychology and current controlled text generation tasks [23], we add control factors to influence the direction of text generation. To prepare questions controlled information, we first classify them like Krishna et al. [12] at the earliest for VQA task. Different from them, we merge the unreasonable classifications and used a more accurate model-based method instead of keywords to label questions' categories.

In QGen task, our CSQG predicts the probability distribution of category and determines the certain category in this turn based on the explicit prior-probability strategy. As we can see, there is currently no work to add the question category as an element to the generation process. Our method is based on human cognition and has better interpretability and logic. The specific effect will also be confirmed in subsequent experiments.

3   Proposed Model

In this section, we will introduce our question generation model CSQG in detail. Denote the dialogue history as H, $H_{t-1} = [q_0, a_0, q_1, a_1...q_{t-1}, a_{t-1}]$ as history before the current turn, and I as the image representation. Our model can be described as $q_t = f(H_{t-1}, I)$ where q_t is the generated question for user in this turn.   The overall structure of the model is shown in Fig. 2. There are three module: (1) a multi-modality encoder with a hierarchical RNN-based neural network and an image-text attention structure, (2) a category predictor and (3) a generation decoder.

3.1   The Design and Implement of Category Info to Questions

In the current most datasets [17,28], there is no specific type annotation for each question. Therefore, before introducing the category-driven policy into visual dialogue generation task, we need to complete the classification and annotation of question categories. As one of our references, Krishna et al. [12] defines 15 category tags(such as objects, attributes and so on) and marks each question

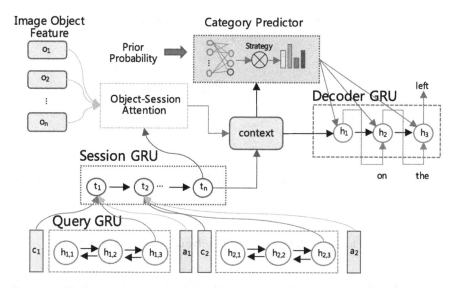

Fig. 2. This is the structure of our CSQG model. The query and session GRU is hierarchical RNN-based dialogue history encoder. For the image object feature, we use object-session attention as shown in the top left of the figure. We then predict the next category and add it to the decoder as shown in the top and right of the figure.

with one or more of these tags in the vqa dataset [1]. We find that 15 tags are too complex and unnecessary for our task. Thus we merge the tags and propose four categories: object, color, location and other.

The method that Krishna et al. [12] tag questions is using key-words by counting the static words of each category. When those words appear in a question, it can be marked with the category tag. This method is simple but not accurate. Especially the key-words of location category are confused with many preposition words in normal questions.

As a result, we propose a supervised Bert-CNN classification model to deal the task which will be at first trained by a part of manual marking data. The model is shown in Fig. 3.

Denote question Q with n tokens as $Q = [x_0, x_1, ..., x_{n-1}]$. We first use pretrained Bert model [7] to extract and get the representation of the sentence. When a list of question tokens is fed into the model, each token will be converted as word embedding $E_{word} \in R^h$ and position embedding $P_{word} \in R^d$ where d is the hidden dimension of the model. The embedding input sequence $[e_0, e_1, ...e_{n-1}]$ is then fed into Bert model which is basically a Bi-Transformer encoder containing several blocks, from which the main structure is Multi-head Self-attention [27]. The final output of Bert model can be expressed as below:

$$BertInput(x_i) = E(x_i) + P(x_i) = e_i \in R^d, \forall 0 \leq i \leq n \tag{1}$$

$$BertOut = BertModel([e_0, e_1, ...e_{n-1}]) = [t_0, t_1, ...t_{n-1}] \tag{2}$$

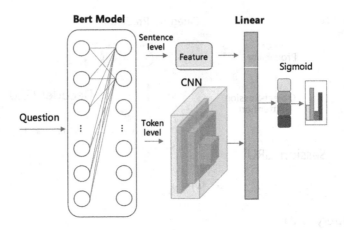

Fig. 3. The Bert-CNN model for classifying category of questions.

We split the output into token-level and sentence-level representation. Sentence-level representation $O_{sentence}$ is the final state corresponding to the [CLS] token which we added in the beginning of input sentence. Token-level representation O_{tokens} is the final state corresponding to each token in the raw sentence. A convolution layer is added behind to deal with token-level representation for more classification information.

Concatenate the convolution output and the sentence-level representation from Bert model, we map features to specific categories through a full-connected layer and use the Sigmoid function to normalize the probability of each category.

In the face of millions of questions, it is not realistic to label all the question categories manually. Therefore, we estimate the minimum number of training data required by our model, and then labeling them manually. After the model is trained completely, we label the category information of the remaining questions by inference process of this model.

3.2   Multi-modality Encoder

The multi-modality encoder is used to get the representation from current context which include the image feature and the dialogue history. The dialogue history contains several turns of question-answer pairs which can be expressed as $D = [q_0, a_0, q_1, a_1, ..., q_{n-1}, a_{n-1}]$. Each question is an independent sentence while the answer is limited into three choices: 'Yes', 'No', 'N/A'. Each question and its corresponding answer will form a turn of game, and there is an obvious progressive relationship between each turn. So combine all the questions and answers into a long list is not a wise approach. Refer to Serban et al. [18], we build a hierarchical recurrent neural network based on GRU [4] as our dialogue history encoder. It can be divided as two layers.

The first layer is query layer, which is Bi-GRU. It process each question and output the last hidden. Suppose the i-th question sentence is $q_i =$

$[x_{i,0}, x_{i,1}, ..., x_{i,n-1}]$, where $x_{i,j} \in R^{d_{emb}}$ means the embedding of j-th token in question sequence q_i. The procedure of each timestamp is shown below:

$$h_n^{forward} = GRU(x_n, h_{n-1}, W) \tag{3}$$

$$h_n^{backward} = GRU(x_n, h_{n+1}, W) \tag{4}$$

$$h_n = h_n^{forward} + h_n^{backward} \tag{5}$$

$h_{n-1} \in R^{d_{query}}$ means the hidden state of last timestamp, $h_{n+1} \in R^{d_{query}}$ means the hidden state of next timestamp, and $W \in R^{d_{emb}, d_{query}}$ means the parameters in this layer. When the last token embedding is processed, we summed each backward and forward hidden state and the last timestamp's hidden state is regarded as the representation of this sequence.

After all the questions are encoded, our model needs to further encode the previous history messages at the session level. The second layer is session layer. Not like the former query layer, this is based on Uni-GRU with only forward direction. For turn i, the input information contains question q_i, answer and the category. Because the types of both answer and category are limited, we transfer them by using one-hot embedding and express answer and category embedding of as a_i and c_i. We concatenate them and input to the session layer.

On the other hand, we introduce the commonly used object detection model Faster-RCNN [16] and get the feature of objects in the image. Besides the representation vector, we concatenate the related position from each object to others. The final object features of image I is expressed as below. R_{pos} is the relative position matrix and num is the number of objects in the image.

$$O_I = Concat(RCNN(I), R_{pos}) \in R^{num \times d_{image}} \tag{6}$$

Considering that the attention of different objects is different in each turn, we introduces the object-session attention mechanism to guide the attention of different objects. For object features list O_I and history representation t_n, a dot-attention is introduced to calculate the similarity score.

$$v_{mid} = Conv1d(O_I) \in R^{num \times d_{mid}} \tag{7}$$

$$t_{mid} = unsqueeze(W \cdot t_n) \in R^{1 \times d_{mid}} \tag{8}$$

$$h_{att} = t_{mid} \cdot v_{mid}^T \in R^{1 \times num} \tag{9}$$

In Eq. 9, h_{att} is the score of each object. After softmax function, we sum all objects weighted and concatenate with history representation t_n to get the context vector C as shown in Eq. 11.

$$V_I = \sum_{i=0}^{num-1} Softmax(h_{att}) * O_i \tag{10}$$

$$C = concat(V_I, t_n) \in R^{d_{context}} \tag{11}$$

3.3 Category Predictor

Category predictor is used to predict the category of the next question to guide the generation. Because we use the explicit way to express the category, the prediction task can be regarded as a multi-labeled classification task. The original method is using a full-connected neural network to get the signal of each category from the context vector and turn it to probability by a sigmoid function.

$$s_i = W^s \cdot C + b^s \tag{12}$$

$$p_i = Sigmoid(s_i) \tag{13}$$

W^s and b^s are parameters for full-connected neural network, and p_i is the probability of category i. When p_i exceeds the threshold, category i is considered to appear in the next question. In the training process, we directly use binary cross-entropy (BCE) loss function to obtain gradient information as below, where y_i is the target of category label i. After backward propagation of gradient and iteration of parameters by optimizer, the predictor will be able to forecast the needed category in the following generation task.

$$Loss = \sum_{i=1}^{C} y_i * log p_i + (1 - y_i) * log(1 - p_i) \tag{14}$$

In addition to simply predicting categories through neural networks, we introduce an explicit prior-probability strategy based on the answers.

When generating questions, there is an interesting phenomenon that the historical answers usually have an important impact on the category of questions to be generated in this turn. Especially the answer from last turn will have a direct and significant impact.

Considering that when the last question receive a positive answer, they tend to change the category of questions to expect more information gain. When the answer has a disposition to negative, the next question can continue to search for other possibilities in this domain. Based on the above, we make a constraint between the last turn's answer & category and the forecast category. The flow chart of the strategy is shown in Fig. 4. In addition to using the classification model to obtain the initial probability value $p_{t,i}$ of category i at turn t, we make statistics on the prior probability in the game. As the prior probability of same category class varies greatly between different turns, we separate them where $q_{t,i}$ means the prior probability of category i in t-th turn. This prior probability is used as benchmark to judge the confidence of predictor result.

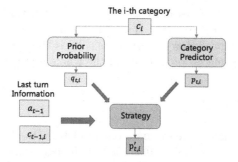

Fig. 4. This is the strategy for generating next question's category. In the figure, c_i means the i-th category. a_{t-1} and $c_{t-1,i}$ means the answer and i-th category of last turn. $p'_{t,i}$ is the final probability for the following generation task.

The principle of discriminator is described in Eq. 15:

$$p'_{t,i} = \begin{cases} min(p_t) & (p_{t,i} < q_{t,i}) \& (a_{t-1} = Yes) \& (c_{t-1,i} = 1) \\ p_{t,i} & else \end{cases} \qquad (15)$$

a_{t-1} and $c_{t-1,i}$ means the answer and i-th category of last turn. When last turn's question has involved such category and answer is "Yes", the category's income is greatly reduced. If predictor gives a high enough confidence which exceeds a given threshold (its prior probability in this turn), we trust it and continue to use $p_{t,i}$. Otherwise, we cut down its probability to the minimum.

For example, the category of last turn is B, and the answer is "Yes". The predicted probability of all categories is [A, B, C, D]. For category at position B, if B is greater than its prior probability, we adopt B. If B is lower than the prior probability, we cut down B to the minimum between A D to transfer the attention to other categories.

3.4 Generation Decoder

The generation decoder is used to generate questions based on context representation and predicted category. Its structure is based on GRU. Similar to common LSTM based sequence-to-sequence models [8,25], the first hidden state of decoder GRU is the context representation C computed in Eq. 11. During each timestamp, we combine the input embedding and the category probability vector to guide the decoder generating preset types of tokens. Suppose the input for GRU is $input_t^{dec}$ which is concatenated by category probability vector p_c and token embedding emb_j^{dec}. The hidden state of last generated token is h_{j-1}^{dec} and the computed progress is as below:

$$h_j^{dec} = GRU(h_{j-1}^{dec}, input_t^{dec}) \qquad (16)$$

$$logits_j = W_{lm}^{dec} \cdot h_j^{dec} \qquad (17)$$

The outputs of each timestamp are updated hidden state h_j^{dec} and current output logits $logits_j$. Cross entropy is used to compute loss and generate gradient. And in inference procedure, beam search [9] is the generating strategy we use to maximize the general probability of the whole question.

4   Experiments

To access the performance, we evaluate our models on the large visual dialogue dataset GuessWhat?!, which contains 155,281 dialogues and 821,955 question-answer pairs. These dialogues are based on 66537 images with 134,074 different objects. To make a fair contrast between ours and the previous, we divided GuessWhat?! dataset according to the same way from others. All dialogues are split into train, dev and test dataset, 70%, 15%, 15%.

In this section, we show the performance of CSQG on the question generation task. The implement details of module Oracle and Guesser will be introduced, and we will report the contrast experiment between ours and related comparison objects. In order to analyze the effect in more detail, an ablation experiment, the accuracy of category in generating questions and representative case analysis will be reported in the next.

4.1   Implementation Details on Guesser and Oracle

There are three parts in a GuessWhat?! game in all. Besides question generation task, we implement the baselines of Oracle and Guesser module that assist to complete the whole game.

The job for an oracle is to answer questions based on the given object. We choose GRU to build a question encoder and the representation is the hidden state of last timestamp. The hidden size of GRU is 2400, with only forward direction and 300 embedding dimension. The representation of the object contains crop vector, spatial and object category information. After concatenate both features, a feed-forward neural network is built with two linear layer and a ReLU [10] activation function beside them. The middle hidden size is 512 and the output is the score of each kind of answers. After a Softmax function, we compute the loss by cross entropy function. After 15 epochs with batch size of 128, the accuracy on the test set is about 78.5%.

The job for an guesser is to guess which object has been selected in advance according to the information of dialogues, image and objects. The dialogue encoder is also built by GRU with only forward direction. Different from Oracle, the image feature is extracted from the fc8 layer of pretrained vgg16 model [22]. A feed-forward neural network is used to transfer the object features into vectors, whose are computed cosine distances with dialogue hidden state. The feed-forward neural network contains two linear layer with 512 hidden size and ReLU activation function. At last we softmax the scores and guess the object. The accuracy of this baseline is about 64.6%.

4.2 Contrast Experiments

Follow the existing studies [20, 28, 30], we choose the comprehensive game success rate as evaluation metrics. The reason we don't choose traditional text generation evaluation indicators such as Bleu [15] and Rouge [13] is that the diversity of questions will affect the accuracy of these evaluation metrics.

We compared the recent representative models in this field, whose are baseline SL [28], VDST-SL [14], ADVSE-QGen [29] and GDSE-SL [21]. Because the implementation of CSQG is only under supervised learning, we compare with other models only on supervised learning part.

Baseline SL [28]: This is an end-to-end QGen model with HRED dialogue encoder and decoder. It uses the output of fc8 layer of VGG16 as image feature.

VDST-SL [14]: An end-to-end QGen model based on visual dialogue state tracking structure. It includes a visual-language-visual multi-step reasoning cycle. Refer to the description from Pang et al. [14], the dimension of word embedding is set to 512. We don't append the reinforce training part.

ADVSE-QGen [29]: An end-to-end model with answer-driven focusing attention, which gets visual state estimation by conditional visual information fusion.

GDSE-SL [21]: It contains a grounded dialogue state encoder which addresses a foundational issue on how to integrate visual grounding with dialogue system components. It is only trained by supervised learning.

CSQG-CD and CSQG are ours. CSQG-CD only includes category driven, without adding specific strategies while CSQG contains both category driven structure and prior probability strategy. We use the vector of objects from faster-RCNN which contains 36 objects per image and each size is 2048. The hidden size of our hierarchical encoder is 800 for query layer and 1000 for session layer. The hidden size of our decoder is 600. The dropout is set as 0.2 for the whole model.

Table 1 shows comparisons among the above models. We conducted comparative experiments on new targets which pictures have been seen by models and on new images which totally new for models. 5 and 8 turns of dialogue are set respectively, with both beam search and greedy search. CSQG achieves the success rate of 54.4% on new objects and 51.7% on new games. It exceeds 9.7% on new objects and 11% on new games from baseline-SL and 2% from the current state of the art GDSE-SL. Ours achieve the highest success rate under supervising learning at present.

The comparative experiments show that our module for controlling question generation category and the strategy of predicting categories effectively improve the key success rate of games. Although we only use the baseline modules of oracle and guess, the overall success rate of the game is still significantly improved. This shows that the conditional introduction of high-level semantic control information still has room for improvement compared with pure end-to-end models whose are actually pure black boxes.

Table 1. Comparison results of QGen on the task success rate. G means Greedy Search and B means Beam Search.

		New object		New game	
		Max turn 5	Max turn 8	Max turn 5	Max turn 8
Baseline SL	G	43.5	–	40.8	40.7
	B	47.1	–	44.6	–
VDST-SL	G	49.49	48.01	45.94	45.03
ADVSE-QGen	G	50.66	–	47.03	–
	B	47.47	–	44.70	–
GDSE-SL	G	–	–	47.8	49.7
CSQG-CD (ours)	G	52.4	52.7	48.0	48.3
	B	51.5	52.7	46.6	48.2
CSQG (ours)	G	53.2	54.4	49.9	51.7
	B	52.4	53.9	48.1	49.7

4.3 Ablation Experiments

In order to evaluate efficiency of the different improvements in our models, we conduct ablation experiments on settings of category-driven structure, category strategy and prior probability of categories.

Baseline is an end-to-end QGen model with RCNN image features and visual attention between dialogue sessions and images. The reason why we don't use the model proposed by de Vries et al. [28] is to eliminate the difference from image features of VGG16 and RCNN.

CSQG-CD includes category driven without adding specific strategies. The next question's category is entirely predicted by neural networks from the context vector.

CSQG-Punish is with different category prediction strategy. Instead of prior probability in each turn, it only punishes the probability when same former categories receive "Yes". The more positive answers, the more punishment for this category.

The rest of settings remain keep up with CSQG where the hidden sizes of encoder are 800 and 1000. The image feature is 36×2048 from RCNN and the visual attention is set with the hidden size of 800 and dropout of 0.2. The ablation experiment result shows in Table 2. RCNN-based baseline achieves 44.3% on new objects and 41.8% on new games.

The improvement of category-driven raises the success rate from 44.3% and 41.8% to 52.7% and 48.3% (comparing Baseline with CSQG-CD). It approves the positive impact of this improvement. Comparing with pure NN-predicted category method, the punishment strategy raises the success rate from 52.7% and 48.3% to 53.7% and 48.4%. This proves that compared with human questioning strategy, simply relying on neural networks can not make the global revenue maximization. Although we can rely on more complex reinforcement learning

Table 2. Ablation results of QGen on the task success rate. The generation strategy is greedy search.

	New object		New game	
	Max turn 5	Max turn 8	Max turn 5	Max turn 8
Baseline	43.9	44.3	41.6	41.8
CSQG-CD	52.4	52.7	48.0	48.3
CSQG-Punish	52.1	53.7	47.0	48.4
CSQG	53.2	54.4	49.9	51.7

optimization, our strategy is obviously more concise and good interpretability. The comparison between CSQG-Punish and CSQG shows that prior probability of categories can adjust the strategy to gain the maximum incomes.

4.4   Good Case Study

Figure 5 shows a couple of examples of dialogues generated by CSQG, whose are sampled from the evaluation set. As can be seen in figure, the sample at the top generates questions according to the categories very effectively. The subsequent categories are adjusted accordingly with the answers. When the answer is "Yes", almost all the next turns have changed the categories, which is also reasonable from the case point of view. These cases proves that our prior probability category strategy is reasonable and well implemented.

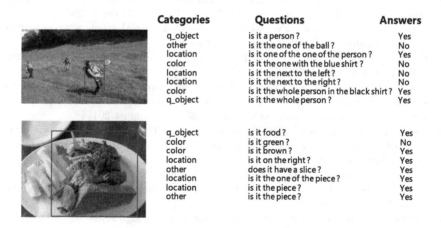

Fig. 5. Generated dialogue examples of CSQG.

5 Conclusions

We introduced CSQG, a category-based strategy-driven question generator which is guided with category information. At First, we predict the initial probability of each class by classifier and joint learning training method. Then we introduce an update strategy based on the prior probability of each category in each turn. The updated category guides the decoder to generate the next question. Experiments shows that our proposed model achieves 51.7% on new games in large visual dialogue dataset GuessWhat?!, which is the state of the art under supervise learning. The ablation experiments further proved that both category guided and prior probability have positive effect on the success rate of the game.

Our CSQG model is training only on supervised learning. However, due to the coherence, interactivity and dynamics of visual dialogue, there is a lot of room for improvement in reinforcement learning. In the future, we will explore how to produce category prediction which is more consistent with the overall goal through reinforcement learning. And this will balance the goals of interpretability and game success rate.

Acknowledgments. We would like to thank anonymous reviewers for their suggestions and comments. The work was supported by the National Natural Science Foundation of China (NSFC62076032) and the Cooperation Poject with Beijing SanKuai Technology Co., Ltd.

References

1. Antol, S., et al.: VQA: visual question answering. In: Proceedings of the IEEE International Conference on Computer Vision (ICCV), December 2015
2. Chattopadhyay, P., et al.: Evaluating visual conversational agents via cooperative human-AI games. arXiv:1708.05122 [cs], August 2017
3. Chen, Q., Zhuo, Z., Wang, W.: BERT for joint intent classification and slot filling. arXiv preprint arXiv:1902.10909 (2019)
4. Cho, K., et al.: Learning phrase representations using RNN encoder-decoder for statistical machine translation. arXiv preprint arXiv:1406.1078 (2014)
5. Das, A., et al.: Visual Dialog. arXiv:1611.08669 [cs], August 2017
6. Deng, C., Wu, Q., Wu, Q., Hu, F., Lyu, F., Tan, M.: Visual grounding via accumulated attention. In: Proceedings of the IEEE Conference on Computer Vision and Pattern Recognition (CVPR), June 2018
7. Devlin, J., Chang, M.W., Lee, K., Toutanova, K.: BERT: pre-training of deep bidirectional transformers for language understanding. arXiv preprint arXiv:1810.04805 (2018)
8. Dušek, O., Jurčíček, F.: A context-aware natural language generator for dialogue systems. arXiv preprint arXiv:1608.07076 (2016)
9. Freitag, M., Al-Onaizan, Y.: Beam search strategies for neural machine translation. arXiv preprint arXiv:1702.01806 (2017)
10. Glorot, X., Bordes, A., Bengio, Y.: Deep sparse rectifier neural networks. In: Proceedings of the 14th International Conference on Artificial Intelligence and Statistics (AISTATS), pp. 315–323 (2011)

11. He, K., Zhang, X., Ren, S., Sun, J.: Deep residual learning for image recognition. In: Proceedings of the IEEE Conference on Computer Vision and Pattern Recognition, pp. 770–778 (2016)
12. Krishna, R., Bernstein, M., Fei-Fei, L.: Information maximizing visual question generation. In: Proceedings of the IEEE Conference on Computer Vision and Pattern Recognition, pp. 2008–2018 (2019)
13. Lin, C.: Recall-oriented understudy for gisting evaluation (rouge) (2005). Accessed 20 Aug 2005
14. Pang, W., Wang, X.: Visual dialogue state tracking for question generation. arXiv:1911.07928 [cs], November 2019
15. Papineni, K., Roukos, S., Ward, T., Zhu, W.J.: BLEU: a method for automatic evaluation of machine translation. In: Proceedings of the 40th Annual Meeting of the Association for Computational Linguistics, pp. 311–318 (2002)
16. Ren, S., He, K., Girshick, R., Sun, J.: Faster R-CNN: towards real-time object detection with region proposal networks. IEEE Trans. Pattern Anal. Mach. Intell. **39**(6), 1137–1149 (2016)
17. Saha, A., Khapra, M., Sankaranarayanan, K.: Towards building large scale multimodal domain-aware conversation systems. arXiv:1704.00200 [cs], January 2018
18. Serban, I., Sordoni, A., Bengio, Y., Courville, A., Pineau, J.: Building end-to-end dialogue systems using generative hierarchical neural network models. In: Proceedings of the AAAI Conference on Artificial Intelligence, vol. 30 (2016)
19. Shan, Y., et al.: A contextual hierarchical attention network with adaptive objective for dialogue state tracking. arXiv preprint arXiv:2006.01554 (2020)
20. Shekhar, R., Baumgartner, T., Venkatesh, A., Bruni, E., Bernardi, R., Fernández, R.: Ask no more: deciding when to guess in referential visual dialogue. arXiv preprint arXiv:1805.06960 (2018)
21. Shekhar, R., et al.: Beyond task success: a closer look at jointly learning to see, ask, and guesswhat. arXiv preprint arXiv:1809.03408 (2018)
22. Simonyan, K., Zisserman, A.: Very deep convolutional networks for large-scale image recognition. arXiv preprint arXiv:1409.1556 (2014)
23. Smith, E.M., Gonzalez-Rico, D., Dinan, E., Boureau, Y.L.: Controlling style in generated dialogue. arXiv preprint arXiv:2009.10855 (2020)
24. Strub, F., De Vries, H., Mary, J., Piot, B., Courville, A., Pietquin, O.: End-to-end optimization of goal-driven and visually grounded dialogue systems. arXiv preprint arXiv:1703.05423 (2017)
25. Sutskever, I., Vinyals, O., Le, Q.V.: Sequence to sequence learning with neural networks. arXiv preprint arXiv:1409.3215 (2014)
26. Tan, Y., Ou, Z., Liu, K., Shi, Y., Song, M.: Turn-level recurrence self-attention for joint dialogue action prediction and response generation. In: Wang, X., Zhang, R., Lee, Y.K., Sun, L., Moon, Y.S. (eds.) Asia-Pacific Web (APWeb) and Web-Age Information Management (WAIM) Joint International Conference on Web and Big Data, pp. 309–316. Springer, Cham (2020). https://doi.org/10.1007/978-3-030-60290-1_24
27. Vaswani, A., et al.: Attention is all you need. arXiv preprint arXiv:1706.03762 (2017)
28. de Vries, H., Strub, F., Chandar, S., Pietquin, O., Larochelle, H., Courville, A.: GuessWhat?! Visual object discovery through multi-modal dialogue. arXiv:1611.08481 [cs], February 2017

29. Xu, Z., Feng, F., Wang, X., Yang, Y., Jiang, H., Wang, Z.: Answer-driven visual state estimator for goal-oriented visual dialogue. In: Proceedings of the 28th ACM International Conference on Multimedia, Seattle, WA, USA, pp. 4271–4279. ACM, October 2020. https://doi.org/10.1145/3394171.3413668

30. Zhang, J., Wu, Q., Shen, C., Zhang, J., Lu, J., Van Den Hengel, A.: Goal-oriented visual question generation via intermediate rewards. In: Ferrari, V., Hebert, M., Sminchisescu, C., Weiss, Y. (eds.) Proceedings of the European Conference on Computer Vision (ECCV), pp. 186–201. Springer, Cham (2018). https://doi.org/10.1007/978-3-030-01228-1_12

31. Zhao, R., Tresp, V.: Learning goal-oriented visual dialog via tempered policy gradient. In: 2018 IEEE Spoken Language Technology Workshop (SLT), pp. 868–875. IEEE (2018)

32. Zhuang, B., Wu, Q., Shen, C., Reid, I., van den Hengel, A.: Parallel attention: a unified framework for visual object discovery through dialogs and queries. In: Proceedings of the IEEE Conference on Computer Vision and Pattern Recognition (CVPR), June 2018

From Learning-to-Match to Learning-to-Discriminate: Global Prototype Learning for Few-shot Relation Classification

Fangchao Liu[1,4], Xinyan Xiao[3], Lingyong Yan[1,4], Hongyu Lin[1], Xianpei Han[1,2(✉)], Dai Dai[3], Hua Wu[3], and Le Sun[1,2]

[1] Chinese Information Processing Laboratory, Institute of Software, Chinese Academy of Sciences, Beijing, China
{fangchao2017,lingyong2014,hongyu}@iscas.ac.cn
[2] State Key Laboratory of Computer Science, Institute of Software, Chinese Academy of Sciences, Beijing, China
{xianpei,sunle}@iscas.ac.cn
[3] Baidu Inc., Beijing, China
{xiaoxinyan,daidai,wu_hua}@baidu.com
[4] University of Chinese Academy of Sciences, Beijing, China

Abstract. Few-shot relation classification has attracted great attention recently, and is regarded as an effective way to tackle the long-tail problem in relation classification. Most previous works on few-shot relation classification are based on learning-to-match paradigms, which focus on learning an effective universal matcher between the query and *one* target class prototype based on inner-class support sets. However, the learning-to-match paradigm focuses on capturing the similarity knowledge between query and class prototype, while fails to consider discriminative information between different candidate classes. Such information is critical especially when target classes are highly confusing and domain shifting exists between training and testing phases. In this paper, we propose the *Global Transformed Prototypical Networks (GTPN)*, which learns to build a few-shot model to directly discriminate between the query and *all* target classes with both inner-class local information and inter-class global information. Such learning-to-discriminate paradigm can make the model concentrate more on the discriminative knowledge between all candidate classes, and therefore leads to better classification performance. We conducted experiments on standard FewRel benchmarks. Experimental results show that GTPN achieves very competitive performance on few-shot relation classification and reached the best performance on the official leaderboard of FewRel 2.0 (https://thunlp.github.io/2/fewrel2_da.html).

Keywords: Few-shot learning · Relation classification · Transformer

---

F. Liu, part of the work was done during an internship at Baidu.

S. Li et al. (Eds.): CCL 2021, LNAI 12869, pp. 193–208, 2021.
https://doi.org/10.1007/978-3-030-84186-7_13

1   Introduction

Few-shot relation classification aims to build relation extractors with only a few instances. Different from supervised learning that requires large scale training data of target classes, few-shot learning-based approaches can effectively build relation extractors with only a few examples (i.e., the support set) of each target class. This property makes it very appealing in the real application, where the training data of new target class can be scarce.

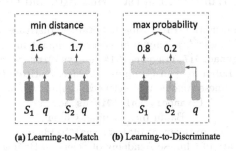

(a) Learning-to-Match   (b) Learning-to-Discriminate

Fig. 1. Illustration of the motivation for our Global Transformed Prototypical Networks (GTPN). (a) is the previous learn-to-match paradigm method, that matches the query instance q to *each* relation support set S_1 and S_2, and measures the similarity seperately. (b) is the learn-to-discriminate paradigm of GTPN, that directly discriminate the query instance q on *all* relation support sets, and give the normalized scores on each relation type for further learning of the prototypes.

Previous few-shot relation classification approaches can be summarized into a learning-to-match paradigm. Specifically, as shown in Fig. 1(a), these methods try to learn a universal matcher between the query and *each* target relation type, and measure their similarity separately to infer the type of queries. Along this line, Gao [7] propose the hybrid attention to attach different importance for each relation feature, and match the query to each relation type with the reweighted features. Ye [33] performs a local matching and aggregation between the query instance and each relation type. Besides, Gao [8] proposes an instance-pair matcher for scoring the similarity between query and each relation instance. Generally, the goal of these methods is to effectively obtain a query-class matcher with the limited given instances in the support set.

However, because such learning-to-match based approaches focus on capturing the similarity information between query and classes, they are unable to consider discriminative knowledge between different candidate classes. This can significantly undermine the model performance when target classes are highly confusing, and domain shifting exists between training and testing phases. For example, a query with relation mention word "contain" may correspond to either a "Component-Whole" relation or a "Member-Collection" relation. Therefore, under the learning-to-match paradigm, this query will achieve high matching scores with both two relation types, which may lead to confusion and misclassification when we determine the instance relation type. Furthermore, the existence of domain shifting can result in the inaccuracy of the similarity measurement on the out-of-domain relation. To tackle these problems, a model needs to take discriminative information into consideration, and focus more on how to distinguish between confusing target relations.

To this end, this paper proposes to resolve few-shot relation classification in a learning-to-discriminate paradigm. The main idea behind, as shown in Fig. 1(b), is to learn a meta classification model which can directly generate a relation classifier based on the query and small support sets of *all* candidate relations, rather than a meta matcher between the query and *one* relation. Motivated by this, we design the *Global Transformed Prototypical Networks* (GTPN), which takes the query and the support sets of all target relations as input, and output the probabilities of the input query correspond to each relation simultaneously. The architecture of GTPN is shown in Fig. 2. Specifically, GTPN first encodes the query and all relation instances in the support sets into the same embedding space. Then we conducted a multi-view global transformation on each relation instance to extract features of different facets, based on the knowledge from both intra-relation and inter-relation instances. After that, all representations of instances of the same relation are summarized to form the class prototypes. Finally, these prototypes, as well as the query representation, are simultaneously sent into a classifier to determine the relation type of the input query. The main advantage of GTPN, compared with previous approaches, is that all candidate relation types are considered jointly rather than independently during relation classification, which makes the model able to leverage discriminative information between classes to more preciously distinguish between confusing relation pairs.

To verify the effectiveness of GTPN, we conduct thorough experiments on FewRel 1.0 [10] and FewRel 2.0 [8], two standard benchmarks for few-shot relation classification. Experimental results demonstrate that GTPN can achieve effective and robust performance on few-shot relation classification, even domain shifting exists. Furthermore, GTPN reaches the best performance on the FewRel 2.0 official leaderboard. These all demonstrate the effectiveness of GTPN and the proposed learning-to-discriminate paradigm.

Generally, the main contributions of this paper can be summarized as:

- We propose to resolve few-shot relation classification in a learning-to-discriminate paradigm, which is able to leverage discriminative knowledge for better distinguishing between confusing type pairs, compared with previous learning-to-match paradigm.
- Based on the paradigm, we propose the Global Transformed Prototypical Networks (GTPN), an effective neural network-based architecture with global transformation for few-shot relation classification.
- The proposed GTPN achieved the new state-of-the-art performance on the official leaderboard of FewRel 2.0, which demonstrates the effectiveness of the proposed paradigm and architecture.

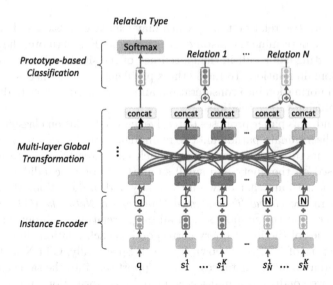

Fig. 2. Framework of GTPN. q is the query instance, S_i^j means the jth support instance of relation i.

2   Background

This section describes the definition and the notations of few-shot relation classification task. Besides, we will briefly illustrate previous learning-to-match based approaches.

Relation Classification. Let $\mathcal{X} = [x_0, ..., x_{n-1}]$ be the sequence which contains n words, $e_1 = [i, j]$ and $e_2 = [k, l]$ indicate the entity pair spans, where $0 \leq i \leq j, j < k \leq l$ and $l \leq n - 1$, a relation instance is defined as $(\mathcal{X}, e_1, e_2)$. For example, (*"Steve Jobs was the co-founder of Apple Inc.", "Steve Jobs", "Apple Inc."*) is a relation instance. The aim of relation classification is to learn a mapping function: $f : r \rightarrow c$, where c is the relation class. For example, we want mapping (*"Steve Jobs was the co-founder of Apple Inc.", "Steve Jobs", "Apple Inc."*) to its relation class *"Founder-of"*.

Few-Shot Relation Classification. Few-shot relation classification task contains a support set S and a query instance q, where $S = \{S_1, .., S_N\}$ are support sets for N relation classes and each

$$S_i = \{s_i^1, ..., s_i^K\}$$

contains K support relation instances for relation i. Then the query instance q is to be classified based on the support relation instances in S. Such a few-shot setting is commonly named as an **N-way K-shot** relation classification.

Learning-to-match Based Methods. Learning-to-match based methods regard few-shot relation classification as a matching problem [25,30]. Generally, it encodes each relation class into a prototype, which is learned from its support set S_i:

$$\mathcal{P}_{S_i} = f(S_i) \in \mathcal{R}^{d_k},$$

where d_k is the dimension of the metric space. Then a matcher is learned to measure the similarity between each prototype and the query. For classification, it first encodes the query instance to the same metric space:

$$\mathcal{P}_q = g(q) \in \mathcal{R}^{d_k}.$$

Then the matcher computes the distances between the query instance and each prototype of different relation classes, and the relation class of the query instance is inferred by the nearest distance:

$$y = \arg\min_i \mathcal{D}(\mathcal{P}_{S_i}, \mathcal{P}_q), \tag{1}$$

where y is the label of the nearest relation class prototype, $\mathcal{D}(.)$ is the distance function to measure the similarity between support set and query instance.

The main drawback of such learning-to-match approaches is that the prototype only contains the information from the support instances of its own class. This results in the absence of discriminative knowledge between different candidate classes, which can significantly undermine the performance on confusing relation type pairs. Besides, the inaccurate measurement of similarities on out-of-domain relations can also lead to a negative impact on system performance.

3   Global Transformed Prototypical Networks

In this section, we introduce our Global Transformed Prototypical Networks (GTPN) for few-shot relation classification. Different from previous work, GTPN follows a learning-to-discriminate paradigm, which directly takes the query and support sets of all classes as input. Figure 2 shows the framework of our method. First, GTPN encodes the query and instances in support sets into a hidden vector space via an instance encoder, as well as a relation marker, which represents whether two support instances belong to the same relation support set. These embeddings are then all synchronously sent to multi-view global transformation layers to learn the discriminative information based on both intra-relation and inter-relation knowledge. Then, we summarize the representations of all instances of the same relation to generate the prototypes of each class. Finally, the query representation and all prototypes are sent into a classifier to predict the relation type. In the following, we will describe each step of GTPN in detail.

3.1   Instance Encoder

The instance encoder module encodes each instance x into the same embedding space:

$$\mathbf{h} = \mathcal{H}(x), \quad \mathbf{h} \in \mathcal{R}^{d_k}, \tag{2}$$

where d_k is the dimension of the embedding space, $\mathbf{h}$ is the embedding of each instance, which will be used for further prototype learning. Recent years, several instance encoders have been proposed, including Convolutional Neural Networks (CNN) [10,34]; Recurrent Neural Networks (RNN) [11,36]; and the recent Transformer architecture [28]. Pre-trained language models like BERT [5] also provide promising encoders for relation instances [2,8]. In this work, we adopt the BERT-based instance encoder similar to [2], which wraps entities in the instance with special markers [ENTITY] and [/ENTITY], and concatenate the representations of the first marker [ENTITY] of each entity as the instance embedding.

3.2   Relation Marker

Apart from the instance encoder, it is necessary to introduce relation markers to encode whether two instances belong to the same relation support set. To this end, we propose to use two kinds of relation markers for GTPN, including:

- **Randomly initialized relation encoding.** For each instance embedding $\mathbf{h}_i, 1 \le i \le N * K$ with relation type r_i, we randomly initialize the relation encoding embedding $\mathbf{h}_{r_i} \in \mathcal{R}^{d_k}$ for it, and add it with the relation encoding embedding: $\mathbf{h}_i = \mathbf{h}_i + \mathbf{h}_{r_i}$ as the new hidden states. For the query instance, we regard it as the instance of relation r_{N+1}: $\mathbf{h}_q = \mathbf{h}_q + \mathbf{h}_{r_{N+1}}$.
- **One-hot relation encoding.** we use an one-hot vector to indicate each relation class, e.g., $[1, 0, ..., 0]$ for the first relation class and $[0, 1, 0, ..., 0]$ for the second relation class. For the query instance, we use the last one-hot vector to indicate the relation class.

The relation markers are directly concatenated to the embeddings from previous instance encoder to form the instance representations. Finally, we represent the output after instance encoder as:

$$\mathcal{X} = [\mathbf{h}_q, \mathbf{h}_{s_1^1}, ..., \mathbf{h}_{s_1^K}, ..., \mathbf{h}_{s_N^1}, ..., \mathbf{h}_{s_N^K}]. \tag{3}$$

3.3   Multi-view Global Transformation

After obtaining the representations of all support instances, we will conduct global transformation among them to learn the knowledge from both intra-relation and inter-relation support sets. This module is based on a multi-layer transformer model [28]. Specifically, given the embeddings of all instances, each instance is firstly mapped into multiple views of semantics by a multi-view projection:

$$\mathbf{v}_j^i = \mathbf{W}_i \cdot \mathbf{h}_j + \mathbf{b}_i, \tag{4}$$

where $\mathbf{v}_j^i$ is the ith view of the jth instance, $\mathbf{W}_i$ is the mapping matrix and $\mathbf{b}_i$ is the bias for view i. Then the model will transform the representation of all instances of all relations from each view via an attention-based mechanism:

$$\mathbf{v}_j^i = \sum_{m=0}^{N*K} \alpha_{jm} \cdot \mathbf{v}_m^i$$

$$\alpha_{jm} = \text{Softmax}(\frac{\mathbf{v}_j^i \cdot {\mathbf{v}_m^i}^T}{\sqrt{d_k}}), \tag{5}$$

where α_{jm} is the normalized attention score between the jth and m^{th} instances, the attention score is divided by dimension d_k to rescale the inner product of two vectors. After that, the output representation of each instance of this transformation layer is obtained by concatenating the representations in its all views. This output representation is then fed into next global transformation layer. We denote the output representation of instance S_i^j as $\mathbf{o}_i^j$. Furthermore, we use $\mathbf{h}_q$ to represent the final output representation of the given query.

3.4   Prototype-Based Classification

After obtaining the global transformed instance representations, we calculate the prototype of each class by averaging the representations of all instances in its support set:

$$\mathbf{p}_i = \frac{1}{K} \sum_{k=1}^{K} \mathbf{o}_i^k, \tag{6}$$

where $\mathbf{p}_i$ is the prototype of relation i. Then we calculate the probability of the query being an instance of relation y_i by sending the query representation $\mathbf{h}_q$ and all prototypes into a softmax-based classifier:

$$\mathcal{P}(y_i|S_i, q; \theta) = \frac{\exp(-\mathcal{D}(\mathbf{p}_i, \mathbf{h}_q))}{\sum_{j=1}^{N} \exp(-\mathcal{D}(\mathbf{p}_j, \mathbf{h}_q))}. \tag{7}$$

Here $\mathcal{D}$ is a score function between the prototype and the query. In this paper, we simply choose $\mathcal{D}$ to be the Euclidean distance function. But it can be easily replaced with other parametrized functions such as MLP. Finally, we choose the relation y_i with maximum probability as the relation class of the given query:

$$y = \arg\max_i \mathcal{P}(y_i|S_i, q, \theta). \tag{8}$$

3.5   Model Learning

Similar to previous work [8,10], all components in GTPN are trained in an end-to-end manner. Specifically, given a training task with N support relations (S_n, y_n) and one query instance (q, y_q), GTPN is learned by minimizing the following loss function:

$$J(\theta) = -\sum_{n=1}^{N} I(y_n) \log P(y_n|S_n, q, \theta), \tag{9}$$

where y_n is the relation of support set S_n and $I(.)$ is an indicator function:

$$I(y_n) = \begin{cases} 1, & y_n = y_q \\ 0, & y_n \neq y_q \end{cases}, \tag{10}$$

which indicates whether the relation corresponds to the golden relation of the given query.

Table 1. Hyperparameter settings.

Hyperparameter	Value
Batch size	1
Layer number	1, 2, 3
View number	1, 2, 4, 8
Learning rate	10^{-5}
Weight decay	10^{-6}
Optimization strategy	Adam
Learning rate decay	0.6
Learning rate decay step	1000
Maximun sequence length	256

4   Experiments

4.1   Experimental Settings

Datasets. We conducted experiments on two standard few-shot relation classification benchmarks: FewRel 1.0 [10] dataset and FewRel 2.0 Domain Adaptation [8] dataset[1]. FewRel 1.0 is constructed based on the articles from Wikipedia. FewRel 2.0 Domain Adaptation task further extends FewRel 1.0 by introducing an additional test data from PubMed footnote https://www.ncbi.nlm.nih.gov/pubmed/, which is a large database of biomedical literature, and is significantly different from Wikipedia train set. Totally, FewRel 1.0 consists of 100 relation classes and 700 instances for each relation class, and the standard 64/16/20 of train/validate/test relation class splits are adopted in our experiments. While FewRel 2.0 shares the same training data with FewRel 2.0, but a different test set including XXX relations in medical domain.

---

[1] https://github.com/thunlp/FewRel.

Table 2. Accuracies(%) on FewRel 2.0 test sets. *is reported in Peng *et al.*. [2020].

Model	FewRel 2.0				
	5-Way 1-Shot	5-Way 5-Shot	10-Way 1-Shot	10-Way 5-Shot	Avg.
Proto-CNN	35.1	49.4	23.0	35.2	35.7
Proto-Bert	40.1	51.5	26.5	36.9	38.8
Proto-Adv (Bert)	41.9	54.7	27.4	37.4	40.4
Proto-Adv (CNN)	42.2	58.7	28.9	44.4	43.6
DaFeC	61.2	77.0	47.6	64.8	62.7
BERT-PAIR	67.4	78.6	54.9	66.9	66.9
CP	79.7	84.9	68.1	79.8	78.1
MTB*	74.7	87.9	62.5	81.1	76.6
GTPN	80.0	**92.6**	69.25	**86.9**	**82.2**

Table 3. Accuracies(%) on FewRel 1.0 test sets. MTB and CP pretrained the matcher on Wikipedia, which is also the source of FewRel 1.0. While Pony introduced additional world knowledge to improve the performance. So there results are not directly comparable with GTPN.

Model	FewRel 1.0				
	5-Way 1-Shot	5-Way 5-Shot	10-Way 1-Shot	10-Way 5-Shot	Avg.
Proto-Adv (CNN)	70.3	84.6	56.3	74.7	71.5
Proto-Adv (Bert)	73.4	82.3	61.5	72.6	72.4
Proto-CNN	74.5	88.4	62.4	80.5	76.4
HATT	–	90.1	–	83.1	–
Proto-Bert	80.7	89.6	71.5	82.9	81.2
MLMAN	83.0	92.7	73.6	87.3	84.1
BERT-PAIR	88.3	93.2	80.6	87.0	87.3
Bert-EM	89.8	93.6	83.4	88.6	88.9
REGRAB	90.3	94.3	84.1	89.9	89.6
CP	**95.1**	**97.1**	**91.2**	**94.7**	**94.5**
MTB	93.9	**97.1**	89.2	94.3	93.6
GTPN	89.4	97.0	84.4	93.8	91.2

Baselines. We compared GTPN with the following published baselines:

- **Bert-Pair** [8]: A few-shot model that utilize the BERT sequence-pair model to measure similarity between two instances.
- **DaFeC** [4]: An inductive unsupervised domain adaptation framework for few-shot relation classification.
- **Proto-Adv(Bert)** [8]: The adversarial trained prototypical networks with Bert as the encoder.
- **Proto-Bert** [8]: The vanilla prototypical networks using Bert as the encoder and the [CLS] token to represent the instance.
- **Proto-Adv(CNN)** [8]: The adversarial trained prototypical networks with CNN as the encoder.
- **Proto-CNN** [8]: The vanilla prototypical networks using CNN as the encoder.
- **REGRAB** [21]: A bayesian meta-learning method that utilize additional global relation graph knowledge.
- **Bert-EM** [2]: A simple Bert-based prototypical networks that wrap entity with special markers.
- **MTB** [2]: The same base model with **Bert-EM**, and uses large-scaled wiki data to pre-train the model, which is of the source with FewRel 1.0.
- **MLMAN** [33]: The multi-level matching and aggregation network that involve intr-class information and support-query interactions.
- **HATT** [7]: The hybrid attention-based prototypical networks that also includes local context to avoid noisy data.
- **CP** [18]: This model utilizes the same model of Bert-EM and proposes a different contrastive-based pretraining on Wikipedia.

Evaluation Metrics. We follow the settings in FewRel [10] and adopt the official evaluation scripts[2] to evaluate the accuracy of all models on tasks including 5-way 1-shot, 5-way 5-shot, 10-way 1-shot, and 10-way 5-shot tasks.

Hyperparameter Settings and Infrastructure specifications. In the experimental period, we fix all the hyper-parameters listed in Table 1 during training stage except for the layer number and view number. We conducted grid search on validation set of FewRel 1.0 to find the best layer number and view number, which will be detailedly analyzed in the following. We train and evaluate our model using one Titan RTX GPU with about 24 GB memory. Each training period costs about 2 h for about 15000 steps of batches. The trainable parameters of GTPN are about 4M, excluding the BERT parameters.

4.2 Overall Results

Table 2 shows the overall results on FewRel 2.0 Domain Adaptation task and Table 3 shows the result on FewRel 1.0. From these tables, we can see that comparing with previous work:

---

[2] https://github.com/thunlp/FewRel.

1. **GTPN achieves the very competitive performance on few-shot relation classification.** From Table 2, we can see that GTPN achieved the best performance on the FewRel 2.0 domain adaptation task. Besides, we can also see that in Table 3, GTPN achieved a very competitive performance on the Fewrel 1.0. Note that the top 2 systems (i.e. MTB and CP) on FewRel 1.0 introduced additional in-domain knowledge into the task, and therefore is not directly comparable with GTPN. Specifically, MTB and CP pretrained the matcher on Wikipedia, which is also the source of FewRel 1.0. While the second-best model introduced additional world knowledge to improve the performance. On the contrast, GTPN achieves very strong performance without introducing any additional knowledge, which demonstrates the effectiveness of GTPN.

2. **GTPN is robust when domain shifting exists.** In FewRel 2.0 domain adaptation task, GTPN outperforms all other baseline models with a large margin, and also reaches the Top 1 in leaderboard of FewRel 2.0 domain adaptation benchmark, which demonstrates that GTPN is very robustness in few-shot relation classification task even without any knowledge of the target domain. We believe that this is because the proposed learning-to-discriminate paradigm can transfer more knowledge between different domains than previous learning-to-match paradigm. Besides, the learning-to-discriminate paradigm also more corresponds to the nature of few-shot relation classification task. These all results in the performance improvements when domain shifting exists.

3. **GTPN is even more effective on 5-shot learning paradigm.** Comparing with the improvement over the 1-shot settings, GTPN achieves more improvements on 5-shot settings. We believe that this is because a little bit more support instances could provide more sufficient discriminative information between different relations, and therefore results in better learning-to-discriminate performance.

4.3 Detailed Analysis

In this section, we conducted detailed analysis on the behavior of GTPN. Since the test sets of both FewRel 1.0 and FewRel 2.0 are unavailable to the public, we choose FewRel 1.0 validation set as the development set to select model, and use FewRel 2.0 validation set as the test set to evaluate the final performance for each model.

Effect of GTPN with Different Instance Encoder. This experiment analyzes the effect of GTPN over the base instance encoder. We reproduce Bert-EM as the base encoder for GTPN and use the Prototypical Networks on Bert-EM as comparison. As we can see from Table 4a, with the same instance encoder, GTPN significantly outperforms vanilla Prototypical Networks, which means that the improvement of GTPN does not stem from the power of instance encoder, but from the effective learning-to-discriminate paradigm.

Table 4. Accuracy (%) of different instance encoders and relation markers.

Model	5-Way 1-Shot	5-Way 5-Shot	10-Way 1-Shot	10-Way 5-Shot	Relation Encoding	5-Way 1-Shot	5-Way 5-Shot	10-Way 1-Shot	10-Way 5-Shot
Bert-EM	79.0	88.4	64.4	84.4	One-hot	**82.8**	**91.4**	71.0	86.0
Bert-EM+GTPN	**82.8**	**91.4**	**71.0**	**86.0**	Random	82.2	90.2	**71.4**	**86.2**
					None	82.2	91.0	68.6	85.8

(a) Accuracies (%) of GTPN over base encoder on FewRel 2.0. Bert-EM is our reimplement of Soares et al., (2019).

(b) Accuracies(%) with different relation encodings of GTPN on FewRel 2.0 validation set.

Table 5. Accuracy (%) of different transformation view and layer numbers.

View Number	5-Way 1-Shot	5-Way 5-Shot	10-Way 1-Shot	10-Way 5-Shot	Layer Number	5-Way 1-Shot	5-Way 5-Shot	10-Way 1-Shot	10-Way 5-Shot
1	81.8	90.4	69.2	84.6	1	**82.8**	**91.4**	**71.0**	**86.0**
2	82.6	90.8	69.8	**86.0**	2	81.8	89.4	69.4	85.8
4	**82.8**	**91.4**	**71.0**	**86.0**	3	80.4	88.2	68.2	83.6
8	82.4	89.6	68.4	85.6					

(a) Accuracies(%) with different view numbers of GTPN on FewRel 2.0 validation set.

(b) Accuracies(%) with different layer numbers of GTPN on FewRel 2.0 validation set.

Effect of Relation Marker. In this experiment, we study the effect of different relation markers we proposed of GTPN. For fair comparison, we fix the layer number to 1 and the view number to 4. As shown in Table 4b, we can see that the performance of different kinds of one-hot and random relation markers are quite similar. One-hot relation marker performs slightly better on 5-way settings, while random marker performs slightly better on 10-way settings. However, the difference between them is not large, which means that both of them are effective relation marker for GTPN.

Effect of Transformation Layer Number. In this experiment, we study the effect of different transformation layer of GTPN. For fair comparison, we fixed all other hyper-parameters the same as the main experimental setting and the view number to 4. The layer number in this experiment is varied from 1 to 3. From the results shown on Table 5b, we can see that model with one transformation layers performs relatively best and more layer number leads the performance to fall. This is perhaps because the global transformation in GTPN is fully-connected with all instances in the task. Hence one layer is enough to capture global information from all other instances. Besides, the limited training data size can also undermine the performance of deep models, because it requires more training data to learn effective parameters.

Effect of View Number. In this experiment, we verify the effect of different view numbers. we vary the view number of GTPN from 1, 2, 4 to 8, and fix the layer number to 1. From the result shown on Table 5a, we can see that with the view number increases, the performance of GTP also improved. The model with 2 and 4 view number reach the relatively best performance on FewRel 2.0, and the model with single view performs the worst. We believe that this is because the relation representation may contain several perspectives of semantics. So with multi-view semantics, the model is easier to find the specific semantic and aggregates more information with other instances with the similar features. Besides, the model with 8 views are marginally worse than model with 4 views, which may indicate that 4 views are enough to capture different aspects information.

5   Related Work

Relation Classification. Relation classification has long been an important information extraction task. Conventional methods commonly employ syntax structure-based representations, e.g., dependency tree [3] and constituent tree [16,20]. In recent years, neural network-based methods dominate relation classification. [34] proposed to encode relation instances via convolutional neural networks (CNN). [36] proposed an attention-based BiLSTM (AttBLSTM) for instance encoding and classification. [31] proposed a multi-level attention CNNs for capturing multi-level lexical and semantic features. [19] tried to incorporate dependency information into neural networks by extending tree LSTM [27]. [29] proposed graph attention networks (GATs) to incorporate the dependency information. The main drawback of supervised methods is that they require a large amount of annotated data, which is costly and cannot be easily obtained when adapting to new relation classes.

Few-Shot Learning. To resolve the annotated data bottleneck problem, few-shot learning is a promising approach. Many few-shot algorithms have been proposed in Computer Vision (CV) and Natural Language Processing (NLP), which can be categorized into two main paradigms, metric-learning based methods [12,25,30] and meta-learning based methods [1,6,14,15,22,23]. Recently, many research interests have been focused on metric-based methods [9,13,17,26,32,35].

Few-Shot Relation Classification. [10] proposed a few-shot learning task in relation classification, and adopted many few-shot learning methods into relation classification, including prototype-based method [25], meta-learning method [14] and graph neural network [24]. Since then, many prototype-based methods have been proposed for relation classification. [2] utilize the pre-trained language model BERT [5] for relation encoding and use prototypes to represent different relation classes. [21] adds knowledge from graph to guide the meta-gradient for bayesian meta-learning. [7] designed hybrid attention to learn a local matcher between intra-class instances, query and each support instances. [33] proposed the multi-level matching and aggregation network which updates support and query instances by matching and aggregating evidence on each support set. [8] directly average the support-query scores using the BERT sequence-pair classification model for relation classification. We can see that the methods in few-shot relation classification are mostly in a learning-to-match paradigm, The matcher learned in these model are difficult to handle confusing reltion types, as well the domain shift. Thus, in this paper, we propose the learn-to-discriminate-based GTPN to tackle these limitations.

6   Conclusions

In this paper, we propose the *Global Transformed Prototypical Networks*, which switches previous learning-to-match paradigm to the learning-to-discriminate paradigm, and therefore can make the model concentrate more on the discriminative knowledge between all candidate relations. GTPN learns to build a few-shot model to directly discriminate between the query and *all* target classes with both inner-class local information and inter-class global information. Experiments on FewRel 1.0 and FewRel 2.0 demonstrate that GTPN achieves very competitive performance on few-shot relation classification, and reached the best performance on the FewRel 2.0 domain adaptation task. Which shows our method can benefit both further study and practice in few-shot relation classification.

Acknowledgements. This work is supported by the National Key R&D Program of China under Grant 2018YFB1005100.

References

1. Andrychowicz, M., et al: Learning to learn by gradient descent by gradient descent. In: NeurIPS (2016). http://dl.acm.org/citation.cfm?id=3157382.3157543
2. Baldini Soares, L., FitzGerald, N., Ling, J., Kwiatkowski, T.: Matching the blanks: distributional similarity for relation learning. In: ACL(2019). https://www.aclweb. org/anthology/P19-1279
3. Bunescu, R., Mooney, R.: A shortest path dependency kernel for relation extraction. In: EMNLP (2005). https://www.aclweb.org/anthology/H05-1091
4. Cong, X., Yu, B., Liu, T., Cui, S., Tang, H., Wang, B.: Inductive unsupervised domain adaptation for few-shot classification via clustering. In: ECML-PKDD (2020). https://arxiv.org/abs/2006.12816
5. Devlin, J., Chang, M.W., Lee, K., Toutanova, K.: BERT: pre-training of deep bidirectional transformers for language understanding. In: NAACL (2019). https:// www.aclweb.org/anthology/N19-1423
6. Finn, C., Abbeel, P., Levine, S.: Model-agnostic meta-learning for fast adaptation of deep networks. In: ICML (2017). http://dl.acm.org/citation.cfm?id=3305381. 3305498
7. Gao, T., Han, X., Liu, Z., Sun, M.: Hybrid attention-based prototypical networks for noisy few-shot relation classification. In: AAAI (2019). https://aaai.org/ojs/ index.php/AAAI/article/view/4604/4482
8. Gao, T., et al.: FewRel 2.0: Towards more challenging few-shot relation classification. In: EMNLP-IJCNLP (2019). https://www.aclweb.org/anthology/D19-1649
9. Gidaris, S., Bursuc, A., Komodakis, N., Perez, P., Cord, M.: Boosting few-shot visual learning with self-supervision. In: Proceedings of the IEEE/CVF International Conference on Computer Vision (ICCV), October 2019

10. Han, X., et al.: FewRel: a large-scale supervised few-shot relation classification dataset with state-of-the-art evaluation. In: EMNLP (2018). https://www.aclweb.org/anthology/D18-1514

11. Hochreiter, S., Schmidhuber, J.: Long short-term memory. Neural Comput. **9**(8), 1735–1780 (1997)

12. Koch, G., Zemel, R., Salakhutdinov, R.: Siamese neural networks for one-shot image recognition. In: ICML Deep Learning Workshop, vol. 2 (2015). https://sites.google.com/site/deeplearning2015/37.pdf?attredirects=0

13. Liu, Y., et al..: Learning to propagate labels: transductive propagation network for few-shot learning. In: ICLR (2019). https://openreview.net/forum?id=SyVuRiC5K7

14. Mishra, N., Rohaninejad, M., Chen, X., Abbeel, P.: A simple neural attentive meta-learner. In: ICLR (2018). https://openreview.net/forum?id=B1DmUzWAW

15. Munkhdalai, T., Yu, H.: Meta networks. In: ICML (2017). http://proceedings.mlr.press/v70/munkhdalai17a.html

16. Nguyen, T.V.T., Moschitti, A., Riccardi, G.: Convolution kernels on constituent, dependency and sequential structures for relation extraction. In: EMNLP (2009). https://www.aclweb.org/anthology/D09-1143

17. Oreshkin, B., Rodríguez López, P., Lacoste, A.: Tadam: Task dependent adaptive metric for improved few-shot learning. In: NeurIPS (2018). http://papers.nips.cc/paper/7352-tadam-task-dependent-adaptive-metric-for-improved-few-shot-learning.pdf

18. Peng, H., et al.: Learning from context or names? An empirical study on neural relation extraction. In: EMNLP (2020). https://www.aclweb.org/anthology/2020.emnlp-main.298

19. Peng, N., Poon, H., Quirk, C., Toutanova, K., Yih, W.t.: Cross-sentence n-ARV relation extraction with graph LSTMs. In: TACL (2017). https://www.aclweb.org/anthology/Q17-1008

20. Qian, L., Zhou, G., Kong, F., Zhu, Q., Qian, P.: Exploiting constituent dependencies for tree kernel-based semantic relation extraction. In: COLING (2008). https://www.aclweb.org/anthology/C08-1088

21. Qu, M., Gao, T., Xhonneux, L.P.A.C., Tang, J.: Few-shot relation extraction via Bayesian meta-learning on relation graphs. In: ICML (2020), https://proceedings.icml.cc/paper/2020/file/99607461cdb9c26e2bd5f31b12dcf27a-Paper.pdf

22. Ravi, S., Larochelle, H.: Optimization as a model for few-shot learning. In: ICLR (2017)

23. Santoro, A., Bartunov, S., Botvinick, M.M., Wierstra, D., Lillicrap, T.P.: One-shot learning with memory-augmented neural networks. ArXiv abs/1605.06065 (2016)

24. Satorras, V.G., Estrach, J.B.: Few-shot learning with graph neural networks. In: International Conference on Learning Representations (2018). https://openreview.net/forum?id=BJj6qGbRW

25. Snell, J., Swersky, K., Zemel, R.: Prototypical networks for few-shot learning. In: NeurIPS (2017). http://papers.nips.cc/paper/6996-prototypical-networks-for-few-shot-learning.pdf

26. Sung, F., Yang, Y., Zhang, L., Xiang, T., Torr, P.H., Hospedales, T.M.: Learning to compare: Relation network for few-shot learning. In: CVPR (2018)

27. Tai, K.S., Socher, R., Manning, C.D.: Improved semantic representations from tree-structured long short-term memory networks. In: Proceedings of the 53rd Annual Meeting of the Association for Computational Linguistics and the 7th International Joint Conference on Natural Language Processing (Volume 1: Long Papers). pp. 1556–1566. Association for Computational Linguistics, Beijing, China, July 2015. https://doi.org/10.3115/v1/P15-1150, https://www.aclweb.org/anthology/P15-1150

28. Vaswani, A., et al.: Attention is all you need. In: NeurIPS (2017), http://papers.nips.cc/paper/7181-attention-is-all-you-need.pdf

29. Veličković, P., Cucurull, G., Casanova, A., Romero, A., Liò, P., Bengio, Y.: Graph attention networks. In: ICLR (2018), https://openreview.net/forum?id=rJXMpikCZ

30. Vinyals, O., Blundell, C., Lillicrap, T., kavukcuoglu, k., Wierstra, D.: Matching networks for one shot learning. In: NeurIPS (2016). http://papers.nips.cc/paper/6385-matching-networks-for-one-shot-learning.pdf

31. Wang, L., Cao, Z., de Melo, G., Liu, Z.: Relation classification via multi-level attention CNNs. In: ACL (2016). https://www.aclweb.org/anthology/P16-1123

32. Yang, S., Liu, L., Xu, M.: Free lunch for few-shot learning: Distribution calibration. In: International Conference on Learning Representations (2021). https://openreview.net/forum?id=JWOiYxMG92s

33. Ye, Z.X., Ling, Z.H.: Multi-level matching and aggregation network for few-shot relation classification. In: ACL (2019). https://www.aclweb.org/anthology/P19-1277

34. Zeng, D., Liu, K., Lai, S., Zhou, G., Zhao, J.: Relation classification via convolutional deep neural network. In: COLING (2014). https://www.aclweb.org/anthology/C14-1220

35. Zhou, L., Cui, P., Jia, X., Yang, S., Tian, Q.: Learning to select base classes for few-shot classification. In: CVPR (2020)

36. Zhou, P., et al. : Attention-based bidirectional long short-term memory networks for relation classification. In: ACL (2016). https://www.aclweb.org/anthology/P16-2034

Multi-strategy Knowledge Distillation Based Teacher-Student Framework for Machine Reading Comprehension

Xiaoyan Yu[1,2,3], Qingbin Liu[1,2(✉)], Shizhu He[1,2], Kang Liu[1,2], Shengping Liu[5], Jun Zhao[1,2], and Yongbin Zhou[3,4]

[1] School of Artificial Intelligence, University of Chinese Academy of Sciences, Beijing 100049, China
[2] National Laboratory of Pattern Recognition, Institute of Automation, Chinese Academy of Sciences, Beijing 100190, China
{xiaoyan.yu,qingbin.liu,shizhu.he,kliu,jzhao}@nlpr.ia.ac.cn
[3] Institute of Information Engineering, Chinese Academy of Sciences, Beijing, China
[4] School of Cyber Security, Nanjing University of Science and Technology, Nanjing, China
zhouyongbin@njust.edu.cn
[5] Beijing Unisound Information Technology Co., Ltd., Beijing, China
liushengping@unisound.com

Abstract. The irrelevant information in documents poses a great challenge for machine reading comprehension (MRC). To deal with such a challenge, current MRC models generally fall into two separate parts: evidence extraction and answer prediction, where the former extracts the key evidence corresponding to the question, and the latter predicts the answer based on those sentences. However, such pipeline paradigms tend to accumulate errors, i.e. extracting the incorrect evidence results in predicting the wrong answer. In order to address this problem, we propose a **M**ulti-**S**trategy **K**nowledge **D**istillation based **T**eacher-**S**tudent framework (**MSKDTS**) for machine reading comprehension. In our approach, we first take evidence and document respectively as the input reference information to build a teacher model and a student model. Then the multi-strategy knowledge distillation method transfers the knowledge from the teacher model to the student model at both feature and prediction level through knowledge distillation approach. Therefore, in the testing phase, the enhanced student model can predict answer similar to the teacher model without being aware of which sentence is the corresponding evidence in the document. Experimental results on the ReCO dataset demonstrate the effectiveness of our approach, and further ablation studies prove the effectiveness of both knowledge distillation strategies.

Keywords: Machine reading comprehension · Knowledge distillation · Evidence sentence

© Springer Nature Switzerland AG 2021
S. Li et al. (Eds.): CCL 2021, LNAI 12869, pp. 209–225, 2021.
https://doi.org/10.1007/978-3-030-84186-7_14

1 Introduction

Machine reading comprehension (MRC) is a task that enables machines to read and understand natural language documents to answer questions. Since it well indicates the ability of machines in interpreting natural language as well as having a wide range of application scenarios, it has attracted extensive attention from academia and industry over the recent years. Prevailing MRC datasets define their tasks as either extracting spans from reference documents to answer questions, such as SQuAD [28] and CoQA [29], or inferring answers based on pieces of evidence from a given document, which is also referred to as non-extractive MRC, including multiple-choice MRC [16,30], open domain question answering [5] and so on.

Current MRC faces the significant challenge of the irrelevant information in documents causing negative impact on answer predicting. Therefore, our aim is to engage the model to focus on evidence sentences in documents and using them to answer corresponding questions accurately. To illustrate, consider the example shown in Fig. 1 (adapted from the ReCO dataset [35]). In this sample document, only the evidence sentences have a significant impact on predicting the answer; the other sentences are irrelevant information that may confuse the model and preventing it from focusing on the evidence sentences, thus affecting the correctness of answer predicting.

Question: 孕妇可以喝红糖水吗？ Can pregnant women drink brown sugar water?

Document: …红糖水是具有补血、活血的作用，很多的女性在月经期间或者是产后会选择喝一些红糖水，对于身体的恢复是有益处的。也有一些孕妇在怀孕期间会喝一些红糖水，觉得这样可以补充营养和身体的所需，那么孕妇能可以红糖水吗？红糖含很多的蔗糖，太多的糖可能会导致高血糖，吃点红枣，炖点乌鸡枸杞汤也是可以的，一般在孕期多喝些白开水就好。孕妇喝红糖水可能导致妊娠糖尿病加重，胎儿畸形，妊娠糖尿病还有可能变成孕妇患有糖尿的隐患。肥胖的孕妇不可以喝红糖水，因为红糖水的糖分易转化为脂肪。孕妇在怀孕要注意养成好的生活习惯和饮食习惯，女性在孕期可以多吃一些清淡的食物，做好孕期的饮食调理和胎儿的护理保健…

…Brown sugar water has the effect of tonifying and invigorating the blood, many women drink brown sugar water during periods or after pregnancy, which is beneficial to their recovery. There are also some women who drink brown sugar water during pregnancy, thinking that it can provide the nutrition and body needs, so can pregnant women can brown sugar water? Brown sugar contains a lot of sucrose, too much sugar may lead to high blood sugar, eat some jujube, stew some wolfberry chicken soup is also acceptable, generally drink more hot water during pregnancy is great. Drinking brown sugar water during pregnancy may lead to aggravation of gestational diabetes, fetal malformation, and gestational diabetes may also become a potential risk of diabetes for pregnant women. Pregnant women who are obese should not drink brown sugar water because the sugar in brown sugar water is easily converted into fat. During pregnancy, women should pay attention to establish good living and eating habits…

Evidence: 孕妇喝红糖水可能会导致胎儿畸形，有概率得妊娠糖尿病从而发展成为糖尿病。

Drinking brown sugar water during pregnancy may lead to fetal malformations and increase the probability of getting gestational diabetes and thus developing diabetes.

Candidate Answers: 可以|不可以|无法确定 Yes | **No** | Uncertain

Fig. 1. Example of multiple-choice machine reading comprehension. The sentence in green is the evidence sentence for answering the given question in this document, which is of great importance. Other sentences contain irrelevant information, while potentially negatively affecting the answer prediction. The sentence in blue is the evidence obtained by manual annotation (summarized or paraphrased by the annotator). (Color figure online)

Previous attempts mainly focused on the pipeline (coarse-to-fine) paradigm [24,36]: first locating or generating the evidence sentences corresponding to the question by an evidence extractor or generator, then the answer is predicted based on it. Unfortunately, such a pipeline paradigm suffers from the problem of error accumulation. Besides, in real-world scenarios, the evidence supporting the answer to the question is often implicitly present in the document and thus not easily extracted or generated. For instance, 46% of the evidence sentences could not be explicitly found in the documents in the ReCO dataset. Once the evidence extractor or generator gets incorrect evidence, the result obtained by the answer predictor is bound to be wrong.

In this paper, we attempt to engage the model to focus more on the evidence sentences in the document rather than extracting them out. Thus we propose a Multi-Strategy Knowledge Distillation based Teacher-Student framework (**MSKDTS**). In the training phase, we first take evidence and document as the reference information to pretrain a teacher model and a student model, respectively. Then, we incorporate multi-strategy knowledge distillation into the teacher-student framework, which is the student model attempts to produce teacher-like features and predicted answers through feature knowledge distillation and prediction knowledge distillation. Subsequently, in the testing phase, the enhanced student model predicts the answer with only the document (unaware of the evidence sentences). Hence, the whole process obviates the process of explicitly evidence extraction, which naturally circumvents the accumulation of errors in the conventional pipeline paradigm.

Our contributions are summarized as follows:

- We propose a teacher-student framework for MRC to address the issue of irrelevant information in reference documents causing a negative impact on answer inference.
- We propose a multi-strategy knowledge distillation approach in the teacher-student framework, which transfers knowledge from the teacher model to the student model at feature level and prediction level through feature knowledge distillation and prediction knowledge distillation.
- We conducted experiments on the two testing sets of the ReCO dataset, the results demonstrate the effectiveness of our approach, and further ablation experiments prove the effectiveness of both knowledge distillation strategies.

2   Related Work

2.1   Machine Reading Comprehension

The task of machine reading comprehension (MRC) can well indicate the ability of the machine to understand texts. Owing to the rapid development of deep learning and the presence of many large-scale datasets, MRC is under the spotlight in the field of natural language processing (NLP) in recent years. Depending on the format of questions and answers, the MRC datasets can be roughly categorized into cloze-style [10,11], multiple-choice [16,30,35], span prediction [15,28], and free form [9,23]. Lately, new tasks have emerged for MRC, such as knowledge-based MRC [25], MRC with unanswerable questions [13,27,32] and multi-passage MRC [37].

To model human reading patterns, pipeline (coarse-to-fine) paradigm have been proposed [2,19]. These models first extract the corresponding evidence from the document and then predict the answer via such evidence. To train a evidence extractor, current methods are mainly unsupervised methods [14,31], weakly supervised methods [22] and reinforcement learning methods [2]. Besides, Niu et al. [24] proposed a self-training method for MRC with soft evidence extraction, which performs great on several MRC tasks. Moreover, there are supervised methods [8,20] for extractive MRC by automatically generating evidence, which can be adopted in non-extractive MRC by first generating the evidence, then predicting the answer based on it. Last, Wang et al. [35] presents ReCO, a multiple-choice dataset which manually annotated the evidence in the document, which allows training the evidence extractor or generator in a supervised manner.

In order to engage the model focus more on the evidence, while excluding the pipeline paradigm that inevitably leads to error accumulation, we propose a end-to-end teacher-student framework in this paper.

2.2 Knowledge Distillation

Knowledge distillation [12] is an effective means of transferring knowledge over from one model to another by mimicking the outputs of the original model. Knowledge Consolidation Network [1] is proposed to address the problem of catastrophic forgetting in the incremental event detection task by utilizing the knowledge distillation method. To deploy huge neural machine translation models on edge devices, Wu et al. [38] combined layer-level supervision into the intermediate layers of the original knowledge distillation framework. To cope with the problem of performance degradation caused by utilizing lifelong language learning on different tasks, Chuang et al. [3] proposes an approach that assigns the teacher model to first learn the new task and then passes the knowledge to the lifelong language learning model via knowledge distillation.

Adversarial feature learning [6] is a method that renders the student model with comparable feature extraction ability to the teacher model via Generative Adversarial Networks (GANs, Goodfellow et al. [7]). In order to tackle the major challenge faced in event detection, namely ambiguity in natural language expressions, Liu et al. [21] proposed an adversarial imitation based knowledge distillation approach to learn the feature extraction ability from the teacher model. Lample et al. [17] adopts adversarial feature learning to align features extracted from different language auto-encoders for unsupervised neural machine translation.

In our work, we incorporate multi-strategy knowledge distillation (feature level with adversarial feature learning and prediction level) into the teacher-student framework, bringing more attention to the evidence.

3   Methods

Figure 2 demonstrates the overall framework of MSKDTS, which aims to cope with the irrelevant information in documents. MSKDTS is composed of three major parts, namely, the teacher model, the student model and the knowledge distillation strategies. Documents and evidence sentences are concatenated with queries and candidate answers respectively as the input to the teacher model and the student model. Following encoding the input sequences and predicting the answers, we utilize knowledge distillation to align the features and predictions of the student model to the teacher model.

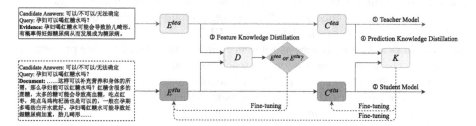

Fig. 2. The overall framework of MSKDTS. The model is composed of six component: the teacher encoder E^{tea}, the student encoder E^{stu}, the discriminator D, the teacher classifier C^{tea}, the student classifier C^{stu} and the prediction knowledge distillation component K. In the training phase, we first take evidence and document as the input reference information to pretrain the teacher model and the student model. Next, E^{stu} and D compete with each other through adversarial imitation strategy. In addition, the probabilities of the answers predicted by C^{stu} and C^{tea} are aligned by a prediction knowledge distillation approach K. In the final testing phase, documents are used as input to the enhanced E^{stu} and C^{stu} for answer prediction.

3.1 MRC Model

Our teacher-student framework mainly oriented towards the multiple-choice MRC problem [16,30,35]. It is composed of a teacher model and a student model, both of which consisting a BERT encoder and a multi-class classifier.

BERT Based Encoder. E^{tea} and E^{stu} are implemented using BERT [4], a multi-layer bidirectional Transformer [34] encoder. Below illustrates several different elements of the input to the BERT encoder and their representations:

- **Document:** A N-token document contains several sentences, distinct parts of which describe different information, denoted as $X^d = \{w_1, w_2, \ldots, w_N\}$, where w_i denotes a word in the document.
- **Evidence:** As the most critical information in inferring the correct answer, the evidence is typically shorter than the document, denoting it by a M-token (where $M \leq N$) sequence as $X^e = \{w_1, w_2, \ldots, w_M\}$, where w_i denotes each word in the evidence sentences.
- **Query:** A L-token query is denoted as $X^q = \{w_1, w_2, \ldots, w_L\}$, where w_i is a word in the query.

- **Alternative Answers:** To predict the correct answer from candidate answers, we denote each candidate answer by A_i. We concatenate A_1 to A_U using [OPT] as the input to the encoder. Note that, in BERT encoder, we use a special token [unused1] as [OPT].

In accordance with the different inputs of the teacher model and the student model, we concatenate these elements above and encode them with the BERT encoder to obtain a context-sensitive representation for the input sequence:

1) For the teacher encoder (E^{tea}), we concatenate candidate answers, query and evidence as input by special tokens of BERT, obtaining an input representation with evidence sentences as reference information. In Fig. 3, we present an example of the input sequence for the teacher encoder.

2) For the student encoder (E^{stu}), analogous to the teacher encoder, except that the reference information is the document rather than the evidence sentence, i.e., candidate answers, query, and document.

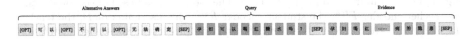

Fig. 3. The BERT input format of the teacher encoder (E^{tea}).

Multi-class Classifier. The softmax classifier is applied as our multi-class classifier, both C^{tea} and C^{stu}, which is used to predict the correct answer from the candidate answers. We take the output of the BERT encoder (E^{tea} and E^{stu}, respectively), i.e., the encoded features, as the input of the classifier. The hidden layer of [OPT] is used as the classification feature f^o for each candidate answer. The multi-class classifier takes these features as input and then computes a prediction probability for each candidate answers as output. The prediction probability $P(A|E, C)$ for each candidate answer is computed as:

$$P(A|E, C) = \text{softmax}(W^o \cdot f^o + b^o) \tag{1}$$

where E is E^{tea} or E^{stu}; C is C^{tea} or C^{stu}; A is the candidate answers; W^o and b^o are trainable parameters of the multi-class classifier (either C^{tea} or C^{stu}).

3.2 Knowledge Distillation Strategies

This section demonstrates in detail of the multi-strategy knowledge distillation in our model, which includes feature-level and prediction-level knowledge distillation, correspondingly, we build a discriminator D and a prediction distiller K. They differ in that the input to D is the extracted feature vector (i.e., the hidden layer of [OPT] mentioned in Sect. 3.1) obtained from either E^{tea} or E^{stu}, whereas the input to K is the logit of the prediction probability from C^{tea} or C^{stu}.

Feature Knowledge Distillation. We adopt an adversarial feature learning approach for feature level knowledge distillation, specifically, we apply a discriminator D, a multi-layer perception (MLP) based binary classifier. It takes the features obtained from E^{tea} and E^{stu} as the input then generates a probability P^D to distinguish the source of the input features. P^D is calculated as:

$$P^D = \text{sigmoid}(W^s(\tanh(W^x f^o + b^x)) + b^s) \tag{2}$$

where sigmoid($*$) is the activation function that maps a scalar to a float number between 0 and 1. A well-trained discriminator would output 1 for the features from E^{tea} and 0 for the features from E^{stu}. W^x, b^x, W^s, and b^s are trainable parameters of the discriminator. We use a two-layer MLP to enhance the representativeness of our discriminator.

The detailed training process of D will be elaborated in Sect. 3.3.

Prediction Knowledge Distillation. Apart from adversarial feature learning for feature level knowledge distillation, we propose a prediction level knowledge distillation. It enables the prediction probability of C^{stu} imitates those of C^{tea}, thereby improving its answer prediction ability. We adopt the knowledge distillation method proposed by Hinton et al. [12], whose specific approach in this framework is demonstrated in Sect. 3.3.

3.3   Overall Training Procedure

Our training process can be summarized into two phases, namely the pretraining phase and the fine-tuning phase. The overall training procedure is demonstrated in Algorithm 1.

Algorithm 1. The Overall Training Procedure

Input: Training Data (x, x^*, y)
1: Pretrain the teacher model (E^{tea}, C^{tea}), the student model (E^{stu}, C^{stu}), and the discriminator (D)
2: Freeze E^{tea} and C^{tea}
3: **repeat**
4:    Freeze D
5:    Unfreeze E^{stu} and C^{stu}
6:    Updata E^{stu} and C^{stu} using Eq.8
7:    **if** the remainder of the batch number to k is 0 **then**
8:        Unfreeze D
9:        Freeze E^{stu} and C^{stu}
10:       Update D using Eq.5
11:   **end if**
12: **until** convergence
Output: An enhanced student model

The Pre-trainig Phase. In the pre-training phase, we first train the teacher model and the student model using the evidence and the documents as reference information, respectively. Then the discriminator is trained using the outputs of E^{tea} and E^{stu}.

First, we train the teacher model (i.e., concatenating E^{tea} and C^{tea}) which is well aware of the evidence sentences. Its loss function is calculated as:

$$\mathcal{L}^{tea} = -\sum_{i=1}^{U} y_i \log(P(A_i|E^{tea}, C^{tea})) \tag{3}$$

where y_i is the label for the i-th answer A_i.

Then, the student model (i.e., concatenating E^{stu} and C^{stu}) is trained, which is not aware of the evidence, it takes the entire document instead of evidence sentences as reference information, thus implicitly introducing considerable irrelevant information. Its loss function is calculated as:

$$\mathcal{L}^{stu} = -\sum_{i=1}^{Y} y_i \log(P(A_i|E^{stu}, C^{stu})) \tag{4}$$

where y_i is the label for the i-th answer A_i.

In the final step, we keep the parameters of E^{tea} and E^{stu} unchanged to train the discriminator by treating the feature vector obtained from E^{tea} as positive examples (label 1) and those from E^{stu} as negative examples (label 0). In this process, the loss function of training the discriminator is calculated as:

$$\mathcal{L}^{D} = \max_{D} \mathbb{E}_{x \sim X}[\log(D(f^{o,tea}))] + \mathbb{E}_{x^* \sim X^*}[\log(1 - D(f^{o,stu}))] \tag{5}$$

where $f^{o,tea}$ is the features of the teacher encoder and $f^{o,stu}$ is the features of the student encoder.

The Fine-Tuning Phase. In the fine-tuning phase, we aim to enhance the feature extraction ability of E^{stu} and the answer prediction ability of C^{stu}, in other words, in document-only cases, we expect the encoder to ignore the irrelevant information as much as possible, focusing more on the evidence sentences.

To enhance the feature extraction ability of E^{stu}, we employ the pretrained D, which can well distinguish between E^{tea} and E^{stu}, to conduct adversarial training with E^{stu}. The loss of E^{stu} is computed as:

$$\mathcal{L}^{afl} = -y \log(D(f^{o,stu})) \tag{6}$$

where y is the label of the output of E^{stu} given to D during adversarial feature learning. Therefore, in order for E^{stu} to produce features similar to those produced by E^{tea}, we set $y = 1$, i.e., we expect the features extracted by E^{stu} to be recognized by D as those extracted by E^{tea}.

After k batches of fine-tuning E^{stu}, the accuracy of D decreases and fails to distinguish well between the outputs obtained from E^{stu} and E^{tea}, then we retrain D using the same loss $\mathcal{L}^D$ as in the pretraining phase. Iteratively fine-tune E^{stu} as well as retrain D until the training process converges. The training procedure is shown in Algorithm 1.

As for prediction level knowledge distillation, the output logit of each sample of C^{tea} and C^{stu} are denoted as v and v^*, respectively. The prediction knowledge distillation is calculated as:

$$\mathcal{L}^{pkd} = -\sum_{i=1}^{U} \tau_i(v^*) \log(\tau_i(v))$$

$$\tau_i(v^*) = \frac{e^{v_i^*/\Omega}}{\sum_{j=1}^{U} e^{v_j^*/\Omega}}, \ \tau_i(v) = \frac{e^{v_i/\Omega}}{\sum_{j=1}^{U} e^{v_j/\Omega}} \tag{7}$$

where Ω is a hyper-parameter, which is usually set to be greater than 1 (e.g. $\Omega = 2$) in our experiments to increase the weights of small values; U is the number of classes; $\mathcal{L}^{pkd}$ is designed to encourage the prediction of the student model to match the prediction of the teacher model.

In the fine-tuning phase, the total loss of the student model is:

$$\mathcal{L}^{stu\_all} = \mathcal{L}^{stu} + \alpha\mathcal{L}^{afl} + \beta\mathcal{L}^{pkd} \tag{8}$$

where α and β are two hyper-parameters. If α and β are very large, the model will focus more on learning knowledge from the teacher model, rather than the ground-truth labels. Noting that, the parameters of D, E^{tea}, C^{tea} are kept unchanged while fine-tuning the components of the student model.

After completing the two knowledge distillation approaches above, we obtained an enhanced student model that has successfully learned the knowledge of the teacher model and is able to predict accurate answers using only the documents as reference information.

4   Experiments

4.1   Datasets

We conduct experiments on a recently proposed MRC dataset, ReCO [35] to evaluate the validity of our model. To the best of our knowledge, this is the only large-scale multiple-choice MRC dataset with manually labeled evidence. ReCO contains 300k document-query pairs, each of them is manually labeled with evidence. It is worth noting that, during the annotation process, for 46% samples, the annotators paraphrase or highly summarize the key sentences according to their understanding, resulting in a situation that not all evidence sentences can be found in its corresponding document.

In ReCO, three candidate answers are available for each query. In order to obtain the correct answer, strong inference capability of the model is required. The dataset contains 280k training samples and 20k test samples, which are further divided into testing set A ($Test_A$) and testing set B ($Test_B$). $Test_B$ is the complement to $Test_A$ in terms of quantity, and can certify the validity of the model more adequately.

4.2 Baseline

To evaluate the capability of MRC models and select well-performing teacher and student models in our framework, we adopt several strong baselines that perform well on many MRC tasks:

BiDAF [31]: BiDAF uses LSTM as its encoder, and models the relationship between the question and the answer by a bidirectional attention mechanism.

BiDAF* [26,31]: BiDAF* replaces the traditional word embedding in BiDAF with ELMO (a language model trained on unsupervised data), which yields better results.

BERT [4]: A multi-layer bidirectional Transformer, which is pretrained on large unlabeled data, has outperformed state-of-the-art models in many NLP tasks.

ALBERT [18]: ALBERT is an improved version of BERT, which reduces the overall number of parameters, speeds up the training process, and is better than BERT in many aspects.

Since the evidence sentences in the 46% samples in the ReCO dataset could not be explicitly found in the corresponding documents, we use generation models as evidence generators in the pipeline baselines instead of extraction models.

Enc2Dec [33]: We designed a coarse-to-fine framework based on the encoder-decoder framework. This model encodes documents with an LSTM encoder and then generates evidence by an LSTM decoder.

Enc2Dec* [33]: In addition to the Enc2Dec model, we adopt the BERT encoder in the encoder-decoder framework.

4.3 Experimental Setup and Evaluation Metrics

We use ALBERT-base from HuggingFace's Transformer library[1] as the encoder for our MRC model. For both teacher model and student model as well as the discriminator D, the learning rate is set to 2e−5, batch size set to 4, hyper-parameters α and β are chosen from [0–100], specified as $\alpha = 0.5$ and $\beta = 20$, with temperature coefficient $\Omega = 2$. Since D can easily learn and distinguish the features obtained from different encoders, we randomly sample 10,000 training samples each time to train D. We retrain the discriminator every $k = 3000$ batches. All hyper-parameters are obtained by grid search in the validation process.

Following the previous work [35], we use accuracy as our metric to evaluate whether each sample is correctly classified.

4.4 Results

We list the following Research Questions (RQ) as guidelines for experimentation in our work:

---

[1] https://huggingface.co/.

– RQ1: How well did the MRC models perform before incorporating the knowledge distillation strategies, and which model we select to be the teacher or student model?
– RQ2: Is there a significant improvement in performance after applying our proposed MSKDTS framework, and does the MSKDTS framework outperform the traditional pipeline paradigm?
– RQ3: Whether the feature knowledge distillation we designed can effectively improve the performance by enabling E^{stu} to imitate the output features of E^{tea}?
– RQ4: Will the prediction knowledge distillation strategy we employ be effective in improving the performance of the student model?

Teacher and Student Models. To answer RQ1, this section shows the performance of several baseline models when inputting documents or evidence as reference information, and compares the performance of different baselines to select the teacher and student models in the MSKDTS framework.

Table 1. Experimental results on the development set (Dev), testing set A (Test$_A$) and testing set B (Test$_B$) of the ReCO dataset. The second/third column shows the result of these models when taking evidence/documents as inference information input in both training and testing phases. Bold indicates the best model. BERT$_b$ and BERT$_l$ denotes BERT base and BERT large, respectively. ALBERT$_{tiny}$ and ALBERT$_b$ denotes ALBERT tiny and ALBERT base, respectively.

	Teacher (Evidence)			Student (Document)		
	Dev	Test$_A$	Test$_B$	Dev	Test$_A$	Test$_B$
Random [35]	33.3	33.3	33.3	33.3	33.3	33.3
BiDAF [31]	68.9	68.3	67.9	55.7	55.8	56.1
BiDAF* [26,31]	70.3	70.9	71.1	58.4	58.9	58.6
BERT$_b$ [4]	73.8	73.4	72.8	61.4	61.1	62.0
BERT$_l$ [4]	76.3	77.0	76.4	65.5	65.3	65.8
ALBERT$_{tiny}$ [18]	70.9	70.4	71.3	63.1	62.7	62.4
ALBERT$_b$ [18]	**77.2**	**77.6**	**77.0**	**68.2**	**68.4**	**69.1**
Human	-	91.5	-	-	88.0	-

Table 1 shows the performance of several baseline models when evidence and documents are used as reference information input, respectively. Comparing the results in Table 1, we can see that irrelevant information in the documents does have a negative effect on answer prediction (for every model except random, the performance with evidence as reference information input is superior to the performance with documents as reference information input), so evidence has a facilitating effect on answer prediction. Also, the results in Table 1 show that there is a gap between predicting answers by documents and by evidence even for human, which proves the importance of evidence in machine reading comprehension.

From the results, we can see that $ALBERT_b$ achieves the best performance and can outperform other BERT-based models when taking evidence and document as reference information input, therefore, we choose $ALBERT_b$ as both our teacher and student models.

The teacher model outperforms the student model by 9.2% and 7.9% on $Test_A$ and $Test_B$, respectively. Since the student model is designed to imitate the behavior of the teacher model in our approach, the performance of the student model cannot exceed that of the teacher model, i.e., the performance of the teacher model is the upper bound of our framework, and the lower bound should be the student model without fine-tuning.

Results on Real Test Scenarios. To answer RQ2, we compared our approach with the pipeline paradigm and the teacher model and student model (which is not fine-tuned with our knowledge distillation strategies) as upper and lower bounds.

Table 2. Experimental results on the development set (Dev), testing set A ($Test_A$), testing set B ($Test_B$). Enc2Dec(*) + $ALBERT_b$ is MRC models with evidence generator (pipeline paradigm).

	Dev	$Test_A$	$Test_B$
Lower bound	68.2	68.4	69.1
Enc2Dec + $ALBERT_b$	68.6	68.9	69.3
Enc2Dec* + $ALBERT_b$	68.9	69.0	69.6
MSKDTS (Ours)	**71.3**	**71.0**	**70.8**
Upper bound	77.2	77.6	77.0

To validate the effectiveness of MSKDTS, we tested the performance of the enhanced student model in real scenarios. In real scenarios, the evidence in the documents is not annotated, and the enhanced student model needs to predict the results directly based on the documents as reference information.

From the experimental results in Table 2, we can see that:

First, our student model achieves the best performance, outperforming all the baseline models that do not use evidence. This demonstrates that the multi-strategy knowledge distillation approach we proposed enables the student encoder to effectively imitate the output features of the teacher encoder, can focus on the evidence sentences in the documents.

Second, to compare with the pipeline model, we train an encoder-decoder model that generates the evidence sentences for each testing sample. The performance of Enc2Dec and Enc2Dec* on the two testing sets is much weaker than our fine-tuned student model. Our model outperforms Enc2Dec* by 2.0% and 1.2% on Test$_A$ and Test$_B$, respectively.

Third, there is still a gap between our approach and the teacher model, which shows that it is still a significant challenge of how to engage the model to focus on the evidence sentences from the documents.

The Effect of Feature Knowledge Distillation. To answer RQ3, we study the effect of feature knowledge distillation in this section.

Table 3. Experimental results on the development set (Dev), testing set A (Test$_A$) and testing set B (Test$_B$) to verify the effect of feature knowledge distillation.

	Dev	Test$_A$	Test$_B$
MSKDTS	**71.3**	**71.0**	**70.8**
MSKDTS (k = 10000)	71.0	70.5	70.6
MSKDTS (k = 50000)	70.6	70.3	70.2
MSKDTS-AFL+COSINE	70.7	70.6	70.4
MSKDTS-AFL	70.2	70.1	69.8

We conducted three experiments to demonstrate the effectiveness of adversarial feature learning as a feature knowledge distillation strategy: 1) Testing the performance of the MSKDTS framework with different hyper-parameters (when the update frequency $k = 10000$ and $k = 50000$ of the discriminator D). 2) Testing the performance of the MSKDTS framework replacing adversarial feature learning with the cosine similarity loss between the features of the teacher model and those of the student model (denoted as MSKDTS-AFL+COSINE), which enables the two features to have the same angle in the high dimensional space. 3) Testing the performance of the MSKDTS framework without any feature knowledge distillation strategies (denoted as MSKDTS-AFL). Table 3 shows the results of these models.

From these results, we can see that: 1) Our approach outperforms all variants, which proves the effectiveness of our feature knowledge distillation strategy. 2) For $k = 10000$ and $k = 50000$, the performance degradation is caused by poor discriminator performance due to updating the discriminator after a larger number of batches. 3) The cosine similarity loss causes a performance degradation with -0.4% and -0.4% on Test$_A$ and Test$_B$ due to features learned based on specific distances is prone to be approximated from a certain aspect (angle) in the high dimensional space, which may result in a loss of semantic information. 4) In the absence of feature knowledge distillation, the performance degrades significantly due to the lack of proximity to the features of the teacher model.

The Effect of Prediction Knowledge Distillation. To answer RQ4, we study the effect of the prediction knowledge distillation in this section.

Table 4. Experimental results on the development set (Dev), testing set A (Test$_A$), testing set B (Test$_B$) to verify the effect of prediction knowledge distillation.

	Dev	Test$_A$	Test$_B$
MSKDTS	**71.3**	**71.0**	**70.8**
MSKDTS-PKD+KL	70.9	70.5	70.3
MSKDTS-PKD	69.7	69.9	70.1

As shown in Table 4, we use KL-divergence (MSKDTS-PKD+KL) to replace the knowledge distillation loss. This variant causes performance degradation due to the absence of the temperature coefficient Ω in the knowledge distillation loss. Therefore, it is difficult for the model to learn small logit values and affects the knowledge distillation ability. Compared to the model without Prediction Knowledge Distillation, our model achieves significant improvement. It demonstrates the effectiveness of prediction knowledge distillation. Compared to the student model trained with document only (without fine-tuning), it verifies that the simultaneous use of feature knowledge distillation and prediction knowledge distillation can effectively improve the performance.

5   Conclusion

We propose a **M**ulti-**S**trategy **K**nowledge **D**istillation based **T**eacher-**S**tudent framework (**MSKDTS**) for MRC to address the challenges posed by irrelevant information in documents for answer prediction. The teacher-student framework naturally circumvents the error accumulation problem in the traditional pipeline paradigm and the knowledge distillation strategies enhance the model capability at the feature and prediction levels. Experiments on the ReCO dataset demonstrate the effectiveness of our approach.

Acknowledgements. This work was supported by the National Key Research and Development Program of China (No. 2020AAA0106400), the National Natural Science Foundation of China (No. 61922085, 61976211, 61632020, U1936209 and 62002353) and Beijing Natural Science Foundation (No.4192067). This work is also supported by Beijing Academy of Artificial Intelligence (BAAI2019QN0301), the Key Research Program of the Chinese Academy of Science (Grant No. ZDBS-SSW-JSC006), the independent research project of National Laboratory of Pattern Recognition, the Youth Innovation Promotion Association CAS and Meituan-Dianping Group.

References

1. Cao, P., Chen, Y., Zhao, J., Wang, T.: Incremental event detection via knowledge consolidation networks. In: Proceedings of the 2020 Conference on Empirical Methods in Natural Language Processing, EMNLP 2020, Online, 16–20 November, 2020, pp. 707–717 (2020)

2. Choi, E., Hewlett, D., Uszkoreit, J., Polosukhin, I., Lacoste, A., Berant, J.: Coarse-to-fine question answering for long documents. In: Proceedings of the 55th Annual Meeting of the Association for Computational Linguistics, ACL 2017, Vancouver, Canada, 30 July – 4 August, Volume 1: Long Papers, pp. 209–220 (2017)

3. Chuang, Y., Su, S., Chen, Y.: Lifelong language knowledge distillation. In: Proceedings of the 2020 Conference on Empirical Methods in Natural Language Processing, EMNLP 2020, Online, 16–20 November, 2020, pp. 2914–2924 (2020)

4. Devlin, J., Chang, M., Lee, K., Toutanova, K.: BERT: pre-training of deep bidirectional transformers for language understanding. In: Proceedings of the 2019 Conference of the North American Chapter of the Association for Computational Linguistics: Human Language Technologies, NAACL-HLT 2019, Minneapolis, MN, USA, 2–7 June, 2019, Volume 1 (Long and Short Papers), pp. 4171–4186 (2019)

5. Dhingra, B., Mazaitis, K., Cohen, W.W.: Quasar: datasets for question answering by search and reading. CoRR abs/1707.03904 (2017)

6. Donahue, J., Krähenbühl, P., Darrell, T.: Adversarial feature learning. In: 5th International Conference on Learning Representations, ICLR 2017, Toulon, France, 24–26 April, 2017, Conference Track Proceedings (2017)

7. Goodfellow, I.J., et al.: Generative adversarial networks. CoRR abs/1406.2661 (2014)

8. Hanselowski, A., Zhang, H., Li, Z., Sorokin, D., Schiller, B., Schulz, C., Gurevych, I.: Ukp-athene: multi-sentence textual entailment for claim verification. CoRR abs/1809.01479 (2018)

9. He, W., et al.: DuReader: a Chinese machine reading comprehension dataset from real-world applications. In: Proceedings of the Workshop on Machine Reading for Question Answering@ACL 2018, Melbourne, Australia, 19 July, 2018, pp. 37–46 (2018)

10. Hermann, K.M., et al.: Teaching machines to read and comprehend. In: Advances in Neural Information Processing Systems 28: Annual Conference on Neural Information Processing Systems 2015, 7–12 December, 2015, Montreal, Quebec, Canada, pp. 1693–1701 (2015)

11. Hill, F., Bordes, A., Chopra, S., Weston, J.: The goldilocks principle: Reading children's books with explicit memory representations. In: 4th International Conference on Learning Representations, ICLR 2016, San Juan, Puerto Rico, 2–4 May, 2016, Conference Track Proceedings (2016)

12. Hinton, G.E., Vinyals, O., Dean, J.: Distilling the knowledge in a neural network. CoRR abs/1503.02531 (2015)

13. Hu, M., Wei, F., Peng, Y., Huang, Z., Yang, N., Li, D.: Read + verify: machine reading comprehension with unanswerable questions. In: The Thirty-Third AAAI Conference on Artificial Intelligence, AAAI 2019, The Thirty-First Innovative Applications of Artificial Intelligence Conference, IAAI 2019, The Ninth AAAI Symposium on Educational Advances in Artificial Intelligence, EAAI 2019, Honolulu, Hawaii, USA, 27 January – 1 February, 2019, pp. 6529–6537 (2019)

14. Huang, H., Choi, E., Yih, W.: Flowqa: Grasping flow in history for conversational machine comprehension. In: 7th International Conference on Learning Representations, ICLR 2019, New Orleans, LA, USA, 6–9 May, 2019 (2019)

15. Joshi, M., Choi, E., Weld, D.S., Zettlemoyer, L.: Triviaqa: a large scale distantly supervised challenge dataset for reading comprehension. In: Proceedings of the 55th Annual Meeting of the Association for Computational Linguistics, ACL 2017, Vancouver, Canada, 30 July – 4 August, Volume 1: Long Papers, pp. 1601–1611 (2017)

16. Lai, G., Xie, Q., Liu, H., Yang, Y., Hovy, E.H.: RACE: large-scale reading comprehension dataset from examinations. In: Proceedings of the 2017 Conference on Empirical Methods in Natural Language Processing, EMNLP 2017, Copenhagen, Denmark, 9–11 September, 2017, pp. 785–794 (2017)

17. Lample, G., Conneau, A., Denoyer, L., Ranzato, M.: Unsupervised machine translation using monolingual corpora only. In: 6th International Conference on Learning Representations, ICLR 2018, Vancouver, BC, Canada, 30 April – 3 May, 2018, Conference Track Proceedings (2018)

18. Lan, Z., Chen, M., Goodman, S., Gimpel, K., Sharma, P., Soricut, R.: ALBERT: a lite BERT for self-supervised learning of language representations. In: 8th International Conference on Learning Representations, ICLR 2020, Addis Ababa, Ethiopia, 26–30 April, 2020 (2020)

19. Li, W., Li, W., Wu, Y.: A unified model for document-based question answering based on human-like reading strategy. In: Proceedings of the Thirty-Second AAAI Conference on Artificial Intelligence, (AAAI-18), the 30th innovative Applications of Artificial Intelligence (IAAI-18), and the 8th AAAI Symposium on Educational Advances in Artificial Intelligence (EAAI-18), New Orleans, Louisiana, USA, 2–7 February, 2018, pp. 604–611 (2018)

20. Lin, Y., Ji, H., Liu, Z., Sun, M.: Denoising distantly supervised open-domain question answering. In: Proceedings of the 56th Annual Meeting of the Association for Computational Linguistics, ACL 2018, Melbourne, Australia, 15–20 July, 2018, Volume 1: Long Papers, pp. 1736–1745 (2018)

21. Liu, J., Chen, Y., Liu, K.: Exploiting the ground-truth: An adversarial imitation based knowledge distillation approach for event detection. In: The Thirty-Third AAAI Conference on Artificial Intelligence, AAAI 2019, The Thirty-First Innovative Applications of Artificial Intelligence Conference, IAAI 2019, The Ninth AAAI Symposium on Educational Advances in Artificial Intelligence, EAAI 2019, Honolulu, Hawaii, USA, 27 January – 1 February, 2019, pp. 6754–6761 (2019)

22. Min, S., Zhong, V., Socher, R., Xiong, C.: Efficient and robust question answering from minimal context over documents. In: Proceedings of the 56th Annual Meeting of the Association for Computational Linguistics, ACL 2018, Melbourne, Australia, 15–20 July, 2018, Volume 1: Long Papers, pp. 1725–1735 (2018)

23. Nguyen, T., et al.: MS MARCO: a human generated machine reading comprehension dataset. In: Proceedings of the Workshop on Cognitive Computation: Integrating Neural and Symbolic Approaches 2016 Co-located with the 30th Annual Conference on Neural Information Processing Systems (NIPS 2016), Barcelona, Spain, 9 December, 2016, vol. 1773 (2016)

24. Niu, Y., Jiao, F., Zhou, M., Yao, T., Xu, J., Huang, M.: A self-training method for machine reading comprehension with soft evidence extraction. In: Proceedings of the 58th Annual Meeting of the Association for Computational Linguistics, ACL 2020, Online, 5–10 July, 2020, pp. 3916–3927 (2020)

25. Ostermann, S., Modi, A., Roth, M., Thater, S., Pinkal, M.: MCScript: a novel dataset for assessing machine comprehension using script knowledge. In: Proceedings of the Eleventh International Conference on Language Resources and Evaluation, LREC 2018, Miyazaki, Japan, 7–12 May, 2018 (2018)

26. Peters, M.E., et al.: Deep contextualized word representations. In: Proceedings of the 2018 Conference of the North American Chapter of the Association for Computational Linguistics: Human Language Technologies, NAACL-HLT 2018, New Orleans, Louisiana, USA, 1–6 June, 2018, Volume 1 (Long Papers), pp. 2227–2237 (2018)

27. Rajpurkar, P., Jia, R., Liang, P.: Know what you don't know: unanswerable questions for squad. In: Proceedings of the 56th Annual Meeting of the Association for Computational Linguistics, ACL 2018, Melbourne, Australia, 15–20 July, 2018, Volume 2: Short Papers, pp. 784–789 (2018)

28. Rajpurkar, P., Zhang, J., Lopyrev, K., Liang, P.: Squad: 100,000+ questions for machine comprehension of text. In: Proceedings of the 2016 Conference on Empirical Methods in Natural Language Processing, EMNLP 2016, Austin, TX, USA, 1–4 November, 2016, pp. 2383–2392 (2016)

29. Reddy, S., Chen, D., Manning, C.D.: CoQA: a conversational question answering challenge. Trans. Assoc. Comput. Linguist. **7**, 249–266 (2019)

30. Richardson, M., Burges, C.J.C., Renshaw, E.: MCTest: a challenge dataset for the open-domain machine comprehension of text. In: Proceedings of the 2013 Conference on Empirical Methods in Natural Language Processing, EMNLP 2013, 18–21 October 2013, Grand Hyatt Seattle, Seattle, Washington, USA, A meeting of SIGDAT, a Special Interest Group of the ACL, pp. 193–203 (2013)

31. Seo, M.J., Kembhavi, A., Farhadi, A., Hajishirzi, H.: Bidirectional attention flow for machine comprehension. In: 5th International Conference on Learning Representations, ICLR 2017, Toulon, France, 24–26 April, 2017, Conference Track Proceedings (2017)

32. Sun, F., Li, L., Qiu, X., Liu, Y.: U-net: machine reading comprehension with unanswerable questions. CoRR abs/1810.06638 (2018)

33. Sutskever, I., Vinyals, O., Le, Q.V.: Sequence to sequence learning with neural networks. In: Advances in Neural Information Processing Systems 27: Annual Conference on Neural Information Processing Systems 2014, 8–13 December, 2014, Montreal, Quebec, Canada, pp. 3104–3112 (2014)

34. Vaswani, A., et al.: Attention is all you need. In: Advances in Neural Information Processing Systems 30: Annual Conference on Neural Information Processing Systems 2017, 4–9 December, 2017, Long Beach, CA, USA, pp. 5998–6008 (2017)

35. Wang, B., Yao, T., Zhang, Q., Xu, J., Wang, X.: Reco: a large scale Chinese reading comprehension dataset on opinion. In: The Thirty-Fourth AAAI Conference on Artificial Intelligence, AAAI 2020, The Thirty-Second Innovative Applications of Artificial Intelligence Conference, IAAI 2020, The Tenth AAAI Symposium on Educational Advances in Artificial Intelligence, EAAI 2020, New York, NY, USA, 7–12 February, 2020, pp. 9146–9153 (2020)

36. Wang, H., et al.: Evidence sentence extraction for machine reading comprehension. In: Proceedings of the 23rd Conference on Computational Natural Language Learning, CoNLL 2019, Hong Kong, China, 3–4 November, 2019, pp. 696–707 (2019)

37. Wang, Y., et al.: Multi-passage machine reading comprehension with cross-passage answer verification. In: Proceedings of the 56th Annual Meeting of the Association for Computational Linguistics, ACL 2018, Melbourne, Australia, 15–20 July, 2018, Volume 1: Long Papers, pp. 1918–1927 (2018)

38. Wu, Y., Passban, P., Rezagholizadeh, M., Liu, Q.: Why skip if you can combine: a simple knowledge distillation technique for intermediate layers. In: Proceedings of the 2020 Conference on Empirical Methods in Natural Language Processing, EMNLP 2020, Online, 16–20 November, 2020, pp. 1016–1021 (2020)

LRRA:A Transparent Neural-Symbolic Reasoning Framework for Real-World Visual Question Answering

Zhang Wan, Keming Chen, Yujie Zhang$^{(\boxtimes)}$, Jinan Xu, and Yufeng Chen

School of Computer and Information Technology,
Beijing Jiaotong University, Beijing 100044, China
{19120413,20120341,yjzhang,jaxu,chenyf}@bjtu.edu.cn

Abstract. The predominant approach of visual question answering (VQA) relies on encoding the image and question with a "black box" neural encoder and decoding a single token into answers such as "yes" or "no". Despite this approach's strong quantitative results, it struggles to come up with human-readable forms of justification for the prediction process. To address this insufficiency, we propose LRRA [Look, Read, Reasoning,Answer], a transparent neural-symbolic framework for visual question answering that solves the complicated problem in the real world step-by-step like humans and provides human-readable form of justification at each step. Specifically, LRRA learns to first convert an image into a scene graph and parse a question into multiple reasoning instructions. It then executes the reasoning instructions one at a time by traversing the scene graph using a recurrent neural-symbolic execution module. Finally, it generates answers to the given questions and makes corresponding marks on the image. Furthermore, we believe that the relations between objects in the question is of great significance for obtaining the correct answer, so we create a perturbed GQA test set by removing linguistic cues (attributes and relations) in the questions to analyze which part of the question contributes more to the answer. Our experiments on the GQA dataset show that LRRA is significantly better than the existing representative model (57.12% vs. 56.39%). Our experiments on the perturbed GQA test set show that the relations between objects is more important for answering complicated questions than the attributes of objects.

Keywords: Visual question answering · Relations between objects · Neural-symbolic reasoning

1 Introduction

Currently, the predominant approach to visual question answering (VQA) relies on encoding the image and question with a black-box transformer encoder [1, 2]. These works carry out complex calculations behind the scenes but only produce a single token as prediction output (for example, "yes", "no") and they

© Springer Nature Switzerland AG 2021
S. Li et al. (Eds.): CCL 2021, LNAI 12869, pp. 226–236, 2021.
https://doi.org/10.1007/978-3-030-84186-7_15

can not provide an easy-to-understand form of justification consistent with their predictions. In addition, recent studies have shown that the end-to-end model can be easily optimized to learn the "shortcut bias" of the data set instead of reasoning (for example, the model uses the implicit fused question representations [3,4], the answer can be directly inferred according to certain language patterns), which tend to undesirably adhere to superficial or even potentially misleading statistical associations [5], so they do not really understand the question, and often perform poorly in the face of complex reasoning problems in the real world. In order to solve the above insuficiencys, we learn the correct problem solving process step-by-step mimicking humans and propose a neural-symbolic approach for visual question answering that fully disentangles vision and language understanding from reasoning. A human would first (1) look at the image, (2) read the question, (3) reason and think (4) answer questions. Following this intuition, our model deploys four neural modules, each mimicking one problem solving step that humans would take: A scene graph generation module first converts an image into a scene graph; A semantic parsing module parses each question into multiple reasoning instructions; A neural execution module interprets reason instructions one at a time by traversing the scene graph in a recurrent manner; Answer generation module predicts the answer with the highest probability. These four modules are connected through hidden states instead of explicit outputs. Therefore, the entire framework can be trained end-to-end from pixels to answers. In addition, since LRRA also produces human-readable output from individual modules during testing, we can easily locate the error by checking the modular output. Our experiments on the GQA dataset show that LRRA is significantly better than the existing representative model (57.12% vs. 56.39%). Furthermore, we believe that the relations between objects in the question is of great significance for obtaining the correct answer, so we create a perturbed GQA test set by removing linguistic cues (attributes and relations) in the questions to analyze which part of the question contributes more to the answer. Ablation experiment further show that the relations between objects is more important for answering complicated questions than the attributes of objects. To summarize, the main contributions of our paper are threefold:

- When we give the answer, we also make the corresponding mark on the image to improve explainability and discourage superficial guess for answering the questions.
- We propose an end-to-end trainable modular VQA framework LRRA. Compared with contemporary black-box methods, it has interpretability and enhanced error analysis capabilities.
- We create a perturbed GQA test set that provides an effficient way to validate our approach on the perturbed dataset. The dataset will be announced soon.

2   Related Work

Visual Reasoning. It is the process of analyzing visual information and solving problems based on it. The most representative benchmark of visual reasoning is GQA [6] a diagnostic visual Q&A dataset for compositional language and elementary visual reasoning. The majority of existing methods on GQA can be categorized into two families: 1) holistic approaches [7–10], which embed both the image and question into a feature space and infer the answer by feature fusion; 2) neural module approaches [11–15], which first parse the question into a program assembly of neural modules, and then execute the modules over the image features for visual reasoning. Our LRRA belongs to the second one but replaces the visual feature input with scene graphs.

Neural Module Networks. They dismantle a complex question into several subtasks, which are easier to answer and more transparent to follow the intermediate outputs. Modules are predefined neural networks that implement the corresponding functions of subtasks, and then are assembled into a layout dynamically, usually by a sequence-to-sequence program generator given the input question. The assembled program is finally executed for answer prediction [12–14]. In particular, the program generator is trained based on the human annotations of desired layout or with the help of reinforcement learning due to the nondifferentiability of layout selection. Recently, Hu et al. [15] proposed StackNMN, which replaces the hard-layout with soft and continuous module layout and performs well even without layout annotations at all. Our LRRA experiments on GQA follows their softprogram generator.

Recently, NS-VQA [16] firstly built the reasoning over the object-level structural scene repre-sentation, improving the accuracy on CLEVR from the previous state-of-the-art 99.1% [14] to an almost perfect 99.8%. Their scene structure consists of objects with detected labels, but lacked the relationships between objects, which limited its application on real-world datasets such as GQA [6]. In this paper, we propose a much more generic framework for visual reasoning over scene graphs, including object nodes and relationship edges represented by either labels or visual features. Our scene graph is more flexible and more powerful than the table structure of NS-VQA.

Scene Graphs. This task is to produce graph representations of images in terms of objects and their relationships. Scene graphs have been shown effective in boosting several vision-language tasks [17–19]. However, scene graph detection is far from satisfactory compared to object detection [20–22]. To this end, our scene graph implementation also supports cluttered and open-vocabulary in real-world scene graph detection, where the nodes are merely RoI features and the edges are their relations.

3   Approach

We build our neural module network over scene graphs to tackle the visual reasoning challenge. As shown in Fig. 1, given an input image andquestion, we first parse the image into a scene graph and parse the question into a module program, an then execute the program over the scene graph. Besides, our approach are totally attention-based, making all theintermediate reasoning steps transparent. The model framework as shown in Fig. 1.

Scene Graph Generation. Given an image I, its corresponding scene graph represents the objects in the image (e.g., girl, hamburger) as nodes and the objects' pairwise relationships (e.g., holding) as edges. The fifirst step of scene graph generation is object detection. We use DETR [22] as the object detection backbone since it removes the need for hand-designed components like non-maximum suppression. DETR [22] feeds the image feature from ResNet50 [23] into a nonau-toregressive transformer model, yielding an orderless set of N object vectors $[o_1, o_2, \cdots, o_N]$, as in (1). Each object vector represents one detected object in the image. Then, for each object vector, DETR uses an object vector decoder (feed-forward network) to predict the corresponding object class (e.g., girl), and the bounding box in a multi-task manner. Since the set prediction of N object vectors is orderless, DETR calculates the set prediction loss by first computing an optimal matching between predicted and ground truth objects, and then sum the loss from each object vector. N is fifixed to 100 and DETR creates a special class label "no object", to represent that the object vector does not represent any object in the image.

$$[o_1, o_2, \cdots, o_N] = DETR \tag{1}$$

The object detection backbone learns object classes and bounding boxes, but does not learn object attributes, and the objects' pairwise relationships. We augment the object vector decoder with an additional object attributes predictor. For each attribute meta-concept (e.g., color), we create a classifier to predict the possi-ble attribute values (e.g., red, pink). To predict the relationships, we consider all $N(N-1)$ possible pairs of object vectors, $[e_1, e_2, \cdots, e_{N(N-1)}]$. The relation encoder transforms each object vector pair to an edge vector through feed-forward and normalization layers as in (2). We then feed each edge vector to the relation decoder to classify its relationship label. Both object attributes and inter-object relationships are supervised in a multitask manner. To handle the object vector pair that does not have any relationship, we use the "no relation" relationship label.

We construct the scene graph represented by N object vectors and $N(N-1)$ edge vectors instead of the symbolic outputs, and pass it to downstream modules.

$$e_{i,j} = LayerNorm(FeedForward(o_i \oplus o_j)) \tag{2}$$

Semantic Parsing. The semantic parser works as a "compiler" that translates the question tokens $(q_1, q_2, \cdots, q_Q)$ into an neural executable program, which consists of multiple instruction vectors. We adopt a hierarchical sequence generation design: a transformer model [23] first parses the question into a sequence of Minstruction vectors, $[i_1, i_2, \cdots, i_M]$. The ith instruction vector will correspond exactly to the ith execution step in the neural execution. engine. To enable human to understand the semantics of the instruction vectors, we further translate each instruction vector to human-readable text using a transformer-based instruction vector decoder. We pass the M instruction vectors rather than the human-readable text to the neural execution module.

$$[i_1, i_2, ..., i_M] = Transformer(q_1, ..., q_Q) \tag{3}$$

Visual Reasoning. The neural execution engine works in a recurrent manner: At the mth time step, the neural execution engine takes the mth instruction vector (i_m) and outputs the scene graph tra-versal result. Similar to recurrent neural networks, a history vector that summarizes the graph traversal states of all nodes in the current time-step would be passed to the next time-step. The neural execution engine operates with graph neural network. Graph neural network generalizes the convolution operator to graphs using the neighborhood aggregation scheme [24, 25]. The key intuition is that each node aggregates feature vectors of its immediate neighbors to compute its new feature vector as the input for the following neural layers. Specifically, at m^{th} time step given a node as the central node, we first obtain the feature vector of each neighbor (f_k^m) through a feed-forward network with the following inputs: the object vector of the neighbor (o_k) in the scene graph, the edge vector between the neighbor node and the central node $(e_{k,central})$ in the scene graph, the $(m-1)^{th}$ history vector (h_{m-1}), and the mth instruction vector (i_m).

$$f_k^m = FeedForward(o_k \oplus e_{k,central} \oplus h_{m-1} \oplus i_m) \tag{4}$$

We then average each neighbor's feature vector as the context vector of the central node

$$c_{central}^m = \frac{1}{K} \sum_{k=1}^{K} f_k^m \tag{5}$$

Next, we perform node classification for the central node, where an "1" means that the corre-sponding node should be traversed at the mth time step and "0" otherwise. The inputs of the node classifier are: the object vector of the central node in the scene graph, the context vector of the central node, and the mth instruction vector.

$$s_{central}^m = Softmax(FeedForward(o_{central} \oplus c_{central}^m \oplus i_m)) \tag{6}$$

where $s_{central}^m$ is the classification confidence score of central node at m^{th} time step. The node classification results of all nodes constitute a bitmap as the scene graph traversal result. We calculate the weighted average of all object vectors as the history vector (h_m), where the weight is each node's classification confidence score.

$$h_m = \sum_{i}^{N} s_i^m \cdot o_i \tag{7}$$

Predict answer. VQA is commonly formulated as a classification problem where the model learns to answers with one token (e.g., "yes" or "no"). To do this, the language output at the last step is passed to a feed-forward network with softmax activation to obtain the distribution for the predicted answers.

$$w = atgmax(softmax(wh_t)) \qquad (8)$$

End-to-End Training: From Pixels to Answers We connect four modules through hidden states rather than symbolic outputs [26]. Therefore, the whole framework could be trained in an end-to-end manner, from pixels to answers. The training loss is simply the sum of losses from all four mod-ules. Each neural module receives supervision not only from the module's own loss, but also from the gradient signals backpropagated by downstream modules. We start from the pretrained weights of DETR for the object detection backbone and all other neural modules are randomly initialized.

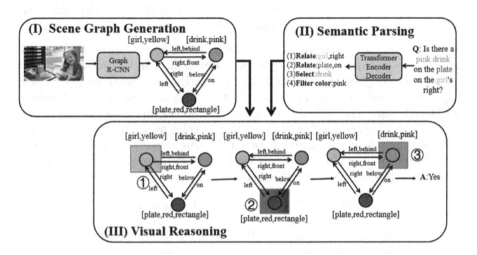

Fig. 1. The framework of LRRA model

4   Experiments

4.1   Dataset

We demonstrate the value and performance of our model on the "balanced-split" of GQA v1.1, which contains 1M questions over 140K images with a more balanced answer distribution. Compared with the VQA v2.0 dataset [27], the questions in GQA are designed to require multi-hop reasoning to test the reasoning skills of developed models. Compared with the CLEVR dataset [28], GQA greatly increases the complexity of the semantic structure of questions, leading to a more diverse function set. The real-world images in GQA also bring in a bigger challenge in visu-al understanding. Following [28], the main evaluation metrics used in our experiments are accuracy, validity and distribution.

4.2 Implementation Details

We first pre-trained DETR for object detection, and then fix the parameters in the backbone to train the scene graph generation model. SGD is used as the optimizer, with initial learning rate 1e–2 for both training stages. For question parser, we train with learning rate $7 \times 10^{-}4$ for 20,000 iterations. The batch size is fixed to be 64.

4.3 Results

We evaluated our method on the GQA dataset [6], which contains 1.5 million questions out of 1.1 million images. We use standard data set splitting. In the training process, we use the basic facts of the scene graph, reasoning explanation, and the traversal result of the scene graph for each step. During the test, we only used images and questions. We will present the state-of-the-art model LXMERT [1] as a baseline. We report the accuracy of the answers of LXMERT and LRRA.

VQA. We compare our performance both with baselines, as appear in [6], as well as with other prior arts of VQA model. Apart from the standard accuracy metric and the more detailed type-based diagnosis (i.e. Binary, Open), we get further insight into reasoning capabilities by reporting three more metrics [6]: Validity, and Distribution. The validity metric checks whether a given answer is in the question scope, e.g. responding some color to a color question. The distribution score measures the overall match between the true answer distribution and the model predicted distribution (for this metric, lower is better). As Table 1 shows, our model achieves competitive accuracy among.

Table 1. VQA results on GQA data sets.

Model	Accuracy	Distribution ↓	Binary	Validity
Language [6]	41.07	16.63	60.39	95.70
BottomUp [29]	49.74	5.60	65.64	94.13
MAC [10]	54.05	5.14	70.49	96.16
LCGN [30]	56.28	4.30	74.87	96.48
Vision [6]	18.93	19.27	36.05	–
LXMERT [1]	56.39	4.80	75.16	96.35
Ours	57.12	3.75	74.87	96.87

Table 2. The accuracy (%) of our question parser and symbolic executor. Program Acc. repre-sents the accuracy of generated program, which is evaluated by the accuracy of operation token, arguments token and the function (It is positive when both operation and arguments in a function are correct). Executor Acc. represents the accuracy of the answers obtained by our deterministic part of program executor executed on the ground-truth scene graph, by using ground-truth (G.T.) and generated (Gen.) program.

Table 2. The accuracy of LRRA model in question parser and symbolic executor

Data Split	Program Acc			Executor Acc	
	Operation	Arguments	Function	G.T	Gen
Testdev	96.65	80.49	81.34	-	-
Val	97.49	82.50	81.75	96.84	90.46

Re-posuton GQA Dataset and Additional Analysis. Finally, we use a comprehensive list of attributes obtained by [28] and mask them using a predefined mask token. For effectively masking relationships, we use Spacy POS-Tagger and mask verbs (VB) and prepositions (PRPN) from the question. The results are reported in Table 3. For attributes, we see that LRRA performance drops by 21.35% as compared to 8.07% drop in LXMERT, while for relations, the margin is more signifficant at 27.73% and 18.33% respectively, thus providing us a strong convergent evidence for our hypothesis that the relations between objects is more important for answering complicated questions than the attributes of objects.

Table 3. Perturbation analysis on testdev set.

Model	Acc drop(from→to)
Relations masked	
LXMERT	18.33% (55.49%→37.16%)
LRRA	27.73% (57.15%→29.39%)
Attributes masked	
LXMERT	8.07% (55.49%→47.42%)
LRRA	21.35% (57.15%→35.80%)

4.4   Example Analysis

Illustrative execution trace generated by our Neuro-Symbolic Concept Learner on the GQA dataset. Execution traces A and B shown in the figure leads to the correct answer to the question.

Our model effectively learns visual concepts from data. The symbolic reasoning process brings transparent execution trace and can easily handle quantities (e.g., object counting in Example A).

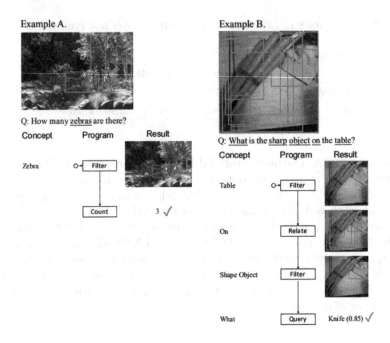

Fig. 2. Example of GQA data set generation

5   Conclusion

We present a transparent neural-symbolic reasoning framework for visual question answering, providing a human-readable form of justification at each step. The modular design of our methodology enables the whole framework to be trainable end-to-end. Our experiments on GQA dataset show that LRRA achieves high accuracy on answer generation task, outperforming the state-of-the-art LXMERT results. In addition, Our experiments on the perturbed GQA test set show that the relations between objects is more important for answering complicated questions than the attributes of objects. Furthermore LRRA performance drops significantly more than LXMERT, when object attributes and relationships are masked, hence indicating that LRRA makes a step forward, towards truly understanding the question.

Acknowledgements. The research work descried in this paper has been supported by the National Nature Science Foundation of China(Contract 61876198, 61976015, 61976016). The authors would like to thank the anonymous reviewers for their valuable comments and suggestions to improve this paper.

References

1. Tan, H., Bansal, M.: LXMERT: Learning cross-modality encoder representations from transformers. In Proceedings of the 2019 Conference on Empirical Methods in Natural Language Processing and the 9th International Joint Conference on Natural Language Processing (EMNLP-IJCNLP), pp. 5100–5111. Hong Kong, China, Association for Computational Linguistics, November 2019

2. Lu, J., Goswami, V., Rohrbach, M., Parikh, D., Lee, S.: 12-in-1: multi-task vision and language representation learning. In: 2020 IEEE/CVF Conference on Computer Vision and Pattern Recognition (CVPR) (2020)

3. Antol, S., et al.: Vqa: visual question answering. In: 2015 IEEE International Conference on Computer Vision (ICCV), pp. 2425–2433 (2015)

4. Goyal, Y., Khot, T., Summers-Stay, D., Batra, D., Parikh, D.: Making the v in VGA matter: elevating the role of image understanding in visual question answering. In: 2017 IEEE Conference on Computer Vision and Pattern Recognition (CVPR), pp. 6325–6334 (2017)

5. Agrawal, A., Batra, D., Parikh, D.: Analyzing the behavior of visual question answering models. In: Proceedings of the 2016 Conference on Empirical Methods in Natural Language Processing, pp. 1955–1960, Austin, Texas, Association for Computational Linguistics November 2016

6. Hudson, D.A., Manning, C.D.: Gqa: a new dataset for real-world visual reasoning and compositional question answering. In: 2019 IEEE/CVF Conference on Computer Vision and Pattern Recognition (CVPR) (2019)

7. Johnson, J., Hariharan, B., Maaten, L., Li, F.F., Girshick, R.: Clevr: a diagnostic dataset for compositional language and elementary visual reasoning. In: 2017 IEEE Conference on Computer Vision and Pattern Recognition (CVPR) (2017)

8. Santoro, A., et al.: A simple neural network module for relational reasoning. In: Guyon, I., et al. (eds.) Advances in Neural Information Processing Systems, vol. 30. Curran Associates Inc., Red Hook (2017)

9. Perez, E., Strub, F., De Vries, H., Dumoulin, V., Courville, A.: Visual reasoning with a general conditioning layer, Courville. Film (2017)

10. Hudson, D.A., Manning. C.D.: Compositional attention networks for machine reasoning. In :International Conference on Learning Representations (ICLR) (2018)

11. Andreas, J., Rohrbach, M., Darrell, T., Klein, D.: Neural module networks. In: 2016 IEEE Conference on Computer Vision and Pattern Recognition (CVPR), pp. 39–48 (2016)

12. Hu, R., Andreas, J., Rohrbach, M., Darrell, T., Saenko. K.: Learning to reason: end-to-end module networks for visual question answering. In: 2017 IEEE International Conference on Computer Vision (ICCV), pp. 804–813 (2017)

13. Johnson, J., et al.: Inferring and executing programs for visual reasoning. In: 2017 IEEE International Conference on Computer Vision (ICCV), pp. 3008–3017 (2017)

14. Mascharka, D., Tran, P., Soklaski, R., Majumdar, A.: Transparency by design: casing the gap between performance and interpretability in visual reasoning. In: 2018 IEEE/CVF Conference on Computer Vision and Pattern Recognition, pp. 4942–4950 (2018)

15. Hu, R., Andreas, J., Darrell, T., Saenko, K.: Explainable neural computation via stack neural module networks. In: European Conference on Computer Vision (2018)

16. Yi, K., et al.: Disentangling reasoning from vision and language understanding. In: Bengio, S., et al. (eds.) Advances in Neural Information Processing Systems, vol. 31. Curran Associates Inc., Red Hook (2018)

17. Johnson, J., et al.: Image retrieval using scene graphs. In: 2015 IEEE Conference on Computer Vision and Pattern Recognition (CVPR), pp. 3668–3678 (2015)
18. Teney, D., Liu, L., Hengel. A.: Graph-structured representations for visual question answering. In: 2017 IEEE Conference on Computer Vision and Pattern Recognition (CVPR) (2017)
19. Yin, X., Ordonez. V.: Obj2text: generating visually descriptive language from object layouts (2017)
20. Xu, D., Zhu, Y., Choy, C., Fei-Fei, L.: Scene graph generation by iterative message passing. In :Computer Vision and Pattern Recognition (CVPR) (2017)
21. Zellers, R., Yatskar, M., Thomson, S., Choi. Y.: Neural motifs: scene graph parsing with global context. In: 2018 IEEE/CVF Conference on Computer Vision and Pattern Recognition, pp. 5831–5840 (2018)
22. Li, Y., Ouyang, W., Zhou, B., Cui, Y., Wang, X.: Factorizable net: an efficient subgraph-based framework for scene graph generation. In: Ferrari, V., Hebert, M., Sminchisescu, C., Weiss, Y. (eds.) Computer Vision– ECCV 2018. ECCV 2018. Lecture Notes in Computer Science, vol. 11205, Springer, Cham (2018). https://doi.org/10.1007/978-3-030-01246-5_21
23. He, K., Zhang, X., Ren, S., Sun. J.: Deep residual learning for image recognition. In: 2016 IEEE Conference on Computer Vision and Pattern Recognition (CVPR), pp. 770–778 (2016)
24. Battaglia, P.W.: Relational inductive biases, deep learning, and graph networks (2018)
25. Xu, K., Hu, W., Leskovec, J., Jegelka. S.: How powerful are graph neural networks? (2018)
26. Liang, W., Tian, Y., Chen, C., Yu. Z.: Moss: end-to-end dialog system framework with modular supervision (2019)
27. Ren, S., He, K., Girshick, R., Sun, J.: Faster r-CON: towards real-time object detection with region proposal networks. IEEE Trans. Pattern Anal. Mach. Intell. **39**(6), 1137–1149 (2017)
28. Chen, W., Gan, Z., Li, L., Cheng, L., Wang, W., Liu, J.: Meta module network for compositional visual reasoning (2019)
29. Anderson, P., et al.: Bottom-up and top-down attention for image captioning and visual question answering. In: 2018 IEEE/CVF Conference on Computer Vision and Pattern Recognition, pp. 6077–6086 (2018)
30. Hu, R., Rohrbach, A., Darrell, T., Saenko, K.: Language-conditioned graph networks for relational reasoning. In: 2019 IEEE/CVF International Conference on Computer Vision (ICCV), pp. 10293–10302 (2019)

Linguistics and Cognitive Science

Meaningfulness and Unit of Zipf's Law: Evidence from Danmu Comments

Yihan Zhou[✉]

University of Illinois at Urbana-Champaign, Champaign, IL 61820, USA
yzhou114@illinois.edu

Abstract. Zipf's law is a succinct yet powerful mathematical law in linguistics. However, the meaningfulness and units of the law have remained controversial. The current study uses online video comments call "danmu comment" to investigate these two questions. The results are consistent with previous studies arguing Zipf's law is subject to topical coherence. Specifically, it is found that danmu comments sampled from a single video follow Zipf's law better than danmu comments sampled from a collection of videos. The results also suggest the existence of multiple units of Zipf's law. When different units including words, n-grams, and danmu comments are compared, both words and danmu comments obey Zipf's law and words may be a better fit. The issues of combined n-grams in the literature are also discussed.

Keywords: Zipf's law · Danmu · Internet language

1 Introduction

1.1 Zipf's Law

Zipf's law is an important empirical law describing the statistical properties of many natural phenomena. The law states that the frequency of a word in a given corpus has an inverse proportion with its frequency rank [53]. Ideally, the word that ranks first will be twice as frequent as the word that ranks second and so forth. This quantitative relation is captured in the following equation where f represents word frequency, r stands for the frequency rank, and C is a constant.

$$f = \frac{C}{r} \tag{1}$$

Mandelbrot proposed a refinement of Zipf's original equation by adding two constants m and β [33]. m is the shifted rank and β is the exponent which is estimated to be 1. The following equation shows Mandelbrot's revision:

$$f = \frac{C}{(r+m)^\beta} \tag{2}$$

Zipf's law is quite common in natural languages and language-related phenomena. It was found that the translated versions of the Holy Bible in one

© Springer Nature Switzerland AG 2021
S. Li et al. (Eds.): CCL 2021, LNAI 12869, pp. 239–253, 2021.
https://doi.org/10.1007/978-3-030-84186-7_16

hundred natural languages approximately follow Zipf's law [36]. Zipf's law also holds for artificial languages such as Esperanto, programming languages such as Python and UNIX [8,32,42].

Beyond languages, Zipf's law also has a wide coverage in physical, biological, and behavioral phenomena. Examples include city sizes, webpage visits, scientific citation numbers, earthquake magnitudes [12,15,39,41].

1.2   Remaining Questions in Zipf's Law

Zipf's law has been proposed for more than 70 years, yet a central question still persists: why the complex language production processes should conform to a mathematically concise equation [40]. While many studies have successfully demonstrated Zip's law in languages, very few explained the underlying cause of word frequency distribution [35]. It is thus crucial for any explanation of Zipf's law to make new predictions and have their assumptions tested with more data [40].

Another question that has been little addressed is the unit in Zipf's law [11]. In the literature, the majority of studies have used word as the frequency unit to derive Zipf's law [11]. However, word may not always be the right unit since the meaningful components in languages are a combination of words and phrases [48]. Moreover, word as an umbrella term can be difficult to define linguistically [14]. Even if we get by with words, empirical data show that other units such as phrases and combined n-grams sometimes fit Zipf's law better than words [17,48]. Therefore, it is important to compare how different units obey Zipf's law before taking word for granted as the default unit.

1.3   Goal of the Current Study

The current study uses danmu comments as data to explore the meaningfulness and unit of Zipf's law. It aims to answer three questions: 1. Do danmu comments follow Zipf's law? 2. Is there an optimal unit of Zipfian distribution in danmu comments? 3. What does the distribution of danmu comments imply for the meaningfulness of Zipf's law?

The study can contribute to research on Zipf's law in three ways. First, it extends Zipf's law to new data. Although it is tempting to assume that Zipf's law is universal, it is advocated that we should examine the data first [25]. In that case, we can identify new data which follow Zipf's law and reject false data that are assumed to exhibit Zipf's law. Second, it can advance our understanding of Zipf's law in internet language. Previous studies have examined how search terms in searching engine and tags in online blogs exhibit Zipf's law [7,28]. However, these items are not very different from words in regular texts in terms of length and composition. In comparison, danmu comments have features that are not commonly found in normal texts such as a large amount of code-mixing and neologisms. Finally, comparing how different units follow Zipf's law may provide indirect evidence to the cause and meaningfulness of Zipf's law.

1.4 Danmu Comments

Danmu comments, or danmaku in Japanese, is an emerging type of commentary system for online videos [50]. Danmu comments first appeared in Niconico, a Japanese video sharing website and spread to China afterwards [51]. Danmu comments are scrolling anonymous comments on the screen that allow participants to express feelings or opinions while watching a video [1]. When danmu comments become dense, they can cover the entire screen and create a visual impression resembling the artillery barrage in warfare. Therefore, this type of comments acquires the name "danmu" which literally means "barrage" in Chinese [9].

There are three characteristics of danmu comments worth pointing out. First, danmu comments are comprised of diverse symbols, including linguistic symbols, digits, punctuation, emoji, etc. [23]. Second, danmu comments often employed homophones called mishearing or soramimi [38]. Some danmu comments sound similar to what is said in the video, but convey different meanings. Third, danmu comments have independent meanings from the video such that users can be as much interested in the danmu comments as in the video itself [38]. Some users just watch a video for the sake of danmu comments.

2   Previous Work

2.1   Meaningfulness of Zipf's Law

Despite its seeming omnipresence, the origin of Zipf's law remains a controversy [6]. In particular, whether Zipf's law is meaningful has been heatedly debated. In the literature, there are mainly four accounts of Zipf's law in natural languages: random account, stochastic account, communication account, and semantic account [40].

The first account is the random account. This view demonstrates mathematically that random texts can exhibit Zipf's law and concludes Zipf's law is purely a statistical phenomenon without linguistic meanings [24,34]. For example, Li (2002) created random texts by inserting M + 1 symbols in each position with one of the symbols being the word boundary [25]. The random "words" were then extracted based on the word boundary symbol. Both mathematical proof and numerical simulation showed that the random "words" follow Zipf's law. Similarly, random sequences consist of symbols from a set of M symbols with equal probability are also shown to conform to Zipf's law [54].

The second view draws on a stochastic model. Simon (1955) postulated that a power-law distribution will take shape if new elements grow at a constant rate and old elements reoccur at a rate proportional to their probability in all elements that have appeared [45]. Similarly, Barabási and Albert (1999) proposed that growth and preferential attachment are the origin of scale-free power-law distribution [3]. On the one hand, new vertices are added as the network grows. On the other hand, new vertices prefer to attach to vertices that already have many connections. Removing either factor will eliminate the scale-free features in the network.

The third view argues the origin of Zipf's law is the results of optimal communication. Zipf (1949) proposed least effort is the fundamental principle governing all human actions [53]. This principle entails two types of economy: the speaker's economy which prefers to express all meanings with one word and the auditor's economy that favors a one-to-one mapping between meanings and word forms. Ultimately, Zipf's law is a vocabulary balance between these two conflicting forces. Cancho et al. (2003) implemented Zipf's idea of least effort with an energy function defined as the sum of the effort for the hearer and the effort for the speaker [6]. The model showed that Zipf's law is the compromise between the needs of the both the hearer and the speaker.

Finally, the semantic view presumes that word frequency is determined by semantics. It is argued that word meanings tend to expand and people are reluctant to use too many synonyms. Under the influence of these two forces, words meanings develop into a semantic space with multiple layers, which give rise to Zipf's law [35].

2.2   Units in Zipf's Law

Studies on Zipf's law in Indo-European languages have predominately used word as the unit. However, as mentioned before, the concept of word can be ambiguous. More importantly, empirical evidence shows that other linguistic units may also fit Zipf's law.

Kwapień and Drożdż (2012) studied the distribution of words and lemmas (i.e. the dictionary form of words) in one English text and one Polish text [22]. It was found that words and lemmas have similar distribution in the English text, but not in the Polish text. It was also pointed out that Zipfian-like scaling covers wider range in words than in lemmas.

In a follow-up study, Corral et al. (2015) conducted a large-scale comparison on how words and lemmas follow Zipf's law in single-authored texts of 4 different languages [11]. These languages range from morphologically poor language to morphologically rich language (i.e. English, Spanish, French, and Finnish). The authors found that Zipf's law holds for both word and lemmas.

Williams et al. (2015) compared different linguistic units with three kinds of text partition: (1) no whitespace serves as the word boundary and clauses remain clauses (2) each whitespace has 50% chance of being the word boundary and clauses are cut into phrases of one or more words (3) every whitespace is treated as the word boundary and clauses are segmented into words [48]. The results showed that both words and phrases yielded β (i.e. exponent of Zipf's law) close to 1, but phrases ($\beta = 0.95$) may be a better fit than words ($\beta = 1.15$).

Ha et al. (2009) extend Zipf's law to n-grams in English, Latin, and Irish with large corpora [17]. It was found that word frequency in English follows Zipf's law only up to the rank of 5000 while Latin and Irish words follow Zipf's law till the rank of 30000. In addition, none of the individual n-grams (i.e. bigrams, trigrams, 4-grams, 5-grams) in the three languages fit the Zipf's curve. However, when all the n-grams from unigrams to 5-grams are combined, the data in the three languages all become close to Zipf's curve for almost all ranks. The study suggested that combined n-grams are a better fit for Zipf's law than words.

For languages like Chinese and Tibetan, there are no whitespaces to mark the word boundary. As a result, multiple linguistic units exist in the written corpora. For example, apart from word, Chinese also has a conventional linguistic unit called "character". Character is the basic unit in Chinese writing system and one character often corresponds to one morpheme and one syllable [13].

Guan et al. (1999) compared Zipf's law in three linguistic units including characters, words, and bigrams and concluded that Zifp's law applies to all three units [16]. Another study also showed that Chinese characters are similar unit to English words in following Zipf's law using short texts [13]. However, neither studies clarified how well each unit fits. Wang et al. (2005) found that Chinese character frequency obeys Zipf's law in texts written before Qin dynasty but not anymore afterwards [46]. The authors attributed the change to the unification of Chinese characters in Qin dynasty, which leaves little room for the growth of new characters.

Ha et al. (2003) showed the distribution of single characters fall below the expected Zipf's curve in Mandarin news corpora [18]. However, bigram curve fits Zipf's law better than any other n-grams. Moreover, when all n-grams are combined, the data approximately followed Zipf's law for nearly all ranks. Chau et al. (2009) found similar patterns in the distribution of Chinese characters in web searching [7]. They found that bigrams fit the Zipfian distribution better than other n-grams. In addition, the combined n-grams also approximately follow Zipf's law.

In Tibetan, there are super character (i.e. a cluster of consonants and vowels), syllable (a combination of one to seven phonemes), and words [27]. It was found that syllable and word fit Zipf's law while super character does not, when n-grams from unigrams to 5-grams are combined.

There are also units beyond characters, words and phrases in internet languages. It was also shown that the distribution of hashtags on twitter follows Zipf's law [9,37]. The tags in Chinese blogs also approximately fit Zipf's law [28]. In addition, the number of microblog reposts on Sina Weibo obeys Zipf's law [52].

Some non-word symbols also obey Zipf's law. For example, the frequency of emoji used in the discussion of a topic on Chinese microblogging platform follows Zipf's law [26]. Punctuation in novels written in six Indo-European languages is very similar to words in obeying Zipf's law [21]. Williams et al. (2017) further showed that whitespace should also be considered as a word in Zipf's law [49]. Furthermore, both studies showed that when punctuation is added to the analysis, the discrepancy between the power-law and the shifted power-law is resolved.

3   Research Method

3.1   Data

The current study used two datasets. The first dataset contains longitudinal danmu comments in a single video and the second dataset includes danmu comments from different videos in 8 categories.

All data come from Bilibili.com, which is the most popular danmu-supported video sharing site in China. The 2020 fourth quarter and fiscal year financial results published by Bilibili Inc. showed that the average monthly active users (MAUs) reached 202 million and the average daily active users (DAUs) rose to 54 million [4].

The first dataset was collected by the author and the video was selected for three reasons (https://www.bilibili.com/video/BV1HJ411L7DP). First, it underwent an abrupt growth. The video was originally uploaded in January 2020, but did not become popular until November 2020. As of March 26, 2021, the video has been watched for more than 30.6 million times and the audience has created more than 90000 danmu comments. Second, the video and its danmu comments have popularized numerous internet buzzwords such as *haoziweizhi* 'mouse tail juice' and *bujiangwude* 'have no martial ethincs'. Finally, the video was relatively recent so the danmu comments can be traced back to the first day the video was published. The Bilibili official API was used to scrape the historical danmu comments between January 5 2020 and March 31 2021. After removing the duplicated items, the author obtained 48459 danmu comments, which is almost half as many as the total danmu comments in the video. Because the Bilibili official API only allows a maximum of 1000 danmu comments for each day and some danmu comments can repeat on different days, it is impossible to get the complete danmu comments.

The second dataset was compiled by the Big Data Lab at University of Science and Technology of China [30, 31]. 7.9 million danmu comments were crawled from 4435 videos in 8 categories: anime, movie, dance, music, play, technology, sport, and show. On average, each video provides 1786 danmu comments. Although the authors did not report, it is very likely that the second dataset only collects live danmu comments instead of the complete historical danmu comments as did in the first dataset.

It is worth noting that the two datasets have three major differences: topical homogeneity/heterogeneity, temporal homogeneity/heterogeneity, and size. First, one dataset comes from a single video while the other is extracted from videos of mixed categories. Second, the first dataset contains diachronic danmu comments but the second dataset only includes synchronic danmu comments. Finally, the second dataset is much larger than the first one.

3.2   Hypotheses

The current study postulates two hypotheses regarding the meaningfulness and unit of Zipf's law. The hypotheses will then be tested on the two danmu datasets.

The first hypothesis states that Zipf's law in languages is not a random process. Instead, Zipf's law must be associated with semantics because we use words to express meanings [35, 40]. This hypothesis is based on three arguments against the random account for Zipf's law.

First, almost all the studies generated random texts with the assumption that each symbol appears with equal probability and thus the frequency of a sequence should decrease monotonically with its length [35]. However, this is not the case for natural languages. It was shown that words with three letters are the most frequent in both English and Swedish [44]. Russian data also showed that words with five to ten letters are used most frequently [35]. Second, even though random texts may exhibit Zipf's law-like distribution, the distribution in random texts is not identical to real texts. When words are restricted to a certain length, random texts no longer have the Zipfian distribution [5]. Random texts can also be easily differentiated from natural texts by vocabulary growth [10]. Third, the words are used by human beings whose behaviors are far from random. Barabási (2005) argued that humans select tasks according to the priority, rather than acting randomly [2]. It was found that human behaviors are characterized by abrupt bursts and long waiting times between the bursts in email communication. This pattern is different from the regular inter-event time predicted by the Possion distribution, which assumes human activities are random. In addition, acting randomly turns out to be a very difficult task for most people [20]. In a psychological experiment, participants were asked to generate 600 random sequences using digits from 1 to 9 [43]. It was shown that even at the eighth sequence, participants' choice of digits can be predicted with an average accuracy rate of 27%, whereas the chance performance is 11%.

The second hypothesis assumes that there is a unit that best fits Zipf's law in linguistic data. The unit must be meaningful and directly observable. According to the first hypothesis, Zipf's law is closely related to meaning. Therefore, the unit in Zipf's law should be carriers of meaningful information. Moreover, it was suggested that direct measurements is more likely to yield Zipfian distribution than derived measurements [25].

3.3 Predictions for Danmu Comments

It is predicted that danmu comments will follow Zipf's law because danmu comments are meaningful and not random. Danmu comments are used to communicate opinions and emotions and may also contain meanings independent of the video content. In addition, He and al. (2017) showed danmu comments in the same video entail a herding effect and multiple-burst phenomena [19]. This pattern is similar to the burst phenomenon in Barabási (2005) which contends that human actions are not random [2]. In addition, danmu comments in a single video will fit Zipf's law better than those in mixed videos because danmu comments in a single video is more topically coherent. Williams et al. (2016) argued that Zipf's law occurs in topically coherent texts. For example, dictionaries, encyclopedias, subjects on questions and answers exhibit poor fit of Zipf's law [47].

In terms of the units, it is predicted that danmu comments or words are the unit of Zipf's law. There are at least three linguistic units: danmu comments, words and n-grams. Both danmu comments and words are meaningful and directly observable. It is thus expected that the frequency of danmu comments or words will conform to Zipf's law.

4   Results and Discussion

4.1   Danmu

In order to examine whether danmu comments follow Zipf's law, raw danmu comments were used for analysis. No prepossessing such as lowercase conversion was conducted. Table 1 shows the top 10 danmu comments in each dataset.

Table 1. Top 10 danmu comments in two datasets

Single video			Mixed video		
Danmu	Translation	Frequency	Danmu	Translation	Frequency
耗子尾汁	Mouse Tail Juice	2682	卧槽	WTF	22771
很快啊	Very fast	882	完结撒花	The end	17132
哈哈哈	Hahaha	579	233333	A loud laugh	15249
婷婷	Tingting	549	23333	A loud laugh	14118
吭	Onomatopoeia	522	2333333	A loud laugh	14117
全文背诵	Full text recitation	494	哈哈哈哈哈哈	Hahaha	13712
哈哈哈哈	Hahaha	390	哈哈哈哈	Hahaha	12904
不讲武德	No martial ethics	374	23333333	A loud laugh	12824
有bear来	A bear comes	309	哈哈哈	Hahaha	12020
万恶之源	The source of all evil	272	233333333	Hahaha	11671

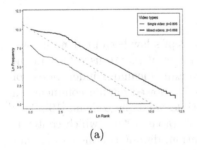

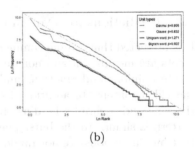

|(a)|(b)|

Fig. 1. Distribution of danmu, clauses, and words in log-log plots. Subfigure (a) shows the distribution of danmu in two datasets. Subfigure (b) shows the distribution of danmu, clause, and words distribution in a single video.

Note that the two datasets differ in diversity and size. The first dataset came from a single video while the second dataset was taken from more than 4000 videos. In addition, the first dataset has only 46754 items but the second dataset contains 7.9 million items. If danmu comments follow Zipf's law in the same way as random texts, we should expect to see the second dataset fits Zipf's law better because it has higher diversity and larger size.

There are two methods to fit Zipf's law: ordinary least squares (OLS) and maximum likelihood estimation (MLE). It was shown MLE fits better than OLS for both Chinese and English data [29]. In the current study, the Zipf's exponent β was obtained by fitting the log-transformed data using MLE. Mandelbrot's revision (Eq. 2) was used to derive the likelihood function below:

$$L(\theta|r) = \prod_{r=1}^{N} f_r * \frac{1/(r+m)^{\beta}}{\sum_{r=1}^{N} 1/(r+m)^{\beta}} \tag{3}$$

The frequency and rank of danmu comments were log-transformed and plotted in Fig. 1a. The exponent in single video danmu is 0.806 and the exponent in mixed video danmu is 0.668. The baseline value for β is 1, as represented by the dashed line.

The results suggest that danmu comments also approximately obey Zipf's law. More importantly, danmu comments in single video fits Zipf's law better. This finding is compatible with Williams et al. (2016), who proposed that Zipf's law is the result of coherent language production and thus topically coherent texts tend to fit Zipf's law better [47].

So far we have some evidence for the Zipfian distribution of raw danmu comments in the first dataset. In the next three sections, other linguistic units including clauses, words, and n-grams will be analyzed. Due to space constraints, only the results in the first dataset will be reported.

4.2   Clauses

Danmu comments have various lengths. Some are composed of clauses or sentences, such as "Did not dodge, it was funny" and "Why are you asking what's going on? Just look at the right eye". Thus, clauses can serve as an intermediate stage between danmu comments and words.

The current study uses comma, period, colon, semi-colon, question mark, exclamation mark, and ellipsis in both Chinese and English are as delimiters to cut danmu into clauses. The frequency of the clauses is fitted to Mandelbrot's revision (Eq. 2) using MLE. The exponent β of clauses is 0.832, slightly closer to 1 than the raw danmu.

4.3   Words

Although methods for Chinese word segmentation are mature, the current study used unigrams and bigrams of Chinese characters as a proxy to Chinese words.

This practice is adopted for two reasons. First, word-segmentation can be difficult to apply to mixed language. Code-mixing and language variants are very common in Internet language. In Table 1, is an example of code mixing and language variants can be found in , , of different lengths. Second, using characters allows the current study to be comparable to previous studies such as Chau et al. (2009), who investigated character usage in Chinese web searching [7].

Before analyzing word frequency, numbers and punctuation were removed from the danmu comments. Strings in Chinese, Japaneses, and Korean were segmented by character or syllable. String in Indo-European languages were segmented by whitespaces. Contractions such as "'s" and "n't" were restored to the original words as "is" and "not".

The unigrams and bigrams are a mixture of different languages, including Chinese, Japanese, Korean, English, French, German, etc. Linear regression was used to calculate Zipf's exponent for each linguistic unit. As shown in Fig. 1b, the exponent β in danmu, clauses, unigrams, and bigrams are 0.806, 0.832 1.271, and 0.922 respectively. Evidently, bigrams best fit Zipf's law. This finding is also consistent with previous studies [7,18,48]. The studies found that bigrams follow Zipf's law better than unigrams in Chinese and phrases fit Zipf's law better than words in English. Furthermore, the exponent in bigrams is less deviated from 1 than that in danmu comments, though both are close to 1. This may suggest that there are different units for Zipf's law in Chinese internet language, but bigrams or words are still the most basic linguistic unit.

The implications of the findings are twofold. It suggests that multiple linguistic units in the same data can fit Zipf's law, similar to the distribution of lemmas and words in Indo-European languages [11,22]. The findings also filled a gap in previous studies on Zipf's law in internet language. The internet language potentially involves different linguistic units, however, those different units were not directly compared in previous studies. For example, character usage in web searching was looked into without comparing it to the query terms [7]. Frequencies of hashtags and blog tags were also examined, but no comparison has been made with regard to characters or words that constitute the hashtags or tags [9,28,37].

4.4   N-grams

Another issue that needs to be addressed is n-grams. According to the hypotheses of unit in Zipf's law proposed in the current study, the unit has to be meaningful and directly observable. However, this is not always the case for n-grams. N-grams are derived data and sometimes do not make sense. Consequently, they should not be considered as the unit for Zipf's law. However, several studies reported that the combined n-grams fit Zipf's law for almost all ranks regardless of the languages and the unit chosen for creating n-grams [7,17,18].

Ha et al. (2009) provided an explanation for the behavior of combined n-grams with randomly generated bits [17]. The paper claimed that the extension of Zipf's law to n-grams may arise from pure probabilities and called for an re-examination of the theoretical motivation of Zipf's law.

To attest the argument, the current study compared separated n-grams and combined n-grams of different symbols, including words, numbers, and punctuation. The n-grams for these symbols were extracted from each danmu comment and then merged together.

The log-log plots for different symbols are shown in Fig. 2 and the Zipf's exponents were presented in Table 2. As can be seen in the table, the combined n-grams have smaller β values than unigrams in all three types of symbols. This characteristic is the same as reported in previous studies.

However, after taking a closer look, the advantage of combined n-grams is not so ubiquitous. In the column of punctuation n-grams, the combined n-grams are only better than the unigrams and bigrams. The β value of combined n-grams of punctuation is also much higher than that of words and number.

Punctuation usually includes one symbol such as "?" and some may contain two or three symbols such as "[ ]" and "!!!". It is hard to interpret what 4-grams and 5-grams of punctuation mean, yet their exponents are all closer to -1 than the combined n-grams. A possible explanation is that the combined n-grams may be a statistical smoothing. Because punctuation data have few 4-grams and 5-grams, the smoothing does not have a strong effect. On the contrary, the combined n-grams for words are much for effective.

In summary, combined n-grams are the best fit for Zipf's law for both words and numbers. However, the somewhat meaningless punctuation 4-grams and 5-grams fit Zipf's law better than the combined n-grams.

Table 2. Zipf's exponent for n-grams of different symbols

Types	Exponent in words	Exponent in numbers	Exponent in punctuation
1-gram	1.271	1.465	2.305
2-gram	0.922	0.764	1.708
3-gram	0.842	0.692	1.255
4-gram	0.775	3.282	1.250
5-gram	0.687	4.997	1.306
Combined	0.956	1.155	1.444

5 Conclusion

The current study explored Zipf's law with novel datasets from danmu comments, which is an emerging type of online video comments and internet language.

Specifically, the study has three findings. First, danmu comments also follow Zipf's law and danmu comments from topically homogeneous video fit Zipf's law better. The findings suggest that Zipf's law may be driven by semantic

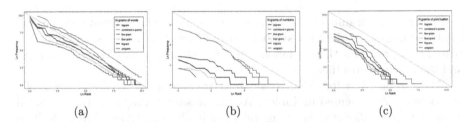

Fig. 2. N-grams of different symbols in log-log plots. Subfigure (a), (b), (c) shows N-grams of words, numbers, and punctuation respectively.

motivations. Second, danmu comments, clauses, and character unigrams and character bigrams in danmu comments all follow Zipf's law. There seems to be a continuum of fitness in the order of bigrams, clauses, and danmu comment. The results indicate that multiple units for Zipf's law may exist even in the same data. Finally, different from previous studies, combined n-grams do not always have the best fit for Zipf's law. It is argued that combined n-grams may be considered as a statistical smoothing rather than a manifestation of Zipf's law.

There are two limitations of the datasets. First, danmu comments are cumulative and may change over time. Therefore, a temporal analysis of danmu comments is necessary in the future. In addition, the danmu comments from a single video seems to fit Zipf's law. However, the dataset is relatively small and larger data from more videos are needed to validate the Zipfian distribution in danmu comments.

Internet has greatly changed the way we use languages and multilingualism is becoming increasingly popular. Studies on Zipf's law in internet language will help us understand the language phenomena better.

References

1. Bai, Q., Hu, Q.V., Ge, L., He, L.: Stories that big danmaku data can tell as a new media. IEEE Access **7**, 53509–53519 (2019)
2. Barabasi, A.L.: The origin of bursts and heavy tails in human dynamics. Nature **435**(7039), 207–211 (2005)
3. Barabási, A.L., Albert, R.: Emergence of scaling in random networks. Science **286**(5439), 509–512 (1999)
4. Bilibili: 2020 fourth quarter and fiscal year financial results (2021). https://ir.bilibili.com/static-files/09f30d5d-5de5-4338-b767-921ce1a07a47. 26 Mar 2021
5. Cancho, R.F.i., Solé, R.V.: Zipf's law and random texts. Adv. Complex Syst. **5**(01), 1–6 (2002)
6. Cancho, R.F.i., Solé, R.V.: Least effort and the origins of scaling in human language. Proc. Natl. Acad. Sci. **100**(3), 788–791 (2003)
7. Chau, M., Lu, Y., Fang, X., Yang, C.C.: Characteristics of character usage in Chinese web searching. Inf. Process. Manage. **45**(1), 115–130 (2009)
8. Chen, Y.S.: Zipf's law in natural languages, programming languages, and command languages: the Simon-Yule approach. Int. J. Syst. Sci. **22**(11), 2299–2312 (1991)

9. Chen, Y., Gao, Q., Rau, P.-L.P.: Understanding gratifications of watching danmaku videos – videos with overlaid comments. In: Rau, P.L.P. (ed.) CCD 2015. LNCS, vol. 9180, pp. 153–163. Springer, Cham (2015). https://doi.org/10.1007/978-3-319-20907-4_14

10. Cohen, A., Mantegna, R.N., Havlin, S.: Numerical analysis of word frequencies in artificial and natural language texts. Fractals **5**(01), 95–104 (1997)

11. Corral, Á., Boleda, G., Ferrer-i Cancho, R.: Zipf's law for word frequencies: word forms versus lemmas in long texts. PLoS ONE **10**(7), e0129031 (2015)

12. Cunha, C.R., Bestavros, A., Crovella, M.E.: Characteristics of www client-based traces. Technical report, Boston University Computer Science Department (1995)

13. Deng, W., Allahverdyan, A.E., Li, B., Wang, Q.A.: Rank-frequency relation for Chinese characters. Eur. Phys. J. B **87**(2), 1–20 (2014)

14. Dixon, R.M., Aikhenvald, A.Y.: Word: a typological framework. In: Dixon, R.M., Aikhenvald, A.Y. (eds.) Word: A Cross-Linguistic Typology, pp. 1–41. Cambridge University Press, Cambridge (2002)

15. Gabaix, X.: Zipf's law for cities: an explanation. Q. J. Econ. **114**(3), 739–767 (1999)

16. Guan, Y., Wang, X., Zhang, K.: Xiandai hanyu jisuan yuyan moxing zhong yuyan danwei de pindu—pinji guanxi [the frequency-rank relation of language units in modern chinese computational language model]. Zhongwen Xinxi Xuebao [J. Chin. Inf. Process.] **13**(2), 9–16 (1999)

17. Ha, L.Q., Hanna, P., Ming, J., Smith, F.: Extending Zipf's law to n-grams for large corpora. Artif. Intell. Rev. **32**(1), 101–113 (2009)

18. Ha, L.Q., Sicilia-Garcia, E.I., Ming, J., Smith, F.J.: Extension of Zipf's law to word and character n-grams for English and Chinese. Comput. Linguist. Chin. Lang. Process. **8**, 77–102 (2003)

19. He, M., Ge, Y., Chen, E., Liu, Q., Wang, X.: Exploring the emerging type of comment for online videos: danmu. ACM Trans. Web (TWEB) **12**(1), 1–33 (2017)

20. Iba, Y., Tanaka-Yamawaki, M.: A statistical analysis of human random number generators. In: Proceedings of the 4th International Conference on Soft Computing (IIZUKA 1996), Iizuka, Fukuoka, Japan, September, pp. 467–472. Citeseer (1996)

21. Kulig, A., Kwapień, J., Stanisz, T., Drożdż, S.: In narrative texts punctuation marks obey the same statistics as words. Inf. Sci. **375**, 98–113 (2017)

22. Kwapień, J., Drożdż, S.: Physical approach to complex systems. Phys. Rep. **515**(3–4), 115–226 (2012)

23. Li, R.: Shipin danmu de yuyanxue yanjiu [A linguistic analysis of danmu comments]. Master's thesis, Shaanxi Normal University (2018)

24. Li, W.: Random texts exhibit Zipf's-law-like word frequency distribution. IEEE Trans. Inf. Theory **38**(6), 1842–1845 (1992)

25. Li, W.: Zipf's law everywhere. Glottometrics **5**, 14–21 (2002)

26. Liu, F., Wang, H., Xu, X.: Shejiaomeiti zhong biaoqingfuhao de shiyong xingwei ji chengyin fenxi [usage behavior and cause of emoji in social media]. Fuza Xitong Yu Fuzaxing Kexue [Complex Syst. Complex. Sci.] **17**(3), 70–77 (2020)

27. Liu, H., Nuo, M., Wu, J.: Zipf's law and statistical data on modern Tibetan. In: Proceedings of COLING 2014, The 25th International Conference on Computational Linguistics: Technical Papers, pp. 322–333 (2014)

28. Liu, Z.: Chinese language networks research and applications. Master's thesis, Tsinghua University (2008)

29. Lu, G., Han, P., Shen, S.: Comparative empirical study on Zipf's law with two fitting methods. Libr. Inf. Serv. **56**(24), 71–76 (2012)

30. Lv, G., Xu, T., Chen, E., Liu, Q.F., Zheng, Y.: Reading the videos: temporal labeling for crowdsourced time-sync videos based on semantic embedding. AAAI, pp. 3000–3006 (2016)
31. Lv, G., Zhang, K., Wu, L., Chen, E., Xu, T., Liu, Q., He, W.: Understanding the users and videos by mining a novel danmu dataset. IEEE Trans. Big Data (2019)
32. Manaris, B.Z., Pellicoro, L., Pothering, G., Hodges, H.: Investigating esperanto's statistical proportions relative to other languages using neural networks and Zipf's law. In: Artificial Intelligence and Applications, pp. 102–108 (2006)
33. Mandelbrot, B.: An informational theory of the statistical structure of language. Commun. Theory **84**, 486–502 (1953)
34. Mandelbrot, B.B., Mandelbrot, B.B.: The Fractal Geometry of Nature, vol. 1. WH Freeman New York (1982)
35. Manin, D.Y.: Zipf's law and avoidance of excessive synonymy. Cogn. Sci. **32**(7), 1075–1098 (2008)
36. Mehri, A., Jamaati, M.: Variation of Zipf's exponent in one hundred live languages: a study of the holy bible translations. Phys. Lett. A **381**(31), 2470–2477 (2017)
37. Melián, J.A.P., Conejero, J.A., Ramirez, C.F.: Zipf's and benford's laws in Twitter hashtags. In: Proceedings of the Student Research Workshop at the 15th Conference of the European Chapter of the Association for Computational Linguistics, pp. 84–93 (2017)
38. Nakajima, S.: The sociability of millennials in cyberspace: a comparative analysis of barrage subtitling in Nico Nico Douga and Bilibili. In: Frangville, V., Gaffric, G. (eds.) China's Youth Cultures and Collective Spaces: Creativity, Sociality, Identity and Resistance, pp. 98–115. Routledge (2019)
39. Newman, M.E.: Power laws, pareto distributions and Zipf's law. Contemp. Phys. **46**(5), 323–351 (2005)
40. Piantadosi, S.T.: Zipf's word frequency law in natural language: a critical review and future directions. Psychon. Bull. Rev. **21**(5), 1112–1130 (2014)
41. Redner, S.: How popular is your paper? An empirical study of the citation distribution. Eur. Phys. J. B Condens. Matter Complex Syst. **4**(2), 131–134 (1998)
42. Sano, Y., Takayasu, H., Takayasu, M.: Zipf's law and Heaps' law can predict the size of potential words. Prog. Theor. Phys. Suppl. **194**, 202–209 (2012)
43. Schulz, M.A., Schmalbach, B., Brugger, P., Witt, K.: Analysing humanly generated random number sequences: a pattern-based approach. PLoS ONE **7**(7), e41531 (2012)
44. Sigurd, B., Eeg-Olofsson, M., Van Weijer, J.: Word length, sentence length and frequency-Zipf revisited. Studia linguistica **58**(1), 37–52 (2004)
45. Simon, H.A.: On a class of skew distribution functions. Biometrika **42**(3/4), 425–440 (1955)
46. Wang, D., Li, M., Di, Z.: True reason for Zipf's law in language. Phys. A Stat. Mech. Appl. **358**(2–4), 545–550 (2005)
47. Williams, J.R., Bagrow, J.P., Reagan, A.J., Alajajian, S.E., Danforth, C.M., Dodds, P.S.: Zipf's law is a consequence of coherent language production. arXiv preprint arXiv:1601.07969 (2016)
48. Williams, J.R., Lessard, P.R., Desu, S., Clark, E.M., Bagrow, J.P., Danforth, C.M., Dodds, P.S.: Zipf's law holds for phrases, not words. Sci. Rep. **5**(1), 1–7 (2015)
49. Williams, J.R., Santia, G.C.: Is space a word, too? arXiv preprint arXiv:1710.07729 (2017)
50. Wu, Q., Sang, Y., Huang, Y.: Danmaku: a new paradigm of social interaction via online videos. ACM Trans. Soc. Comput. **2**(2), 1–24 (2019)

51. Yao, Y., Bort, J., Huang, Y.: Understanding danmaku's potential in online video learning. In: Proceedings of the 2017 CHI Conference Extended Abstracts on Human Factors in Computing Systems, pp. 3034–3040 (2017)
52. Zhang, N., Zhang, S., Chen, H., Luo, Y.: Xinlang weibo zhuanfashu de milüfenbu xianxiang [the power-law distribution in the number of reposts in sina microblog]. Jisuanji Shidai [Comput. Era] **3**, 33–35 (2015)
53. Zipf, G.K.: Human Behavior and the Principle of Least Effort: An Introduction to Human Ecology. Addison-Wesley Press, Cambridge (1949)
54. Zörnig, P.: Zipf's law for randomly generated frequencies: explicit tests for the goodness-of-fit. J. Stat. Comput. Simul. **85**(11), 2202–2213 (2015)

Language Resource and Evaluation

Unifying Discourse Resources
with Dependency Framework

Yi Cheng[1], Sujian Li[1(✉)], and Yueyuan Li[2]

[1] MOE Key Lab of Computational Linguistics, School of EECS, Peking University,
Beijing, China
yicheng@pku.edu.cn, lisujian@pku.edu.cn
[2] School of Electronic Information and Electrical Engineering, Shanghai Jiao Tong
University, Shanghai, China
rowena_lee@sjtu.edu.cn

Abstract. For text-level discourse analysis, there are various discourse schemes but relatively few labeled data, because discourse research is still immature and it is labor-intensive to annotate the inner logic of a text. In this paper, we attempt to unify multiple Chinese discourse corpora under different annotation schemes with discourse dependency framework by designing semi-automatic methods to convert them into dependency structures. We also implement several benchmark dependency parsers and research on how they can leverage the unified data to improve performance (The data is available at https://github.com/PKU-TANGENT/UnifiedDep.).

Keywords: Discourse parsing · Discourse resources · Discourse dependency framework

1 Introduction

Discourse parsing aims to construct the logical structure of a text and label relations between discourse units, as shown in Fig. 1 and Fig. 2. Various discourse corpora have been developed to promote the discourse parsing technique. There exist multiple discourse schemes such as Rhetorical Structure Theory (RST) [1] and PDTB [13], which act as guidelines of various discourse corpora. At the same time, text-level discourse annotation is complicated and laborious, so the scale of a single discourse corpora is often much smaller compared with other NLP tasks.

One way to conquer data sparsity is to use different discourse corpora through multi-task learning. One way to conquer the data-sparsity problem is to multi-task learning multiple corpora simultaneously. Prior efforts on that mainly focus on discourse relation classification between two text spans, without considering the whole discourse structure [7,8], since one principle of multi-task learning is that the tasks should be closely related, but the discourse structures of different

© Springer Nature Switzerland AG 2021
S. Li et al. (Eds.): CCL 2021, LNAI 12869, pp. 257–267, 2021.
https://doi.org/10.1007/978-3-030-84186-7_17

corpora vary a lot, e.g. shallow predicate-arguments structure (e.g. the relations in Fig. 1) and deep tree structure (e.g. the CDTB tree in Fig. 2).

In this work, we explore another possible way to simultaneously leverage different corpora: we unify the existing discourse corpora under one same framework to form a much larger dataset. Our choice for this unified framework is dependency discourse structure (DDS), where elementary discourse units (EDUs) are directly related to each other without consideration of intermediate text spans. Because as [9] have argued, DDS is a very general discourse representation framework, and they adopt a dependency perspective to evaluate the English discourse corpora RST-DT [1] and the related parsing techniques.

Our work explore the feasibility of unifying discourse resources under dependency framework with consideration of the following two questions: (i) How to convert other discourse structures into DDS? (ii) How to make the best use of the unified data to improve discourse parsing techniques?

Oriented by the questions above, we unify three Chinese discourse corpora under dependency framework: HIT-CDTB [18], SU-CDTB [6] and Sci-CDTB [2]. HIT-CDTB adopts the predicate-argument structure similar to PDTB, with a connective as predicate and two text spans as arguments. Following rhetorical structure theory (RST), SU-CDTB uses a hierarchical tree to represent the inner structure of each text, with EDUs as its leaves and connectives as intermediate nodes. Sci-CDTB is a small-scale DDS corpus composed of 108 scientific abstracts. It is the only Chinese DDS corpus as far as we know.

The primary obstacle of unifying these corpora is inconsistency of the representation schemes, such as granularity of EDU and definition of relation types. Besides, the predicate-argument structure of HIT-CDTB leads to the problem that some discourse relations between adjacent text spans are absent, so that the information provided by the original annotation might be insufficient to form a complete dependency structure. To tackle these problems, we redefine granularity of EDU, conduct mapping among different relation sets, and design semi-automatic methods to convert other discourse structures into DDS. As the automatic part, we design the dependency tree transformation method for each corpora. As the manual part, we proofread all the segmentation of EDUs to follow the same definition, complement necessary information, and correct the transformed dependency trees.

Different from [9] who only consider conversion between RST and DDS, we attempt to unify more discourse schemes into dependency framework and explore whether discourse parsing techniques can be promoted by the unified dataset. Here we implement several discourse dependency parsers and research through experiment on how they can leverage the unified data to improve performance. Then we give out our findings about how to make better use of the unified discourse data.

Contributions of this paper are summarized as follows:

- we propose to integrate the existing discourse resources under a unified framework;

- we design a unified DDS framework and convert three Chinese discourse corpora into dependency structure in a semi-automatic way and get a unified large-scale dataset;
- we implement several discourse dependency parsers and explore how the unified data can be leveraged to improve parsing performance.

2 Background

In this section we mainly introduce the three Chinese discourse corpora: HIT-CDTB, SU-CDTB and Sci-CDTB, which we use in this work.

HIT-CDTB. Borrowing the discourse scheme of PDTB, HIT-CDTB adopts one-predicate two-arguments structure, where a connective serves as the predicate and two text spans as two arguments, as shown in Fig. 1. The connective can be either explicitly identified if it already exists in the original text, or implicitly added by annotators, while the arguments can be phrases, clauses, sentences, or sentence groups. Each connective corresponds to a relation type. In total, there are 4 coarse-grained types (i.e., *temporal causal, comparative* and *extension*) and 22 fine-grained types (e.g., *temporal* is further divided as *synchronous* and *asynchronous*). The documents cover multiple domains, such as news, editorials and popular science articles. When labeling each document, the discourse relations are labeled locally by only considering adjacent text spans, so a complete logical structure of the text may not be obtained.

SU-CDTB. Similar to RST, SU-CDTB represents the inner structure of a text with a hierarchical tree with EDUs as leaves, and uses connectives as intermediate nodes to indicate rhetorical relations. The whole text is first divided into several text spans which are recursively divided until getting EDUs. In SU-CDTB, EDUs are segmented according to the punctuation marks. The nucleus-satellite relation structure in RST is retained through arrows in the trees. Connectives in SU-CDTB are also given some relation attributes. 500 news documents from the Chinese Treebank [15] are annotated. A discourse tree is constructed for each paragraph rather than the entire document, with an average of 4.5 EDUs per tree, which are relatively shallow. Besides, the top-down constructed tree cannot cover some particular discourse structures, such as non-adjacent relations. So parsers trained only with this corpus are not very likely to analyze more complex discourse logic.

Sci-CDTB. Sci-CDTB is a small Chinese discourse dependency corpus, where EDUs are directly connected with discourse relations without intermediate levels. Its definition of EDU refers to RST-DT, with some modifications based on the linguistic characteristics of Chinese. The head of each EDU and the relations between them are annotated. Sci-CDTB is a small-scale corpus, only composed of 108 annotated scientific abstracts, so it is hard to support the training of a competitive discourse parser. It is the only Chinese DDS corpus to the best of our knowledge.

3 Unifying Discourse Corpora

In this section, we introduce how to convert the three discourse corpora into one unified framework, which mainly involves two aspects, i.e., EDU segmentation and dependency tree construction. Here, we adopt the EDU segmentation guideline of Sci-CDTB, which is similar to RST-DT [1]. Basic discourse units of SU-CDTB and HIT-CDTB are divided mainly based on punctuation marks, which is inconsistent with our EDU definition, so we manually re-segment their EDUs to ensure the same segmentation rules.

As for dependency tree construction, we should ensure correctness of both the tree structure and the relations between them. In a dependency tree, each relation connects a *head* EDU to a *dependent* EDU. Each EDU should have one and only one *head*. As the three corpora adopt different relation sets (22 relation types in HIT-CDTB$_{dep}$, 18 in SU-CDTB$_{dep}$[1] and 26 in Sci-CDTB), during conversion we keep the relations unchanged for each corpus, and then map them into 17 predefined relation types, which are basically the same as the ones of SU-CDTB, except that relation *example illustration* is merged into *explanation*. It requires to be further investigated whether there is a more appropriate mapping, and there exists the possibility that these relation sets have inherent incompatibility.

Experiments are conducted both before and after relation mapping.

3.1 Conversion of HIT-CDTB

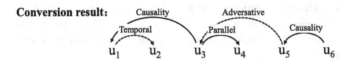

Inner-sentence relation:

S_1: [一渔船昨晚出海时]u_1 [冷动机突然发生故障。]u_2
When a fishing boat was sailing out at sea last night, its cold motor broke down.

S_2: [结果造成机长和 3 名渔工眼睛受伤,]u_3 [还$_{and}$合并有呕吐现象。]u_4
As a result, the captain and three fishermen suffered eye injuries, and kept vomiting.

S_3: [不过当时渔船已经失去了动力,]u_5 （所以$_{causality}$） [受伤的人员只能在渔船上等待救援。]u_6
But by then the boat had lost power, (so) the injured can only wait on the boat to be rescued.

Inter-sentence relation: $\underline{S_1}$ 结果$_{causality}$ $\underline{S_2}$

Conversion result:

Causality   Adversative
Temporal   Parallel   Causality
u_1 u_2 u_3 u_4 u_5 u_6

Fig. 1. An example text in HIT-CDTB and its conversion result of DDS. Underlined texts are arguments of labeled HIT-CDTB discourse relations. The dependency relations derived from the original corpus are represented with solid lines in the conversion result, while the ones complemented during conversion are with dotted lines.

---

[1] HIT-CDTB$_{dep}$ and SU-CDTB$_{dep}$ refer to the converted HIT-CDTB and SU-CDTB before relation mapping.

Text: [张三三十出头，]ₐ [既没有什么学历，]ᵦ [又没有多少工作经验，]ᵪ [但无论是什么的工作，]d
[他都尽力做好，]ₑ [_所以只要有重要的任务_f₁ _处长总是会交给他。_ f₂]f

[Zhang is in his early thirties.]ₐ [He has neither competive educational background]ᵦ [nor rich experience.]ᵪ [But
whatever task assigned to him,]d [he always tries his best to accomplish it.]ₑ [So as long as there is an important
job f₁ the director will delegate it to him. f₂]f

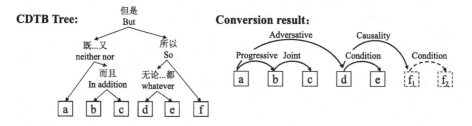

Fig. 2. An example SU-CDTB tree and its conversion result of DDS

Formally, a discourse relation in HIT-CDTB can be denoted as $R = <\overline{rel}_{HIT},$ $con,\ ARG1,\ ARG2>$, where $\overline{rel}_{HIT}$ is a relation type, con a connective, and $ARG1$, $ARG2$ are two arguments. The arguments can be phrases, clauses, sentences or sentence groups. We will use the example in Fig. 1 to elaborate the conversion process of HIT-CDTB.

For the example, HIT-CDTB labels two inner-sentence relations and one inter-sentence relation, which can not form a complete tree for the text. As preparation for DDS construction, EDU segmentation is conducted with an pre-trained segmenter [17], which are then manually checked to ensure quality. With EDUs as basic units, we utilize the original relations labeled in HIT-CDTB and complement some relations to construct a complete dependency tree. For instance, the relations derived from the original corpus and the ones complemented during conversion are respectively represented with solid lines and dotted lines in the conversion result of Fig. 1.

Specifically, assume that U_i is the EDU set covered by Arg_i $(i = 1, 2)$, the absent relations are **first** complemented to construct a complete dependency subtree. We generate these complemented relations in a rule-based way, by summarizing common discourse markers for each relation type. For example, discourse markers for _Temporal_ relation include "时" (when), "前" (before), "后" (after). **Then**, with two subtrees for U_1 and U_2, we replace U_1 and U_2 with their root EDUs as the arguments of R' (e.g., _Causality_ relation between u_1 and u_3 in Fig. 1). **Finally**, all the automatically-added relations are manually checked by two annotators.

3.2   Conversion of SU-CDTB

Like RST, SU-CDTB represents the inner structure of a text with a hierarchical tree, with EDUs as its leaves and connectives as its intermediate nodes. The "nucleus-satellite" structure in RST is retained through the arrows on the tree. In SU-CDTB, EDU segmentation is conducted according to punctuation marks. Figure 2 shows an example discourse tree in SU-CDTB and its conversion result of DDS.

The conversion process consists of three steps. **First**, as the annotation scheme of SU-CDTB is similar to that of RST-DT, we conduct conversion in a way similar to the algorithm of transforming RST-style discourse trees into DDS [4,5,10]. **Then**, we further subdivide some of the EDUs because EDU granularity in SU-CDTB is larger than our definition. **Finally**, we manually label the relations between newly-segmented EDUs, such as f_1 and f_2 in Fig. 2. As SU-CDTB constructs a complete tree for each text, the conversion process is relatively labor-saving.

3.3   Corpus Statistics

Size of HIT-CDTB$_{dep}$, SU-CDTB$_{dep}$ and Sci-CDTB are presented in Table 1. We also list the statistics of RST-DT and SciDTB [16], an English dependency corpus, for comparison. We can see that Sci-CDTB has a much smaller scale than HIT-CDTB$_{dep}$ and SU-CDTB$_{dep}$. HIT-CDTB$_{dep}$ and SU-CDTB$_{dep}$ have similar number of relations, but the average length of each document (531.7 characters, 28.6 EDUs) in HIT-CDTB$_{dep}$ is larger than that in SU-CDTB$_{dep}$ (94.8 characters, 4.8 EDUs). Size of the unified DDS corpus Unified$_{dep}$ is comparable to the two English discourse corpora.

Two annotators have participated in the manual labeling work. It takes them about 3 months and 3 weeks respectively to do the manual annotation and checking for HIT-CDTB and SU-CDTB. In comparison, it takes two annotators 3 months to build Sci-CDTB from scratch, which has a much smaller scale than the converted data, showing that converting existing resources is a relatively efficient way to generate dependency discourse dataset. Conversion of SU-CDTB is less time-consuming because its discourse structure is more complete than HIT-CDTB. For HIT-CDTB, annotators take much more time on EDU segmentation and tree completion, since this corpus is constructed bottom up and its scheme is more different from our discourse dependency framework. Table 2 shows distribution of the five most frequent relations in each corpus. We can see that the relation distribution in Sci-CDTB is quite different from the other corpora.

4   Dependency Discourse Parsing

Baselines. Following the work of [16], we implement four Chinese discourse dependency parsers:

Table 1. Corpus size comparison

Corpus	HIT-CDTB$_{dep}$	SU-CDTB$_{dep}$	Sci-CDTB	Unified$_{dep}$	SciDTB	RST-DT
#Doc	353	2332	108	**2793**	798	385
#Rel	9796	8181	1392	**19369**	18978	21787

Table 2. Distribution of the 5 most frequently-used relations in HIT-CDTB$_{dep}$, SU-CDTB$_{dep}$ and Sci-CDTB

Corpus	HIT-CDTB$_{dep}$		SU-CDTB$_{dep}$		Sci-CDTB		Unified$_{dep}$	
	Rel.	%	Rel.	%	Rel.	%	Rel.	%
Relation	Joint	46.4	Joint	53.2	Elaboration	29.3	Joint	52.7
Distribution	Explanation	8.7	Explanation	10.9	Joint	17.0	Explanation	16.7
	Progressive	8.5	Causality	8.3	Enablement	9.9	Causality	6.9
	Expression	8.1	Continuation	6.8	Bg-general	9.7	Continuation	4.3
	Causality	6.2	Goal	4.1	Evaluation	6.1	Progressive	4.1

- **Graph-based Parser** adopts Eisner algorithm to predict the most possible dependency tree structure [6]. It refers to the graph-based method of syntax parsing, which adopts Eisner algorithm and MST algorithm to predict the most possible dependency tree structure. For simplicity, an averaged perceptron is implemented to train weights.
- **Vanilla Transition-based Parser** adopts the transition method of dependency parsing proposed by [12], employing the action set of arc-standard system [11]. An SVM classifier is trained to predict transition action for a given configuration.
- **Two-stage Transition-based Parser** [14] first adopts the transition-based method to construct a naked dependency tree, and then uses another SVM to predict relations, which can take the tree-structure as features.
- **Bert-based Parser** also conducts parsing in a two-stage transition way, but in the second stage it uses a bert-based model, which incorporates BERT [3] with one additional output layer, to identify relation types. It keeps the pre-trained parameters[2] and is fine-tuned on our task using Adam.

Metrics. As for metrics, we use UAS and LAS to measure the dependency tree labeling accuracy without and with relation labels respectively. LAS$_O$ and LAS$_U$ denote using the original relation set of each corpus or the predefined unified relation set.

Results. Table 4 compares performance of the benchmark parsers. To give a rough upper bound of the parsing performance, the last row lists the consistency of human annotation by comparing two annotators' labelling results on

---

[2] https://github.com/huggingface/transformers.

Table 3. Division of training, validation and test set for the three corpora

	Train	Dev	Test
HIT-CDTB$_{dep}$	250	50	53
SU-CDTB$_{dep}$	1600	400	332
Sci-CDTB	68	20	20
Unified$_{dep}$	1918	470	405

Table 4. Performance comparison of benchmark parsers on different dependency discourse corpora

	HIT-CDTB$_{dep}$			CDTB$_{dep}$			Sci-CDTB		
	UAS	LAS$_O$	LAS$_U$	UAS	LAS$_O$	LAS$_U$	UAS	LAS$_O$	LAS$_U$
Graph-based	0.353	0.237	0.255	0.585	0.415	0.428	0.338	0.175	0.199
Vanilla Trans	**0.835**	0.551	0.588	**0.803**	0.580	0.588	**0.525**	0.276	0.299
Two-stage	**0.835**	**0.565**	**0.600**	**0.803**	0.587	0.597	**0.525**	0.276	0.299
Bert-based	**0.835**	0.564	0.576	**0.803**	**0.767**	**0.783**	**0.525**	**0.358**	**0.408**
Two-stage$_{Uni}$	0.813	-	**0.614**	0.802	-	0.591	**0.603**	-	0.393
Bert-based$_{Uni}$	0.813	-	0.606	0.802	-	0.637	**0.603**	-	0.220
Human	0.872	0.723	-	0.897	0.774	-	0.806	0.627	-

30 documents from each corpus respectively. Division of training, validation and test set is shown in Table 3. The first four methods in the first block show the results of the benchmark parsers which are trained and tested respectively on HIT-CDTB$_{dep}$, SU-CDTB$_{dep}$ and Sci-CDTB.

Graph-based method is less effective than the others, but it could probably be improved by using other training methods like MIRA. Among the transition-based methods, *Bert-based* performs the best on SU-CDTB$_{dep}$ and Sci-CDTB$_{dep}$ with respect to LAS, but is 2.4% lower than *Two-stage* on HIT-CDTB$_{dep}$, probably because HIT-CDTB$_{dep}$ is a multi-domain corpus, and feature-based methods are more robust to changes in different domains. Comparing the three corpora, parsing results on Sci-CDTB are the worst because Sci-CDTB is too small for supervised learning. As we expect, SU-CDTB performs the best since it has relatively shallow tree structure and its labeling is highly consistent. With original relations mapped to the unified relation set, LAS$_U$ results of all the methods on the three corpora have been improved compared with LAS$_O$, proving that the unified relation set we use is acceptable. Two-stage transition-based method performs better than the vanilla one with respect to LAS due to the addition of tree structural features in relation type prediction.

Two-stage$_{Uni}$ and Bert-based$_{Uni}$ use Unified$_{dep}$ as training data, and are tested on each corpus respectively. From Table 4, we can see that Sci-CDTB has a significant promotion of 7.8% on UAS, while HIT-CDTB$_{dep}$ and SU-CDTB$_{dep}$ slightly decline. One possible explanation is that documents in HIT-CDTB$_{dep}$

are much longer than in SU-CDTB$_{dep}$ so their structures are not similar enough to improve each other's performance. For Sci-CDTB, however, it also has a small numbers of EDUs per text so parsing performance can be improved by learning the augmented data of short trees from SU-CDTB.

For LAS$_U$, both methods increase on HIT-CDTB$_{dep}$, but drop on SU-CDTB$_{dep}$. As HIT-CDTB$_{dep}$ covers multiple domains and has an average of 28.6 EDUs per text, its annotation is more difficult and the DDS-conversion process is also more complicated. In comparison, with an average of 4.5 EDUs per tree, SU-CDTB only focuses on news domain and its DDS-conversion is relatively easy. As a result, SU-CDTB$_{dep}$ keeps a better data consistency than HIT-CDTB$_{dep}$, which may explain why the augmented data promotes the results on HIT-CDTB$_{dep}$, but introduces noise data for SU-CDTB$_{dep}$. As Sci-CDTB is from scientific domain and its relations are much different from those of HIT-CDTB and SU-CDTB, the augmented labeled relations do not bring much improvement on relation labeling.

Overall, Unified$_{dep}$ can serve as a discourse dataset for researching cross-domain or long text discourse parsing. However, our unification method may also introduce noise due to difference in text length and domain, so it remains to be considered how to better leverage this unified large corpus to improve the Chinese discourse parsing techniques.

5   Conclusions

In our work, we design semi-automatic methods to unify three Chinese discourse corpora with dependency framework. Our methods of converting PDTB-style and RST-style discourse annotation into DDS can be used to convert more other corpora and further enlarge our unified dataset. At the same time, by implementing several benchmark parsers on the converted data, we find that augmenting training set with the unified data can to some extent improve performance, but may also introduce noise and bring performance loss due to difference in text length and domain. This unified dataset is potentially helpful to research on cross-domain and long text discourse parsing, which will be our future work.

Acknowledgments. We thank the NLP Lab at Soochow University and HIT-SCIR for offering the discourse resources, *i.e.* CDTB and HIT-CDTB, to conduct the research. This work was partially supported by National Key Research and Development Project (2019YFB1704002) and National Natural Science Foundation of China (61876009). The corresponding author of this paper is Sujian Li.

References

1. Carlson, L., Marcu, D., Okurowski, M.E.: Building a discourse-tagged corpus in the framework of Rhetorical structure theory. In: Proceedings of the Second SIGDIAL Workshop on Discourse and Dialogue-Volume 16, pp. 1–10, September 2001
2. Cheng, Y., Li, S.: Zero-shot Chinese discourse dependency parsing via cross-lingual mapping. In: Proceedings of the 1st Workshop on Discourse Structure in Neural NLG, pp. 24–29, November 2019
3. Devlin, J., Chang, M. W., Lee, K., Toutanova, K.: BERT: pre-training of deep bidirectional transformers for language understanding. In: Proceedings of the 2019 Conference of the North American Chapter of the Association for Computational Linguistics: Human Language Technologies, Volume 1 (Long and Short Papers), pp. 4171–4186, June 2019
4. Hirao, T., Yoshida, Y., Nishino, M., Yasuda, N., Nagata, M.: Single-document summarization as a tree knapsack problem. In: Proceedings of the 2013 Conference on Empirical Methods in Natural Language Processing, pp. 1515–1520, October 2013
5. Li, S., et al.: Text-level discourse dependency parsing. In: Proceedings of the 52nd Annual Meeting of the Association for Computational Linguistics (Volume 1: Long Papers), pp. 25–35 (2014)
6. Li, Y., et al.: Building Chinese discourse corpus with connective-driven dependency tree structure. In: Proceedings of the 2014 Conference on Empirical Methods in Natural Language Processing (EMNLP), pp. 2105–2114 (2014)
7. Liu, Y., Li, S., Zhang, X., Sui, Z. (2016, February). Implicit discourse relation classification via multi-task neural networks. In Proceedings of the Thirtieth AAAI Conference on Artificial Intelligence (pp. 2750–2756)
8. Li, H., Zhang, J., Zong, C.: Implicit discourse relation recognition for English and Chinese with multiview modeling and effective representation learning. ACM Trans. Asian Low Resour. Lang. Inf. Process. (TALLIP) **16**(3), 1–21 (2017)
9. Morey, M., Muller, P., Asher, N.: A dependency perspective on RST discourse parsing and evaluation. Comput. Linguist. **44**(2), 197–235 (2018)
10. Muller, P., Afantenos, S., Denis, P., Asher, N.: Constrained decoding for text-level discourse parsing. In: Proceedings of the 24th International Conference on Computational Linguistics (COLING), December 2012
11. Nivre, J., Hall, J., Nilsson, J.: Memory-based dependency parsing. In: Proceedings of the Eighth Conference on Computational Natural Language Learning (CoNLL) at HLT-NAACL 2004, pp. 49–56 (2004)
12. Nivre, J.: An efficient algorithm for projective dependency parsing. In: Proceedings of the Eighth International Conference on Parsing Technologies, pp. 149–160, April 2003
13. Prasad, R., et al.: The Penn Discourse TreeBank 2.0. In: LREC, May 2008
14. Wang, Y., Li, S., Wang, H.: A two-stage parsing method for text-level discourse analysis. In: Proceedings of the 55th Annual Meeting of the Association for Computational Linguistics (Volume 2: Short Papers), pp. 184–188, July 2017
15. Xue, N., Xia, F., Chiou, F.D., Palmer, M.: The penn Chinese TreeBank: phrase structure annotation of a large corpus. Nat. Lang. Eng. **11**(2), 207 (2005)
16. Yang, A., Li, S.: SciDTB: discourse dependency TreeBank for scientific abstracts. In: Proceedings of the 56th Annual Meeting of the Association for Computational Linguistics (Volume 2: Short Papers), pp. 444–449, July 2018

17. Yang, J., Li, S.: Chinese discourse segmentation using bilingual discourse commonality. arXiv preprint arXiv:1809.01497 (2018)
18. Zhang, M., Qin, B., Liu, T.: Chinese discourse relation semantic taxonomy and annotation. J. Chin. Inf. Process. **28**(2), 28–36 (2014)

A Chinese Machine Reading Comprehension Dataset Automatic Generated Based on Knowledge Graph

Hanyu Zhao[1], Sha Yuan[1(✉)], Jiahong Leng[1], Xiang Pan[2], Zhao Xue[1],
Quanyue Ma[1], and Yangxiao Liang[1]

[1] Beijing Academy of Artificial Intelligence, Beijing, China
{hyzhao,yuansha,jhleng,xuezhao,maqy,yxliang}@baai.ac.cn
[2] New York University, New York, USA
xiangpan@nyu.edu

Abstract. Machine reading comprehension (MRC) is a typical natural language processing (NLP) task and has developed rapidly in the last few years. Various reading comprehension datasets have been built to support MRC studies. However, large-scale and high-quality datasets are rare due to the high complexity and huge workforce cost of making such a dataset. Besides, most reading comprehension datasets are in English, and Chinese datasets are insufficient. In this paper, we propose an automatic method for MRC dataset generation, and build the largest Chinese medical reading comprehension dataset presently named CMedRC. Our dataset contains 17k questions generated by our automatic method and some seed questions. We obtain the corresponding answers from a medical knowledge graph and manually check all of them. Finally, we test BiLSTM and BERT-based pre-trained language models (PLMs) on our dataset and propose a baseline for the following studies. Results show that the automatic MRC dataset generation method is considerable for future model improvements.

Keywords: Machine reading comprehension · Knowledge graph · PLMs

1 Introduction

Medical care is closely related to people's lives and helps people keep healthy in several different aspects, including disease prevention, medical examination, disease diagnosis, and treatment. Generally, various medical services are provided by those professionals who have specialized knowledge and rich experience. However, many countries now face severe medical personnel shortages,

The work is supported by the National Natural Science Foundation of China (NSFC) under Grant No. 61806111, NSFC for Distinguished Young Scholar under Grant No. 61825602 and National Key R&D Program of China under Grant No. 2020AAA010520002.

S. Li et al. (Eds.): CCL 2021, LNAI 12869, pp. 268–279, 2021.
https://doi.org/10.1007/978-3-030-84186-7_18

which means the demand for medical services dramatically exceeds the limit that professionals can supply. The rapidly developed artificial intelligence (AI) technology is a potential solution to the doctor shortage problem. AI medical experts are expected to offer various kinds of service for patients after learning enough knowledge from human experts and thus, reduce doctors' burden to a great extent.

One of the potential uses of AI experts is to provide medical consultation for patients. The technical nature of such kinds of applications is the automated question answering technology. Recently, the automatic question answering system is a significant development direction of the intelligent medical industry. Retrieval-based question answering and knowledge base question answering (KBQA) are two primary forms of present medical QA systems. Retrieval type system selects the candidate in the existing collection of QA pairs with the highest similarity to the user's input question. It returns the corresponding answer as the final response. As for KBQA, the system first extracts a topic entity in the user's question, then searches it in the knowledge base and returns the most suitable neighbor of a topic entity as the final answer. However, the above two methods both have poor performances when answering new questions that have not been appeared in the dataset or knowledge base of the QA system. Thus, they are not adaptive to the medical industry, where new things appear almost all the time. MRC provides a new approach to construct a QA system. It enables the machine to understand the meaning of contexts and answer related questions. The difference is that the MRC QA system answers questions based on its knowledge learned from a large volume of data rather than directly matching one from the dataset or knowledge base.

Although MRC systems have many advantages in question answering tasks, they are not developed a lot recently due to the lack of large-scale and high-quality datasets. At present, most MRC datasets are generated and annotated by humans, which is time-consuming and non-objective sometimes. In the medical industry, some data annotation work can only be done by those annotators with specialized knowledge, which also means a high economic cost. In this case, it is helpful to generate and annotate datasets automatically. Here, we propose a knowledge-graph-based automatic method to generate Q&A datasets and construct a medical dataset. We will show the details of the method in Sect. 3.

The main contributions of our work are as follows:

- We propose a reading comprehension dataset automatic generation method based on knowledge graph (KG) and pretrained language models (PLMs).
- By using this method, we release the largest Chinese Medical Reading Comprehension dataset.
- We propose several baseline models based on our medical dataset CMedRC, such as BERT, ERNIE, etc.

2   Related Work

2.1   English Machine Reading Comprehension Datasets

There are many reading comprehension datasets for Machine reading comprehension research. Though these datasets concentrate on different question types and refer to different corpora domains, most of them are in English. The following paragraph introduces several famous English datasets. CNN/Daily Mail [6] is a large one in the news, which contains approximately 1.4M fill-in questions. NewsQA [16] is another corpus that focuses on the news field, but it is relatively small, and most questions are the span of words. RACE [10] collects around 870K multiple-choice questions that appeared in Chinese middle school students' English reading comprehension examination. SQuAD2.0 [12] and TriviaQA [8] are collections of contents and question-answer pairs in Wikipedia. Besides the span of words, some questions in those two datasets are more complex, which need to be inferred by summarizing multiple sentences in corresponding contexts. CoQA [13] gathers its data mainly from children's storybooks. It contains various types of questions, including the span of words, yes/no, and even some unanswerable questions whose answers cannot be obtained according to the given context.

2.2   Chinese Machine Reading Comprehension Datasets

As for the Chinese dataset, HFL-RC [3] and C3 [14] are the first Chinese cloze-style and free-form multiple-choice MRC datasets respectively. The former is collected from fairy tales and newspapers, while the latter is based on documents in Chinese-as-a-second-language examinations. In the field of law, CJRC [2] is a famous dataset that contains more than 10K documents published by the Supreme People's Court of China. It contributes a lot to the research of judicial information retrieval and factor extraction. DuReader [5] is a large-scale and open-domain Chinese MRC dataset, which gathers 200K questions from Baidu Search and Baidu Zhidao with manually generated answers.

2.3   Machine Reading Comprehension Models

Considering the characteristics of Chinese MRC tasks, various models have been proposed for solving them. By using attention mechanism [9], AoA (attention over attention) [1], Gated Attention [4], Hierarchical Attention [19] and ConvAtt [22], some models have achieved considerable performance increase. From MRC task perspective, there are also various specific models that have been proposed recently [7,11,17,20,21].

However, most of the proposed methods and their corresponding datasets rely on humans. There are several disadvantages in efficiency and accuracy. Those methods involve resource-intensive processes such as the pre-defining question template and the linkage between question and their paragraph in documents. Those proposed fully automated annotation methods still require humans for

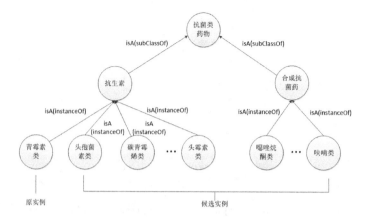

Fig. 1. Example of entity replacement using conceptual knowledge graph

evaluation and dataset cleaning. For those professional areas such as medical, the annotation and evaluation require professional knowledge. Those proposed methods are not usable. Furthermore, present methods are mostly straightforward and do not involve multiple-step reasoning and inferences.

3   CMedRC: A Chinese Medical MRC Dataset

CMedRC dataset is generated based on a chinese medical knowledge graph[1]. Knowledge graph is a summary and abstraction of existing human knowledge. With the upper and lower relationship in the knowledge graph, new questions can be generated by conducting replacement keywords in seed questions. During the generation process, SimBert is applied to create synonymous sentences of questions to guarantee the corpus' richness. Answers to questions are obtained from a knowledge graph, and corresponding documents that contain answers are crawled from Baidu Baike and Wikipedia. In the following subsections, we will introduce the dataset construction in detail.

3.1   Question and Answer Generation

For a given cypher (seed question), we can extract topic entities and relations from the sentence. Here, a relation means a piece of knowledge related to a topic entity in our medical knowledge graph. Thus, we can replace entities and/or relations in those seed questions to generate new questions automatically.

Entity Replacement. We conduct entity replacing based on a medical conceptual knowledge graph. For each topic entity in our knowledge graph, it has a hypernym and a class. Those topic entities that belong to the same hypernym

---

[1] http://cmekg.pcl.ac.cn/.

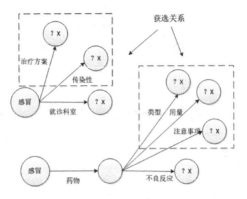

Fig. 2. Examples of relation replacement using conceptual knowledge graph

or other hypernyms under the same class form a candidate set. By selecting a suitable entity from the candidate set for substitution, a new question can be generated. For example, in Fig. 1, the seed question is "青霉素类药品的作用是什么? (What is the effect of penicillin?)". The original entity of cypher is "青霉素类 (penicillin)", which is an instance of its hypernym "抗生素 (antibiotics)". The original hypernym is a subclass of class "抗菌类药物 (antibacterial drug)". When conducting replacement, the candidate set includes other entities of original hypernym "抗生素 (antibiotics)" and entities of the hypernym that belongs to the same class "抗菌类药物 (antibacterial drug)". After entity substitution, one example of the new question is "头孢菌素类药品的作用是什么? (What is the effect of cephalosporin?)".

Relation Replacement. Similarly, a conceptual knowledge graph is used for relation replacement. After extracting a topic entity and its relation from a cypher, we can choose other relations of that entity in the knowledge graph to replace the original one. Thus, a new question can be generated. If there is a multi-step relation in the seed question, we substitute for the last step relation. Figure 2 shows relation replacement examples of the above two cases. For single-step relation case, the seed question is "感冒的就诊科室是? (Which clinic department should we go when having a cold?)". We can extract the entity "感冒 (cold)" and its relation "就诊科室 (clinic department)" from cypher. To make a substitution, we randomly choose one relation from the candidate set: the rectangle area enclosed with a dotted line. One example of a generated question is "感冒的治疗方案是? (What is the treatment method for cold?)". As for the multi-step case, our cypher is "感冒药物双黄连的副作用是? (What are the side effects of cold medicine ShuangHuangLian?)". Here, we have "感冒 (cold)", and "双黄连 (ShuangHuangLian)" two entities, their corresponding relations are "药物 (medicine)" and "副作用 (side effect)" respectively. When conducting replacement, only the last step relation "副作用 (side effect)" is substituted by a new relation of entity "双黄连 (ShuangHuangLian)". Thus, a new question can be "感冒药物双黄连的用量是? (What is the dosage of cold medicine ShuangHuangLian?)".

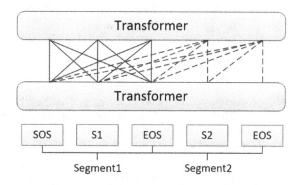

Fig. 3. The structure of SimBERT model

3.2 Synonymous Sentences Generation

To guarantee question corpus diversity, we use SimBERT to produce a synonymous sentence for those new questions after entity and/or relation replacement. SimBERT is a model trained by many synonymous sentence pairs and can predict labels according to the cosine similarity between contextual token embeddings. The structure of SimBERT is shown in Fig. 3. In our method, we generate ten synonymous sentences for each new question by SimBERT and randomly select one from those sentences to be the final version of the question.[2] For example, we use a previously generated question "感冒的治疗方案是? (What is the treatment method for cold?)" as the input, the modified new question can be "感冒的治疗方法都有哪些? " (What treatment method can be applied for cold?).

3.3 Answer Matching and Document Linkage

We need to match their answers for newly-generated questions and collect those documents containing QA pairs. First, we obtain answers directly from our medical conceptual knowledge graph. We exclude those questions without answers in the dataset. We can crawl the related paragraph from web pages such as Baidu Baike and Wikipedia for those well-matched QA pairs. Suppose the answer to a question cannot match the crawled paragraph exactly, we calculate the Levenshtein Distance between answer and each sliding window of that paragraph. If there is at least one distance larger than 0.7, we record corresponding window with largest distance and submit it for manual check. Otherwise, we delete that question from our dataset.

---

[2] https://github.com/ZhuiyiTechnology/simbert.

Table 1. Comparison of CMedRC with existing reading comprehension datasets

DataSets	Language	Domain	Answer type
RACE	ENG	English exam	Multi choise
SQuAD	ENG	Wikipe	Span of word, Unanswerable
NewsQA	ENG	CNN	Span of word
CNN	ENG	News	Fill in entity
HFRC	CHN	Fairy/News	Fill in word
CMRC	CHN	Wikipe	Span of word
CJRC	CHN	Law	Span of word, Yes/No, unanswerable
CMedRC	CHN	Medical	Span of word, Yes/No, unanswerable

Table 2. Dataset statistics of CMedRC

	Docments	Q&A pairs	Span of word	Yes/No	Unanswerable
Train	6000	13000	10002	2255	743
Test	2000	4000	3261	693	46

3.4 Dataset Cleaning

After answer and document matching, we can obtain a set of triples (Q, A, P). To guarantee the quality of the corpus, we conduct data cleaning before adding them to the dataset. Specifically, we check the generated questions and remove those with faulty or misleading wording. Even though our method is more labor-efficient, evaluation is more knowledge-reliable than the human generation.

3.5 Data Statistics

Our dataset consists of more than 8k medical documents and 17k QA pairs. It is the largest medical reading comprehension dataset. Table 1 compares our dataset with some other datasets mentioned in Sect. 2, mainly considering three dimensions: language, domain, and answer type. Table 1 shows that CMedRC contains more problematic types of questions, thus can evaluate the ability of algorithms from more perspectives.

In order to carry out the research better, we divide the dataset into training set and test set, and the data distribution is shown in Table 2. Additionally, Table 3 gives an example of a piece of data.

4   Experiment

4.1 Evaluation Metric

In this paper, we adopt two different evaluation metrics to measure the performance of the question-answering task. The metrics are Exaction Position Match and F1-Score.

Table 3. Example of data in CMedRC

Content	鼠疫 (plague) 是由鼠疫耶尔森菌感染引起的烈性传染病，属国际检疫传染病，也是我国法定传染病中的甲类传染病，在法定传染病中位居第一位。鼠疫为自然疫源性传染病，主要在啮齿类动物间流行，鼠、旱獭等为鼠疫耶尔森菌的自然宿主。鼠蚤为传播媒介。临床表现为发热、毒血症症状、淋巴结肿大、肺炎、出血。本病传染性强，病死率高。鼠疫在世界历史上曾有多次大流行，我国在解放前也曾发生多次流行，目前已大幅减少，但在我国西部、西北部仍有散发病例发生。肺鼠疫可以挂传染科。(Plague is a virulent infectious disease caused by Yersinia pestis. It is an international quarantinable infectious disease and a Class A infectious disease in China. Among all Chinese statutory infectious diseases, it ranks in the first place. Plague is a natural infectious disease, which is mainly prevalent among rodents. Mice and marmots are the natural hosts of Yersinia pestis. Rat fleas are the medium of transmission. The clinical manifestations are fever, toxemia, lymph node enlargement, pneumonia, and hemorrhage. The disease is highly infectious and has a high fatality rate. Plague has had many epidemics in world history, and many epidemics occurred in China before liberation. At present, the epidemic has been dramatically reduced, but sporadic cases still appear in the west and northwest of China. Pneumonic plague can be linked to the department of infection in the hospital.)
QA Pairs	{"id": "36_0","question": "肺鼠疫有哪些临床上的表现?" (What are the clinical manifestations of pneumonic plague?),"answers": "发热、毒血症症状、淋巴结肿大、肺炎、出血" (fever, symptoms of toxemia, lymph node enlargement, pneumonia, bleeding),} {"id": "36_1", "question": " 肺鼠疫的病因都有什么?" (What are the causes of pneumonic plague?), "answers": " 鼠疫耶尔森菌感染" (the infection of Yersinia pestis)}, {"id": "36_2","question": " 肺鼠疫通过什么途径传播?" (How does pneumonic plague spread?),"answers": " 鼠蚤为传播媒介" (Rat fleas are the medium of transmission.)}, {"id": "36_3","question": " 肺鼠疫翻译成英文是?" (What is the Engish translation of " 肺鼠疫"?), "answers": "plague"},

Extraction Position Match. For the extracted machine reading comprehension(MRC) task, we measure the position match using the exact match (EM). When the extracted beginning position and the end position are the same, the score is one. Otherwise, the score is zero. Such a measure can be regarded as a strict performance indicator.

F1-Score. For MRC tasks, if the starting or ending position is close but not the same, the answer is still valuable sometimes. Therefore, we also use F1-score, which can measure the characters' overlap between the prediction and ground truth. Specifically, the formula of F1-score is as follow:

$$Pre = \frac{TP}{LP} \qquad Rec = \frac{TP}{LG} \qquad F1 = \frac{2*Pre*Rec}{Pre+Rec}$$

Here, TP is the overlap length between prediction and ground truth, LP and LG are the length of prediction and ground truth, respectively. Moreover, if there are successive non-Chinese tokens, they are counted as one unit length rather than segmented into characters.

4.2 Baselines

For MRC tasks, there is usually a sequence model for contextual encoding and a task-specific layer for index prediction. We adopt traditional BiLSTM as our baseline model. For BERT-based PLMs, we test the normal BERT-based model and BERT-WWM (Whole Word Masking), which considers the Chines language characteristic. We also test the ERNIE [15] as the SOTA model for our evaluation. Finally, we provide estimated human performance as the human performance for reference.

Sequential Model. Traditional MRC model uses BiLSTM as the input sequence encoding layer and the final linear layer for the answer index prediction [18]. BiLSTM considering two directions of information flow, and it is lightweight and classic to be the baseline model.

BERT. As the representation of pre-training language models(PLMs), BERT is pre-trained on a large text corpus (like Wikipedia). Thus it can capture the general language features and be used as a general contextual embedding layer for downstream tasks. In MRC, BERT is used as the contextual encoding layer to get the sentence embedding. There are various versions of BERT, and we chose the word piece MLM version and whole word masking version (WWM) to be the Bert-based models in our experiment.

ERNIE. ERNIE (Enhanced Language Representation with Informative Entities) can learn abundant information from broad knowledge resources, including knowledge graphs. Thus it is more potent in knowledge-oriented downstream tasks, such as MRC. We use it as a SOTA model in the experiment.

4.3 Evaluation

Experimental results on the test set are shown in Table 4. From this table, it is obvious that the pre-trained language models (PLMs) such as BERT is about 20% points better than BiLSTM in EM and F1, which indicates that PLMs on large corpora can help it learn general language representation. Fine-tuning based on PLMs can improve downstream tasks' effects and avoid training models of downstream tasks from scratch. Besides, we also experiment with two improved BERT models—BERT_WWM and ERNIE. The table shows that BERT_WWM can be one percentage point higher than BERT_Base in EM by using Whole Word Masking. And ERNIE can be one percentage point higher in

Table 4. Experiment results

Model	EM	F1
Human	0.861	0.912
BiLSTM	0.532	0.623
BERT_Base	0.737	0.832
BERT_WWM	0.746	0.837
ERNIE	0.740	0.841

F1 by introducing knowledge into BERT. Finally, we also evaluate the results of Human Performance, which is approximately seven percentage points higher than PLMs. It implies that the automatically generated data can markedly improve models in future research.

5 Conclusion and Further Research

5.1 Conclusion

We present an automatic generation method for machine reading comprehension dataset. The method merely relies on an existing conceptual knowledge graph and a small number of seed questions to automatically construct a Chinese reading comprehension dataset. According to this approach, we construct the largest Chinese medical reading comprehension dataset presently, which contains more than 17k QA pairs and 8k documents. In comparison with those human-annotated datasets, our dataset's construction cost is reduced to a great extent. Next, we propose baseline models based on our dataset for BiLSTM and BERT-based PLMs. After comparing with human answers, the results show that those models have room for further improvement. Although our method is practiced in the medical area, such a method can be easily adopted and generalized into other areas.

5.2 Further Research

In our method, we only consider the one-step replacement. We can use similar methods for generated multi-hop reasoning reading comprehension datasets, which is a more challenging task for evaluating the MRC model and machine intelligence.

References

1. Cui, Y., Chen, Z., Wei, S., Wang, S., Liu, T., Hu, G.: Attention-over-attention neural networks for reading comprehension. In: Proceedings of the 55th Annual Meeting of the Association for Computational Linguistics (Volume 1: Long Papers), pp. 593–602. Association for Computational Linguistics (2017). https://doi.org/10.18653/v1/P17-1055. http://aclweb.org/anthology/P17-1055

2. Cui, Y., et al.: A span-extraction dataset for Chinese machine reading comprehension, p. 7

3. Cui, Y., Liu, T., Chen, Z., Wang, S., Hu, G.: Consensus attention-based neural networks for Chinese reading comprehension. In: Proceedings of COLING 2016, the 26th International Conference on Computational Linguistics: Technical Papers, pp. 1777–1786. The COLING 2016 Organizing Committee (2016). http://aclweb.org/anthology/C16-1167

4. Dhingra, B., Liu, H., Yang, Z., Cohen, W., Salakhutdinov, R.: Gated-attention readers for text comprehension. In: Proceedings of the 55th Annual Meeting of the Association for Computational Linguistics (Volume 1: Long Papers), pp. 1832–1846. Association for Computational Linguistics (2017). https://doi.org/10.18653/v1/P17-1168. http://www.aclweb.org/anthology/P17-1168

5. He, W., et al.: DuReader: a Chinese machine reading comprehension dataset from real-world applications. arXiv preprint arXiv:1711.05073 (2017)

6. Hill, F., Bordes, A., Chopra, S., Weston, J.: The goldilocks principle: reading children's books with explicit memory representations. arXiv preprint arXiv:1511.02301 (2015)

7. Hu, M., Peng, Y., Huang, Z., Qiu, X., Wei, F., Zhou, M.: Reinforced mnemonic reader for machine reading comprehension. In: Proceedings of the Twenty-Seventh International Joint Conference on Artificial Intelligence, IJCAI-18, pp. 4099–4106. International Joint Conferences on Artificial Intelligence Organization, July 2018. https://doi.org/10.24963/ijcai.2018/570

8. Joshi, M., Choi, E., Weld, D., Zettlemoyer, L.: TriviaQA: a large scale distantly supervised challenge dataset for reading comprehension. arXiv preprint arXiv:1705.03551 (2018)

9. Kadlec, R., Schmid, M., Bajgar, O., Kleindienst, J.: Text understanding with the attention sum reader network. In: Proceedings of the 54th Annual Meeting of the Association for Computational Linguistics (Volume 1: Long Papers), pp. 908–918. Association for Computational Linguistics (2016). https://doi.org/10.18653/v1/P16-1086. http://aclweb.org/anthology/P16-1086

10. Lai, G., Xie, Q., Liu, H., Yang, Y., Hovy, E.: RACE: large-scale reading comprehension dataset from examinations. In: Proceedings of the 2017 Conference on Empirical Methods in Natural Language Processing, pp. 785–794. Association for Computational Linguistics (2017). http://aclweb.org/anthology/D17-1082

11. Liu, T., Cui, Y., Yin, Q., Zhang, W.N., Wang, S., Hu, G.: Generating and exploiting large-scale pseudo training data for zero pronoun resolution. In: Proceedings of the 55th Annual Meeting of the Association for Computational Linguistics (Volume 1: Long Papers), pp. 102–111. Association for Computational Linguistics (2017). https://doi.org/10.18653/v1/P17-1010. http://www.aclweb.org/anthology/P17-1010

12. Rajpurkar, P., Zhang, J., Lopyrev, K., Liang, P.: Squad: 100,000+ questions for machine comprehension of text. In: Proceedings of the 2016 Conference on Empirical Methods in Natural Language Processing, pp. 2383–2392. Association for Computational Linguistics (2016). https://doi.org/10.18653/v1/D16-1264. http://www.aclweb.org/anthology/D16-1264

13. Reddy, S., Chen, D., Manning, C.: CoQA: a conversational question answering challenge. arXiv preprint arXiv:1808.07042 (2018)

14. Sun, K., Yu, D., Yu, D., Cardie, C.: Investigating prior knowledge for challenging Chinese machine reading comprehension. Trans. Assoc. Comput. Linguist. 8, 141–155 (2020)

15. Sun, Y., et al.: ERNIE: enhanced representation through knowledge integration. arXiv:1904.09223 [cs], April 2019. http://arxiv.org/abs/1904.09223. arXiv: 1904.09223

16. Trischler, A., Wang, T., Yuan, X., Harris, J., Sordoni, A., Bachman, P., Suleman, K.: NewsQA: a machine comprehension dataset. arXiv preprint arXiv:1611.09830 (2016)

17. Wang, S., Jiang, J.: Machine comprehension using match-LSTM and answer pointer. arXiv preprint arXiv:1608.07905 (2016)

18. Wang, S., Jiang, J.: Machine Comprehension Using Match-LSTM and Answer Pointer. arXiv:1608.07905 [cs], November 2016. http://arxiv.org/abs/1608.07905. arXiv: 1608.07905

19. Wang, W., Yan, M., Wu, C.: Multi-granularity hierarchical attention fusion networks for reading comprehension and question answering. In: Proceedings of the 56th Annual Meeting of the Association for Computational Linguistics (Volume 1: Long Papers), pp. 1705–1714. Association for Computational Linguistics (2018). http://aclweb.org/anthology/P18-1158

20. Wang, W., Yang, N., Wei, F., Chang, B., Zhou, M.: Gated self-matching networks for reading comprehension and question answering. In: Proceedings of the 55th Annual Meeting of the Association for Computational Linguistics (Volume 1: Long Papers), pp. 189–198. Association for Computational Linguistics (2017). https://doi.org/10.18653/v1/P17-1018. http://www.aclweb.org/anthology/P17-1018

21. Xiong, C., Zhong, V., Socher, R.: Dynamic coattention networks for question answering. arXiv preprint arXiv:1611.01604 (2016)

22. Yu, A.W., et al.: QANet: combining local convolution with global self-attention for reading comprehension. arXiv preprint arXiv:1804.09541 (2018)

Morphological Analysis Corpus Construction of Uyghur

Gulinigeer Abudouwaili[1,3], Kahaerjiang Abiderexiti[1,3], Jiamila Wushouer[1,3(✉)],
Yunfei Shen[2,3], Turenisha Maimaitimin[1,3], and Tuergen Yibulayin[1]

[1] School of Information Science and Engineering, Xinjiang University,
Urumqi, Xinjiang, China
107556518131@stu.xju.edu.cn, {jiamila,turgun}@xju.edu.cn
[2] School of Software, Xinjiang University, Urumqi, Xinjiang, China
shenyunfei@stu.xju.edu.cn
[3] Laboratory of Multi-Language Information Technology, Urumqi, Xinjiang, China
kaharjan@aliyun.com

Abstract. Morphological analysis is a fundamental task in natural language processing, and results can be applied to different downstream tasks such as named entity recognition, syntactic analysis, and machine translation. However, there are many problems in morphological analysis, such as low accuracy caused by a lack of resources. In this paper, to alleviate the lack of resources in Uyghur morphological analysis research, we construct a Uyghur morphological analysis corpus based on the analysis of grammatical features and the format of the general morphological analysis corpus. We define morphological tags from 14 dimensions and 53 features, manually annotate and correct the dataset. Finally, the corpus provided some informations such as word, lemma, part of speech, morphological analysis tags, morphological segmentation, and lemmatization. Also, this paper analyzes some basic features of the corpus, and we use the models and datasets provided by SIGMORPHON Shared Task organizers to design comparative experiments to verify the corpus's availability. Results of the experiment are 85.56%, 88.29%, respectively. The corpus provides a reference value for morphological analysis and promotes the research of Uyghur natural language processing.

Keywords: Morphological analysis · Corpus · Uyghur

1 Introduction

Morphological analysis is the process of dividing words into different morphologies or morphemes and analyzing their internal structure to obtain grammatical

Supported by the National Natural Science Foundation of China (grant numbers 61762084); and the Scientific Research Program of the State Language Commission of China (grant number ZDI135-54); the Opening Foundation of the Key Laboratory of Xinjiang Uyghur Autonomous Region of China (grant number 2018D04019).

S. Li et al. (Eds.): CCL 2021, LNAI 12869, pp. 280–293, 2021.
https://doi.org/10.1007/978-3-030-84186-7_19

information. It is an essential step in lexical analysis. Therefore, [1] showed that the related research of morphological analysis has also attracted the attention of most researchers. In natural language processing based on deep learning, researchers have found that if modeling to words directly, it is easy to ignore the relationship within the words, which will also bring limitations to the model. Thus, the model's input has changed from a single word to a character, subword, and morpheme or morphology. When the word is split into different granularities [2–4], the performance of the model is improved. Among these words' segmentation methods, morphology or morpheme relies on the morphological analysis tagger or manual annotation. In languages with mature natural language processing technologies such as English, Finnish, and Chinese, there are many manually annotated morphological analysis corpus, and the lexical analysis (including morphological analysis) technology of these languages has reached a high level. To further promote the lexical analysis of minority languages in Xinjiang, many researchers have obtained preliminary results in lexical analysis. [5] constructed a Uyghur lexical tagging corpus of 1.23 million words; [6] built a Uyghur stemming corpus of 10,000 sentence sets; [7] annotated about 30,000 Kazakh sentences and studied a lexical analysis of Kazakh; [8] constructed a character-level Uyghur morphological collaborative word corpus and annotated morphological analysis for 3500 sentences; for the first time, [9] released the Uyghur language based on morpheme sequence for morphological segmentation corpus; the corpus includes about 20,000 Uyghur words, including word-level and sentence-level corpus; based on the statistics of Uyghur noun affixes, [10] proposed a hybrid strategy for the Uyghur noun morphological Re-Inflection model; [11] constructed a small-scale stemming corpus for Uzbek, which includes 7435 words and 568 sentences in total.

Up until now, no reports have been published about the publicly available Uyghur of morphological analysis corpus. In addition, because there is no unified reference standard, the private morphological analysis datasets created by previous studies have fewer or incomplete features. This is not only directly affecting the development of the language's morphological analysis technology but also indirectly affecting the performance of downstream tasks, such as named entity recognition [12], syntactic analysis [13], text classification [14], and machine translation [15]. Therefore, to alleviate the shortage of Uyghur language morphological analysis resources, this paper refers to the format of universal morphological feature schema (UniMorph Schema), analyzes Uyghur words from 14 dimensions and 53 features, constructs a Uyghur language morphological analysis corpus, and provides lemma, part of speech. Morphological analysis tags, morphological segmentation, and lemmatization. This can be used for various lexical analysis tasks, like morphological analysis, stemming, and other downstream tasks of natural language processing.

2   Related Work

2.1   Agglutinative Language and Morphological Analysis

The world's languages can be roughly classified into four types [16], based on their morphology: isolated language, agglutinative language, fusional language, and polysynthetic language. The main feature of agglutinative language is that there is no inflection inside the word. A word is composed of several morphemes, and a morpheme is the smallest grammatical unit [17]. According to the different roles of morphemes in words, they are divided into root and affix. Among them, affix can be subdivided into word-forming affix and inflectional affix. Root and word-forming affix can form new words; the inflectional affix to words can change the grammatical category. The following will take Uyghur as an example:

ئوقۇ	+	غۇچى	=	ئوقۇغۇچى
(read)		(morphological affix)		(student)
ئوقۇغۇچى	+	لار	=	ئوقۇغۇچىلار
(student)		(configuration affix)		(students)

In the example, "ئوقۇ" is the root of the word, combined with the word-forming affix "غۇچى" will form a new word "ئوقۇغۇچى", and then combined with the plural affix "لار" will change the grammatical category and meaning of "ئوقۇغۇچىلار" change from a student to students.

Morphology [18] refers to the study of the internal structure of words and how they are formed. It refers to recognition of words' lemma [19], part of speech (POS), and morphological features of a word. Lemma refers to the condition where the word is not connected to affixes. For example, in English, words "write", "writes", "written" and "writing", all have the same lemma "write"; POS is defined according to syntactic function and morphological function. If words have a similar syntactic function, they can appear in similar contexts. Or they have affixes with similar morphological functions, and then they can be classified into one category. The morphological feature of a word is the grammatical categories (related grammatical information) attached to the lemma, such as number, case, tense, aspect, mood, person, and so on. The example is shown in Table 1.

Table 1. Lemma, POS, Morphological analysis for "shadows"

Word	Shadows
Lemma	Shadow
POS	V (verb)
Morphological analysis	Shadow; V; SG; 3; PRS

2.2   Morphological Analysis Corpus

The corpus is integral part of natural language processing tasks and is a collection of written or spoken material stored [20]. Early corpus research dates back before the 1950s, but its rise and development period can trace on the 1980s. In 1964, W. Nelson Francis and Henry Kučera of Brown University released the first machine-readable corpus and the first parallel corpus: the Standard Corpus of Present-Day Edited American English (the Brown Corpus)[1]. However, the Brown corpus was built early and roughly, it has always been the standard for English parallel corpus. Geoffrey Leech and Stig Johansson released the original version of the Lancaster Oslo/Bergen corpus (LOB corpus)[2] and the annotated POS version[3] in 1976 and 1986, respectively. In 1993, the University of Pennsylvania released the Penn Treebank (PTB) [21], which mainly annotated POS and syntactic component analysis; NEGRA corpus[4] is a German syntax annotated corpus constructed by Saarland University in Germany, and the second version has been released. There are about 350,000 words (20,603 sentences), mainly annotated POS, MSDs (morphosyntactic descriptions), the grammatical function in the directly dominating phrase, and the category of nonterminal nodes edge labeling. In 2013, Jan Hajič[5] released the Czech Morphological Dictionary, a spelling checker and lemmatization dictionary. The description contains not only traditional morphological categories but also some semantic, stylistic, and derivational information. Almaty Corpus of Kazakh language[6] is an online corpus containing about 40 million words. The corpus texts were marked utilizing the automatic morphological analyzer, 86% of word forms of the corpus were parsed. In terms of other agglutinative languages, Finnish has also released different dependency treebanks, which also annotate the morphological tag of words. For example, Turku Dependency Treebank (TDT)[7] and Finn Treebank (FTB), etc. TDT collects more than 10,000 sentences and about 180,000 words in different fields, respectively annotating lemma, POS, and MSDs; there are three versions of the FTB dependency treebank, the first version FTB1 is a manually annotated corpus, including 19,000 sentences or sentence fragments and about 160,000 words (including punctuations), and can be provided Online services. The corpus is mainly marked with lemma, POS, MSDs, dependency relationship, and sentence component analysis; the second version FTB 2 improves the first version, sentences and words are more than the previous version. For these, uses the same labeling format as the first version is adopted, the sentence components are manually annotated, and the MSDs uses three different analyzers, and finally, the results are manually verified; the third version, FTB 3, contains about 4.36 million sentences and about 76.36 million words. It has the functions

---

[1] http://icame.uib.no/brown/bcm.html.

[2] http://korpus.uib.no/icame/manuals/LOB/INDEX.HTM.

[3] http://korpus.uib.no/icame/manuals/LOBMAN/INDEX.HTM.

[4] http://www.coli.uni-saarland.de/projects/sfb378/negra-corpus/.

[5] http://ufal.mff.cuni.cz/morfflex.

[6] http://web-corpora.net/KazakhCorpus/search/index.php.

[7] https://bionlp.utu.fi/fintreebank.html.

of automatic morphology and dependent syntax analysis. The main difference between TDT and FTB is that [22] FTB includes various grammatical examples, while TDT is more of daily Expressions.

In 1994, the Association for Computational Linguistics (ACL) established the Special Interest Group on Computational Morphology and Phonology group (SIGMORPHON)[8] and regularly held morphology-related share tasks to promote further the basic research of natural language processing, which mainly models to words, lemma, and MSDs. From the share tasks in the past five years, it is found that most of their data sets are provided by Universal Morphology (UniMorph) and Surrey Morphology Group; among them, UniMorph is used widely. This corpus, published by the Center for Language and Speech Processing (CLSP) of Johns Hopkins University, currently contains a morphological analysis corpus of 118 languages and annotated more than 23 dimensions of meaning with over 212 features, and UniMorph3.0 [23] released in May 2020 is by far the largest high-quality morphological analysis corpus. Also, other share tasks selected the Universal Dependencies (UD) as datasets. UD provides 92 languages of POS, MSDs, and syntactic dependencies. It is an improvement based on Universal Stanford dependencies, Google universal part-of-speech tags, and the Interset interlingua for morphosyntactic tagsets, using CoNLL-X [24] format to annotate each word (CoNLL-X format mainly includes word ID, word form or punctuation, lemma, UPOSTAG, XPOSTAG, morphological features FEATS, the central word HEAD of the current word, dependency relationship DEPREL, second-level dependency (head-deprel pair) DEPS and others label MISC). Since the corpus's existence related to morphological analysis, researchers have also released many morphological analyzers, such as Morfessor[9], UDPipe[10], Omorfi, and MorphoDiTa[11]. Morphological analysis corpus currently available for Uyghur, Kazakh, and Uzbek languages. But there are some problems of those languages' corpus, such as the corpus has a one kind of POS, a small amount of data, or incomplete tags, and so on.

3   Construction of Uyghur Morphological Analysis Corpus

This chapter will introduce the construction process of the Uyghur morphological analysis corpus. Firstly, a more suitable morphological annotation guidelines is proposed based on [25] and [26]. Secondly, the collected dataset is preprocessed. Finally, considering the context, the whole dataset is annotated by human-machine interaction.

---

[8] https://sigmorphon.github.io/.
[9] http://morpho.aalto.fi/projects/morpho/.
[10] https://bnosac.github.io/udpipe/en/.
[11] http://ufal.mff.cuni.cz/morphodita.

3.1 Data Preparation and Preprocessing

We crawled Uyghur news articles from TianshanNet[12] and NurNet[13] as raw corpus and preprocesses them. It is including finance, lottery, technology, tourism, society, fashion, sports, entertainment, and other fields. The preprocessing mainly includes removing web page tags, to punctuate sentences, and word segmentation, and finally selecting 5014 sentences with high quality. These 5014 sentences are used to construct the morphological analysis corpus. These sentences are used to construct the morphological analysis corpus.

3.2 The Annotation Scheme

This paper makes a more fine-grained morphological analysis and annotation of each word based on the previous annotation guidelines. The corpus' annotation scheme will be explained below: the annotation takes sentences as the context environment and words as the basic unit. The annotation mainly includes the current word, lemma, POS, MSDs, morphological segmentation, and lemmatization. The grammatical categories and morphological tags are shown in Table 2.

Table 2. Grammatical features and morphological tags

Grammatical features	Morphological tags
Parts of speech	N, ADJ, NUM, CLF, ADV, PRO, IMI, V, ADP, CONJ, PART, INTJ, AUX, Y, X, LW;
Case	NOM, GEN, ACC, ALL, LOC, ABL, LQ, LMT, SML, EQUI;
Possession	PSSPNO;
Person	1, 2, 3;
Number	PL, SG
Aspect	PROG, PROSP, ABIL, ITER, DISP, SELF;
Voice	CAUS, PASS, REFL, PECP;
Polarity	NEG;
Participle	PTCP;
Masdar	MSDR;
Converb	CVB;
Tense	PST, NPST;
Mood	IND, INT, COND, IMP, OPT;
Politeness	POL;

– Lemma: give the valid word;
– POS: mainly divided into noun, adjective, numeral, classifier, adverb, pronoun, imitation word, verb, postposition, conjunction, particles, interjection, auxiliary verb, in addition to punctuation, additional ingredients, and Latin;

---

[12] http://uy.ts.cn/.
[13] https://www.nur.cn/index.shtml.

- MSDs: including POS and grammatical features. The grammatical features are expressed through inflectional affixes, mainly including plural affix, possession affixes, case affix, voice affix, aspect affix, negative affix, masdar affix, participle affix, converb affix, tense affix, and mood affix;
- Morphological segmentation: according to different grammatical features, each type of affix is segmented in a fine-grained way;
- Lemmatization: restore the affix obtained by morphological segmentation to obtain a valid affix.

When formulating tags, this article refers to the morphological tags proposed by [25]. However, these tags are universal tags and do not include special grammatical features. Therefore, new tags are also made for grammatical features that are not in UniMorph.

3.3   Corpus Construction Process

After preprocessing the data crawled from news websites, a dataset based on sentences is obtained. To reduce manual annotation workload, POS tagging [27], and stemming [17] are performed on the dataset. Since POS used in this article is not the same as the first-level part-of-speech proposed by [27], the tagging set is modified. The morphological analysis corpus does not provide stemming result, but it is valuable for constructing the corpus.

After the labeled data are initially obtained, the data is manually annotated. Three students took part in the manual annotation. First, students should understand annotation scheme. Secondly, they conduct small-scale annotation and check each other. Then refer to the annotation scheme, they discuss the inconsistencies or ambiguous annotations and achieve a consensus. Finally, annotate the data in batches. To ensure the consistency of the annotated data, after the annotation, each student checks the annotation result of the other two students and submits the annotated data after it is true. An example of annotation is shown in Fig. 1.

4   Corpus Information Statistics and Evaluation

The statistical distribution of words and sentences in each domain is shown in Fig. 2, which including 5014 sentences, 152669 words, 24631 tokens. The average word repetition rate is 1:6, and it can be found that each sentence contains an average of 30 words. Compared with other categories, the number of news in finance is most, sport news has the largest number of words, and entertainment news has more tokens. Figure 3(a) and (b) respectively represent the distribution of sentence length and word length in the corpus. The abscissa represents the length of the statistics, and the ordinate represents the number of statistics. Figure 3(a) sentence length statistics figure, most of the sentence length is in the (40,140), the shortest sentence length is 3, the longest is 195. Figure 3(b) word length statistics figure, most of the word length is (1, 10), the shortest word length is 1, and the longest word length is 33.

sentence	سانلىق مەلۇماتتا سىنا تورىنىڭ پىرىۋوت سانلىق مەلۇماتى ئاساس قىلىندى .
Translation	The data is based on Sina's exchange data .

word	stem	POS	MSDs	stemming	lemmatization
سانلىق	سانلىق	ADJ	ADJ;SG;NOM	سانلىق	سانلىق
مەلۇماتتا	مەلۇمات	N	N;SG;ON	مەلۇمات+تا	مەلۇمات+تا
سىنا	سىنا	N	N;SG;NOM	سىنا	سىنا
تورىنىڭ	تور	N	N;SG;PSS3S;GEN	تور+ى+نىڭ	تور+ى+نىڭ
پىرىۋوت	پىرىۋوت	N	N;SG;NOM	پىرىۋوت	پىرىۋوت
سانلىق	سانلىق	ADJ	ADJ;SG;NOM	سانلىق	سانلىق
مەلۇماتى	مەلۇمات	N	N;SG;PSS3S;NOM	مەلۇمات+ى	مەلۇمات+ى
ئاساس	ئاساس	N	N;SG;NOM	ئاساس	ئاساس
قىلىندى	قىل	V	V;PASS;PST;3;SG;IND	قىل+ىن+دى	قىل+ىن+دى
.	—	Y	Y	—	—

Fig. 1. An example of annotation

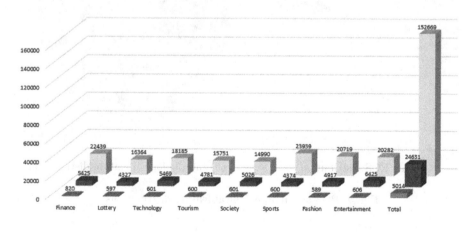

Fig. 2. The number of words and sentences in each field

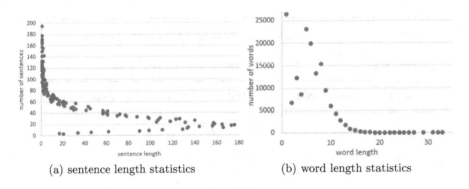

(a) sentence length statistics   (b) word length statistics

Fig. 3. Sentence and word length statistics

In general, a sentence contains at least one verb and several nouns, and there are a small number of conjunctions or interjections, etc. Each type of word has a different grammatical category, and as the number of words increases, its grammatical features will increase. According to the morphological labels, the POS distribution, morphological feature distribution, and morphological label distribution of some words are respectively counted, as shown in Figs. 4(a), (b), and 5.

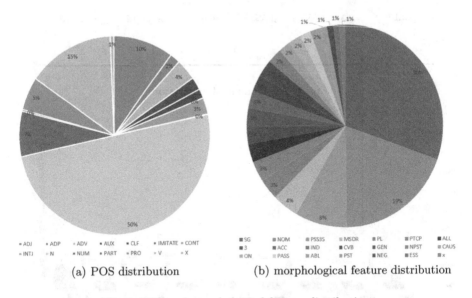

(a) POS distribution   (b) morphological feature distribution

Fig. 4. POS and morphological feature distributions

From the part of speech distribution in Fig. 4(a), we found that the proportions of noun (N), adjective (ADJ), pronoun (PRO), verb (V), and auxiliary verb (AUX) are more than the proportions of classifier (CLF), interjection (INTJ)

and imitate (I). From the distribution of lexical features in Fig. 4(b) and the morphological distribution label in Fig. 5, we observed that the proportion of grammatical labels that modify nominal words is the largest, such as SG, NOM, and PSS3S, etc., followed by verbs or auxiliary verbs, such as MSDR, PTCT, and so on. This statement can be further verified from Fig. 5. Also, it proves that the morphology of verbal words is more complex and richer than nominal words.

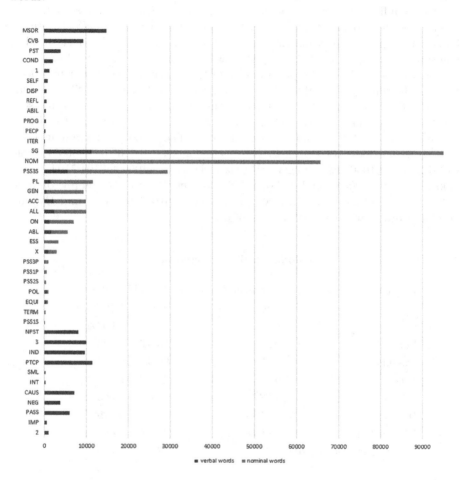

Fig. 5. MSDs distribution

To verify the effectiveness of the morphological analysis dataset constructed in this paper, we select part of the agglutinative or morphologically complex languages and use baseline models provided by the SIGMORPHON Shared Task. And we design a comparative experiment.

Evaluation Task Definition. Given the word, lemma, and morphological label to train model, so that the model predicts the word that by the lemma and morphological labels, such as given the lemma "shadow" and morphological label "V;SG;3;PRS" to predict words "shadows".

The experiment uses two modes provided by the task organizers, a non-neural baseline [28] and a neural baseline [29]. The neural baseline is a multilingual transformer. The version of this model adopted for character-level tasks currently holds the state-of-the-art on the 2017 SIGMORPHON shared task data. The transformer takes the lemma and morphological tags as input and outputs the target inflection. The non-neural baseline heuristically extracts lemma-to-form transformations; it assumes that these transformations are suffix-or prefix-based. A simple majority classifier is used to apply the most frequent suitable transformation to an input lemma, given MSDs, yielding the output form.

Table 3 shows the model performance of the Czech, Polish, Russian, Sakra, Eibela, Hebrew, and Uyghur datasets on the SIGMORPHON Share Task, with the first six languages provided by the organizers, the last two datasets constructed in this paper, and nominal words in the dataset filtered out to build the Uyghur nominal words dataset. The number of training sets, test sets, and development sets is Czech datasets (94169:6659:6659), Polish datasets (100039:8023:6023), Russian datasets (100002:7104:7104), and Sakra datasets (100046:6971:7098), Eibela dataset (918:64:66), Hebrew dataset (23204:1640:1642), Uyghur ALL Word dataset (19704:2463:2463) and Uyghur nominal words dataset (13652:1706:1706).

Table 3. Result of experiments

Languages	Non-neural baseline	Neural baseline	
	Test	Test	Development
Czech	92.81%	96.95%	96.92%
Polish	94.13%	99.54%	99.33%
Russian	88.72%	96.33%	96.34%
Sakha	90.54%	95.51%	95.49%
Eibela	4.68%	4.68%	13.63%
Hebrew	36.16%	99.02%	99.09%
Uyghur (ALL)	85.56%	88.29%	94.00%
Uyghur (nominal words)	89.25%	92.77%	96.00%

Experimental results can be found that (1) the performance of the neural model is superior to the performance of the non-neural model, such as Czech, Hebrew, and Uyghur, which are respectively 4.14%, 62.86%, and 2.76% higher than the non-neural model. (2) In the languages with large datasets, such as Czech, Polish, Russian, and Sakha, the accuracy rates of the non-neural model are 92.81%, 94.13%, 88.72%, and 90.54%, and the accuracy rate of the neural model respectively 96.95%, 99.54%, 96.33%, 95.51%. However, for languages with small datasets, such as Eibela, the effect of the two models was not noticeable. Because the training dataset is more extensive, and coverages is broader, the model has a strong learning ability; when the scale of the training set is large enough, the learning ability of the neural network model will be significantly higher than that of the statistical learning model, which is also more in line with the actual situation and our expectations. (3) From the overall experimental results, regardless of the size of the dataset or the language feature, the Uyghur language dataset in the two models are similar to other languages. It also reflects the dataset effectiveness constructed in this paper. (4) From two Uyghur datasets, the results of nominal dataset higher than all words dataset by 3.69% and 4.48%. Because the nominal dataset is relatively single part of speech, and the morphological changes are not too complicated. Compared with the nominal dataset, all words dataset is richer in morphological feature and have more POS.

5 Summary

In this paper, firstly, we mainly introduce the related work of constructing the Uyghur language morphological analysis corpus, including data preparation and preprocessing, making morphological analysis tags (POS and morphological tags), and corpus construction process. To reduce manual work, we used a POS tagger and stemming tool. Secondly, we statistically analyzed the basic information of the dataset, such as the distribution of the number and length of words and sentences, the distribution of POS, morphological feature, and MSDs. Finally, we designed comparative experiments using the models and datasets provided by the SIGMORPHON Shared task, to analyze and verify the validity of the dataset. The result of experiment shows that the dataset constructed on this paper is similar to other datasets. The morphological analysis corpus provides information about the lemma, POS, MSDs, morphological segmentation, and lemmatization of words. In the following research, we will continue to expand the data set, increase the coverage of words, enrich the language, and study morphological analyzers suitable for agglutinative language, to provide high-quality data sets for downstream tasks.

References

1. Straková, J., Straka, M., Hajič, J.: Open-source tools for morphology, lemmatization, POS tagging and named entity recognition. In: Proceedings of 52nd Annual Meeting of the Association for Computational Linguistics: System Demonstrations, pp. 13–18 (2015)
2. Zhuang, H., Wang, C., Li, C., Li, Y., Wang, Q., Zhou, X.: Chinese language processing based on stroke representation and multidimensional representation. IEEE Access 6, 41928–41941 (2018)
3. Üstün, A., Kurfalı, M., Can, B.: Characters or morphemes: how to represent words? In: Proceedings of the 3rd Workshop on Representation Learning for NLP, pp. 144–153 (2019)

4. Zhu, Y., Vulić, I., Korhonen, A.: A systematic study of leveraging subword information for learning word representations. In: Conference of the North (2019)
5. Ibrahim, T., Baoshe, Y.: A survey on minority language information processing research and application in Xinjiang. J. Chin. Inf. Process. **25**(6), 149 (2011)
6. Enwer, S., Lu, X., Zong, C., Pattar, A., Hamdulla, A.: A multi-strategy approach to Uyghur stemming. J. Chin. Inf. Process. **29**(5), 204 (2015)
7. Altenbek, G., Wang, X., Haisha, G.: Identification of basic phrases for Kazakh language using maximum entropy model. In: Proceedings of COLING 2014, the 25th International Conference on Computational Linguistics: Technical Papers, pp. 1007–1014 (2014)
8. Osman, T., Yating, Y., Tursun, E.: Collaborative analysis of Uyghur morphology based on character level. Acta Scientiarum Naturalium Universitatis Pekinensis **55**(1), 47 (2019)
9. Abudukelimu, H., Sun, M., Liu, Y., Abulizi, A.: THUUyMorph: an Uyghur morpheme segmentation corpus. J. Chin. Inf. Process. **32**(2), 81 (2018)
10. Munire, M., Li, X., Yang, Y.: Construction of the Uyghur noun morphological re-inflection model based on hybrid strategy. Appl. Sci. **9**(4), 722 (2019)
11. Maimaitiming, W., Gulinigeer, A., Maimaiti, M., Abiderexiti, K., Yibulayin, T.: A comparative study of Uzbek stemming algorithms. J. Chin. Inf. Process. **34**(1), 45 (2020)
12. Güngör, O., Güngör, T., Üsküdarli, S.: The effect of morphology in named entity recognition with sequence tagging. Nat. Lang. Eng. **25**(1), 147–169 (2019)
13. Vania, C., Grivas, A., Lopez, A.: What do character-level models learn about morphology? The case of dependency parsing (2018)
14. Parhat, S., Ablimit, M., Hamdulla, A.: A robust morpheme sequence and convolutional neural network-based Uyghur and Kazakh short text classification. Inf. **10**(12), 387 (2019)
15. Bisazza, A., Tump, C.: The lazy encoder: a fine-grained analysis of the role of morphology in neural machine translation. In: Proceedings of 2018 Conference on Empirical Methods in Natural Language Processing, EMNLP 2018, pp. 2871–2876 (2020)
16. Ye, F., Xu, T.: Essentials of Linguistics. Peking University Press, Beijing (2006)
17. Abudouwaili, G., Tuergen, Y., Kahaerjiang, A., Wang, L.: Research on Uyghur stemming based on Bi-LSTM-CRF model. J. Chin. Inf. Process. **33**(08), 65–71 (2019)
18. Vania, C.: On understanding character-level models for representing morphology (2020)
19. Jurafsky, D., Martin, J.H.: Speech and language processing: an introduction to natural language processing, computational linguistics, and speech recognition (2000)
20. Zong, C.: Statistical Natural Language Processing. Tsinghua University Press, Beijing (2013)
21. Marcus, M.P., Marcinkiewicz, M.A., Santorini, B.: Building a large annotated corpus of English: the Penn Treebank. Comput. Linguist. **19**(2), 313–330 (1993)
22. Silfverberg, M., Ruokolainen, T., Linden, K., Kurimo, M.: FinnPos: an open-source morphological tagging and lemmatization toolkit for Finnish. Lang. Resour. Eval. **50**(4), 863–878 (2016)
23. McCarthy, A.D., et al.: UniMorph 3.0: universal morphology. In: 2020–12th International Conference on Language Resources and Evaluation Conference Proceedings, no. May, pp. 3922–3931 (2020)

24. Buchholz, S., Marsi, E.: CoNLL-X shared task on multilingual dependency parsing. In: Proceedings of the Tenth Conference on Computational Natural Language Learning, CoNLL-X (2006)

25. Sylak-Glassman, J., Kirov, C., Post, M., Que, R., Yarowsky, D.: A universal feature schema for rich morphological annotation and fine-grained cross-lingual part-of-speech tagging. In: Mahlow, C., Piotrowski, M. (eds.) Systems and Frameworks for Computational Morphology. SFCM 2015. CCIS, vol. 537, pp. 72–93. Springer, Cham (2015). https://doi.org/10.1007/978-3-319-23980-4_5

26. Tuohuti, L.: Modern Uyghur Reference Grammar. China Social Sciences Press, Beijing (2012)

27. Maimaiti, M., Wumaier, A., Abiderexiti, K., Yibulayin, T.: Bidirectional long short-term memory network with a conditional random field layer for Uyghur part-of-speech tagging. Inf. **8**(4), 157 (2017)

28. Cotterell, R., et al.: CoNLL-SIGMORPHON 2017 shared task: universal morphological reinflection in 52 languages. In: CoNLL 2017 - Proceedings of CoNLL SIGMORPHON 2017 Shared Task: Universal Morphological Reinflection, pp. 1–30 (2017)

29. Vaswani, A., et al.: Attention is all you need. In: Advances in Neural Information Processing Systems, NIPS, pp. 5999–6009 (2017)

Knowledge Graph and Information Extraction

Improving Entity Linking by Encoding Type Information into Entity Embeddings

Tianran Li, Erguang Yang, Yujie Zhang$^{(\boxtimes)}$, Jinan Xu, and Yufeng Chen

School of Computer and Information Technology, Beijing Jiaotong University,
Beijing 100044, China
yjzhang@bjtu.edu.cn

Abstract. Entity Linking (EL) refers to the task of linking entity mentions in the text to the correct entities in the Knowledge Base (KB) in which entity embeddings play a vital and challenging role because of the subtle differences between entities. However, existing pre-trained entity embeddings only learn the underlying semantic information in texts, yet the fine-grained entity type information is ignored, which causes the type of the linked entity is incompatible with the mention context. In order to solve this problem, we propose to encode fine-grained type information into entity embeddings. We firstly pre-train word vectors to inject type information by embedding words and fine-grained entity types into the same vector space. Then we retrain entity embeddings with word vectors containing fine-grained type information. By applying our entity embeddings to two existing EL models, our method respectively achieves 0.82% and 0.42% improvement on average F1 score of the test sets. Meanwhile, our method is model-irrelevant, which means it can help other EL models.

Keywords: Entity Linking · Entity embedding · Entity disambiguation

1 Introduction

Entity Linking (EL) refers to the task of linking entity mentions in the text to correct entities in the Knowledge Base (KB) to help text comprehension by making use of rich semantic information in KB. The state-of-the-art entity linking models ignore fine-grained entity type information which causes the type of the linked entity is incompatible with the mention context. For example, among the error cases in AIDA-CoNLL [7] dataset produced by Le and Titov [10], the mention "Japan" in the sentence "Japan began the defense of their Asian Cup" is linked to the entity "Japan" *(country)* while it should be linked to the entity "Japan national football team" *(sports team)*. And such errors account for more than 1/2 of the total errors which show that their model cannot correctly distinguish different types of candidate entities.

Entity embeddings are especially important to entity linking task for the sake of distinguishing entities with similar characters. We hope the entity embeddings

S. Li et al. (Eds.): CCL 2021, LNAI 12869, pp. 297–307, 2021.
https://doi.org/10.1007/978-3-030-84186-7_20

can have type information so as to differentiate entities in terms of categories. However, previous pre-trained entity embeddings are learned by using plenty of texts, which ignores fine-grained type information [5,18]. As a result, the similar entity vectors in the vector space are of different types. In order to alleviate this problem, we firstly pre-train word vectors to inject type information by embedding words and fine-grained entity types into the same vector space. Then, we follow Ganea and Hofmann [5] to retrain the representation of each entity by making it closer to the word vectors of its context, using the word vectors containing type information. Finally, we apply the entity embeddings into the existing EL models without modifying the model architectures. Compared with two baseline models, our method respectively achieves 0.82% and 0.42% improvement on average F1 score of the test sets and obtains new state-of-the-art results on multiple datasets. Besides, detailed experimental analysis confirms the hypothesis that encoding type information in entity embeddings can reduce the type error cases. Our contributions can be summarized as follows:

- We propose a simple but effective method to encode fine-grained entity type information into entity embeddings. Moreover, our method does not need to change the model structure thus can help other entity linking models.
- Experiment results on the EL task show that our method achieves substantial improvements and obtains new state-of-the-art results on multiple datasets.
- Detailed experimental analysis shows that introducing type information can correct substantial type error cases produced by the baseline model.

2   Background and Related Works

2.1   Local and Global Entity Linking Models

The local entity linking model disambiguates each mention separately and measures the relevance scores between each candidate entity and mention context designed by Ganea and Hofmann [5]:

$$\Psi(e_i, c_i) = \mathbf{x}_{e_i}^T \mathbf{B} f(c_i) \tag{1}$$

where $\mathbf{B}$ is a learnable diagonal matrix, $\mathbf{x}_{e_i}$ is the embedding of entity e_i, $f(c_i)$ applies hard attention mechanism to obtain the feature representation of c_i which is the local context around mention m_i.

The global entity linking model encourages all entities in a document to be consistent with the topic. Ganea and Hofmann [5] define a pairwise global score by calculating the consistency between each pair of entities in document D:

$$\Phi(e_i, e_j | D) = \frac{1}{n-1} \mathbf{x}_{e_i}^T \mathbf{C} \mathbf{x}_{e_j} \tag{2}$$

where $\mathbf{C}$ is a diagonal matrix, n is the number of mentions in document D. Based on Formula 2, Le and Titov [10] propose a pairwise global score considering K potential relations as follows:

$$\Phi(e_i, e_j | D) = \sum_{k=1}^{K} \alpha_{ijk} \mathbf{x}_{e_i}^T \mathbf{R}_k \mathbf{x}_{e_j} \tag{3}$$

where $\mathbf{R}_k$ is a diagonal matrix to measure potential relationship k between entities e_i and e_j, and α_{ijk} is the weight corresponding to relation k.

2.2 Related Works

Entity Embeddings: Entity embeddings encode the structural or textual information of candidate entities. According to the different learning contents, the existing pre-trained entity embeddings are mainly categorized into the following three types:

(1) Context-based. Embeddings are typically trained using canonical Wikipedia articles that contain a lot of contextual information [4,5,16,18].
(2) Graph-based. This method builds entity graph based on KB and generates graph embeddings to capture structured information [1,17,21].
(3) Description-based. This approach makes use of descriptions, names, redirects of entities in Wikipedia, which are dictionary-based data [2,6,8,13].

Our approach is similar to Hou et al. [8] that generates semantic type words chosen from the first sentence of entity description pages in Wikipedia and combine them with entity embeddings of Ganea and Hofmann [5] through linear aggregation. Our approach is also similar to Chen et al. [2] that uses the pretrained BERT [3] to capture the latent type information in the mention context. Compared with their methods, the main advantage of our approach is that we explicitly encode fine-grained type information into the entity embeddings, while they implicitly capture the underlying type information.

Fine-grained Entity Type: Existing entity type labeling systems classify entities in texts into fine-grained types. Our work utilizes ZOE [20] system, which predicts the type in the FIGER [12] type set containing 112 two-dimensional entity types, such as *"/location/country"* and *"/location/city"*.

3   Our Method

Our method mainly consists of four steps: (1) Labeling named entities in Wikipedia articles as corresponding types. (2) Pre-training word and entity type embeddings by employing Skip-gram model [15]. (3) Training the representation for each entity using the word vectors obtained in the previous step. (4) Combining entity vectors trained by Genea and Hofmann [5] (hereafter referred to as "Wikitext entity embeddings") and our entity embeddings. The word vectors obtained in step (2) are also fused with the word vectors they use in the same way. Finally, we apply the result word and entity embeddings to the existing entity linking models. The overall architecture of our method is illustrated in Fig. 1.

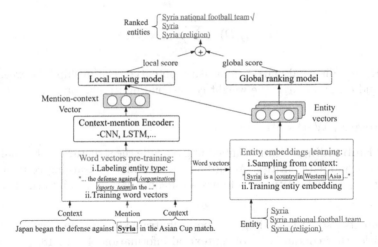

Fig. 1. Overall framework of our method. The main component of our method is painted in red. We first pre-train word vectors to encode entity type information using Wikipedia articles with entity type tags. Then we utilize the word vectors with type information to train entity embedding. Both of the local and global ranking model can leverage the type information in entity embeddings and word vectors. (Color figure online)

(a)**Beaverton** is a community in **Brock Township** in the **Regional Municipality of Durham** , **Ontario** , **Canada**.

(b) */location/city* is a community in */location/city* in the */location* , */location/province* , */location/country*.

Fig. 2. Example of preprocessed sentences.

3.1 Entity Labeling

To embed words and entity types into the same vector space, we adopt a simple but effective preprocessing method: labeling named entities in Wikipedia articles as corresponding fine-grained 2-dimensional entity types in the FIGER type set as shown in Fig. 2. We believe that the same type entities tend to appear in the similar contexts. To train embedding for each entity type, we utilize quite a few contexts of entities with the same type. For example, the embedding of "/location/country" would be trained from the contexts of all the country entities in Wikipedia articles. Specifically, we utilize hyperlinks in annotated Wikipedia corpus to extract the entity mention and map Freebase type of each entity mention into a set of fine-grained types.

3.2 Word and Type Vectors Pre-training

After processing the Wikipedia document, we employ the Skip-gram [15] model to learn uncased 300-dimensional embeddings for both words and entity types with a minimum word frequency cut-off of 5 and a window size of 5. We project 1200 words vectors of 6 different types and corresponding type vectors using the T-SNE algorithm [14], as shown in Fig. 3. It can be observed that the word vectors are clustered according to the fine-grained entity type. For instance, "mouse", "monkey" and "dog" are close to the type */livingthing/animal*. These results indicate our method can encode the type information into word vectors. We also observe that there are some overlaps between the entities of type */location/country* and */language* which is possibly due to the similar contexts or annotation errors.

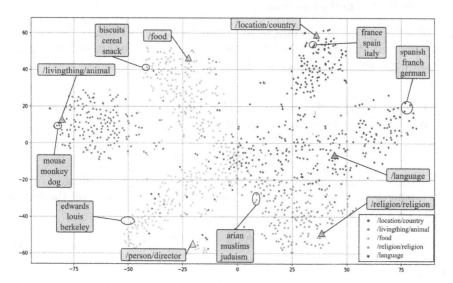

Fig. 3. Two-dimensional representation of the vector space containing word and entity types embeddings. Type embeddings are represented by triangle, while words are represented by dot, and color represents the type to which they belong. (Color figure online)

3.3 Entity Embeddings Training

We utilize the distribution hypothesis to learn the entity embeddings, where we use the word vectors obtained in Sect. 3.2 as the input. During training, we sample 20 context words around each entity as positive samples and 5 words at random as negative samples each iteration. Thus we train the entity embedding by making it close to the context words and further from other words in the vector space. Finally, the embeddings are normalized to obtain the joint distribution of entities and words on the unit sphere:

$$J(\mathbf{z}; e) = E_{\omega^+|e}E_{\omega^-}[h(\mathbf{z}; \omega^+, \omega^-)] \tag{4}$$

$$h(\mathbf{z}; w, v) = [\gamma - <z, x_w - x_v>]_+ \tag{5}$$

$$x_e = \arg\min_{\mathbf{z}:||\mathbf{z}||=1} J(\mathbf{z}; e) \tag{6}$$

where x_e is the final entity embedding, γ is a margin parameter, $<, >$ denotes dot product, $[.]_+$ denotes the RELU function, (ω^+, ω^-) denotes the sample positive word and negative word respectively, E denotes the process of sampling word vectors according to the given entity, $\mathbf{z}$ is the entity representation. The main idea is to find a $\mathbf{z}$ on the unit sphere that is closer to the positive samples and farther from the negative samples.

3.4 Entity Embeddings Fusing

In order to exploit the advantages of different entity embeddings, we combine our entity embeddings with Wikitext entity embeddings for entity linking, and the word vectors are fused in the same way. Specifically, we adopt three fusion methods:

(i) Linearly Interpolation. Similar to Hou et al. [8], we use a parameter α to control the weight of the two entity embeddings.

$$\mathbf{e}^\alpha = (1 - \alpha)\mathbf{e}^\omega + \alpha\mathbf{e}^t \tag{7}$$

where $\mathbf{e}^\omega$ is the Wikitext entity embedding, $\mathbf{e}^t$ is our embedding.

(ii) Nonlinear Interpolation. We also use a parameter β to control the weight of the two entity embeddings.

$$\mathbf{e}^n = (1 - \beta) * \mathbf{e}^\omega + \beta * Tanh(\mathbf{e}^t) \tag{8}$$

(iii) Gate Interpolation.

$$r, z = \sigma(W[\mathbf{e}^\omega; \mathbf{e}^t]) \tag{9}$$

$$\mathbf{e}^g = r \cdot \mathbf{e}^\omega + z \cdot \mathbf{e}^t \tag{10}$$

where W is a learnable matrix and σ is the sigmoid function, $[;]$ denotes the concatenation process.

4   Experiments

4.1 Datasets and Evaluation Metric

We use standard benchmark dataset AIDA-train in AIDA-CoNLL [7] for training, AIDA-A for validating, AIDA-B for testing. We also use MSNBC, AQUQINT, ACE2004, WNED-WIKI (WIKI), WNED-CWEB (CWEB) which come from different domains for evaluation just like Ganea and Hofmann [5] and Le and Titov [10,11]. It should be noted that CWEB and WIKI are believed to be less reliable. The standard micro-F1 score is employed as evaluation metric.

Table 1. Comparison of experimental results of different models. The first column represents models and entity embeddings they use. The last column is the average of F1 scores on the five test sets. Underlined scores show cases where we outperforms the baseline models **wnel** and **mulnel**.

Embeddings	Models	AIDA-B	MSNBC	AQUAINT	ACE2004	CWEB	WIKI	Avg
Weakly-supervised								
-	Plato	86.4	-	-	-	-	-	-
G and H	wnel	89.66 ± 0.16	92.2 ± 0.2	90.7 ± 0.2	88.1 ± 0.0	78.2 ± 0.2	81.7 ± 0.1	86.18
Hou et al.	wnel	89.23 ± 0.31	92.15 ± 0.24	91.22 ± 0.18	88.02 ± 0.15	78.29 ± 0.17	81.92 ± 0.36	86.32
Ours ($\alpha = 0.1$)	wnel	89.44 ±0.13	92.46 ± 0.08	91.69 ± 0.10	88.13 ± 0.00	78.90 ± 0.05	80.82 ± 0.44	86.39
Ours ($\beta = 0.1$)	wnel	90.40 ± 0.14	92.49 ± 0.01	91.64 ± 0.15	88.21 ± 0.22	78.77 ± 0.01	81.90 ± 0.15	86.60
Ours (*gate*)	wnel	89.28 ± 0.12	92.55 ± 0.08	91.58 ± 0.15	87.89 ± 0.27	78.77 ± 0.06	81.10 ± 0.44	86.38
Fully-supervised								
Yamada et al.	JEWE	91.5	-	-	-	-	-	-
G and H	Deep-ed	92.22 ± 0.14	93.7 ± 0.1	88.5 ± 0.4	88.5 ± 0.3	77.9 ± 0.1	77.5 ± 0.1	85.22
G and H	mulnel	93.07 ± 0.27	93.9 ± 0.2	88.3 ± 0.6	89.9 ± 0.8	77.5 ± 0.1	78.0 ± 0.1	85.5
G and H	DCA	**93.73 ± 0.2**	93.8 ± 0.0	88.25 ± 0.4	90.14 ± 0.0	75.59 ± 0.3	78.84 ± 0.2	85.32
Chen et al.	BERT-Sim	93.54 ± 0.12	93.4 ± 0.1	89.8 ± 0.4	88.9 ± 0.7	77.9 ± 0.4	**80.1 ± 0.4**	86.02
Hou et al.	mulnel	92.63 ± 0.14	94.26 ± 0.17	88.47 ± 0.23	90.7 ± 0.28	77.41 ± 0.21	77.66 ± 0.23	85.7
Ours ($\alpha = 0.1$)	mulnel	92.41 ± 0.32	93.59 ± 0.10	90.57 ± 0.45	90.22 ± 0.82	77.86 ± 0.26	77.94 ± 0.60	86.04
Ours ($\beta = 0.1$)	mulnel	92.85 ± 0.13	93.96 ± 0.27	90.15 ± 0.38	90.87 ± 0.65	78.07 ± 0.27	78.67 ± 0.27	86.32
Ours (*gate*)	mulnel	92.97 ± 0.07	94.29 ± 0.35	90.10 ± 0.35	90.30 ± 0.44	77.80 ± 0.06	77.83 ± 0.40	86.06

4.2 Experimental Settings

We use canonical Wikipedia dumps data[1] published on 2020-07-01 which consists of 1.3G tokens to train word and type vectors. We totally annotate 157.4M entity mentions with their types. We use the method of Sect. 3.3 to train our entity vectors on the entity similarity task shared by Genea and Hofmann [5], and finally we obtain 276,030 entity vectors containing type information according to the entity vocabulary.

The parameter α in Eq. 7 and β in Eq. 8 greatly affect the experimental results, we present the best group of results. To demonstrate the effectiveness of our entity embeddings, we use the existing state-of-the-art entity linking models **mulrel** [10] (relations number K = 3) and **wnel** [11] as our baseline models. To make our results comparable, we just replace the entity embeddings [5] and the word vectors [15] that they used without modifying their model architectures. Besides, we follow Ganea and Hofmann [5] to only consider the in-KB mentions and use the same candidate generation strategy. Similar to Le and Titov [10], we run each combination of entity embedding and linking model for 5 times, record the mean and 95% confidence intervals of micro F1 score.

4.3 Results

As shown in Table 1, we compare the experimental results on the six test sets with several state-of-the-art models: Plato [9], JEWE [18], Deep-ed [5], DCA [19], BERT-Sim [2] etc. The models are divided into two groups: (1) models using

---

[1] https://dumps.wikimedia.org/enwiki/latest/.

Wikipedia and Unlabelled document which are weakly-supervised; (2) models using Wikipedia and labelled dataset AIDA-train [7] which are fully-supervised.

For the **mulrel** model, we achieve new state-of-the-art results on MSNBC, AQUAINT, ACE2004 and CWEB datasets and the average F1 score of the five out-domain test sets. Compared to the baseline model, both of our approaches using nonlinear interpolation and gate interpolation can improve performance on four of the six datasets. For the **wnel** model, our entity embeddings obtain new state-of-the-art results on five of the six test sets, and obtain the optimal average F1 score of the five out-domain test sets. The experimental results can further draw the following three conclusions:

(1) In the weakly-supervised setting, our method achieves 0.74% performance improvement on the in-domain dataset and the improvement of the out-domain test sets is relatively solid. It can be seen that our embeddings perform better in the scenario with low resources.
(2) In most cases, three fusion methods can improve multiple datasets which come from different domains. And we respectively achieve 0.82% and 0.42% improvement on average F1 score of the five test sets. This demonstrates our method performs well on cross-domain issues.
(3) For different fusion ways of entity embeddings, nonlinear interpolation reveals excellent and solid results in both of the fully-supervised and weakly-supervised settings, gate interpolation and linearly interpolation also produce competitive results. Comparatively speaking, nonlinear interpolation is the most appropriate to combine the advantages of the two kinds of embeddings.

Compared with the entity embeddings of Hou et al. [8] and Ganea and Hofmann [5], our embeddings achieve solid performance improvement on most of the datasets, proving the effectiveness of the introduction of type information. Compared with the BERT-based entity embeddings of Chen et al. [2], we achieve 0.32% performance improvement on five out-domain datasets using non-linear interpolation, while we achieve lower performance on the in-domain dataset AIDA. This experimental result also shows the advantages of making use of the latest pre-trained language models which have powerful encoding capability.

4.4 Analysis

Type Errors Correction: We respectively analyzed the error cases again generated by **mulrel** model ($\alpha = 0.1$) on the AQUAINT dataset (totally 727 mentions) and **wnel** model ($\beta = 0.1$) on the AIDA-B dataset (totally 4485 mentions). Type error cases specifically refer to the cases where the type of predicted entity is inconsistent with the ground-truth entity's type and doesn't match the corresponding mention context. For the **mulrel** model, there are 21 type errors cases (about 34.2% of the total errors), 17 of which are corrected by our method. And for the **wnel** model, it generates totally 259 type error cases (about 56.8% of the total errors), we correct 54 of them. As listed in Table 2, our entity embeddings can correct a large number of type error cases. We further analyzed the remaining type error cases and find they are mainly due to the insufficient helpful local

Table 2. Comparison of results between beseline model and our model. Mentions in texts are bold, the type of each entity is italic.

Contexts	Mulrel or Wnel	Our model
(1) **Palestinians** await word from Israel	Palestine (*/location/country*)	Palestinian people (*/people/ethnicity*)
(2) The victims of the poisoning were in stable condition, but three suffered mild **liver disorders**	Liver (*/body part*)	Liver disease (*/disease*)
(3) That would allow it to use clean burning **hydrogen** to generate electricity	Hydrogen vehicle (*/product*)	Hydrogen (*/chemistry*)
(4) ...Who has joined Espanyol of Barcelona from **Nantes** since missing the European championship finals through injury	Nantes (*/location/city*)	FC Nantes (*/organization/sports league*)

Table 3. Examples of nearest entities in Ganea and Hofmann [5] and our entity representation space. The entity whose type is inconsistent with the entity being queried is bold.

Methods	George Washington (*/person/politician*)	Beijing (*/location/city*)	On the Origin of Species (*/art/written work*)
G and H	Abraham lincoln	Seoul	**Charles Darwin**
	Gilbert Du Motier, Marquis de Lafayette	Shanghai	**Evolution**
	American revolutionary war	**China**	**Thomas Henry Huxley**
Ours	Abraham lincoln	Athens	History of evolutionary thought
	Ulysses S. Grant	Seoul	Introduction to evolution
	Jefferson Davis	Tokyo	Theistic evolution

context and the low prior probability, which are hard to address. We leave these problems to our future work.

Nearest Entities: In addition, we compared the closest entity embeddings in Ganea and Hofmann [5] and our entity representation space as shown in Table 3. When we query the entity "George Washington", the first three most similar entities in our vector space are all of the */person/politician* category, which is consistent with the type of queried entity "George Washington". While, among the result of Ganea and Hofmann [5], only one entity remains the same type. It can be seen that our entity embeddings are more sensitive to type.

5   Conclusion

This paper proposes a simple yet effective method to encode fine-grained entity type information into the entity embeddings, enabling models to better capture the fine-grained entity type information. We obtain solid improvement compared with the previous models and the detailed experimental analysis shows our method can correct a fair portion of type error cases produced by the baseline model. Moreover, our method doesn't need to modify the EL model architecture, thus can be easily applied to other EL models.

In the future, we will consider injecting the entity information and fine-grained type information into the latest pre-trained language model (e.g., BERT [3]) to obtain entity representation and further improve the performance of entity linking.

Acknowledgements. We are grateful for helpful comments and suggestions from the anonymous reviewers. This work is supported by the National Nature Science Foundation of China (Contract 61876198, 61976015, 61976016).

References

1. Cao, Y., Hou, L., Li, J., Liu, Z.: Neural collective entity linking. In: Proceedings of the 27th International Conference on Computational Linguistics, pp. 675–686. Association for Computational Linguistics, Santa Fe (2018). https://www.aclweb.org/anthology/C18-1057
2. Chen, S., Wang, J., Jiang, F., Lin, C.: Improving entity linking by modeling latent entity type information. CoRR abs/2001.01447 (2020). http://arxiv.org/abs/2001.01447
3. Devlin, J., Chang, M., Lee, K., Toutanova, K.: BERT: pre-training of deep bidirectional transformers for language understanding. CoRR abs/1810.04805 (2018). http://arxiv.org/abs/1810.04805
4. Eshel, Y., Cohen, N., Radinsky, K., Markovitch, S., Yamada, I., Levy, O.: Named entity disambiguation for noisy text. In: Proceedings of the 21st Conference on Computational Natural Language Learning (CoNLL 2017), pp. 58–68. Association for Computational Linguistics (2017). https://doi.org/10.18653/v1/K17-1008, https://www.aclweb.org/anthology/K17-1008
5. Ganea, O.E., Hofmann, T.: Deep joint entity disambiguation with local neural attention. In: Proceedings of the 2017 Conference on Empirical Methods in Natural Language Processing, pp. 2619–2629. Association for Computational Linguistics, Copenhagen (2017). https://doi.org/10.18653/v1/D17-1277, https://www.aclweb.org/anthology/D17-1277
6. Gupta, N., Singh, S., Roth, D.: Entity linking via joint encoding of types, descriptions, and context. In: Proceedings of the 2017 Conference on Empirical Methods in Natural Language Processing, pp. 2681–2690. Association for Computational Linguistics, Copenhagen (2017). https://doi.org/10.18653/v1/D17-1284, https://www.aclweb.org/anthology/D17-1284
7. Hoffart, J., et al.: Robust disambiguation of named entities in text. In: Proceedings of the 2011 Conference on Empirical Methods in Natural Language Processing, pp. 782–792. Association for Computational Linguistics, Edinburgh (2011). https://www.aclweb.org/anthology/D11-1072
8. Hou, F., Wang, R., He, J., Zhou, Y.: Improving entity linking through semantic reinforced entity embeddings. In: Proceedings of the 58th Annual Meeting of the Association for Computational Linguistics, pp. 6843–6848. Association for Computational Linguistics, Online (2020). https://doi.org/10.18653/v1/2020.acl-main.612, https://www.aclweb.org/anthology/2020.acl-main.612

9. Lazic, N., Subramanya, A., Ringgaard, M., Pereira, F.: Plato: a selective context model for entity resolution. Trans. Assoc. Comput. Linguist. **3**(1), 503–515 (2015)
10. Le, P., Titov, I.: Improving entity linking by modeling latent relations between mentions. In: Proceedings of the 56th Annual Meeting of the Association for Computational Linguistics (Volume 1: Long Papers), pp. 1595–1604. Association for Computational Linguistics, Melbourne (2018). https://doi.org/10.18653/v1/P18-1148, https://www.aclweb.org/anthology/P18-1148
11. Le, P., Titov, I.: Boosting entity linking performance by leveraging unlabeled documents. In: Proceedings of the 57th Annual Meeting of the Association for Computational Linguistics, pp. 1935–1945. Association for Computational Linguistics, Florence (2019). https://doi.org/10.18653/v1/P19-1187, https://www.aclweb.org/anthology/P19-1187
12. Ling, X., Weld, D.S.: Fine-grained entity recognition. In: Proceedings of the 26th AAAI Conference on Artificial Intelligence (2012)
13. Logeswaran, L., Chang, M.W., Lee, K., Toutanova, K., Devlin, J., Lee, H.: Zero-shot entity linking by reading entity descriptions. In: Proceedings of the 57th Annual Meeting of the Association for Computational Linguistics, pp. 3449–3460. Association for Computational Linguistics, Florence (2019). https://doi.org/10.18653/v1/P19-1335, https://www.aclweb.org/anthology/P19-1335
14. Maaten, L.J.P.V.D.: Accelerating t-SNE using tree-based algorithms. J. Mach. Learn. Res. **15**, 3221–3245 (2014)
15. Mikolov, T., Chen, K., Corrado, G., Dean, J.: Efficient estimation of word representations in vector space. Computer Science (2013)
16. Rijhwani, S., Xie, J., Neubig, G., Carbonell, J.: Zero-shot neural transfer for cross-lingual entity linking. In: Proceedings of the AAAI Conference on Artificial Intelligence, vol. 33, pp. 6924–6931 (2019)
17. Sevgili, Ö., Panchenko, A., Biemann, C.: Improving neural entity disambiguation with graph embeddings. In: Proceedings of the 57th Annual Meeting of the Association for Computational Linguistics: Student Research Workshop, pp. 315–322. Association for Computational Linguistics, Florence (2019). https://doi.org/10.18653/v1/P19-2044, https://www.aclweb.org/anthology/P19-2044
18. Yamada, I., Shindo, H., Takeda, H., Takefuji, Y.: Joint learning of the embedding of words and entities for named entity disambiguation. In: Proceedings of The 20th SIGNLL Conference on Computational Natural Language Learning, pp. 250–259. Association for Computational Linguistics, Berlin (2016). https://doi.org/10.18653/v1/K16-1025, https://www.aclweb.org/anthology/K16-1025
19. Yang, X., et al.: Learning dynamic context augmentation for global entity linking. In: Proceedings of the 2019 Conference on Empirical Methods in Natural Language Processing and the 9th International Joint Conference on Natural Language Processing (EMNLP-IJCNLP), pp. 271–281. Association for Computational Linguistics, Hong Kong (2019). https://doi.org/10.18653/v1/D19-1026, https://www.aclweb.org/anthology/D19-1026
20. Zhou, B., Khashabi, D., Tsai, C.T., Roth, D.: Zero-shot open entity typing as type-compatible grounding. In: Proceedings of the 2018 Conference on Empirical Methods in Natural Language Processing, pp. 2065–2076. Association for Computational Linguistics, Brussels (2018). https://doi.org/10.18653/v1/D18-1231, https://www.aclweb.org/anthology/D18-1231
21. Zhou, S., Rijhwani, S., Wieting, J., Carbonell, J., Neubig, G.: Improving candidate generation for low-resource cross-lingual entity linking. Trans. Assoc. Comput. Linguist. **8**, 109–124 (2020)

A Unified Representation Learning Strategy for Open Relation Extraction with Ranked List Loss

Renze Lou[1], Fan Zhang[2], Xiaowei Zhou[3], Yutong Wang[4], Minghui Wu[1], and Lin Sun[1(✉)]

[1] Department of Computer Science, Zhejiang University City College,
Hangzhou, China
{mhwu,sunl}@zucc.edu.cn
[2] Faculty of Economics, Hitotsubashi University, Tokyo, Japan
[3] Zhejiang Qianyue Digital Technology Co., Ltd., Hangzhou, China
[4] Tsinghua Shenzhen International Graduate School, Shenzhen, China
wangyt19@mails.tsinghua.edu.cn

Abstract. Open Relation Extraction (OpenRE), aiming to extract relational facts from open-domain corpora, is a sub-task of Relation Extraction and a crucial upstream process for many other NLP tasks. However, various previous clustering-based OpenRE strategies either confine themselves to unsupervised paradigms or can not directly build a unified relational semantic space, hence impacting down-stream clustering. In this paper, we propose a novel supervised learning framework named MORE-RLL (Metric learning-based Open Relation Extraction with Ranked List Loss) to construct a semantic metric space by utilizing Ranked List Loss to discover new relational facts. Experiments on real-world datasets show that MORE-RLL can achieve excellent performance compared with previous state-of-the-art methods, demonstrating the capability of MORE-RLL in unified semantic representation learning and novel relational fact detection.

Keywords: Open-domain · Relation extraction · Deep metric learning

1 Introduction

Relation Extraction (RE) aims to extract pre-defined relational facts from plain text (e.g., *"Mary gave birth to Keller in 1989s."*, RE can extract ***"gave_birth_to"*** between two named entities *"Mary"* and *"Keller"*). It is an important task that can structure a large amount of text data. Therefore, it can benefit for unstructured text data storing and the procedure of many other down-stream NLP tasks or applications, such as knowledge graph construction [29], information retrieval [33], and logic reasoning [26]. Nevertheless, with the rapid

---

R. Lou and F. Zhang—Equal contribution.

© Springer Nature Switzerland AG 2021
S. Li et al. (Eds.): CCL 2021, LNAI 12869, pp. 308–323, 2021.
https://doi.org/10.1007/978-3-030-84186-7_21

development of social media and human civilization, novel relationships and new knowledge in open-domain text data are also increasing. Accordingly, the relation types in the open-domain corpora may not be pre-defined, which is hard for RE to handle. To meet the rapid emergence of such novel knowledge, OpenRE emerged as the times required [2]. The goal of OpenRE is to detect novel relational facts from open-domain datasets. It is a crucial task for updating the human knowledge base and the study of human civilization.

Existing OpenRE methods are divided into two main categories: tagging-based and clustering-based. The tagging-based strategies treat OpenRE as a sequence labeling problem [2,3]. Still, these methods often extract surface forms that can not be utilized for downstream tasks (i.e., some sequences have the same relational semantic type, but their phrases generated from tagging-based methods are different because of overly-specific). Comparatively, clustering-based methods aim to identify the rich semantic features in the text, then cluster them into certain relation types. Recently, many efforts have been devoted to exploring clustering-based methods, such as [8,34,36]. Yet, those schemes are laborious and time-consuming because of the high dependence on well-designed features created by hand.

Profited from the substantial improvement of computing power in recent years, neural networks begin to be exploited in clustering-based OpenRE tasks to alleviate the above issues, such as [10,15,24]. Even so, these strategies confine themselves to unsupervised or self-supervised paradigms and can not fully benefit from current high-quality human-labeled corpora. Although several unconventional works have gained phenomenal performance, such as [36], the reliance on the extra knowledge for these strategies make it hard for us to compare with. Besides, another supervised scheme learned the similarity metrics from labeled instances and further transferred the relational knowledge to the open-domain scene, namely Relational Siamese Networks (RSNs) [32]. However, RSNs target learning a similarity classifier rather than building relational representations directly. Thereby, this may impact the efficiency and effect of down-stream clustering.

In order to address these issues, we propose a novel supervised learning framework via a clustering-based scheme driving neural encoder to build rich semantic representation directly. From our insight view, the essential target of the clustering-based OpenRE algorithm is to construct a reasonable semantic space on the open-domain corpora, where all different relational facts can be distinguished clearly. Therefore, the learning of semantic representation is a fundamental part of the whole task. It can not only extend the functionality of the neural encoder (i.e., the semantic space construction ability of the neural model can be used in other scenes, such as classification, etc.) but also bring benefits to downstream clustering.

As a result, we pay attention to the unified semantic representation learning ability of neural encoders. Specifically, we employ deep metric learning to drive the neural encoder to build a distinguishable semantic space on open-domain datasets. However, most prevailing deep metric learning methods, such as triplet loss [14], N-pair-mc [27], or Proxy-NCA [20], always bring low yield due to the poor supervision signals from the limited number of training data points. Inspired by [30], we chose Ranked List Loss (RLL) instead, which can capture set-based rich supervision signals. Meanwhile, RLL can preserve a better intraclass similarity structure within a hypersphere than other set-based schemes, hence constructing a more desirable semantic space.

Additionally, considering that the open scene corpora is usually full of noise, hence directly transferring knowledge may not be an ideal choice. To enhance the model's robustness, we also design virtual adversarial training for our semantic space construction algorithm. Experiments demonstrate that MORE-RLL can build more distinguishable semantic representations and obtain excellent performances on real-world datasets.

To sum up, the main contributions of this work are as follows:

* We propose a novel clustering-based OpenRE framework, namely MORE-RLL. The MORE-RLL combines deep metric learning and neural encoder to build a unified relational semantic space to discriminate samples rather than utilize an additional classification layer. Thus, it can handle the enormous undefined relation types in the open-domain corpora and facilitate downstream clustering to discover valuable novel types. Meanwhile, we adopt a Ranked List Loss to gain more prosperous supervision signals and construct a more desirable semantic space than other prevailing metric learning losses.
* Considering the noise and bias present in the text of open scenarios, we also design virtual adversarial training to enhance the robustness of MORE-RLL instead of directly transferring the knowledge that comes from clean RE datasets to the open-domain corpora.
* Experiments illustrate that the proposed MORE-RLL achieves state-of-the-art performance on real-world datasets, even if the imbalance distribution presents in the test set. Moreover, the visual analysis also demonstrates its excellent ability of relational representation learning and novel knowledge detecting.

2 Methodology

In this section, we will introduce our framework in detail. As shown in Fig. 1, we exploit a neural encoder to extract relational representations from a batch of training samples. These sentence-level representations can be taken as relational semantic vectors, indicating the relative locations of facts in the semantic feature space. Then, we use them to calculate the Ranked List Loss (RLL) and gain rich supervision signals to train the encoder. Besides, we set virtual adversarial training to smooth the feature space to overcome noise in the open scenes. We repeat these steps until the encoder is well-trained and then transfer the prior knowledge from the training samples to the open-domain corpora.

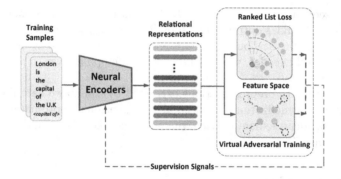

Fig. 1. Overall architecture of MORE-RLL.

2.1 Neural Encoders

As a vital component of MORE-RLL, the neural encoder aims at extracting semantic representations of relation types from natural language sentences. In this paper, we mainly use CNN as the encoder. Meanwhile, we also experiment with the pre-trained language model to demonstrate the expansibility of our framework.

CNN+GloVe. Following [32,35], we take the CNN encoder as our primary choice and utilize the pre-trained GloVe embedding [22]. To be specific, we firstly use the pre-trained word embedding layer and a randomly initialized position embedding layer to transform the original text sequences. Both these embedding layers are trainable. Then, the outputs of these two embedding layers will be concatenated and passed to a one-dimensional CNN followed by a max-pooling layer. After that, we employ a linear layer to map these raw representations to a high-dimensional semantic space. So far, the structure of our model is the same as that of RSNs [32], so we don't detail it. However, unlike RSNs, which utilize an additional linear classifier to predict the similarity of the extracted representation pair, we simply construct such a feature space and prepare for the next step.

BERT. Inspired by SelfORE [15], which exploited the pre-trained language model, we also choose BERT [7] as our contextual neural encoder. Following the operation proposed by [25], we take the relational hidden states of BERT as representations rather than the output of $[CLS]$ token. More specifically, for a sentence $\mathcal{S} = \{s_1, ..., s_T\}$ (where s indicates the token and T is the length of $\mathcal{S}$), we insert four special tokens before and after each entity mentioned in a sentence and get a new sequence:

$$\mathcal{S} = [s_1, .., [E1_{start}], s_p, .., s_q, [E1_{end}],$$
$$.., [E2_{start}], s_k, .., s_l, [E2_{end}], .., s_T] \tag{1}$$

We use this sequence as the input of BERT, and then we concatenate the last hidden states of BERT's outputs corresponding to $[E1_{start}], [E2_{start}]$, take these relational hidden states as our raw representations. Same as what we have mentioned in the CNN encoder, we then use a linear mapping layer to process these representations.

After obtaining the representations extracted by the above neural encoders, we perform L_2 normalization on these high dimensional representations, thus construct a Euclidean semantic space where we can predict the similarity metrics of relations conveniently.

2.2 Ranked List Loss

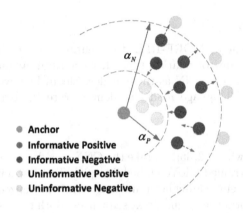

Fig. 2. Illustration of the ranked list loss.

So far, we have introduced how to built a Euclidean semantic space. Therefore, the next problem is the optimization of this space, which we will detail in this section.

As we have introduced in the Subsect. 1, our essential objective is to make the neural encoder gain a unified representation learning ability on high-quality RE corpora. Thus, we adopt deep metric learning as the optimization algorithm on relational semantic space. What's more, considering the limited information from point-based or pair-based metric learning methods (e.g., triplet loss [14], N-pair-mc [27]), we try to use a set-based or group-based scheme instead. Inspired by [30], we finally choose Ranked List Loss (RLL) to explore set-based similarity structure from a training batch and gain richer supervision signals.

For an anchor selected from the training batch, RLL rank the similarity of all the same type (positive) points before the different categories (negative) points and preserve a preset margin between them. To be specific, given a batch of normalized relation representations $\mathcal{B} = \{r_1, ..., r_m\}$ generated from the neural encoder, and an instance (anchor) r_i in $\mathcal{B}$ (where m indicates the batch size), we expect that the positive points for the anchor in $\mathcal{B}$ can be gathered together while those negative points are the opposite. So we calculate the following formula:

$$\mathcal{L}(r_i, \mathcal{B}; f) = \sum_{r_j \in \mathcal{B}, j \neq i} [(1 - y_{ij})[\alpha_N - d_{ij}]_+$$

$$+ y_{ij}[d_{ij} - \alpha_P]_+] \tag{2}$$

where f is model parameters; y indicates the relation type, $y_{ij} = 1$ if r_j is a positive point, $y_{ij} = 0$ otherwise; d_{ij} denotes the Euclidean distance between two points; α_P, α_N represent the positive and negative boundary respectively; $[.]_+$ denote the hinge function.

Intuitively, as shown in Fig. 2, those positive instances outside the α_P will be pulled closer, while those negative points within the α_N will be pushed further. The remaining uninformative points which have already met our objective will not be taken into count because of the hinge function.

More concisely, for an anchor r_i in $\mathcal{B}$, let's define $\mathcal{L}_P(r_i, \mathcal{B}; f)$ as the total loss of all informative positive points, and $\mathcal{L}_N(r_i, \mathcal{B}; f)$ is the sum of all informative negative samples loss, thus the optimization objective function can be summarized as below:

$$\mathcal{L}_{RLL}(\mathcal{B}; f) = \sum_{r_i \in \mathcal{B}} [(1 - \lambda)\mathcal{L}_P(r_i, \mathcal{B}; f) + \lambda\mathcal{L}_N(r_i, \mathcal{B}; f)] \tag{3}$$

Here, the λ is the balance factor between $\mathcal{L}_P(r_i, \mathcal{B}; f)$ and $\mathcal{L}_N(r_i, \mathcal{B}; f)$. Usually, we set it as 0.5.

Additionally, given r_i as an anchor, there are always numerous informative negative points that can be found in $\mathcal{B}$. To deal with the magnitude difference laying in the negative loss, we follow [30], weight the negative examples according to the values of their loss:

$$w_{ij} = \exp[T_n * (\alpha_N - d_{ij})], r_j \in \mathcal{R}_i^{\mathcal{N}} \tag{4}$$

where $\mathcal{R}_i^{\mathcal{N}}$ represents the set of informative negative samples of r_i, and T_n indicates the temperature factor which controls the degree of weighting these negative samples (because the temperature factor of positive samples, namely T_p, is always set to 0, we don't formalize it). When T_n is 0, every instance will be treated equally. But if set T_n to $+\infty$, the loss function will devote almost all attention to the hardest sample.

Consequently, the $\mathcal{L}_N(r_i, \mathcal{B}; f)$ in (3) can be updated as:

$$\mathcal{L}_N(r_i, \mathcal{B}; f) = \sum_{r_j \in \mathcal{R}_i^{\mathcal{N}}} [\frac{w_{ij}}{\sum_{r_j \in \mathcal{R}_i^{\mathcal{N}}} w_{ij}} (\alpha_N - d_{ij})] \tag{5}$$

After the neural encoder retrieves the supervision signals of $\mathcal{B}$, we sample next m instances sequentially from the training corpus and performed the above operation iteratively.

2.3 Virtual Adversarial Training

Different from many other tasks, there is always bias (e.g., the extreme imbalance of labels) and noise (e.g., spelling mistakes) present in the text of open scenarios. As a result, directly transfer the knowledge from the standardized corpus to an open-domain setting is not an ideal scheme. To address this issue, we design virtual adversarial training (VAT) [19] to smooth the semantic space, hence enhance the model's robustness.

Specifically, for any given sentence S and its original representation r, we first generate a normalized perturbation ξ on the word embedding within S randomly, add it to the original word embedding, and then take this disturbed embedding as the input of encoder to build a new representation $\tilde{r}$. Next, we calculate the gradient g of the Euclidean distance between r and $\tilde{r}$ with respect to the ξ. Then, we regard ϵ times normalized g as the worst-case perturbation $\tilde{\xi}$, where ϵ is a small decimal number we set as 0.02 in all our experiments. Finally, we use $\tilde{\xi}$ to disturb the original embedding of S. In a word, given a batch of samples $\mathcal{B}$, we penalize neural encoder with the following VAT loss:

$$\mathcal{L}_{adv}(\mathcal{B}; f) = \frac{1}{m} \sum_{i=1}^{m} D(F(S_i; f), F(S_i + \tilde{\xi}_i; f)) \qquad (6)$$

Where S_i indicates the i sequence, $F(S_i + \tilde{\xi}; f)$ denotes the distributed representation encoded by the neural model, while $F(S_i; f)$ is the original one (namely r_i) and D calculates the Euclidean distance between two representations. Intuitively, we expect that any representation r_i encoded by model in $\mathcal{B}$ is stable as possible under such worst-case perturbations.

Thence, the final objective loss function can be written as:

$$\mathcal{L}(\mathcal{B}; f) = \mathcal{L}_{RLL}(\mathcal{B}; f) + \beta \mathcal{L}_{adv}(\mathcal{B}; f) \qquad (7)$$

Where β is a factor that indicates the weight of virtual adversarial training. Same as [19], we set it as 1 practically.

3 Experiment

3.1 Datasets

FewRel is derived from Wikipedia and annotated by crowd workers [12]. Different from most other datasets, the entity pair of each instance in FewRel is unique, which makes the model unable to obtain shortcuts by memorizing the entities. Following the paper [32], we choose 64 relations as the train set and randomly select 16 relations with 1600 instances as the test set; the remaining sentences are validation set (as can be seen in Table 1).

NYT+FB-sup is generated from NYT+FB. The original NYT+FB is built by distant supervision, and its text sequences come from New York Times corpus [23] while the relational types are extracted from Freebase [5]. Following [15,24], we process the raw data and get the original NYT+FB. Since the whole dataset is built via distant supervision, its labels are full of noise and bias. In order to fit the supervision setting and to better simulate open scenes, we then divide the original dataset again, hence obtaining NYT+FB-sup. Usually, the relations which occur frequently are common categories, and those relations with rare instances are insufficient to be regarded as novel types. Therefore, we select relations with the number of instances between 20 and 2000 as novel relations and append them to the test set. As shown in Table 1, we finally obtained 72 novel relations equally divided between the test and validation set, leaving 190 relations as the train set, which contains both common and scarce types to simulate a real unbalanced environment.

Table 1. The statistical information on FewRel and NYT+FB-sup

Partition	FewRel		NYT+FB-sup	
	Relation type	Instance	Relation type	Instance
Train	64	44800	190	25521
Dev	16	9600	72	8100
Test	16	1600	72	8063

3.2 Settings

In all our experiments, we use the NVIDIA RTX2080 graphics card. We choose Adam [16] for our optimization and fix the learning rate with 3e-4 and 1e-5 on CNN and BERT, respectively. Since the batch size m is a significant factor in our metric learning-based framework, we use the conclusion that comes from 3.5.3. Expressly, we set m to 100, fix both the relation types C and the number of instances for each type K to 10. To solve out-of-memory problems when utilizing BERT, we use parallel training on 4 graphics cards. For the hyperparameter α_P, α_N, we follow the original paper [30] and set them as 0.8 and 1.2 separately, hence preserving a margin of 0.4 between these two boundaries. The temperature factor T_n in this work is 10, which is also the same as the original paper. We train our framework with 4 epochs on the training set and adopt early stopping.

The clustering algorithm is a general factor in the OpenRE task, so there can be multiple choices. Since the semantic space constructed by our framework is a normalized Euclidean space, the distance metric described by our model is linear, which is different from [32]. Therefore, we utilize two commonly used algorithms: K-Means [13] and Mean-Shift [6]. On FewRel, we choose K-means as our downstream clustering algorithm and set the number of clusters as 24. And on NYT+FB-sup, we choose Mean-Shift instead of K-means to deal with the imbalance, which can automatically find clusters based on spatial density. We don't use Louvain [4] or HAC [31] in our framework. The reason is that the Louvain is a graph-based algorithm that is not very compatible with our normalized Euclidean semantic space, while the HAC is highly time-consuming.

What's more, we adopt B^3 F_1 score [1] as our metrics, which is widely used in previous works [8,15,18,32]. F_1 calculates the harmonic mean of precision and recall, and its value is more affected by the lower one, which can fairly demonstrate the performance of the model.

3.3   Main Results

Table 2. The results on FewRel and NYT+FB-sup.

Method	FewRel			NYT+FB-sup		
	Prec.	Rec.	F1	Prec.	Rec.	F1
VAE [18]	17.9	69.7	28.5	20.3	40.7	27.1
RW-HAC [8]	31.8	46.0	37.6	25.2	33.9	28.9
SelfORE [15]	50.8	51.6	51.2	30.9	46.4	37.1
RSNs [32]	48.9	77.5	59.9	31.1	**52.0**	38.8
MORE-RLL (GloVe+CNN)	57.1	68.0	62.0	39.1	49.1	43.5
MORE-RLL (BERT)	**70.1**	**79.6**	**74.5**	**48.7**	50.8	**49.7**

We compare our MORE-RLL with four previous state-of-the-art baselines [8,15, 18,32] on two datasets. All these models are evaluated on the test set to show their performance. The main results can be seen in Table 2, the scores of all algorithms are the highest among the statistical testings (some borrowed from the original paper). From Table 2, we can draw the following conclusions:

* Benefiting by the rich supervision signals come from the labeled RE corpora (even the distant-supervised annotations), MORE-RLL(GloVe+CNN) outperforms all unsupervised or self-supervised methods on both datasets. The results indicate the effectiveness of prior knowledge transfer and our unified semantic representation learning strategy, which can be conducive to novel type-detection in open scenarios.
* MORE-RLL (GloVe+CNN) outperforms RSNs on precision and F_1. However, compared with RSNs, the superiority of MORE-RLL is not obvious on recall. Because the clustering method adopted by RSNs is Louvain [4], which constantly produces coarse-grained clustering results, as mentioned in [32]. Since the essential objective of OpenRE is to detect valuable novel relations, the quality of relation types detected by the model is more significant than the quantity. Therefore, the impressive precision of MORE-RLL also indicates its capability of high-quality knowledge discovery.
* The F_1 scores of all methods on NYT+FB-sup are lower than the results on FewRel. Even SelfORE [15], which has achieved admirable performance on NYT+FB, also has a poor performance. This phenomenon shows that NYT+FB-sup can simulate the real open scene and presents a challenging problem for all models. However, even in a hard setting, MORE-RLL can still

maintain a better performance than others. This result proves that MORE-RLL is robust to noise in the dataset and can distinguish those ambiguous novels and rare classes in open-domain corpora.

* To demonstrate the expansibility of our framework, we also adopt BERT as our neural encoder. Owing to the powerful extracting ability of the pre-trained language model, the performance of our framework has been greatly improved. In fact, there are multiple encoders that can be adopted in our framework. Due to the limitation of the space, we do not extend it here.

3.4 Visualization Analysis

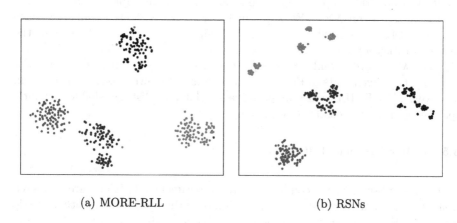

(a) MORE-RLL   (b) RSNs

Fig. 3. The t-SNE visualization on FewRel.

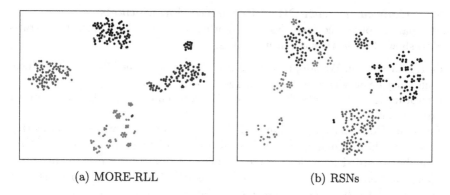

(a) MORE-RLL   (b) RSNs

Fig. 4. The t-SNE visualization on NYT+FB-sup.

In order to intuitively demonstrate the capability of MORE-RLL on semantic representation learning, we visualize the semantic space of relational representations with t-SNE [17]. Specifically, We randomly sample 4 relation types from the test set of FewRel and NYT+FB-sup, respectively, 100 instances per type, and construct representations from these instances on both MORE-RLL(GloVe+CNN) and RSNs, then color these representations according to their ground-truth types.

The Fig. 3 illustrates the visualization on FewRel. The semantic space of MORE-RLL is more distinguishable that almost all four types can preserve the intraclass similarity within a hypersphere while leaving a distinct margin between any two categories. In contrast, RSNs attempt to shrink the representations of the same type into one point. Thus, the distribution of points for RSNs in each cluster is denser than MORE-RLL. However, excessive attention to the similarity between point pairs may drop the intraclass similarity structure, so it is easier for RSNs to divide samples of the same category into multiple subcategories.

Visualization on NYT+FB-sup can be seen in Fig. 4. The semantic space on NYT+FB-sup is harder to construct compared with Fig. 3. Because of the distant-supervision labels and the extreme imbalance in the training set, it is easy for both MORE-RLL and RSNs to be confused with some relation types. For example, the olive points in Fig. 4 cannot maintain the intraclass structures well. However, MORE-RLL can still preserve a relatively distinguishable semantic space even in such a challenging setting.

3.5   Other Empirical Studies

In this subsection, we conduct several experiments: 1). Compare the VAT with other regularization strategies; 2). Compare the RLL with the other metric losses; 3). Explore the vital factor for the batch content. In all the following experiments, we use CNN as the neural encoder.

3.5.1   Comparing with Other Regularization Methods

In this paragraph, we compare VAT with other widely used regularization strategies to show the effectiveness of it, that is:

* **L_2 regularization.** We adopt L_2 regularization on both the CNN and the linear mapping layer with the 2e−4 and 1e−3 weight decay ratio, respectively.
* **Dropout** [28]. We apply Dropout on the word embedding, the drop rate we use here is 0.3.
* **Random Perturbation Training (RPT)** is a naive smoothing scheme that disturbs the original input with an isotropic distribution. Since the RPT can be regarded as a downgraded version of our VAT, we take the initial perturbation ξ (we have mentioned in Sect. 2.3) as a random perturbation and add it to the original word embedding within each sequence.

The hyparameters of each strategy above are chosen via greedy trials. For each of them, we report the best score among 10 experiments.

As can be seen in Table 3, with the help of VAT, MORE-RLL can achieve better performance. In contrast, some other widely used regularization methods (e.g., L_2, Dropout) can't bring phenomenal improvement due to the shallow structure of our framework. Meanwhile, suffering from the influence of isotropic distribution, the RPT also achieve barely satisfactory compared with VAT. Besides the theoretical and performance promise, the VAT calculates "virtual adversarial direction" (as mentioned in the original paper [19]), so there is no dependency on the ground-truth information for VAT. Consequently, it is more suitable to use VAT in the open domain than other label-dependency methods, e.g., adversarial training [11].

Table 3. The results of adopting different regularization methods.

Method	FewRel			NYT+FB-sup		
	Prec.	Rec.	F1	Prec.	Rec.	F1
MORE-RLL w/o VAT	56.6	60.3	58.4	36.4	48.9	41.8
MORE-RLL w/o VAT + L_2	52.9	66.4	58.9	39.0	45.5	42.0
MORE-RLL w/o VAT + Dropout	53.9	66.8	59.6	**39.5**	45.2	42.1
MORE-RLL w/o VAT + RPT	**57.7**	62.7	60.1	38.7	43.9	41.1
MORE-RLL	57.1	**68.0**	**62.0**	39.1	**49.1**	**43.5**

3.5.2 Comparing with Other Metric Learning Losses

In this paragraph, we try to demonstrate the effectiveness of the Ranked List Loss. We compare RLL with other prevailing metric losses:

* **Triplet Loss (TL)** [14] is a pair-based metric loss, which aims to pull an anchor closer to a positive point while pushing further from a negative point.
* **N-Pair-Mc Loss (NPML)** [27] is similar to triplet loss which increases the number of data points used to calculate. Further, it utilizes an efficient batch composition method.
* **Proxy-NCA Loss (PNL)** [20] is a proxy-based loss, which aims at selecting proxies that represent the desirable cluster center of those positive or negative samples.
* **Facility Location Loss (FLL)** [21] is another outstanding set-based metric loss. The author takes the global embedding structure into account and proposes a better optimization strategy on FLL than the greedy algorithm.

In this experiment, We fix the batch size m of all schemes to 100. All the results reported here are the highest one during 10 experiments. To avoid the influence of other factors, we don't use VAT on MORE-RLL here. The result is presented in Table 4. It shows that these set-based metric losses do capture richer supervision signals than those pair-based methods. Moreover, RLL performs better than FLL, this mainly owing to the superiority of RLL on the intraclass similarity structure-preserving. At the same time, FLL also suffers from the excessive pursuit of the similarity metric. We also note that the FLL is much more time-consuming than RLL and is sensitive to its hyparameters, making it hard to optimize.

Table 4. The results of adopting different metric losses.

Method	FewRel			NYT+FB-sup		
	Prec.	Rec.	F1	Prec.	Rec.	F1
MORE-TL	37.2	44.2	40.4	32.3	30.2	31.2
MORE-NPML	40.6	48.6	44.2	32.5	33.6	33.1
MORE-PNL	46.3	55.4	50.4	31.7	33.1	32.4
MORE-FLL	50.0	57.0	53.3	32.7	41.3	36.5
MORE-RLL w/o VAT	**56.6**	**60.3**	**58.4**	**36.4**	**48.9**	**41.8**

3.5.3 The Vital Factor for Batch Content

Unlike many other OpenRE methods, the training batch content is a significant part of our metric learning-based strategy. As mentioned in [9], almost all metric losses can benefit from a larger batch size due to the more prosperous signals. However, there are two critical factors for the batch content in RLL, that is, the number of relation types C and the number of instances K in each relation type (i.e., $m = C \times K$, m is the batch size). Though Wang et al. [30] has experimented with the impact of batch content on RLL, it is still ambiguous which factor for batch content is more contributing.

Accordingly, we change the training batch size in a larger range on both datasets to reveal the impact of C and K more clearly. To be specific, we fix one of these two factors to 10 and range the other from 3 to 13, then plot the results to show each factor's impact. All scores we report in this paragraph are averaged from 10 experiments.

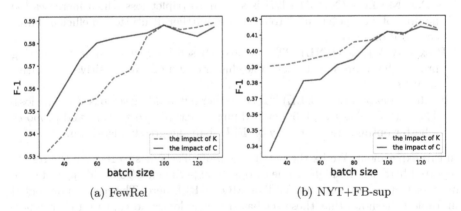

(a) FewRel   (b) NYT+FB-sup

Fig. 5. The influence of training batch content on FewRel and NYT+FB-sup. The solid blue line represents the impact of changing in K while the dashed red line indicates the influence of changing in C

As can be seen in Fig. 5, we can draw the following conclusions:

* The batch size is a significant hyparameters in our framework. With a larger batch size, the neural encoder does capture more prosperous supervision signals. However, this improvement will not be noticeable when the batch size is sufficiently large. Hence, we recommend to take 100 ($C = K = 10$) as an ideal batch size on both FewRel and NYT+FB-sup.
* The result on FewRel illustrates that K seems to have a greater impact than C, i.e., the dashed red line rises more rapidly with batch size growth. Meanwhile, the batch content with larger K (the solid blue line) usually brings more performance improvement. As we have mentioned in Sect. 2.2, there is a vast gap between positive and negative loss present in RLL. Since the FewRel is a human-labeled corpus, the larger K is, the more high-quality positive supervision signals RLL can bring. Thus K is a more vital factor when training for these gold-label corpora.
* On the contrary, C plays a vital role when the training dataset is NYT+FB-sup. Unlike FewRel, the NYT+FB-sup corpus is full of noise, so it is difficult for RLL to generate beneficial positive signals. Taking another route, increasing the diversity of relationship types can dramatically increase the negative signals. Though there is still noise present, RLL is more likely to find those informative and beneficial negative points in it. Hence the neural encoder can have more opportunities to obtain instructive semantic signals. In this case, a larger C may be a desirable choice.

4 Conclusion

In this paper, we propose a novel supervised learning framework for open-domain relation extraction, namely MORE-RLL. It aims to make the neural network gain a unified relational representation encoding ability and handle the open-domain relational instances. We utilize deep metric learning to drive the neural model to learn relational representations directly, thereby conducive to downstream clustering efficiency. Moreover, we set virtual adversarial training to enhance the robustness of the neural encoder. Our experiments show that MORE-RLL achieves state-of-the-art performance on real-world RE corpora comparing with previous methods and can build a more desirable semantic space. These all demonstrate the capability of our scheme on relational representation learning and novel relation detection.

Acknowledgements. This work was supported by the Key Research and Development Project of Zhejiang Province (No. 2021C01164) and the National Innovation and Entrepreneurship Training Program for College Students (No. 202113021002, No. 202113021003).

References

1. Bagga, A., Baldwin, B.: Algorithms for scoring coreference chains. In: The First International Conference on Language Resources and Evaluation Workshop on Linguistics Coreference, vol. 1, pp. 563–566. Citeseer (1998)

2. Banko, M., Cafarella, M., Soderland, S., Broadhead, M., Etzioni, O.: Open information extraction from the web. In: Proceedings of the 20th International Joint Conference on Artificial Intelligence (2007)

3. Banko, M., Etzioni, O.: The tradeoffs between open and traditional relation extraction. In: Proceedings of ACL-08: HLT, pp. 28–36 (2008)

4. Blondel, V.D., Guillaume, J.L., Lambiotte, R., Lefebvre, E.: Fast unfolding of communities in large networks. Journal of statistical mechanics: theory and experiment 2008(10), P10008 (2008)

5. Bollacker, K., Evans, C., Paritosh, P., Sturge, T., Taylor, J.: Freebase: a collaboratively created graph database for structuring human knowledge. In: Proceedings of the 2008 ACM SIGMOD International Conference on Management of Data, pp. 1247–1250 (2008)

6. Cheng, Y.: Mean shift, mode seeking, and clustering. IEEE transactions on pattern analysis and machine intelligence 17(8), 790–799 (1995)

7. Devlin, J., Chang, M.W., Lee, K., Toutanova, K.: Bert: pre-training of deep bidirectional transformers for language understanding. arXiv preprint arXiv:1810.04805 (2018)

8. Elsahar, H., Demidova, E., Gottschalk, S., Gravier, C., Laforest, F.: Unsupervised open relation extraction. In: European Semantic Web Conference. pp. 12–16. Springer (2017)

9. Fehervari, I., Ravichandran, A., Appalaraju, S.: Unbiased evaluation of deep metric learning algorithms. arXiv preprint arXiv:1911.12528 (2019)

10. Gao, T., et al.: Neural snowball for few-shot relation learning. In: AAAI, pp. 7772–7779 (2020)

11. Goodfellow, I.J., Shlens, J., Szegedy, C.: Explaining and harnessing adversarial examples. arXiv preprint arXiv:1412.6572 (2014)

12. Han, X., et al.: Fewrel: a large-scale supervised few-shot relation classification dataset with state-of-the-art evaluation. arXiv preprint arXiv:1810.10147 (2018)

13. Hartigan, J.A., Wong, M.A.: Algorithm as 136: a k-means clustering algorithm. J. Roy. Stat. Soc. Ser. C (Appl. Stat.) 28(1), 100–108 (1979)

14. Hoffer, E., Ailon, N.: Deep metric learning using triplet network. In: International Workshop on Similarity-Based Pattern Recognition. pp. 84–92. Springer (2015)

15. Hu, X., Wen, L., Xu, Y., Zhang, C., Yu, P.S.: Selfore: self-supervised relational feature learning for open relation extraction. arXiv preprint arXiv:2004.02438 (2020)

16. Kingma, D.P., Ba, J.: Adam: a method for stochastic optimization. arXiv preprint arXiv:1412.6980 (2014)

17. Maaten, L., Hinton, G.: Visualizing data using T-SNE. J. Mach. Learn. Res. 9, 2579–2605 (2008)

18. Marcheggiani, D., Titov, I.: Discrete-state variational autoencoders for joint discovery and factorization of relations. Transactions of the Association for Computational Linguistics 4, 231–244 (2016)

19. Miyato, T., Maeda, S.i., Koyama, M., Ishii, S.: Virtual adversarial training: a regularization method for supervised and semi-supervised learning. IEEE Trans. Pattern Anal. Mach. Intell. 41(8), 1979–1993 (2018)

20. Movshovitz-Attias, Y., Toshev, A., Leung, T.K., Ioffe, S., Singh, S.: No fuss distance metric learning using proxies. In: Proceedings of the IEEE International Conference on Computer Vision, pp. 360–368 (2017)

21. Oh Song, H., Jegelka, S., Rathod, V., Murphy, K.: Deep metric learning via facility location. In: Proceedings of the IEEE Conference on Computer Vision and Pattern Recognition, pp. 5382–5390 (2017)

22. Pennington, J., Socher, R., Manning, C.D.: Glove: global vectors for word representation. In: Proceedings of the 2014 Conference on Empirical Methods in Natural Language Processing (EMNLP), pp. 1532–1543 (2014)
23. Sandhaus, E.: The New ork times annotated corpus. Linguistic Data Consortium, Philadelphia **6**(12), e26752 (2008)
24. Simon, É., Guigue, V., Piwowarski, B.: Unsupervised information extraction: regularizing discriminative approaches with relation distribution losses. In: ACL 2019–57th Annual Meeting of the Association for Computational Linguistics, pp. 1378–1387. Association for Computational Linguistics, Florence, Italy, Jul 2019. https://doi.org/10.18653/v1/P19-1133, https://hal.archives-ouvertes.fr/hal-02318233
25. Soares, L.B., FitzGerald, N., Ling, J., Kwiatkowski, T.: Matching the blanks: distributional similarity for relation learning. arXiv preprint arXiv:1906.03158 (2019)
26. Socher, R., Chen, D., Manning, C.D., Ng, A.: Reasoning with neural tensor networks for knowledge base completion. In: Advances in Neural Information Processing Systems, pp. 926–934 (2013)
27. Sohn, K.: Improved deep metric learning with multi-class n-pair loss objective. In: Advances in Neural Information Processing Systems, pp. 1857–1865 (2016)
28. Srivastava, N., Hinton, G., Krizhevsky, A., Sutskever, I., Salakhutdinov, R.: Dropout: a simple way to prevent neural networks from overfitting. The journal of machine learning research 15(1), 1929–1958 (2014)
29. Suchanek, F.M., Kasneci, G., Weikum, G.: Yago: a core of semantic knowledge. In: Proceedings of the 16th International Conference on World Wide Web, pp. 697–706 (2007)
30. Wang, X., Hua, Y., Kodirov, E., Hu, G., Garnier, R., Robertson, N.M.: Ranked list loss for deep metric learning. In: Proceedings of the IEEE Conference on Computer Vision and Pattern Recognition pp. 5207–5216 (2019)
31. Ward Jr., J.H.: Hierarchical grouping to optimize an objective function. Journal of the American statistical association 58(301), 236–244 (1963)
32. Wu, R., et al.: Open relation extraction: Relational knowledge transfer from supervised data to unsupervised data. In: Proceedings of the 2019 Conference on Empirical Methods in Natural Language Processing and the 9th International Joint Conference on Natural Language Processing (EMNLP-IJCNLP), pp. 219–228 (2019)
33. Xiong, C., Power, R., Callan, J.: Explicit semantic ranking for academic search via knowledge graph embedding. In: Proceedings of the 26th International Conference on World Wide Web, pp. 1271–1279 (2017)
34. Yao, L., Riedel, S., McCallum, A.: Unsupervised relation discovery with sense disambiguation. In: Proceedings of the 50th Annual Meeting of the Association for Computational Linguistics (Volume 1: Long Papers), pp. 712–720 (2012)
35. Zeng, D., Liu, K., Lai, S., Zhou, G., Zhao, J.: Relation classification via convolutional deep neural network. In: Proceedings of COLING 2014, the 25th International Conference on Computational Linguistics: Technical Papers, pp. 2335–2344 (2014)
36. Zhang, K., et al.: Open hierarchical relation extraction. In: Proceedings of the 2021 Conference of the North American Chapter of the Association for Computational Linguistics: Human Language Technologies, pp. 5682–5693 (2021)

NS-Hunter: BERT-Cloze Based Semantic Denoising for Distantly Supervised Relation Classification

Tielin Shen, Daling Wang$^{(\boxtimes)}$, Shi Feng, and Yifei Zhang

School of Computer Science and Engineering, Northeastern University,
Shenyang, China
{shentielin,wangdaling,fengshi,zhangyifei}@cse.neu.edu.cn

Abstract. Distant supervision can generate large-scale relation classification data quickly and economically. However, a great number of noise sentences are introduced which can not express their labeled relations. By means of pre-trained language model BERT's powerful function, in this paper, we propose a BERT-based semantic denoising approach for distantly supervised relation classification. In detail, we define an entity pair as a source entity and a target entity. For the specific sentences whose target entities in BERT-vocabulary (one-token word), we present the differences of dependency between two entities for noise and non-noise sentences. For general sentences whose target entity is multi-token word, we further present the differences of last hidden states of [MASK]-entity (MASK-lhs for short) in BERT for noise and non-noise sentences. We regard the dependency and MASK-lhs in BERT as two semantic features of sentences. With BERT, we capture the dependency feature to discriminate specific sentences first, then capture the MASK-lhs feature to denoise distant supervision datasets. We propose NS-Hunter, a novel denoising model which leverages BERT-cloze ability to capture the two semantic features and integrates above functions. According to the experiment on NYT data, our NS-Hunter model achieves the best results in distant supervision denoising and sentence-level relation classification.

Keywords: Distant supervision · Relation classification · Semantic denoising

1 Introduction

Relation classification (RC) is a fundamental task in natural language processing. The goal of RC [26] is to identify the relation type in a sentence for a given entity pair, which is particularly important for the construction of knowledge bases. In recent years, deep learning has performed very well in relation extraction, but the technique needs a great number of labeled data, which is very expensive for manual tagging. In order to obtain a large amount of labeled RC data quickly and cheaply, distant supervision (DS) [16] was proposed to automatically generate

© Springer Nature Switzerland AG 2021
S. Li et al. (Eds.): CCL 2021, LNAI 12869, pp. 324–340, 2021.
https://doi.org/10.1007/978-3-030-84186-7_22

data by aligning a knowledge base with an unlabeled corpus. It is built on a weak assumption that if an entity pair has a relation in a knowledge base, all the sentences containing this entity pair will express the corresponding relation and exist in the dataset as a bag [16]. Based on such an assumption, a large number of noise sentences are generated by DS because many sentences can not express their labeled relation in fact. For example, in Fig. 1, sentence S-2 can not express the labeled relation "founder". These noise sentences such as S-2 will cause error propagation and may significantly reduce the performance of RC model.

Entity Pair and Labeled Relation

Entity1	Entity2	Relation
Jobs	Apple	founder

Two Example Sentences

S-1	Jobs was the co-founder and CEO of Apple.
S-2	During Jobs' tenure, Apple released four iPhones.

Fig. 1. Two sentences generated by DS method.

Since the pre-trained language model BERT was put forward [4], it has performed very well in the fully supervised RC datasets such as SemEval 2010 task 8 [22,25]. Different from DS, it is manually labeled, so there is no noise-sentences in it. Following these two works, even if we only keep two entities and delete the other parts of the sentence in test set, we can still get the F1-value of 49.99% (89.2% in MTB [22]) in 19-class fully supervised RC task. Only keeping entity pair in every sentence of test set can get such a high F1-value, which shows that BERT-based RC model will pay more attention to the entity pair itself rather than other words of the sentence. So, the model can not effectively discriminate noise and non-noise sentences, because in a bag generated by DS-method, noise and non-noise sentences have the same entity pair. Therefore, this method can not be directly used in sentence-level DS-RC.

Predicting masked word is one of the two pre-training tasks of BERT. Cui et al. [3] showed that there is a great deal of commonsense knowledge in BERT. After our verification, BERT can predict most blanks in general texts, which enables us to identify the noise-sentences in DS dataset.

In this paper, we propose NS-Hunter, a novel denoising model for DS-RC, which leverages BERT-cloze to capture semantic features of noise sentences in DS dataset (Short for **N**oise **S**entence **Hunter**). We define the entity pair in each sentence as a **source entity** and a **target entity**, and the rest of the sentence as **relation pattern**. Here the source entity is known, but the target entity needs to be predicted which may be head entity or tail entity. Our NS-Hunter is based on the following assumption: **in a non-noise sentence, the correct prediction of the target entity requires the both attendance of source entity and relation pattern.** The assumption restricts the dependency of the source entity and target entity based on the relation pattern marked, which means for a sentence, if the target entity can be predicted only based on one of the either source entity or relation pattern, it will be regarded as a noise sentence. We regarded the dependency of the source entity and the target entity as the first semantic feature for detecting noise sentence in DS datasets.

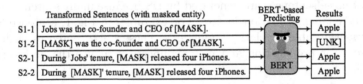

Fig. 2. An example of identifying noisy sentences.

According to the function of BERT, if a word in a sentence was replaced by [MASK] and then the revised sentence was input to BERT, BERT can make a reasonable prediction for the word at the position of [MASK] according to other words in the sentence. To explain the rationality of our assumption, we transformed the S-1 in Fig. 1 into S1-1 and S1-2 in Fig. 2, and also the S-2 in Fig. 1 into S2-1 and S2-2 in Fig. 2. For S1 in Fig. 1, when we do not mask Jobs, i.e. S1-1 in Fig. 2, BERT can predict Apple. When we mask Jobs with [MASK], i.e. S1-2 in Fig. 2, BERT can not predict Apple. That means Jobs and Apple have closely related dependency in S-1, so we think S-1 can express the labeled relation. We use the same method to judge S-2, whether we mask Jobs (i.e. S2-2) or not (i.e. S2-1), BERT can predict Apple, which shows that in S-2, Jobs and Apple are loosely related, so we think S-2 can not express the labeled relation, and it is a noise sentence. It is based on the first semantic feature, i.e. dependency feature, we can recognize a part of noise-sentences.

However, this method is only applicable to the sentences whose target entities in BERT-vocabulary (one-token word). In order to make our model can deal with multi-token word, we build some noise reducers, which are actually binary classifiers based on BERT. Their training sets and development sets are from the part that can be discriminated in the original training sets with our dependency feature, while their test sets are the whole original training set, so we can denoise the original training set for RC. For S2-1 and S2-2 in Fig. 2, the prediction results of [MASK] are all Apple, but the prediction results of [MASK] in S1-1 and S1-2 are different. So, we consider that there are semantic difference in the [MASK]-representation between noise and non-noise sentences, which can be captured by fully-connected layer. For the reason, we do not use the commonly used [CLS] feature, but concatenate last hidden states of [MASK] in transformed sentences as the feature (we call it as MASK-lhs), and regard MASK-lhs as the second semantic feature. We utilize MASK-lhs feature to denoise general sentences.

To the best of our knowledge, this is the first work of presenting semantic feature differences between noise and non-noise sentences, and implementing semantic denoising based on the features. We use the sentence-level test set in ARNOR tagged by Jia et al. [9] from NYT dataset for sentence-level evaluation of our NS-Hunter.

Overall, our contributions can be summarized as follows:

- We propose a novel model NS-Hunter, which denoises the datasets in DS for RC by leveraging BERT-cloze ability to capture our proposed two noise semantic features (dependency feature and MASK-lhs feature).
- The NS-Hunter we proposed is independent of RC, and it is a plug and play denoiseing model, which can be applied to any existing RC model. We verify the denoising ability of NS-Hunter on CNN [27] and BERT [4].
- We conduct experiments by using ARNOR dataset from NYT. The results show our NS-Hunter model achieves state-of-the-art results.

2   Related Work

Neural network based models have performed very well in RC [23]. However, training effective neural classifiers requires a large amount of labeled data, which is usually hard to obtain. DS [16] provides a way to create massive weakly labeled data for RC.

Many studies train RC model in DS by applying multi-instance learning (MIL) to reduce the impact of noise-sentences [8,12,13], which relaxes the label of each instance to a bag of sentences containing the same entity pair. Some MIL-based studies introduce adversarial training [7]. MIL assumes at least one sentence in a bag was labeled correctly. When all sentences in a bag are noise-sentences, MIL still suffers from noise [11,19]. Moreover, these MIL-based approaches are designed and tested for a pair of entities [10,20], and they are not suitable for sentence-level RC. Alternatively, some studies evaluate and select training instances individually without relying on the at-least-one assumption [5,9,18,28]. They usually rely on the classification effect to denoise. It is difficult to measure the denoising ability alone, and some of them need the pure data labeled manually [1,17].

Our approach uses the commonsense knowledge in the pre-training language model (PLM) to measure whether the two entity pairs are closely related, does not rely on assumption of MIL and needs no manually labeled data. Our denoising model is independent of classification process. It is plug and play, so can be used with all the above models.

Recently, more and more PLMs have adopted the method of predicting masked word to learn grammar and semantics, such as Roberta [15], Electra [2], ERNIE [29] etc. With more and more parameters and larger corpus for training, it can be expected that their cloze-accuracy will be higher and higher and the DS-RC model based on these PLMs will have stronger denoising ability.

Moreover, CASREL [24] is BERT-based RC model, they extract entities and relations jointly. This model use DS datasets, but the sentences labeled as NA were deleted. So, we can not compare the denoising ability of NS-Hunter with it.

3 Model

In this paper, we propose NS-Hunter, a novel BERT-based denoising model for DS-RC shown as Fig. 3. For a DS dataset with n classes and a NA (no relation) class, we construct a noise reducer with BERT for every class except the NA. Finally, the denoised training set is composed of n classes after denoising and a NA class in the original training set. In Fig. 3, the left is the overall framework and the right is the detail of training k-th noise reducer for Class k ($1 \leq k \leq n$) dataset.

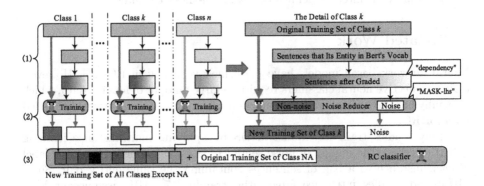

Fig. 3. NS-Hunter model. For each class, the lighter color on top means dataset that mixed with noise sentences and non-noise sentences, the gradual color in the middle means the sorted sentences, and the dark color below means the non-noise sentences we extracted. We do the same operation for each class and take the non-noise sentences of each class and the sentences labeled NA in the original dataset as the new training set. (Color figure online)

NS-Hunter consists of three parts (see (1)–(3) on the left side of Fig. 3). The first part is to discriminate the sentences whose target entities in BERT-vocabulary, then we get n small-scale pure binary datasets, which is labeled as noise and non-noise. The second part is to train noise reducers with n small-scale datasets. In this part, we design a novel feature MASK-lhs that can effectively capture the semantic differences between noise and non-noise sentences. It should be noted that the first two parts are for the single relation class, that is why we train a noise reducer for each class. The third part is the application of the denoised training set. After denoising, we get a training set with the same form as the original training set, but more pure. We introduce each part in Sect. 3.1–Sect. 3.3, and finally give the algorithm description of NS-hunter.

3.1 Source Entity and Target Entity

The source entity and the target entity are mentioned above, and here we need the target entity existing in BERT-vocabulary when discriminating noise sentences. We assume that in a non-noise sentence, the correct prediction of the target entity requires the both attendance of source entity and relation pattern. Therefore, the target entity must be semantically predictable compared to the source entity. For the example entity pair "Europe" and "Norway" labeled "contains", there is a sentence "Norway is a country in northern Europe" including this entity pair. The relation pattern of this sentence is "* is a country in northern *". However, from the perspective of semantics, it is very difficult to predict "Norway" based only on the relation pattern and "Europe". Obviously, the relation of "Norway" and "Europe" is 1-to-many, here "many" means "Europe" contains many countries, such as "Finland" and "Sweden", which are also semantically reasonable. If "Norway" is the target entity, our NS-Hunter will judge this sentence as noise sentence, but this sentence can actually express "contains" relation. Therefore, "Europe" is the correct target entity, that is to say, we will select a entitiy with a larger scope in the 1-to-many relation as the target entity. According to this method, we can label a source entity and a target entity for each class of dataset. In addition, if the relation between the two entities is 1-to-1 or many-to-many, we can select any one as the target entity.

3.2 Discrimination on Dependency Features

There are many entity-pairs with at least two relations (EPO) in the dataset. For example, Biden and the United States have both "place of birth" and "president" relation. According to the DS method, the sentence "Biden is the president of the United States" is labeled as the relation "place of birth" in the dataset. If we only build a noise reducer for the training set, this sentence will be considered as non-noise sentence because this sentence can express the relation of "president". However, according to its "place of birth" label, it is a noise sentence because it can not express its labeled relation "place of birth". As shown in Fig. 3, we need to build a noise reducer for each class to avoid the influence of EPO on denoising. According to DS method, there is no noise sentence in the NA class, so we do not denoise the NA.

According to dependency features, in a non-noise sentence, the correct prediction of the target entity requires the attendance of both source entity and relation pattern, so we believe the three parts (source entity, target entity and relation pattern) of this sentence are closely related. If the target entity can be predicted only based on either relation pattern (mask source entity with [UNK]) or source entity (delete relation pattern), we think that the three parts of this sentence are loosely related and this sentence is noisy. When predicting masked word in the pre-training of BERT, the last hidden states of [MASK] will pass through a fully connected layer shaped (768, 30522), the numbers in the output represent the possibility that each word in the BERT-vocabulary may appear in the [MASK] position. Therefore, when we predict target entity, we take the corresponding number of the target entity as the prediction score shown as Formula 1. That is why we can only discriminate the sentence whose target entity is one-token word at the beginning.

$$G = g\left(s, en_t\right) \qquad (1)$$

where s is transformed sentence, en_t is target entity, g denotes function based on BERT and G is corresponding number of the target entity.

As mentioned above, we can discriminate a part of original training set and train the noise reducers with the pure datasets after discriminating. Here we believe that if the higher G is when source entity and relation pattern attend together, and the lower G is when source entity or relation pattern attend alone, the higher the possibility that the sentence is not a noise sentence. We use Formula 2 to quantify this possibility.

$$G_s = g\left(en_s + rp, en_t\right) - g\left(rp, en_t\right) - f \qquad (2)$$

where G_s is the possibility that a sentence is not a noise sentence, en_s is source entity, $en_s + rp$ is transformed sentence such as S1-1 in Fig. 2 (target Apple) and means the attendance of both source entity and relation pattern. Moreover, rp represents the attendance of only relation pattern such as S1-2 in Fig. 2 (target Apple), and f represents the probability of predicting the target entity based only on the source entity's attendance.

$$f = max(g\left(en_s^{mp}, en_t\right), g\left(en_s^{mr}, en_t\right)) \qquad (3)$$

where en_s^{mp} and en_s^{mr} are artificial sentences, en_s^{mp} is "en_s [MASK]" and en_s^{mr} is "[MASK] en_s". In Fig. 1, if we target Apple, en_s^{mp} is "Jobs [MASK]" and en_s^{mr} is "[MASK] Jobs".

In this way, we can grade and sort the sentences whose target entities in BERT-vocabulary. In order to improve the confidence of binary datasets, we discard the middle parts of the sentence sets and take the first n and last n of the sentence sets as positive and negative samples to train the noise reducer of each class.

3.3   Denoising and Classification

After discriminating, in each class of the training set, we get a binary dataset including noise sentences and non-noise sentences. These datasets only contain one-token entities, so we trained a noise reducer for each class to discriminate noise sentences with multi-token entities. We stated that BERT can't be directly used for sentence level denoising in Sect. 1, so we designed a novel feature, MASK-lhs.

In Fig. 4, we use the non-noise sentence "Jobs was the co-founder and CEO of Apple" to illustrate our MASK-lhs. First, we construct S1-1 by masking the entity 'Apple' with [MASK] (record as MASK-1), and construct S1-2 by deleting the entity 'Jobs' and masking the entity 'Apple' with [MASK] (record as MASK-2). Then v_1 is used to represent the last hidden states of [MASK-1], and v_2 is used to represent the last hidden states of [MASK-2]. From the semantic point of view, v_1 is closer to 'Apple', while v_2 is farther from 'Apple', so we think that v_1 is very different from v_2. In contrast, in noise sentence like "During Jobs' tenure, Apple released four iPhones.", whether "Jobs" is deleted or not, BERT can predict "Apple", so we think that v_1 and v_2 in this noise sentence should be

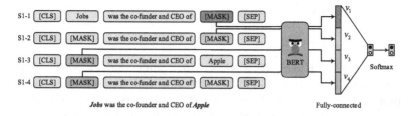

Fig. 4. The training process of our noise reducers. This figure shows the MASK-lhs feature of our training process with "**Jobs** was the co-founder and CEO of **Apple**".

Algorithm 1. Overview of NS-Hunter

Input : original training set *train*, BERT-base with its 110M parameters B, sentence nums n

Output: noise reducer for each class NS_i, relation classifier RC

1: split *train* into $N+1$ classes as $train_i$ according to the label
2: **for** $i = 1,2...N$, **do**
3:     divide entity pairs in $train_i$ into source entities and target entities according to the method in Sect. 3.1;
4:     filter out the sentences whose target entity exists in the Bert-vocabulary from $train_i$ as t_i;
5:     sort t_i according to the dependency feature;
6:     In t_i, the first n sentences are selected as positive examples, and the last n sentences are taken as negative examples to form a denoise dataset nt_i;
7:     training NS_i based B with nt_i;
8:     apply NS_i to denoise $train_i$, remove noise-sentences, get a pure dataset of class i pt_i;
9: training RC based B with $train_0$ and $\sum_{j=1}^{N} pt_j$;

---

more similar. We expect the model to capture this semantic feature to reduce noise.

To avoid missing some information, we perform the opposite operation on two entities to construct S1-3 and S1-4 (in Fig. 4). Signing the last hidden states of [MASK] in the two sentences as v_3, v_4. We concatenate v_1, v_2, v_3, v_4 and add a fully connected layer. In our noise reducer, both one-token entiey and multi-token entity are replace by [MASK], so our noise reducer can deal with a sentence whether its entity is a one-token word or a multi-token word.

After each noise reducer is trained, all the sentences in the original training set will be used as the test set, and the noise sentences will be found out and eliminated. After denoising of all classes, we get a new training set. We use the data pre-processing method in MTB [22], that is, to mark the position of two entities with special symbols. For example, S-1 in Fig. 1 is transformed into "#Jobs# was the co-founder and CEO of $Apple$". The steps of denoising and RC are shown in Algorithm 1.

4 Experiments

4.1 Dataset and Evaluation

We evaluate our NS-Hunter on a widely-used public dataset NYT, which is a news corpus sampled from 294k 1989–2007 New York Times news articles [16]. Most previous works commonly generate their test sets by DS method. Such a test set can only provide an approximate measure because there are many of noise sentences in it. In contrast, Jia et al. [9] published a complete dataset ARNOR 2.0.0[1] on the basis of the one released by Ren et al. [21] including a training set, develop set, test set and denoising dataset, in which the develop set, test set and denoising dataset are manually labeled. Jia et al. [9] removed some of the relation types which are overlapping and ambiguous or are too noisy to obtain a non-noise test sample. ARNOR 2.0.0 is the largest and most accurate dataset of sentence-level annotation at present. The denoising dataset could detect whether the model can recognize the noise sentence. We evaluate NS-Hunter on sentence-level (or instance-level) through this dataset and the details of this dataset are shown in Table 1 and Table 2.

Table 1. Statistics of the dataset in our experiments.

NYT	Training	Dev	Test
#Instances	353,650	4567	4484
#Positive instances	92707	975	1050

Table 2. The 11 relation types retained by Jia et al. [9] and statistics of them.

NYT	Training	Dev	Test
location/location/contains	51766	479	611
business/person/company	5595	113	105
people/person/place_lived	7197	198	185
people/person/nationality	8079	117	91
people/person/place_of_birth	3173	15	13
people/location/place_of_death	1936	14	8
location/country/capital	7690	15	14
business/company/place_founded	412	0	4
location/location/neighborhood_of	5553	7	3
business/company/founders	800	6	10
people/person/children	506	11	6

---

[1] https://github.com/PaddlePaddle/Research/tree/master/NLP/ACL2019-ARNOR.

Sentence-level RC is more friendly to reading comprehension tasks such as question answering and semantic analysis [5]. Different from the commonly used bag level evaluation, sentence-level evaluation directly calculates precision, recall and F1-values for all instances except NA in the dataset [21]. We think this evaluation method is more practical and suitable for a real world application.

4.2 Implementation Details

As we mentioned above, our NS-Hunter consists of three parts. In the first part (shown in Fig. 3), we separate the entity pairs of each class except NA, and the results are shown in Table 3.

After grading and sorting the sentences, for Class k, we set:

$$n_k = min(150, 0.3 \times l_k) \tag{4}$$

and take the first n_k sentences as positive samples, and the last n_k as negative samples. Where l_k is the number of sentence whose target entity in BERT-vocablary in Class k. We separate 30% of $2n_k$ sentences into a development set.

Table 3. Target entity of 11 relation types.

NYT	Head	Tail	Target
contains	location	location	head
company	person	business	tail
place_lived	people	location	tail
nationality	people	nation	tail
place_of_birth	people	location	tail
place_of_death	people	location	tail
capital	country	location	head
place_founded	business	location	tail
neighborhood_of	location	location	head
founders	company	person	head
children	people	person	head

Our noise reducers and relation classifier are based on BERT, the proportion of the one-token-entity datasets to the original dataset is 68169/92707 after looking up the BERT vocabulary. We use BERT-base-uncased with 110M parameters and set learning rate to 2e−5, the batchsize to 4 and utilize Adam for optimization. Generally, the denoising of each class can be completed in 8 epochs. The relation classification will be completed in 1 epoch and we test every 1000 batches. We set max sentence length to 450 and use Nvidia GeForce RTX 2080 Ti for training. The whole experiment will be finished in two hours.

4.3 Baselines

We compare NS-Hunter with several denoising baselines including CNN + RL$_1$ [19], CNN + RL$_2$ [5], PCNN + ATT [12] and ARNOR [9]. The experimental results of these baselines are all from the implementation of Jia et al. [9]. In addition, we use the training data without denoising to classify the relation, so that we can see the good performance of our denoising method more intuitively.

4.4 Main Results

We compare NS-Hunter model with four denoising baselines. As shown in Table 4, NS-Hunter achieves state-of-the-art results in F1 metric. Results of baselines are from Jia et al.'s [9] implementation. Moreover, after denoising, we delete 55634 in 92403 relational sentences and significantly improve the precision without reducing the recall, which shows that NS-Hunter can effectively reduce the impact of noise sentences.

Table 4. Comparison of our NS-Hunter and other baselines. The first four methods are models for DS-RC.

Method	Dev			Test		
	Pre.	Rec.	F1	Pre.	Rec.	F1
CNN+RL$_1$ [19]	42.50	71.62	53.34	43.70	72.34	54.49
CNN+RL$_2$ [5]	42.69	72.56	53.75	44.54	73.40	55.44
PCNN+ATT [12]	**82.41**	34.10	48.24	**81.00**	35.50	49.37
ARNOR [9]	78.14	59.82	67.77	79.70	62.30	69.93
BERT without denoising [4]	43.97	**77.32**	56.06	48.20	**78.85**	60.13
NS-Hunter (our model)	67.53	71.90	**69.65**	69.92	74.38	**72.08**

4.5 Denoising

Our NS-Hunter reduces noise by capturing semantic differences between noise and non-noise sentences. In ARNOR 2.0.0 [9], there is a denoising dataset which includes 466 non-NA sentences labeled as noise and non-noise manually.The experimental results in Table 5 show that the effect of our NS-Hunter improved by 9.72% compared with ARNOR.

Table 5. The experimental results of our NS-Hunter and two baselines on the denoise dataset.

Denoise	Pre.	Rec.	F1
CNN+RL$_2$	41.35	**94.83**	57.59
ARNOR	72.04	74.01	73.01
NS-Hunter	**81.31**	84.19	**82.73**

4.6   Effects of Our MASK-lhs Feature

We illustrated in Sect. 1 that commonly-used [CLS] feature is not suitable for denoising DS dataset because the noise and non-noise sentences in a bag have the same entity pair. For the reason, we design a novel MASK-lhs feature (Fig. 4), which can reduce noise by capturing the semantic differences between noise and non-noise sentences. In order to verify the superiority of the MASK-lhs feature, when training 11 classifiers in ARNOR dataset, we take the [CLS] feature as the baseline, and compare the denoising and RC effect of the model on the development set, test set and denoise set.

Table 6. Experimental results of our MASK-lhs and commonly used CLS features.

Feature	Pre.	Rec.	F1
CLS	**84.33**	67.21	74.80
MASK-lhs	81.31	**84.19**	**82.73**

Table 7. Experimental results of our MASK-lhs and commonly used CLS features on the RC dataset.

Features		Pre.	Rec.	F1
Dev	CLS	52.88	71.23	60.70
	MASK-lhs	**67.53**	**71.90**	**69.65**
Test	CLS	56.12	73.24	63.55
	MASK-lhs	**69.92**	**74.38**	**72.08**

The experimental results in Table 6 show the denoising effect of [CLS] feature and MASK-lhs feature on the denoise dataset. The experimental results in Table 7 show the RC effect of two features on the RC dataset. It can be seen that our MASK-lhs feature has increased by 7.29% in noise reduction and about 9% in RC task. Experiments show that our MASK-lhs feature can actually capture the semantic differences between noise and non-noise sentences and can better denoise DS-RC dataset than [CLS].

4.7   Apply NS-Hunter to CNN

In our NS-Hunter, the denoising part and the RC part are separated. The denoising method is plug and play for any other RC model sucn as CNN, and can also improve the classification effect. We train CNN-RC model with the denoised training set. The experimental results of the original training set come from ARNOR [9] and we use the same settings on the experiments of denoised training set. The comparison is shown in Table 8.

Table 8. Experiment results of CNN-RC model before and after using our NS-Hunter.

CNN RC		Pre.	Rec.	F1
Dev	Original	39.27	**73.80**	51.26
	Denoising	**66.56**	64.51	**65.52**
Test	Original	42.41	**76.64**	54.60
	Denoising	**67.73**	64.95	**66.31**

4.8 Effect of Entity on Denoising

Our NS-Hunter reduces noise by capturing semantic differences between noise and non-noise sentences. For example, although entity "New York" is very common, it is not in BERT-vocabulary, so "New York" corresponds to two vectors in BERT's hidden state. However, in the process of training noise reducers, "New York" is replaced by [MASK], and we hope to obtain the semantic features of "New York" from a single vector corresponding to [MASK]. According to our method, this will bring some errors. Therefore, for a single class, the percentage of entities in the BERT-vocabulary should be related to the improvement of the F1-value after denoising.

In the NYT development set and test set we used, there are only four classes with more than 50 sentences. As we all know, the percentage of location in the BERT-vocabulary is far greater than that name of people. So, we show the impact of denoising module in these four classes respectively in Table 9 where **Pct.** is the percentage of original training set entities in each class included in the BERT-vocabulary.

Table 9. The improvement of F1-value in four classes after denoising.

Class	Pct.	F1 increase	
		Dev	Test
Contains	59%	**24.31**	**21.24**
Company	11%	4.76	5.32
Place-lived	40%	4.28	9.19
Nationality	48%	16.41	10.44

It can be seen from the Table 9 that the class "contains" with the largest percentage gets the best improvement after denoising, followed by the "nationality". The percentage of the class "company" is the smallest, and its improvement is the worst. This shows that our NS-Hunter can reduce noise according to the design principle.

5   Conclusion

After carefully observing the RC dataset generated by the DS method, we present the two semantic features, i.e. dependency and MASK-lhs feature, and propose a BERT-based denoising model NS-Hunter and a denoising approach based on the two semantic features for DS-RC. We present the dependency feature of the entity pair and use BERT-cloze to discriminate some specific sentences with BERT-vocabulary based on the dependency feature, which has strong interpretability. For general sentences generated by the DS method, we designed a novel MASK-lhs feature to capture the semantic differences between noise and non-noise sentences for denoising. The performance of NS-Hunter is better than several other denoising baselines based on CNN, PCNN and BiLSTM. Significant improvements have been made in denoising and RC task. Our denoising method can also be easily combined with other RC methods.

Because we train noise reducers for each class, even if the knowledge is updated (such as the president of a country changes), our noise reducers are still robust according to the existing similar relation patterns in the training set. However, our model may not perform well in some professional domain such as biological, because BERT generates language representation from general corpus and lacks domain-specific knowledge [14]. It may get better results if we continue to pre-training BERT in a large scale of professional texts before applying our denoising model [6].

Acknowledgements. The work was supported by the National Key R&D Program of China under grant 2018YFB1004700 and National Natural Science Foundation of China (61772122, 61872074).

References

1. Beltagy, I., Lo, K., Ammar, W.: Combining distant and direct supervision for neural relation extraction. In: Burstein, J., Doran, C., Solorio, T. (eds.) Proceedings of the 2019 Conference of the North American Chapter of the Association for Computational Linguistics: Human Language Technologies, NAACL-HLT 2019, Minneapolis, MN, USA, June 2–7, 2019, Volume 1 (Long and Short Papers), pp. 1858–1867. Association for Computational Linguistics (2019). https://doi.org/10.18653/v1/n19-1184

2. Clark, K., Luong, M., Le, Q.V., Manning, C.D.: ELECTRA: pre-training text encoders as discriminators rather than generators. In: 8th International Conference on Learning Representations, ICLR 2020, Addis Ababa, Ethiopia, April 26–30, 2020. OpenReview.net (2020). https://openreview.net/forum?id=r1xMH1BtvB

3. Cui, L., Cheng, S., Wu, Y., Zhang, Y.: Does bert solve commonsense task via commonsense knowledge? (2020)

4. Devlin, J., Chang, M., Lee, K., Toutanova, K.: BERT: pre-training of deep bidirectional transformers for language understanding. CoRR abs/1810.04805 (2018). http://arxiv.org/abs/1810.04805

5. Feng, J., Huang, M., Zhao, L., Yang, Y., Zhu, X.: Reinforcement learning for relation classification from noisy data. In: McIlraith, S.A., Weinberger, K.Q. (eds.) Proceedings of the Thirty-Second AAAI Conference on Artificial Intelligence, (AAAI-18), the 30th innovative Applications of Artificial Intelligence (IAAI-18), and the 8th AAAI Symposium on Educational Advances in Artificial Intelligence (EAAI-18), New Orleans, Louisiana, USA, 2–7 February 2018, pp. 5779–5786. AAAI Press (2018). https://www.aaai.org/ocs/index.php/AAAI/AAAI18/paper/view/17151

6. Gururangan, S., et al.: Don't stop pretraining: adapt language models to domains and tasks (2020)

7. Han, X., Liu, Z., Sun, M.: Denoising distant supervision for relation extraction via instance-level adversarial training. CoRR abs/1805.10959 (2018). http://arxiv.org/abs/1805.10959

8. Ji, G., Liu, K., He, S., Zhao, J.: Distant supervision for relation extraction with sentence-level attention and entity descriptions. In: Singh, S.P., Markovitch, S. (eds.) Proceedings of the Thirty-First AAAI Conference on Artificial Intelligence, 4–9 February 2017, San Francisco, California, USA, pp. 3060–3066. AAAI Press (2017). http://aaai.org/ocs/index.php/AAAI/AAAI17/paper/view/14491

9. Jia, W., Dai, D., Xiao, X., Wu, H.: ARNOR: attention regularization based noise reduction for distant supervision relation classification. In: Korhonen, A., Traum, D.R., Màrquez, L. (eds.) Proceedings of the 57th Conference of the Association for Computational Linguistics, ACL 2019, Florence, Italy, 28 July– 2 August 2019, Volume 1: Long Papers, pp. 1399–1408. Association for Computational Linguistics (2019). https://doi.org/10.18653/v1/p19-1135

10. Li, P., Zhang, X., Jia, W., Zhao, H.: GAN driven semi-distant supervision for relation extraction. In: Burstein, J., Doran, C., Solorio, T. (eds.) Proceedings of the 2019 Conference of the North American Chapter of the Association for Computational Linguistics: Human Language Technologies, NAACL-HLT 2019, Minneapolis, MN, USA, 2–7 June 2019, Volume 1 (Long and Short Papers), pp. 3026–3035. Association for Computational Linguistics (2019). https://doi.org/10.18653/v1/n19-1307

11. Li, Y., et al.: Self-attention enhanced selective gate with entity-aware embedding for distantly supervised relation extraction. CoRR abs/1911.11899 (2019). http://arxiv.org/abs/1911.11899

12. Lin, Y., Shen, S., Liu, Z., Luan, H., Sun, M.: Neural relation extraction with selective attention over instances. In: Proceedings of the 54th Annual Meeting of the Association for Computational Linguistics, ACL 2016, 7–12 August 2016, Berlin, Germany, Volume 1: Long Papers. The Association for Computer Linguistics (2016). https://doi.org/10.18653/v1/p16-1200

13. Liu, T., Wang, K., Chang, B., Sui, Z.: A soft-label method for noise-tolerant distantly supervised relation extraction. In: Palmer, M., Hwa, R., Riedel, S. (eds.) Proceedings of the 2017 Conference on Empirical Methods in Natural Language Processing, EMNLP 2017, Copenhagen, Denmark, 9–11 September 2017, pp. 1790–1795. Association for Computational Linguistics (2017). https://doi.org/10.18653/v1/d17-1189

14. Liu, W., et al.: K-bert: enabling language representation with knowledge graph (2019)

15. Liu, Y., et al.: Roberta: a robustly optimized BERT pretraining approach. CoRR abs/1907.11692 (2019). http://arxiv.org/abs/1907.11692

16. Mintz, M., Bills, S., Snow, R., Jurafsky, D.: Distant supervision for relation extraction without labeled data. In: Su, K., Su, J., Wiebe, J. (eds.) ACL 2009, Proceedings of the 47th Annual Meeting of the Association for Computational Linguistics and the 4th International Joint Conference on Natural Language Processing of the AFNLP, 2–7 August 2009, Singapore, pp. 1003–1011. The Association for Computer Linguistics (2009). https://www.aclweb.org/anthology/P09-1113/

17. Pershina, M., Min, B., Xu, W., Grishman, R.: Infusion of labeled data into distant supervision for relation extraction. In: Proceedings of the 52nd Annual Meeting of the Association for Computational Linguistics, ACL 2014, 22–27 June 2014, Baltimore, MD, USA, Volume 2: Short Papers, pp. 732–738. The Association for Computer Linguistics (2014). https://doi.org/10.3115/v1/p14-2119

18. Qin, P., Xu, W., Wang, W.Y.: DSGAN: generative adversarial training for distant supervision relation extraction. In: Gurevych, I., Miyao, Y. (eds.) Proceedings of the 56th Annual Meeting of the Association for Computational Linguistics, ACL 2018, Melbourne, Australia, July 15–20, 2018, Volume 1: Long Papers, pp. 496–505. Association for Computational Linguistics (2018). https://doi.org/10.18653/v1/P18-1046

19. Qin, P., Xu, W., Wang, W.Y.: Robust distant supervision relation extraction via deep reinforcement learning. In: Gurevych, I., Miyao, Y. (eds.) Proceedings of the 56th Annual Meeting of the Association for Computational Linguistics, ACL 2018, Melbourne, Australia, 15–20 July 2018, Volume 1: Long Papers, pp. 2137–2147. Association for Computational Linguistics (2018). https://doi.org/10.18653/v1/P18-1199. https://www.aclweb.org/anthology/P18-1199/

20. Qu, J., Hua, W., Ouyang, D., Zhou, X., Li, X.: A fine-grained and noise-aware method for neural relation extraction. In: Zhu, W., et al. (eds.) Proceedings of the 28th ACM International Conference on Information and Knowledge Management, CIKM 2019, Beijing, China, 3–7 November 2019, pp. 659–668. ACM (2019). https://doi.org/10.1145/3357384.3357997

21. Ren, X., et al.: Cotype: joint extraction of typed entities and relations with knowledge bases. In: Barrett, R., Cummings, R., Agichtein, E., Gabrilovich, E. (eds.) Proceedings of the 26th International Conference on World Wide Web, WWW 2017, Perth, Australia, 3–7 April 2017, pp. 1015–1024. ACM (2017). https://doi.org/10.1145/3038912.3052708

22. Soares, L.B., FitzGerald, N., Ling, J., Kwiatkowski, T.: Matching the blanks: Distributional similarity for relation learning. In: Korhonen, A., Traum, D.R., Màrquez, L. (eds.) Proceedings of the 57th Conference of the Association for Computational Linguistics, ACL 2019, Florence, Italy, 28 July–2 August 2019, Volume 1: Long Papers, pp. 2895–2905. Association for Computational Linguistics (2019). https://doi.org/10.18653/v1/p19-1279

23. Wang, L., Cao, Z., de Melo, G., Liu, Z.: Relation classification via multi-level attention CNNs. In: Proceedings of the 54th Annual Meeting of the Association for Computational Linguistics, ACL 2016, 7–12 August 2016, Berlin, Germany, Volume 1: Long Papers. The Association for Computer Linguistics (2016). https://doi.org/10.18653/v1/p16-1123

24. Wei, Z., Su, J., Wang, Y., Tian, Y., Chang, Y.: A novel cascade binary tagging framework for relational triple extraction (2019)

25. Wu, S., He, Y.: Enriching pre-trained language model with entity information for relation classification. In: Zhu, W., et al. (eds.) Proceedings of the 28th ACM International Conference on Information and Knowledge Management, CIKM 2019, Beijing, China, 3–7 November 2019, pp. 2361–2364. ACM (2019). https://doi.org/10.1145/3357384.3358119

26. Zelenko, D., Aone, C., Richardella, A.: Kernel methods for relation extraction. J. Mach. Learn. Res. **3**, 1083–1106 (2003) http://jmlr.org/papers/v3/zelenko03a.html

27. Zeng, D., Liu, K., Lai, S., Zhou, G., Zhao, J.: Relation classification via convolutional deep neural network. In: Hajic, J., Tsujii, J. (eds.) COLING 2014, 25th International Conference on Computational Linguistics, Proceedings of the Conference: Technical Papers, 23–29 August 2014, Dublin, Ireland, pp. 2335–2344. ACL (2014). https://www.aclweb.org/anthology/C14-1220/

28. Zeng, X., He, S., Liu, K., Zhao, J.: Large scaled relation extraction with reinforcement learning. In: McIlraith, S.A., Weinberger, K.Q. (eds.) Proceedings of the Thirty-Second AAAI Conference on Artificial Intelligence, (AAAI-18), the 30th innovative Applications of Artificial Intelligence (IAAI-18), and the 8th AAAI Symposium on Educational Advances in Artificial Intelligence (EAAI-18), New Orleans, Louisiana, USA, 2–7 February 2018, pp. 5658–5665. AAAI Press (2018). https://www.aaai.org/ocs/index.php/AAAI/AAAI18/paper/view/16257

29. Zhang, Z., Han, X., Liu, Z., Jiang, X., Sun, M., Liu, Q.: ERNIE: enhanced language representation with informative entities. In: Korhonen, A., Traum, D.R., Màrquez, L. (eds.) Proceedings of the 57th Conference of the Association for Computational Linguistics, ACL 2019, Florence, Italy, 28 July 28–2 August 2019, Volume 1: Long Papers, pp. 1441–1451. Association for Computational Linguistics (2019). https://doi.org/10.18653/v1/p19-1139

A Trigger-Aware Multi-task Learning for Chinese Event Entity Recognition

Yangxiao Xiang[1(✉)] and Chenliang Li[2(✉)]

[1] School of Electrical and Electronic Engineering, Huazhong University
of Science and Technology, Wuhan, China
[2] School of Cyber Science and Engineering, Wuhan University, Wuhan, China
`cllee@whu.edu.cn`

Abstract. This paper tackles a new task for event entity recognition (EER). Different from named entity recognizing (NER) task, it only identifies the named entities which are related to a specific event type. Currently, there is no specific model to directly deal with the EER task. Previous named entity recognition methods that combine both relation extraction and argument role classification (named NER+TD+ARC) can be adapted for the task, by utilizing the relation extraction component for event trigger detection (TD). However, these technical alternatives heavily rely on the efficacy of the event trigger detection, which have to require the tedious yet expensive human labeling of the event triggers, especially for languages where triggers contain multiple tokens and have numerous synonymous expressions (such as Chinese). In this paper, a novel **trigger-aware m**ulti-task learning framework (TAM), which jointly performs both trigger detection and event entity recognition, is proposed to tackle Chinese EER task. We conduct extensive experiments on a real-world Chinese EER dataset. Compared with the previous methods, TAM outperforms the existing technical alternatives in terms of $F1$ measure. Besides, TAM can accurately identify the synonymous expressions that are not included in the trigger dictionary. Moreover, TAM can obtain a robust performance when only a few labeled triggers are available.

Keywords: Event entity recognition · Chinese event trigger detection

1 Introduction

In this paper, we introduce a variant of named entity recognition (NER) task, which is event entity recognition (EER). Different from NER task, EER aims at identifying the named entities corresponding to a specific event type. Figure 1 demonstrates some examples of EER. In example a, text contains both the event type '交易违规' (illegal trading) and '涉嫌传销' (illegal pyramid selling). Obviously, the corresponding entity of event type '交易违规' (illegal trading) is the 'organization B'. Hence, the 'organization B' is defined as the event entity of the query event '交易违规' (illegal trading). Similarly, the event entity of query

© Springer Nature Switzerland AG 2021
S. Li et al. (Eds.): CCL 2021, LNAI 12869, pp. 341–354, 2021.
https://doi.org/10.1007/978-3-030-84186-7_23

event '涉嫌传销' (illegal pyramid selling) is 'organization C'. Clearly, EER can be seen as a selective NER task, which treats the 'text' and the 'query event' as the input, and identify the 'event entity' as output. EER is widely useful in many semantic applications such as public opinion analysis in financial domain, disease extraction in medical domain, etc.

Currently, however, there is no specific model to deal with EER task. To the best of our knowledge, mainly three types of techniques can be used to possibly achieve EER: vanilla NER models, question-answer (QA) models and NER+TD+ARC models. For the vanilla NER models, solving EER task may be difficult due to the lack of event information. QA models can effectively incorporate the event information in the form of 'question' input. However, how to derive the appropriate question with respect to EER is still unknown. The NER methods that incorporate both the relation extraction and argument role classification are the most relevant solutions to tackle EER. Specifically, we could adapt the relation extraction component for event trigger detection. These solutions extract all the named entities and the event triggers (*i.e.*, trigger detection (TD)) in a sentence, and then identify the relation between each named entity and the target event trigger respectively (*i.e.*, argument role classification (ARC)). Here, an event trigger is a word indicating the event type mentioned in the text. There are various forms of NER+TD+ARC models. [2] proposed a pipelined DMCNN model to perform TD and ARC tasks separately. However, the pipelined method shows poor performance for both TD and ARC, because it cannot capture the inner dependency between TD and ARC tasks. Afterwards, some joint models (JRNN [16], JMEE [15], etc.), which jointly perform TD and ARC tasks, are proposed. These models significantly improve the performance of the TD and ARC tasks. In addition, models which jointly perform NER, TD and ARC tasks are also proposed, which further improves performance [18]. Although NER+TD+ARC methods are applicable for EER tasks, however, these existing solutions have the following three defects.

- The three subtasks (NER, TD, and ARC) require enormous efforts for human labeling. In detail, NER requires to label all the entities in the text. TD requires to label event triggers as much as possible, and ARC requires to label the relationship between each entity and each event trigger.
- The ARC subtask is severely dependent on the event trigger detection. However in practice, event triggers are not always identifiable. On the one hand, event triggers do not always exist in the texts. In some cases, the event is expressed implicitly. Looking at example *b* demonstrated in Fig. 1, this sentence expresses an event '交易违规' (illegal trading) with no explicit event trigger. On the other hand, for some languages (such as Chinese) where some event triggers are typically composed by multiple tokens and have numerous synonymous expressions, it is challenging to compile all the synonymous expressions of these event triggers with limited human efforts. For example, there are many synonymous expressions for an event trigger '犯罪' (crime), such as '犯了罪' (crime), '犯下罪行' (crime), '犯法' (crime), etc. Last but not least, some triggers can be associated with multiple events. Takeing example

c in Fig. 1 as an example, trigger word '违法' (break the law) indicates both the event '涉嫌违法' (crime) and the event '高管负面' (negative news of the executive) happen. Therefore, multi-label classification for event triggers is needed, which is mainly overlooked in the existing works.

- Previous methods are not directly aimed at dealing with EER task, some redundant process may cause unnecessary training cost and result in poor performance, such as all the named entities have to be recognized.

Example a:公司A举报公司B违规减持公司C债券,而公司C法人此前曾因卷入传销案被捕入狱

  Company B was reported by Company A due to the illegal reduction of the bonds of Company C, the legal person of the Company C was previously arrested for Pyramid selling case.

Example b:公司A董事长提前获取了一些信息，并精准抛售了公司B1000万股

  The chairman of Company A got some inside information in advance, and accurately sold 10 million shares of the Company B

Example c:公司A执行董事被指控违法

  The executive director of Company A was accused by breaking the laws.

Fig. 1. The examples of EER task.

To this end, this paper proposes a novel trigget-aware multi-task learning framework (named TAM) to achieve EER task. TAM joints two networks to deal with the two subtasks. In the first stage of the TAM, a trigger detection network is proposed to perform the Chinese TD. It uses a local attention mechanism to catch the context information of each token, and then perform multi-label classification of each single token. The trigger detection network effectively improves the performance in intractable Chinese event trigger identification. On the one hand, the trigger detection network is capable of identifying the synonymous expressions of event triggers with limited human-labeling. On the other hand, it effectively solves the multi-label classification of the event triggers, which is rarely considered in the previous TD works. In the second stage, we introduce an event-featured transformer-CRF network to identify the event entities directly. It discards the previous entity-trigger-relation classification workflow. Instead, it identifies the event entities by full-text understanding with the help of the three event type relevant features. This alteration effectively increases the robustness of the model with different amount of trigger labeling, and significantly outperforms the NER-TD-ARC models in EER task where event triggers are not explicitly mentioned or more than one event are included in the sentence. The main contribution of this paper can be summarized as follows:

- We propose a novel trigger-aware multi-task learning framework for event entity recognition. TAM jointly combine a trigger detection network and an event-featured transformer-CRF to achieve the trigger detection and event entity recognition simultaneously.

- Experiments demonstrate that TAM significantly outperforms the existing technical alternatives for EER tasks with single event, multiple events and with no explicit event triggers. Besides, TAM also is effective at identifying the synonymous expressions of a Chinese event trigger, and is robust against the availability of the labeled triggers.

2   The Proposed Algorithm

The proposed TAM mainly consists of two components: *trigger detection network* for event trigger detection and *event-featured transformer-CRF* for event entity recognition. The architecture of the proposed TAM is illustrated in Fig. 2. At first, we briefly describe the process to build the event trigger dictionary.

2.1   Event Trigger Dictionary Construction

Formally, an event trigger is a word which indicates a specific event mentioned in a text. For the complex sentence involving multiple events and multiple entities (ref. Figure 1), the corresponding event triggers naturally provide semantic segmentation between the sub-sentences related to different event types. Therefore, event trigger detection is very helpful to solve EER task for the sentences with multiple events and multiple entities.

However, a Chinese event trigger is typically composed by several tokens and would have many synonymous expressions. Therefore, labeling Chinese event triggers is a tedious yet expensive task. We first utilize a simple discriminative measure based on the frequency statistics to obtain the trigger candidates. Specifically, we first perform Chinese word segmentation with an external Chinese NLP toolkit Jieba[1]. Then, for each word w, we calculate the relatedness $\beta_e(w)$ towards an event type e as follows:

$$\beta_e(w) = \frac{c(w,e)}{c(e)} \frac{c(\neg e)}{c(w, \neg e)} \tag{1}$$

where $c(w,e)$ is the number of sentences containing word w and belonging to event type e, $c(e)$ is the number of sentences of event type e. $c(\neg e)$ is the number of sentences of the other event types, $c(w, \neg e)$ is the number of the sentences containing word w but not belonging to event type e. We manually identify the words as the triggers for event type e by examining the words with high $\beta_e(w)$. Note that our trigger labeling is not exhaustive due to the numerous synonymous expressions of event triggers. That is, the built event trigger dictionary is limited on its coverage.

---

[1] https://github.com/fxsjy/jieba.

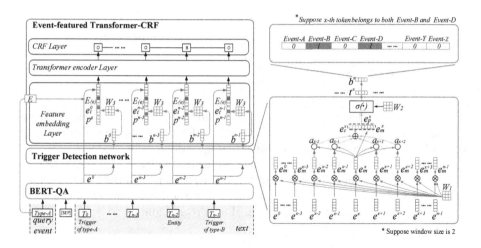

Fig. 2. The network architecture of TAM. (Color figure online)

2.2   Trigger Detection Network

BERT is one of the most popular pretrained language models recently, the resultant contextual embeddings have achieved the state-of-the-art performance in various NLP tasks. Here, we utilize a Chinese character based BERT-QA to obtain the contextual word embedding of each token in the given sentence. The structure of the BERT-QA is shown as the pink block in Fig. 2. Specifically, the sentence and query event are fed into BERT-QA, where text and query are separated by a special token '[SEP]'. Then, the contextual token embedding $\mathbf{e}^x \in R^{d_1}$ of x-th token in the sentence is obtained through the pre-trained Chinese Simplified character based BERT model[2], where d_1 is the dimension size. Note that our trigger labeling is not exhaustive due to the numerous synonymous expressions of event triggers. That is, the built event trigger dictionary is limited on its coverage. In the proposed TAM, we first learn a trigger detection network to identify the plausible triggers which will be utilized in entity recognition for the target event.

We apply a linear mapping by transferring each token embedding $\mathbf{e}^x$ into a feature vector $\mathbf{e}_m^x$ with a learnable weight matrix $\mathbf{W}_1 \in R^{d_2 \times d_1}$, in order to better adapt to the event trigger detection task. Note that a Chinese event trigger often contains several tokens. Here, we utilize a context window of length w centering at each token, $[\mathbf{e}_m^{x-w}, \cdots, \mathbf{e}_m^{x+w}]$, to help identify whether the token is a constituent of a trigger. This local context window could help us judge whether the token belongs to an event trigger and what event type it is. In detail, we utilize an attention network to help $\mathbf{e}_m^x$ find the most related contextual information:

---

[2] https://github.com/google-research/bert.

$$\alpha_k = \frac{\exp(\mathbf{e}_m^{x\top} \mathbf{e}_m^k)}{\sum_{j=x-w}^{x-1} \exp(\mathbf{e}_m^{x\top} \cdot \mathbf{e}_m^j) + \sum_{j=x+1}^{x+w} \exp(\mathbf{e}_m^{x\top} \cdot \mathbf{e}_m^j)} \qquad (2)$$

$$\mathbf{e}_c^x = \sum_{j=x-w}^{x-1} \alpha_j \cdot \mathbf{e}_m^j + \sum_{j=x+1}^{x+w} \alpha_j \cdot \mathbf{e}_m^j \qquad (3)$$

where $\mathbf{e}_c^x$ denotes the contextual information around x-th token. Then, the contextual information is concatenated with $\mathbf{e}_m^x$ to form the phrase embedding $\mathbf{e}_p^x$: $\mathbf{e}_p^x = \mathbf{e}_m^x \oplus \mathbf{e}_c^x$, where $\oplus$ is the concatenation operation.

With the phrase embedding, we then perform a multi-label classification to identify the event types of the current token. The structure of the event mapping layer is shown as the orange block in Fig. 1. Let the total number of the event types is d_3, the classification is performed as a regression:

$$\hat{\mathbf{t}}^x = \sigma(\mathbf{W}_2 \mathbf{e}_p^x + \mathbf{b}) \qquad (4)$$

where $\sigma(\cdot)$ is the nonlinear sigmoid function, $\mathbf{W}_2 \in R^{d_3 \times 2 \cdot d_2}$ is a learnable weight matrix. Each dimension of the $\hat{\mathbf{t}}^x \in R^{d_3}$ is the probability of the x-th token belonging to the corresponding event type, which is exhibited in the green block of Fig. 1. Finally, by using the event trigger dictionary as the supervision, the loss of the trigger detection networks is calculate as the sum of the cross-entropy loss of each token:

$$\mathcal{J}_{TD} = -\sum_{i=1}^{d_3} \sum_{j=1}^{n} \mathbf{t}^j(i) \log(\hat{\mathbf{t}}^j(i)) \qquad (5)$$

where $\mathbf{t}^j \in R^{d_3}$ is the groundtruth multi-hot vector of j-th token, n is the sentence length.

2.3   Event-Featured Transformer-CRF

In order to reduce the strong dependency between TD and ARC subtasks existed in the current technical alternatives, here, we introduce an event-featured transformer-CRF network (shown as the blue block in Fig. 1) to directly identify the event entities. Specifically, besides the contextual token embeddings produced by BERT-QA, the event-featured transformer-CRF incorporates three additional features (ie event type feature, trigger feature and position feature) related to the target event type to perform event entity recognition.

Event Type Feature. We utilize another learnable event type embedding matrix $\mathbf{E} \in R^{d_3 \times d_4}$ to encode the features related to each event type. Given the target event type e, we can adopt look-up operation to extract the corresponding feature embedding $\mathbf{E}(e)$. The utilization of the $\mathbf{E}(e)$ could enhance the model's awareness towards the target event type.

Trigger Feature. Since the trigger information provides both the information of what events occur in the sentence, we utilize two sets of trigger feature embeddings to inject the trigger information detected by the trigger detection network mentioned above. We first binarize the predicted event type vector $\hat{\mathbf{t}}^x$ of each token into vector $\mathbf{b}^x$ by utilizing 0.5 as the threshold. When $\hat{\mathbf{t}}^x(i) \geq 0.5$, the corresponding value $\mathbf{b}^x(i)$ is set to 1; otherwise, $\mathbf{b}^x(i)$ is set to 0. Then, we transform the $\mathbf{b}^x$ into trigger embedding $\mathbf{e}_t^x \in R^{d_5}$ by looking up a learnable matrix $\mathbf{W}_3$: $\mathbf{e}_t^x = \mathbf{W}_3 \mathbf{b}^x$. Here, $\mathbf{e}_t^x$ encodes the event type information of the x-th token in the sentence.

Position Feature. The position embeddings have been widely adopted in many relevant NLP tasks, such as relation extraction [2]. Here, we further encode the position of each token as a d_6 dimensional embedding $\mathbf{p}^x$, using the work presented in [4]. Under the combination of both position embeddings and trigger embeddings, the information of semantic segmentation between sub-sentences of different events could be extracted by the transformer-CRF network.

The above three kinds of event type relevant features are concatenated together with the contextual token embeddings produced by the BERT-QA to form the new token embeddings. Then, these updated token embeddings are then fed into a transformer-CRF layer for event entity recognition. It includes two sub-layers. The first layer is a transformer encoder model [4], which is a powerful feature extractor various NLP tasks. In the second layer, a conditional random field (CRF) is utilized to learn the correlation between the tag correlations. As for the labeling, we utilize the BIO scheme in CRF. The maximum conditional likelihood loss is used to update the parameters as follows:

$$\mathcal{J}_{EER} = -\sum_{i=x}^{n} \log p(y_x | t_x) \tag{6}$$

where $p(y_x | t_x)$ is the probability of assigning label y_x to token t_x, n is the length of the sentence, and y_x is the groundtruth label for token t_x.

2.4   Loss Function

Since TAM is devised with a multi-task learning paradigm, for model training, we minimize the joint negative log-likelihood loss function by combining $\mathcal{J}_{EER}$ and $\mathcal{J}_{TD}$ as:

$$\mathcal{J}_{total} = \mathcal{J}_{EER} + \lambda \mathcal{J}_{TD} \tag{7}$$

where λ is the hyper-parameter which controls the training weight of the two subtasks. We use the Adam algorithm to optimize the parameters with mini-batches. The gradients are computed with back-propagation.

3   Experiments

3.1   Experimental Setup

Dataset. We evaluate the proposed TAM on the CCKS2019-task4 (named CCK19) dataset[3]. CCKS19 dataset contains in total $19,500$ sentences from the financial news. After excluding the overlapping sentences in both training and test sets, There are $16,000$ sentences in training set and $2,400$ sentences in test set. It defines 22 negative event types (including type 'None'). For each item in training set, the event entity and the corresponding event type are annotated. For each item in test set, only query event type is used as an input, and the event entity are used for performance evaluation. It is worth mentioning that since the event triggers are not provided in CCKS19 dataset, we utilize the event trigger dictionary constructed in Sect. 2.1.

Baseline Methods. Currently, there is no direct model to deal with EER task. We compare TAM with the following related methods:

- BiLSTM+CRF was proposed by [8]. It achieves a good performance on vanilla NER tasks. We use the sentence and the annotated event entities while training, and only use the sentence while testing.
- BERT-QA was proposed by [4]. It achieves the state-of-the-art performance in question answer tasks. We use the sentence (paragraph) and the event type (question) as the input of the BERT-QA, and treat the event entity as the answer.
- DMCNN was proposed by [2]. It is a typical pipelined relation extraction method (TD-ARC). In our settings, The NER is realized by BiLSTM+CRF, TD and ARC are realized by DMCNN.
- JRNN was proposed by [16]. It is a typical joint relation extraction method (TD-ARC). In our settings, The NER is realized by BiLSTM+CRF, TD and ARC are realized by JRNN.
- Joint3EE was proposed by [18]. It is a typical joint relation extraction method which combines three subtasks (*i.e.*, NER-TD-ARC).

We further introduce several variants of TAM by excluding the trigger detection network (named TAM/TD), event type feature (named TAM/EE), position feature (named TAM/PE), both event type feature and position feature (named TAM/EE&PE), both trigger detection network and position feature (named TAM/TD&PE), both trigger detection network and event type feature (named TAM/TD&EE) to validate their impacts in an individual and combination manner. Note that when trigger detection network is excluded, trigger embedding e_t^x is equivalent to the zero vector.

---

[3] http://www.ccks2019.cn/?page_id=62.

Parameter Settings. For BiLSTM+CRF, DMCNN, JRNN and Joint3EE, word embeddings are obtained using vanilla BERT with a dimension size 768 for a fair comparison. The same dimension size applies for BERT-QA and TAM. For the trigger detection network used in TAM, the window size is set to 5. For the event-featured transformer-CRF network, the dimension size of each feature embedding is set to 48 (*i.e.,* $d_4 = d_5 = d_6 = 48$). The transformer encoding module utilizes 3 layers and 6 attention heads. Loss weight λ in Eq. 7 is set to 0.05. The dropout rate is 0.5, and we apply the Adam optimizer for parameter update. Three metrics: precision (P), recall (R) and $F1$, are utilized for performance comparison. We report the optimal $F1$ performance of each method on the test set.

Table 1. The performance of different methods in EER task. The best results are highlighted in boldface.

Method	With trigger						Without trigger			Total		
	Single event			Multiple event								
	P	R	$F1$	P	R	$F1$	P	R	$F1$	P	R	$F1$
BiLSTM-CRF	0.935	0.937	0.936	0.850	0.862	0.856	0.750	0.750	0.750	0.896	0.902	0.899
BERT-QA	0.949	0.945	0.947	0.920	0.923	0.922	0.848	**0.875**	**0.862**	0.935	0.935	0.935
DMCNN	0.952	0.949	0.951	0.895	**0.955**	0.924	**1.00**	0.094	0.171	0.926	0.940	0.933
JRNN	0.954	0.938	0.946	0.890	0.949	0.919	0.591	0.406	0.481	0.922	0.936	0.929
Joint3EE	**0.960**	0.940	0.951	0.890	0.948	0.918	0.647	0.344	0.449	0.926	0.936	0.931
TAM	0.958	**0.958**	**0.958**	**0.935**	0.941	**0.938**	0.848	**0.875**	**0.862**	**0.947**	**0.950**	**0.948**
TAM/TD	0.948	0.946	0.947	0.919	0.929	0.924	0.848	**0.875**	**0.862**	0.947	0.950	0.948
TAM/TD&PE	0.949	0.945	0.947	0.921	0.931	0.926	0.848	**0.875**	**0.862**	0.935	0.938	0.937
TAM/TD&EE	0.949	0.947	0.948	0.918	0.924	0.921	0.844	0.844	0.844	0.935	0.936	0.935
TAM/EE&PE	0.957	0.943	0.955	0.926	0.930	0.928	0.897	0.813	0.852	0.943	0.941	0.942
TAM/EE	0.955	0.952	0.953	0.927	0.936	0.931	0.867	0.813	0.839	0.942	0.944	0.943
TAM/PE	0.956	0.956	0.956	0.932	0.936	0.934	0.871	0.844	0.857	0.945	0.946	0.945

Table 2. The performance of different methods in TD task.

Method	Trigger detection		
	P	R	$F1$
DMCNN	0.929	0.840	0.882
JRNN	0.929	0.844	0.884
Joint3EE	0.928	0.844	0.884
TAM	**0.930**	**0.865**	**0.895**

3.2 Results and Discussion

Event Entity Recognition. The results of different models on EER task are reported in Table 1. In order to better demonstrate the traits of each kind of model, the test set are divided into three subsets: 1) sentences with a single event (1, 366 instances); 2) sentences with multiple explicit events (1, 005 instances) and sentences with no event trigger (29 instances).

According to Table 1, it can be seen that TAM outperforms the other baselines in all the three subsets. The BiLSTM-CRF model for vanilla NER delivers the worst performance in dealing with EER task. It is reasonable since no event information is incorporated.

By comparing TAM with TAM/TD and BERT-QA, we can see that since TD effectively provides the event type information and trigger position information, such as the three event type relevant features for transformer-CRF, incorporating the trigger detection network and the additional event type relevant features is notably beneficial for the downstream EER task with explicit event triggers, especially for the sentences with multiple events. However, for the sentences with no trigger, the utilization of TD shows trivial effect, BERT-QA, TAM/TD and TAM all show good performance in sentences with no event trigger. Note that we take the target event type as input into BERT-QA. The observation suggests that explicitly modeling event type information is very useful for EER task.

The three NER-TD-ARC based models (DMCNN, JRNN and Joint3EE) all show good performance in dealing sentence with single event and multiple events, because these models are all relying on event trigger detection. However, TAM shows better performance, especially in EER task with multiple events, due to its good performance in TD task. Since NER-TD-ARC based models are strongly dependent on event trigger detection, for the sentences where event is expressed implicitly (no event trigger exists), these models perform much worse instead. Since DMCNN performs relatively better overall, we take this model as a reference to examine the impact of different amounts of labeled triggers. Figure 3(a) plot the performance patterns of DMCNN and TAM with different rates of dropping out the labeled event triggers for supervision. The results shows that with few trigger labeling, the performance of DMCNN model experiences a large deterioration. In contrast, TAM experience only a very small performance decrease, which is much desired for real-world applications.

As to the several variants of TAM, we can clearly see that excluding each feature or their combinations lead to performance deterioration to some extent. Seperately applying the position feature (i.e., TAM/TD&EE) or event type feature (i.e., TAM/TD&PE) shows little effect on EER task. However, when solely applying TD, the performance of EER is much improved (i.e., TAM/EE&PE), which verifies that the upstream TD task is helpful for the downstream EER task. In addition, it is observed that the inclusion of the event type feature and position feature both effectively enhance the performance of the TD based EER task, especially for sentence with multiple events (i.e., TAM/EE and TAM/PE). It may because that the event type feature helps the model know which event trigger should be attentioned and the position feature offers the position information between entity candidates and event triggers.

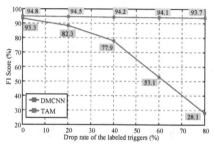

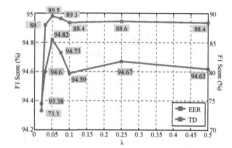

(a) The performance pattern of different numbers of labled triggers.

(b) The performance pattern of varying λ values.

Fig. 3. Ratios at different ranks (a), and performance with λ values (b).

Trigger Detection. To evaluate the performance of the trigger detection network, we randomly pick 240 sentences from the test set and check the detected event trigger manually. Some event triggers are not existing in the constructed trigger dictionary because they rarely appear in the training set, but they are the synonymous expressions of the triggers covered by the dictionary. The classification performance are listed in Table 2. As can be seen, the trigger detection network outperforms the other methods in terms of $F1$. Note that the synonymous expressions of the event triggers covered by the dictionary comprise only a small proportion of the test set. Hence, the improvement over the baseline methods for TD task is significant. Figure 4 gives two examples in which our model outperforms. It can be seen that TAM can effectively identify the synonymous expressions of the trigger but other methods fail. By including '财务造假' (financial fraud) and '非法吸收公众存款' (illegal absorption of public deposits) in the trigger dictionary, TAM successfully predicts their synonymous expression '财务弄虚作假' (financial fraud) and '非法吸储' (illegal absorption of public deposits) as triggers. Furthermore, the results also show that TAM precisely predicts the boundary of trigger phrases. These abilities are very useful for Chinese trigger detection, where numerous synonymous expressions exist and these triggers usually contains multiple tokens.

It is worth mentioning that the loss weight parameter λ has a vital influence on the performance of TAM. As can be seen in Fig. 3, the performance of the downstream task EER is highly related with the upstream TD task. When λ is too large, TAM losts its generation capability and becomes the same as the other TD models. Also, if λ is set too small, nothing will be learned. In our experiment, the trigger detection network shows generation capability among $[0.03, 0.1]$.

4 Related Work

Event entity recognition (EER) is a novel task which has wide applications in various scenarios such as public opinion analysis, etc. Since our work jointly

Example a：

中兴通讯(000063)财务弄虚作假疑问重重 夏草指其信息披露避重就轻

There exists many doubts for the financial fraud of ZTE(000063), Xiacao accused ZTE for intencially evading the crucial point of information disclosure

TAM：财务弄虚作假(Finacial Fraud)   Others：无 (None)

Example b：

北京恒昌柯桥分公司涉及受害人134人，合计追回非法吸储资金3300万元

134 victims are affected by Beijing Hengchang Keqiao branch, and totally 33 million yuan of illegal absorption of public deposits is recovered

TAM：非法吸储(Illegal absorption of public deposits)   Others：无 (None)

Fig. 4. Examples of trigger detection by TAM.

perform EER and TD, we will introduce the related works in these two tasks separately. For the TD task, various methods have been proposed to realize TD task by identifying the event trigger with supervision. In the early years, feature-based methods, which incorporated diverse semantic clues (such as lexical features, dependency features, etc.) into feature vectors, have long been used for TD task [1,7,11]. Recently, representation based methods have achieved the state-of-art performance. These methods represented the event mentions into embeddings, and then be sent to neural networks to classify the event type [2,13,17]. However, current TD models are mainly used in English corpus. For some languages where event triggers are generally in the form of nugget (composed by several tokens), the trigger boundary mismatch problem severely deteriorates the effectiveness of the TD models [3]. To solve that, [9,10] defined manually character compositional patterns for Chinese event triggers. [6] first applied a neural network based model FBRNNs to identify event trigger nugget. Besides, [14] proposed a nugget proposal network to realize Chinese TD by exploiting character compositional structure of Chinese event triggers. Although supervised TD models achieves a good performance in most TD tasks, these methods also require too many labeling labors. Some semi-supervised methods were proposed to relieve the cost of labeling [5,12].

For the EER task, to the best of our knowledge, there is no previous works directly dealing with it. Previous NER methods are a clue for solving EER since they both aim at identifying named entities. However, NER method lacks event information. Previous NER+TD+ARC models are another clues for dealing with EER. Difference lies on that NER+TD+ARC aims at classifying the role of the arguments for an event type, but EER aims at recognizing the named entities that are related to a specified event. Previous methods mainly focus on the TD and ARC, and the named entities were considered as known. [2] proposed DMCNN based method to pipeline realize TD and ARC, [15,16,19] jointly realized ED+ARC to catch the inner relation between the two tasks. [18] pointed that only concerning the TD+ARC had neglected the error propagation from the NER, and further improved the model by jointly realizing NER, TD and ARC in one model.

5 Conclusion

This paper proposed a trigger-aware multi-task model (TAM) to deal with a novel event entity extraction (EER) task. TAM is composed by two subtasks:a trigger detection network for event trigger identification and an event-featured transformer-CRF for event entity recognition. The former effectively tackles the intractable Chinese event trigger detection, where triggers usually contain multiple tokens and have numerous synonymous expressions. Experiments show that the trigger detection network is effective for identifying the synonymous expressions of the labeled event triggers, which is useful to the downstream EER subtask, which is also robust against the availability of labeled triggers. The latter incorporates three kinds of event type relevant features and avoid the dependency modeling between the event trigger and ARC existed in the relevant techniques. The experimental results demonstrate the superiority of the proposed TAM against the existing SOTA technical alternatives. Future work will be dedicated in further improving the performance of trigger detection.

References

1. Ahn, D.: The stages of event extraction. In: ARTE (2006)
2. Chen, Y., Xu, L., Liu, K., Zeng, D., Zhao, J.: Event extraction via dynamic multi-pooling convolutional neural networks. In: ACL, pp. 167–176 (2015)
3. Chen, Z., Ji, H.: Language specific issue and feature exploration in Chinese event extraction. In: NAACL-HLT, pp. 209–212 (2009)
4. Devlin, J., Chang, M., Lee, K., Toutanova, K.: BERT: pre-training of deep bidirectional transformers for language understanding. In: NAACL-HLT, pp. 4171–4186 (2019)
5. Ferguson, J., Lockard, C., Weld, D.S., Hajishirzi, H.: Semi-supervised event extraction with paraphrase clusters. In: NAACL-HLT, pp. 359–364 (2018)
6. Ghaeini, R., Fern, X.Z., Huang, L., Tadepalli, P.: Event nugget detection with forward-backward recurrent neural networks. In: ACL (2016)
7. Gupta, P., Ji, H.: Predicting unknown time arguments based on cross-event propagation. In: ACL-IJCNLP (2009)
8. Huang, Z., Xu, W., Yu, K.: Bidirectional LSTM-CRF models for sequence tagging, vol. abs/1508.01991 (2015)
9. Li, P., Zhou, G.: Employing morphological structures and Sememes for Chinese event extraction. In: COLING, pp. 1619–1634 (2012)
10. Li, P., Zhou, G., Zhu, Q., Hou, L.: Employing compositional semantics and discourse consistency in Chinese event extraction. In: EMNLP-CoNLL, pp. 1006–1016 (2012)
11. Liao, S., Grishman, R.: Using document level cross-event inference to improve event extraction. In: ACL, pp. 789–797 (2010)

12. Liao, S., Grishman, R.: Can document selection help semi-supervised learning? A case study on event extraction. In: ACL, pp. 260–265 (2011)
13. Liu, S., Chen, Y., He, S., Liu, K., Zhao, J.: Leveraging framenet to improve automatic event detection. In: ACL (2016)
14. Liu, S., Li, Y., Zhang, F., Yang, T., Zhou, X.: Event detection without triggers. In: NAACL-HLT, pp. 735–744 (2019)
15. Liu, X., Luo, Z., Huang, H.: Jointly multiple events extraction via attention-based graph information aggregation. In: EMNLP, pp. 1247–1256 (2018)
16. Nguyen, T.H., Cho, K., Grishman, R.: Joint event extraction via recurrent neural networks. In: NAACL-HLT, pp. 300–309 (2016)
17. Nguyen, T.H., Grishman, R.: Event detection and domain adaptation with convolutional neural networks. In: ACL, pp. 365–371 (2015)
18. Nguyen, T.M., Nguyen, T.H.: One for all: neural joint modeling of entities and events. In: AAAI, pp. 6851–6858 (2019)
19. Sha, L., Qian, F., Chang, B., Sui, Z.: Jointly extracting event triggers and arguments by dependency-bridge RNN and tensor-based argument interaction. In: AAAI, pp. 5916–5923 (2018)

Improving Low-Resource Named Entity Recognition via Label-Aware Data Augmentation and Curriculum Denoising

Wenjing Zhu, Jian Liu, Jinan Xu[✉], Yufeng Chen, and Yujie Zhang

School of Computer and Information Technology, Beijing Jiaotong University,
Beijing 100044, China
{18120461,jianliu,jaxu,chenyf,yjzhang}@bjtu.edu.cn

Abstract. Deep neural networks have achieved state-of-the-art performances on named entity recognition (NER) with sufficient training data, while they perform poorly in low-resource scenarios due to data scarcity. To solve this problem, we propose a novel data augmentation method based on pre-trained language model (PLM) and curriculum learning strategy. Concretely, we use the PLM to generate diverse training instances through predicting different masked words and design a task-specific curriculum learning strategy to alleviate the influence of noises. We evaluate the effectiveness of our approach on three datasets: CoNLL-2003, OntoNotes5.0, and MaScip, of which the first two are simulated low-resource scenarios, and the last one is a real low-resource dataset in material science domain. Experimental results show that our method consistently outperform the baseline model. Specifically, our method achieves an absolute improvement of 3.46% F_1 score on the 1% CoNLL-2003, 2.58% on the 1% OntoNotes5.0, and 0.99% on the full of MaScip.

Keywords: Deep learning · Named entity recognition · Data augmentation

1 Introduction

Named entity recognition (NER) is a fundamental natural language processing (NLP) task aiming to identify the names of people, places, organizations, and proper nouns in texts, which supports a wide range of downstream applications [12,15]. The current state-of-the-art methods for NER rely on abundant training data. However, manual annotation is expensive, which limits the effectiveness of the model, especially in bio-medicine and material chemistry domains [7]. Many studies have investigated NER in low-resource scenarios, by transferring pre-trained language representations on self-supervised or rich-resource domains to target domains [10,28]. Others use the knowledge base to

---

W. Zhu and J. Liu—Equal contribution.

S. Li et al. (Eds.): CCL 2021, LNAI 12869, pp. 355–370, 2021.
https://doi.org/10.1007/978-3-030-84186-7_24

semi-automatically label extra data for training [36]. Nevertheless, these methods usually require huge expertise knowledge to obtain good performance.

Data augmentation has been proven effective to alleviate data scarcity in many NLP tasks, including machine translation [8,31], text classification [33,34], question answering [27], etc. [21]. However, most existing studies focus on sentence-level tasks, which generate sentences via word replacement, swap, and deletion [21,33] or generative models [13,35]. Different from these sentence-level NLP tasks, NER predict entities on the token level. That is, for each token in the sentence, NER models predict a label indicating whether the token belongs to a mention and which entity type the mention has. Therefore, applying transformations to tokens may also change their labels. Due to this difficulty, data augmentation for NER is comparatively less studied.

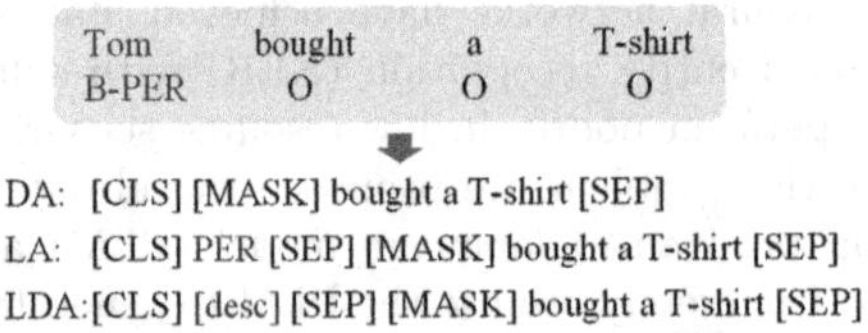

Tom	bought	a	T-shirt
B-PER	O	O	O

⬇

DA: [CLS] [MASK] bought a T-shirt [SEP]

LA: [CLS] PER [SEP] [MASK] bought a T-shirt [SEP]

LDA:[CLS] [desc] [SEP] [MASK] bought a T-shirt [SEP]

Fig. 1. The input format of different data augmentation methods. DA, LA, LDA correspond to the basic method, label additional method, and label description additional method. As shown in blue, LA puts the label at the beginning of the sentence, while LDA uses the description of the label. (Color figure online)

In this work, we propose a novel data augmentation framework for NER in low-resource scenarios, which generates examples with consistent labels and filters noises in the generated data. Concretely, our approach contains two complementary components: 1) *data augmentation via pre-trained BERT*, which uses contextualized information to predict a masked word for data augmentation, and 2) *data denoising via curriculum learning*, which filter noises in the augmented data for further boost learning. In *data augmentation via pre-trained BERT*, our basic idea is to predict the masked words through pre-trained language models (we use BERT [4] in this paper), and then replace original words with predicted words to generate new sentences. However, directly using BERT for prediction may generate some words that mismatch the original labels. As shown in Fig. 1, given a sentence *"Tom bought a T-shirt"*, we replace *"Tom"* with [MASK] and predict it by BERT. The predicted words may be third-person pronouns like *"he"*, *"she"* or wrong words, which causes mismatch between original labels and generated words. In response to this issue, we propose a label-aware data augmentation method, which considers the label information to make the predicted words match the original label more closely. In *data denoising via curriculum learning*, we propose a new method based on curriculum learning to filter augmented data. Considering that the synthetic data still contain noises, we design three evaluation metrics to measure the generated examples by confidence, and then use curriculum learning strategy to filter noises. Consequently, the performance is significantly improved.

To evaluate the effectiveness of our approach, we conduct experiments in both simulated low-resource scenarios and real-world low-resource scenarios. In the former scenarios, we use two standard NER datasets: CoNLL-2003 [30] and OntoNotes5.0 [26], which are randomly sampled to simulate a low-resource scenarios. In the latter scenarios, we use a dataset from the material science domain: MaScip [22]. The results show that our method obtains a significant performance improvement over baseline model (i.e., Bi-LSTM-CRF for NER)

Our contributions are summarized as follows:

- We propose a novel data augmentation method for low-resource NER task, which uses pre-trained BERT to generate label-aware synthetic data and curriculum learning strategy on generated data denoising to improve data quality.
- We conduct experiments on two standard NER datasets and a real-world low-resource NER dataset. Experimental results demonstrate the effectiveness of our methods in low-resource scenarios. Moreover, our methods can be easily applied to other token-level tasks.

2   Related Work

2.1   Low-Resource NER

Deep learning-based methods achieve good performance for NER with abundant annotated data but encounter various challenges when the labeled data is scarce. Therefore, more and more works pay attention to improving the performance of NER in low-resource scenarios. Kruengkrai et al. [14] use sentence-level information in auxiliary tasks to improve model performance on low-resource languages. Peng et al. [23] use dictionaries to directly label data, ignoring entities that are not in the dictionaries. This method greatly reduce the requirements on the quality of the dictionaries. Shang et al. [29] propose a revised fuzzy CRF layer to handle tokens with multiple possible labels and a neural model AutoNER with a new Tie or Break scheme. Han and Eisenstein [11] propose domain adaptive fine-tuning with unlabeled data to reduce the discrepancy between different domains. All of these methods focus on existing resources and do not consider using synthetic data for data augmentation.

Several works have studied using data augmentation for NER. Mathew et al. [18] train a weak tagger to annotate unlabeled data through weak supervision. Dai and Adel [3] summarize the sentence-level and sentence-pair level data augmentation methods on the NLP tasks and apply some of them to the NER task, including the token replacement, synonym replacement, mention replacement, and shuffle within segments. Synonym replacement often relies on external knowledge, e.g. WordNet [20], which is a manually designed dictionary that may have low coverage (or not available) for low-resource languages. Ding et al. [5]

propose an augmentation method with language models trained on the linearized labeled sentences.

Different from the above methods, our approach uses pre-trained BERT to predict the masked words, which contains rich contextual information. Then the masked words are replaced with the predicted words to generate new synthetic sentences, and the original label sequence remains unchanged.

2.2 Curriculum Learning

Curriculum learning [1] is a particular learning paradigm in machine learning, which starts from easy instances and then gradually handles harder ones. Liu et al. [16] propose a natural answer generation framework based on curriculum learning for question answering tasks. Pentina et al. [24] use curriculum learning to study the best order of learning tasks in multitasking problems. Platanios et al. [25] propose a neural machine translation framework based on curriculum learning. It decides which training samples to show to the model at different times during the training process according to the estimated difficulty of the samples and the current capabilities of the model, which greatly reduce the training time. Matiisen et al. [19] propose Teacher-Student Curriculum Learning, a framework for automatic curriculum learning, where the Student try to learn a complex task, and the Teacher automatically choose subtasks from a given set for the Student to train on. Gong et al. [9] employ the curriculum learning methodology by investigating the difficulty of classifying every unlabeled image. The reliability and the discriminability of these unlabeled images are particularly investigated for evaluating their difficulty. As a result, an optimized image sequence is generated during the iterative propagations, and the unlabeled images are logically classified from simple to difficult. Wang et al. [32] propose a unified framework called Dynamic Curriculum Learning (DCL) to adaptively adjust the sampling strategy and loss weight in each batch, which achieves the better ability of generalization and discrimination.

In our work, we propose three different strategies to determine the difficulty of the augmented data and add them to the original data for training in an easy-to-difficult order. As a result, the model can preferentially learn from high-confidence data which contains more accurate information, and thus obtain better performance.

3   Proposed Method

3.1   Overview

Figure 2 shows the overview of our approach, which effectively deals with insufficient data for low-resource NER. The framework consists of two parts: data augmentation and denoising. Data augmentation mainly uses BERT [4] to predict the masked words according to the context and synthesizes new sentences by replacing the words in the original masked position to augment the training set. Furthermore, we denoise the augmented data through curriculum learning [1] to obtain higher quality data. We will introduce the details of each part.

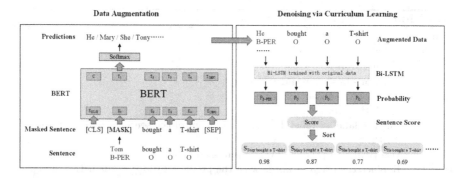

Fig. 2. The overall framework. The left is the data augmentation part using pre-trained BERT, and the right is the denoising part using curriculum learning.

3.2   Data Augmentation via Pre-trained BERT

In this section, we propose a data augmentation method based on pre-trained BERT for low-resource conditions, which predict words based on the context to generate new synthetic sentences. We apply two methods: *basic method*, which directly uses the BERT model for prediction, and *label-aware method*, which attaches label or label description.

The Basic DA Method. Let a labeled sentence be $\mathbf{x} = (x_1, \cdots, x_m)$ and the corresponding labels be $\mathbf{y} = (y_1, \cdots, y_m)$, where x_i denotes the i-th token and y_i denotes the i-th label. If x_i is inside an entity, which can be judged by its gold label (e.g. B-PER, I-PER, ...), we first mask x_i in this sentence, i.e. masked token $\{x_i\}$ and its context S. Then we use the masked sentence as the input of BERT. The final hidden representation corresponding to the masked token is fed into a softmax layer to generate a sequence of vocabulary size with a probability distribution $p(\cdot|S\{x_i\})$. Then we replace x_i with the k words which have the highest probability. For each sentence in the corpus, we perform the above procedure. Especially note that we only mask the words inside the entity, not the non-entity temporarily. After substituting the masked words with predicted words, our method generates some new sentences, which share the same label sequence with the original sentence. Then these sentences are added to the original low-resource dataset.

Label-Aware Data Augmentation. Although, pre-trained BERT encodes the context information, there is still a lot of noises, such as pronouns and wrong words, in the synthetic sentences. Considering that, we propose a label-aware data augmentation method. Different from basic DA method, this method fine-tunes the BERT before prediction to incorporate label information in the prediction process and improve the matching degree between predicted words and their corresponding labels. We elaborate this process in two steps: *BERT fine-tuning* and *Synthetic Sentence Generation*.

Algorithm 1. Label-Aware DA Method

Input: Labeled dataset D
Output: Augmented dataset D'

1: **for** each sentence **do**
2:     **for** each word in sentence **do**
3:         **if** this word in an entity **then**
4:             Mask this word.
5:             Put the label or label description in front of the sentence and separate with [SEP].
6:             Use BERT to predict words on masked position.
7:             Replace the original words with the predicted top k words with the highest probability to generate new sentences.
8:         **end if**
9:     **end for**
10: **end for**
11: Add new sentences to the original dataset D to generate an augmented dataset D'.
12: Perform NER task on augmented dataset D'.

BERT Fine-Tuning. At first step, we fine-tune the BERT with label-aware original data, which allows us to further train the hidden feature representation with label information. Here we consider two strategies for label-aware data generation:

- Label Additional (LA), where we define all entity types (e.g. PER, ORG, ...) as the training signal.
- Label Description Additional (LDA), where we use descriptions[1] of entity types as the training signal.

As shown in Fig. 1, the label (or label description) of masked entity words with [SEP] token is inserted between [CLS] and the first word of the sentence. Then we fine-tune BERT with data obtained from above steps.

Synthetic Sentence Generation. At the second step, we use fine-tuned model to make predictions on original data. Note that input data is also transformed into label-aware format, which is the same as fine-tuning data shown in Fig. 1 (LA and LDA). Given a masked word x_i and its label y_i, we define its label description as d_i. Different from basic DA method in Sect. 3.2 calculating $p(\cdot|S\{x_i\})$, we calculate $p(\cdot|y_i, S\{x_i\})$ or $p(\cdot|d_i, S\{x_i\})$ in label-aware DA method. The concrete process of label-aware method is shown in Algorithm 1.

3.3 Denoising via Curriculum Learning

BERT is a powerful pre-trained language model, which make full use of contextualized information to generate context-sensitive words. However, directly using

---

[1] We obtain the label descriptions from https://spacy.io/api/annotation#named-entities.

original label sequence may cause mismatch between predicted words and original labels, so it is necessary to denoise augmented data via curriculum learning.

Synthetic Example Ranking. As the right part of Fig. 2 shows, we train a Bi-LSTM model on original data. Then we use it to predict augmented data without original data and take the predicted probability of the gold label corresponding to each word, i.e. P_i, where i represents the i-th word in a sentence, as the basis for scoring. Based on this process, we artificially formulate three curriculum indicators described in detail as follows:

- **Average.** We calculate the sentence score S_{sent} by averaging the predicted probabilities P_i of all words in it. The lower the value, the more mismatch between the whole sentence and the original gold label, and the more incorrect information contained, which may hurt the model training.
- **Entity.** Different from *average*, we only consider entity words and average their predicted probabilities P_i as sentence score S_{sent}. Entities are more important than other words, because the named entity recognition task is mainly to recognize entities in sentences. Same as above, the higher the value, the more the predicted word matches the original label.
- **Length.** Using sentence length L as the score S_{sent}, we believe that the longer the sentence, the more information it contains, which is more instructive for the training of the model.

Then we sort the sentences in descending order of sentence scores S_{sent}, corresponding to the easy to difficult in the curriculum learning. We believe that prioritizing model to learn more correct information can lead to an improvement in model performance.

Incremental Training. We sample the sorted data according to the ratio of $0\%, 5\%, 10\%, \cdots, 100\%$, and add them to the original data gradually. In the training process, we save the best model on each scale, and the next scale of data uses the model parameters of the previous scale to train. That is to say, we only use a part of synthetic examples to train our model, i.e., samples with high confidence, to reduce the impact of noises in the data augmentation process.

4 Experimental Setups

4.1 Datasets

We consider both the simulated and real-world low-resource scenarios. In the simulated low-resource scenarios, we conduct our experiments on two English NER datasets of different granularities: CoNLL-2003 [30] and OntoNotes5.0 [26]. CoNLL-2003 is composed of newswire from the Reuters RCV1 corpus and contains four types of named entities: location, organization, person, and miscellaneous. OntoNotes5.0 contains 18 entity types for the CoNLL-2012 task, which includes broadcast conversations, broadcast news, newswire, magazines, telephone conversations, and web texts. Pradhan et al. [26] comply a core portion of the OntoNotes5.0 dataset and describe a standard train/dev/test split, which we use for our evaluation. As our work mainly focuses on low-resource NER, we randomly select four ratios for CoNLL-2003 and OntoNotes5.0 to simulate the low-resource situation, as shown in Table 1.

Table 1. The sampling ratio of the two datasets and the corresponding sentence number.

CoNLL-2003	0.2%	1%	2.5%	5%
Sentence Num	29	149	374	749
OntoNotes5.0	0.05%	0.1%	1%	2%
Sentence Num	57	115	1158	2316

In the real-world low-resource scenarios, we conduct our experiments on a dataset from material science: MaScip [22]². This dataset contains 230 synthesis procedures annotated by domain experts with labeled graphs that express the semantics of the synthesis sentences, and 21 entity types (e.g., Material, Number, Operation, Amount-Unit, etc.). We use the train/dev/test split provided by the authors, which contains 1901/109/158 sentences respectively. To simulate a low-resource setting, we also randomly select 50, 150, 500 sentences that contain all entity types from the training set to create the corresponding small, medium, and large training sets (denoted as S, M, L, whereas the complete training set is denoted as F) for each dataset. Note that we only apply data augmentation to the training set, and the development set and test set remain unchanged.

4.2 Implementations

We regard the NER task as a sequence labeling task: given a token sequence, the model needs to predict the label corresponding to each token, which includes position indicators (BIO schema) and entity types. In our study, we use the Bi-LSTM-CRF model [17] commonly used in NER tasks as the experimental model. It consists of two parts: a neural-based encoder that creates context-sensitive embedding for each token, whose weights are learned from scratch, the other is a condition random field output layer, which captures the dependencies between adjacent labels. Besides, CNN is used to obtain the character representation of each token, which is then concatenated with the word representation and sent to the bidirectional LSTM layer. The hidden states of the forward and backward LSTM are concatenated together as the final representation. We use a single-layer BiLSTM with a hidden state size of 200. Dropout layers are applied before and after the BiLSTM layer with a dropout rate of 0.5. This model is trained

---

² https://github.com/olivettigroup/annotated-materials-syntheses.

using SGD [2] with an initial learning rate of 0.015 and batch size of 10. The learning rate of each epoch decays proportionally. We use randomly initialized word embeddings with a dimension of 100. We stop training when the F_1 score of the development set has not been updated for 10 epochs. We use the F_1 score to evaluate the effectiveness of the models. The best model saved on the development set is measured using the F_1 score, and finally evaluated on the test set.

4.3   Experimental Results

Impact of Data Augmentation. We compare our methods (DA, LA, LDA) with the following models: None, the original Bi-LSTM-CRF model without data augmentation; EDA [33], which includes the substitution of synonyms, random insertion, random exchange, and random deletion of words. DA, LA, LDA correspond to the basic method, label additional method, and label description additional method. For each augmentation method, we take $k = 20$ predicted words to replace the masked words.

Table 2. Evaluation results in F_1 score. $\triangle$ column shows the averaged improvement due to data augmentation. **Bold** means the result is significantly better than the baseline model without data augmentation.

Method	CoNLL-2003				$\triangle$	OntoNotes 5.0				$\triangle$
	0.2%	1%	2.5%	5%		0.05%	0.1%	1%	2%	
None	25.72	38.74	50.65	60.65		2.64	12.43	46.71	56.00	
EDA	11.14	33.67	41.86	51.69	−9.3500	7.23	13.82	40.46	48.60	−1.9175
DA	27.11	40.11	**53.23**	59.25	0.9850	12.37	21.43	47.63	**55.39**	4.7600
LA	**29.03**	**42.20**	51.88	**61.48**	2.2075	11.91	**21.46**	49.29	54.25	4.7825
LDA	27.90	41.38	52.46	59.91	1.4725	**13.39**	20.19	47.84	**55.39**	4.7575

Table 2 provides the evaluation results on the test set. We can first conclude that our augmentation framework improves over the baseline where no augmentation is used in most cases, and superior to EDA in any ratio. For the CoNLL-2003 dataset, the four proportions increased by 3.31%, 3.46%, 2.58%, and 0.83% respectively compared to the baseline. For the OntoNotes5.0 dataset, the first three proportions have increased by 10.75%, 9.03%, and 2.58% respectively, while the performance of the last proportion has decreased slightly. This situation may reflect the trade-off between the diversity and validity of augmented instances. On the one hand, we use BERT to generate different training instances to prevent overfitting. This positive effect is especially useful when the training set is small. On the other hand, it may also increase the risk of generating invalid instances. For larger training sets, this negative effect may be dominant.

Second, the label additional method outperforms other augmentation on average, i.e. 2.21% for CoNLL-2003 and 4.78% for OntoNotes5.0, although there is no single clear winner across both datasets. However, there is little difference in the performance of the three methods on OntoNotes5.0, which may be due to its more fine-grained entity types. For the label description additional method, the performance is slightly lower than the label additional method. We consider that the label description contains more information compared with the label, which leads to more fixed words predicted by the model and causes a slightly negative impact.

Third, data augmentation techniques are more effective when the training sets are small. In both datasets, data augmentation methods achieve more significant improvements when the training sets are small, such as 0.2% of CoNLL-2003 and 0.05% OntoNotes5.0. In contrast, when the larger training sets are used, the augmentation methods achieve less improvements and some even decrease the performance. This has also been observed in previous work on machine translation tasks [6].

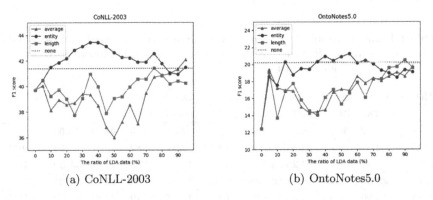

(a) CoNLL-2003   (b) OntoNotes5.0

Fig. 3. The result of *average*, *entity* and *length* strategies. The black dashed line represents the original LDA method without curriculum learning. (Color figure online)

Impact of Curriculum Learning. Figure 3 shows the results of three different indicators via curriculum learning on CoNLL-2003 and OntoNotes5.0 respectively. We take 1% of CoNLL-2003 LDA data and 0.1% of OntoNotes5.0 LDA data as examples and use the method described in Sect. 3.3 for data denoising[3].

As can be seen from the figure, the *entity* indicator is more effective on both datasets. On CoNLL-2003, the best result of 43.43% is achieved on 35% of the LDA augmented data, which is 2.05% higher than the original LDA method. On OntoNotes5.0, the best result of 21.23% is achieved on 55% of the LDA augmented data, which outperforms the original LDA method by 1.04%. It can be explained that when we use data augmentation technologies mentioned in the

---

[3] For OntoNotes5.0, we do not save the previous scale model, and all start training from scratch.

Sect. 3.2, we also introduce noise in the augmented dataset at the same time. The purpose of denoising method is to reduce the negative impact of noise. Experiment results demonstrate the effectiveness of using curriculum learning for data denoising. By giving priority to training high-confidence data except the noise part, the model gets better performance.

Real Low-Resource Scenarios. We use the methods on MaScip[4], which include data augmentation method via BERT model (DA), label additional data augmentation method (LA), and denoising LA data via curriculum learning (LA-CL), but except the method of additional label description, because this dataset does not have an official description of all labels. Note that when denoising, we only use the *entity* indicator among the above three indicators as it has a best performance.

Table 3. The result of MaScip. None represent the baseline with unaugmented data.

Method	S	M	L	F
None	59.95	70.05	72.59	76.30
DA	62.08	69.36	72.62	75.03
LA	**62.36**	68.65	73.35	76.31
LA-CL	62.02	**70.41**	**74.31**	**77.29**

Table 3 presents comparisons of our methods with the baseline. What we see is that our methods are significantly better than the baseline on the real-world low-resource scenarios. Our methods outperforms the baseline with 62.36%, 70.41%, 74.31% and 77.29% in terms of F1 score. It can be seen that the denoising results have been improved to a certain extent in most of the experiments, including the full amount of data. However, in the small-scale experiment, the denoising result is lower than the LA result. We consider the reason is that the amount of data used to train the Bi-LSTM model for scoring is too small, which makes the model learning less information. This situation leads us to fail to select the most correct predicted words, which negatively feeds back on performance. Therefore, when the training data is particularly small, it is enough to use the data augmentation methods without denoising, by which can achieve an obvious improvement. In summary, the experimental results on the MaScip dataset confirm the effectiveness of our methods in real world low-resource situations.

4.4   Discussion

Performance on Entity Types. To further understand the effectiveness of our method, we investigate the performances for each entity type. Table 4 and Table 5 describe the performance on F1 score for each entity type in 1% of CoNLL-2003 data and 0.1% of OntoNotes5.0 data.

---

[4] We leverage GloVe embedding for these experiments.

Table 4. The F_1 score of each label of CoNLL-2003 (1%).

Label	Num	None	DA	LA	LDA
LOC	72	46.04	47.71	**48.98**	47.43
MISC	38	11.04	17.37	**21.78**	20.33
ORG	131	38.93	39.10	**43.55**	40.64
PER	127	39.78	44.50	**45.39**	44.99

In CoNLL-2003, it is clear that our methods both outperform the baseline where no augmentation is used on every entity types. Comparing these three methods, we can see that the label additional method is significantly better than the other two. These results first reflect the effectiveness of the label-aware method mentioned in Sect. 3.2. Second, regarding the reason why the LA's score is higher than the LDA's, we consider that the predicted words by LA method are more diverse, which can prevent the model from overfitting.

Table 5. The F_1 score of each label of OntoNotes5.0 (0.1%).

Label	Num	None	DA	LA	LDA
CARDINAL	16	25.50	**29.79**	27.19	24.68
DATE	41	15.75	32.72	**33.06**	28.13
EVENT	2	0.00	0.00	0.00	**2.11**
FAC	9	0.00	**1.14**	0.00	0.00
GPE	28	13.35	25.05	23.17	**25.31**
LANGUAGE	2	16.67	0.00	8.70	**30.30**
LOC	7	2.52	**10.77**	8.70	9.16
MONEY	12	25.42	**29.76**	28.65	28.49
NORP	4	0.00	7.44	**8.85**	6.05
ORDINAL	2	24.54	**50.24**	30.87	40.25
ORG	82	6.81	10.15	**11.16**	9.51
PERCENT	18	35.12	43.21	56.83	**56.85**
PERSON	37	1.28	8.83	9.69	**12.21**
QUANTITY	2	0.00	0.00	0.00	**0.82**
TIME	6	0.81	1.55	**2.01**	0.76

In OntoNotes5.0, due to the complexity of entity types, the three methods have different manifestations in each entity type. It can be seen that the score of some entity types are zero, because the number of sampled sentences is small, and some entity types have almost never appeared. Among them, we omitted some entity types with all zero F_1 score. Summarized as follows, the basic method is better on 'CARDINAL', 'FAC', etc., the label additional method is more effective on 'DETA', 'NORP', 'ORG', and the label description additional method improves more significantly on entity types with lower scores, like 'EVENT', 'PERSON', 'QUANTITY'. This may be due to the words predicted by the LDA method more closely match the gold label for entity types that appear less frequently.

Case Study. We examine the effect of different methods on generating words. We use 2% of CoNLL-2003 to fine-tune BERT and Table 6 lists some examples of generated words by the basic method and LA method with different settings. All the examples can fill in at least two different entity types of words in the masked position.

Table 6. The table shows the change of the predicted word after using additional label information, where DA means basic method, LA means label additional method, and the brackets indicate the label placed before the sentence. **Bold font** indicates words that match the label. The words from left to right indicate that the predicted probability is from high to low.

Example	Method	Generated Word
[MASK] buys a gaming company	DA	He She It he **Michae . Joe Jack David Tony**
	LA(PER)	He **Sinclair Warner Simon Blackburn Morgan Hamilton Fox Anderson Harry**
	LA(ORG)	**Sony Blackburn** Sinclair He **Dell** Britain **Hamilton Leeds** Fischer **Intel**
[MASK] went bankrupt	DA	It He They and it he . She which they
	LA(PER)	**Blackburn** Leeds Middlesbrough Barrow Yorkshire York **Hamilton** Italy **Sheffield Milan**
	LA(ORG)	**Salzburg** Switzerland **Zürich** Austria **Blackburn Bavaria Zurich Leeds** Italy **Juventus**
[MASK] plans this event	DA	Who He Currently She It **FIFA** who **India Nike Israel**
	LA(LOC)	**Canada Ireland Australia** WHO **Bermuda** FIFA UEFA **Argentina Scotland** Yorkshire
	LA(ORG)	**WHO UCI UEFA** Slovenia Slovakia Canada **Yorkshire** Britain Azerbaijan Schedule

As known to all, BERT encodes semantic features from the input sentences for extracting global contexts. In the fist example, when directly using BERT for prediction, we can see that there are some person names generated, which prove the ability of BERT to obtain contextual information. However, one problem with BERT is that the predicted words have a higher probability to be third-person pronouns or even wrong words, which cannot increase the diversity of the augmented data and may hurt the performance. Therefore, we propose the label-aware method described in Sect. 3.2 to minimize this weakness. When we use label additional method with 'PER', more person names appeared, whose probability of being predicted increased. Then when we change the label to 'ORG', some organization names are predicted by the pre-trained model. The situation goes to show the effectiveness of our methods. The same conclusion can be drawn from the second and third examples. In summary, the LA method considers more label information than the DA method, and makes the generated words contain less impurities.

5   Conclusion

In this paper, we propose a novel framework to generate high-quality synthetic data for low-resource NER. We use pre-trained BERT to fully integrate contextual information to generate diverse synthetic sentences, and leverage curriculum learning to denoise synthetic sentences for obtaining higher quality augmented data. Our framework shows superior performance in both simulated low-resource scenarios and real-world low-resource scenarios. In the future, we will explore the performance of our framework when customizing label descriptions and on other token-level NLP tasks.

Acknowledgements. The research work described in this paper has been supported by the National Key R&D Program of China (2020AAA0108001) and the National Nature Science Foundation of China (No. 61976015, 61976016, 61876198 and 61370130). The authors would like to thank the anonymous reviewers for their valuable comments and suggestions to improve this paper.

References

1. Bengio, Y., Louradour, J., Collobert, R., Weston, J.: Curriculum learning. In: ICML (2009)
2. Bottou, L.: Stochastic gradient descent tricks. In: Montavon, G., Orr, G.B., Müller, K.R. (eds.) Neural Networks: Tricks of the Trade. LNCS, vol. 7700, pp. 421–436. Springer, Heidelberg (2012)
3. Dai, X., Adel, H.: An analysis of simple data augmentation for named entity recognition. In: COLING (2020)
4. Devlin, J., Chang, M.W., Lee, K., Toutanova, K.: BERT: pre-training of deep bidirectional transformers for language understanding. In: NAACL (2019)
5. Ding, B., et al.: DAGA: data augmentation with a generation approach for low-resource tagging tasks. In: EMNLP (2020)

6. Fadaee, M., Bisazza, A., Monz, C.: Data augmentation for low-resource neural machine translation. In: ACL (2017)
7. Friedrich, A., et al.: The SOFC-EXP corpus and neural approaches to information extraction in the materials science domain. In: ACL (2020)
8. Gao, F., et al.: Soft contextual data augmentation for neural machine translation. In: ACL (2019)
9. Gong, C., Tao, D., Maybank, S.J., Liu, W., Kang, G., Yang, J.: Multi-modal curriculum learning for semi-supervised image classification. IEEE Trans. Image Process. 3249–3260 (2016)
10. Gururangan, S., et al.: Don't stop pretraining: adapt language models to domains and tasks. In: ACL (2020)
11. Han, X., Eisenstein, J.: Unsupervised domain adaptation of contextualized embeddings for sequence labeling. In: EMNLP (2019)
12. Huang, Z., Xu, W., Yu, K.: Bidirectional LSTM-CRF models for sequence tagging. CoRR (2015)
13. Iyyer, M., Wieting, J., Gimpel, K., Zettlemoyer, L.: Adversarial example generation with syntactically controlled paraphrase networks. In: NAACL (2018)
14. Kruengkrai, C., Nguyen, T.H., Aljunied, S.M., Bing, L.: Improving low-resource named entity recognition using joint sentence and token labeling. In: ACL (2020)
15. Kuru, O., Can, O.A., Yuret, D.: Charner: character-level named entity recognition. In: COLING (2016)
16. Liu, C., He, S., Liu, K., Zhao, J.: Curriculum learning for natural answer generation. In: IJCAI (2018)
17. Ma, X., Hovy, E.: End-to-end sequence labeling via bi-directional LSTM-CNNs-CRF. In: ACL, August 2016
18. Mathew, J., Fakhraei, S., Ambite, J.L.: Biomedical named entity recognition via reference-set augmented bootstrapping. arXiv preprint arXiv:1906.00282 (2019)
19. Matiisen, T., Oliver, A., Cohen, T., Schulman, J.: Teacher-student curriculum learning. IEEE Trans. Neural Netw. Learn. Syst. 3732–3740 (2020)
20. Miller, G.A.: WordNet: a lexical database for English. Commun. ACM, 39–41 (1995)
21. Min, J., McCoy, R.T., Das, D., Pitler, E., Linzen, T.: Syntactic data augmentation increases robustness to inference heuristics. In: ACL (2020)
22. Mysore, S., et al.: The materials science procedural text corpus: annotating materials synthesis procedures with shallow semantic structures. In: Proceedings of the 13th Linguistic Annotation Workshop (2019)
23. Peng, M., Xing, X., Zhang, Q., Fu, J., Huang, X.: Distantly supervised named entity recognition using positive-unlabeled learning. In: ACL (2019)
24. Pentina, A., Sharmanska, V., Lampert, C.H.: Curriculum learning of multiple tasks. In: CVPR (2015)
25. Platanios, E.A., Stretcu, O., Neubig, G., Poczos, B., Mitchell, T.: Competence-based curriculum learning for neural machine translation. In: NAACL (2019)
26. Pradhan, S., et al.: Towards robust linguistic analysis using ontonotes. In: CoNLL (2013)
27. Raiman, J., Miller, J.: Globally normalized reader. In: EMNLP (2017)
28. Ruder, S.: Neural transfer learning for natural language processing. Ph.D. thesis (2019)
29. Shang, J., Liu, L., Gu, X., Ren, X., Ren, T., Han, J.: Learning named entity tagger using domain-specific dictionary. In: EMNLP (2018)
30. Tjong Kim Sang, E.F., De Meulder, F.: Introduction to the CoNLL-2003 shared task: language-independent named entity recognition. In: CoNLL (2003)

31. Wang, X., Pham, H., Dai, Z., Neubig, G.: SwitchOut: an efficient data augmentation algorithm for neural machine translation. In: EMNLP (2018)
32. Wang, Y., Gan, W., Yang, J., Wu, W., Yan, J.: Dynamic curriculum learning for imbalanced data classification. In: Proceedings of the IEEE/CVF International Conference on Computer Vision (ICCV) (2019)
33. Wei, J., Zou, K.: EDA: easy data augmentation techniques for boosting performance on text classification tasks. In: EMNLP (2019)
34. Xie, Q., Dai, Z., Hovy, E., Luong, T., Le, Q.: Unsupervised data augmentation for consistency training. In: NIPS (2020)
35. Yu, A.W., et al.: QANet: combining local convolution with global self-attention for reading comprehension. CoRR (2018)
36. Zeng, D., Liu, K., Chen, Y., Zhao, J.: Distant supervision for relation extraction via piecewise convolutional neural networks. In: EMNLP (2015)

Global Entity Alignment with Gated Latent Space Neighborhood Aggregation

Wei Chen[1], Xiaoying Chen[1,2(✉)], and Shengwu Xiong[1,3]

[1] Wuhan University of Technology, Wuhan, China
`xiongsw@whut.edu.cn`
[2] Hubei Credit Information Center, Wuhan, China
[3] Sanya Science and Education Innovation Park of Wuhan University of Technology, Sanya, China

Abstract. Existing entity alignment models mainly use the topology structure of the original knowledge graph and have achieved promising performance. However, they are still challenged by the heterogeneous topological neighborhood structures, which could cause the models to produce different representations of counterpart entities. In the paper, we propose a global entity alignment model with gated latent space neighborhood aggregation (LatsEA) to address this challenge. Latent space neighborhood is formed by calculating the similarity between the entity embeddings, it can introduce long-range neighbors to expand the topological neighborhood and reconcile the heterogeneous neighborhood structures. Meanwhile, it uses vanilla GCN to aggregate the topological neighborhood and latent space neighborhood respectively. Then, it uses an average gating mechanism to aggregate topological neighborhood information and latent space neighborhood information of the central entity. In order to further consider the interdependence between entity alignment decisions, we propose a global entity alignment strategy, i.e., formulate entity alignment as the maximum bipartite matching problem, which is effectively solved by Hungarian algorithm. Our experiments with ablation studies on three real-world entity alignment datasets prove the effectiveness of the proposed model. Latent space neighborhood information and global entity alignment decisions both contributes to the entity alignment performance improvement.

Keywords: Knowledge graph · Entity alignment · Latent space neighborhood aggregation

1 Introduction

Knowledge graph (KG) is an important tool to store knowledge, which provides support for many applications, such as question answering systems, recommender systems. Many KGs have been constructed for particular applications.

Supported by the Major project of IoV, Technological Innovation Projects in Hubei Province (Grant No. ZDZX2020000027, 2019AAA024) and Sanya Science and Education Innovation Park of Wuhan University of Technology (Grant No. 2020KF0054).

S. Li et al. (Eds.): CCL 2021, LNAI 12869, pp. 371–384, 2021.
https://doi.org/10.1007/978-3-030-84186-7_25

Sometimes single KG cannot meet the needs of certain applications, so it is necessary to integrate multiple complementary KGs. Entity alignment is non-trivial to integrating different KGs. Traditional entity alignment methods are mainly divided into two categories, one is based on the entity's label information [1,2], the other is based on manually defining features. Recently, the embedding-based representation learning has attracted more and more attention. Given a KG, embedding-based representation learning can map entities to a low-dimensional vector space, then entity alignment model can find counterpart entities according to the similarity of the embeddings. The embedding-based entity alignment approaches show its superiority. Such methods are all built based on KG embedding models (such as TransE [3]). Then they use a transformation matrix to convert the embedding space and complete the entity alignment by calculating the similarity between entity embeddings, such as MTransE [4]. In recent researches, graph convolutional neural networks (GCN) have shown amazing results when processing graph structure data. GCN-based Entity alignment models all hope that counterpart entities have similar neighborhood structures and have similar vector representations. Several recent GCN-based alignment models (such as GCN-Align [5], AVR-GCN [6], GMNN [7], AliNet [8], RDGCN [9]) have achieved good results.

However, existing embedding-based entity alignment models still face a critical problem. The counterpart entities usually have dissimilar neighborhood structures in between KGs, which will produce different embedding. The statistics on DBpedia datasets the percentages of such entity pairs reach 89.97% between Chinese-English, 86.19% between Japanese-English and 90.71% between French-English, respectively [8].

The difficulty in tackling this problem is how to alleviate the difference in topological neighborhood structure. Therefore, we propose a global entity alignment model with gated latent space neighborhood aggregation (LatsEA). Latent space neighborhood is formed in the embedding space by calculating the similarity between the entity embeddings, which can introduce long-distance neighbors to expand the topological neighborhood. The basic idea is to first learn KGs embeddings through KG embedding models and calculate the similarity between entity embeddings to generate latent space neighborhood. So, the model can be combined with many existing KG embedding models. Then, we use the GCN to aggregate topological neighborhood and latent space neighborhood respectively. Finally, we use an average gating mechanism to aggregate topological neighborhood information and latent space neighborhood information.

State-of-the-art models treat entities autonomously when determining entity alignment results. Intuitively, if a target entity is aligned to a source entity with higher confidence, it has a smaller chance to be aligned with another source entity. So, we formulate entity alignment as the maximum bipartite matching problem, which can be solved by Hungarian algorithm, so that the interdependence between entity alignment decisions can be captured to make global entity alignment decisions. Our experiments with ablation studies prove the effectiveness of the proposed model (LatsEA) and the global entity alignment strategy. In summary, our main contributions as follows:

- We propose a novel global entity alignment model, which introduces latent space neighborhood to reconcile the heterogeneous neighborhood structures.
- We propose an average gating mechanism to aggregate topological neighborhood information and latent space neighborhood information, so that the entity embeddings contain the entities information in topological and latent space neighborhood.
- We consider the interdependence between entity alignment decisions, and propose a global entity alignment strategy, i.e., formulate entity alignment as the maximum bipartite matching problem and use the classic Hungarian algorithm to solve.
- We perform experiments with ablation studies on three real-world datasets prove the effectiveness of the proposed model and the global entity alignment strategy.

2 Related Work

In our proposed entity alignment model, the most relevant works are KG embedding and embedding-based entity alignment, so we introduce some of the work related.

2.1 KG Embedding

In the past few years, the researches on KG embedding is mainly divided into three categories: translational distance models, neural network based models, and bilinear models [10]. TransE [3] is a classic model in the translational distance models. It believes that in a triple, the relation is the translations of the head entities to the tail entities. TransE can solve the relation type of 1-1 well, but cannot solve the relation type of 1-N, N-1, N-N. Therefore, subsequent researches have proposed many improved versions based on the TransE, such as TransH [11], TransD [12].

With the rise of neural networks in recent years, neural networks are used to many KG embedding researches, such as ConvE [13] and CapsE [14]. KGs have three widely spread types of relation patterns: symmetry, inversion, and composition [4]. However, models mentioned above can model one or two type of relation patterns. RotatE [15], which is a translational distance models, can model all relation patterns. It maps the entities and relations to the complex vector space and defines each relation as a rotation from the head entities to the tail entities. But real-world KGs usually have semantic hierarchies that RotatE fails to model. So, to tackle this challenge, the HAKE [10] model is based on the RotatE, and retains the rotation characteristics of the RotatE. It contains modulus part and phase part. The modulus part models the entities at different levels of the hierarchy in KGs, and the phase part models the entities at the same level of the semantic hierarchy. In the paper, we use the HAKE [10] model to learn KGs embedding and generate latent space neighborhood by calculating the cosine similarity between entity embeddings.

2.2 Embedding-Based Entity Alignment

Embedding-based entity alignment models employ the embedding models to learn entity embeddings, then learn a mapping to transform the embedding space and find entity alignment through the similarity between these embeddings, such as MTransE [4]. IPTransE [16] iteratively adds the newly discovered entity alignment to the training data to improve the performance. There are also some models that use auxiliary information to improve the performance. For example, JAPE [17] combines attribute information to find entity alignment. Graph convolutional network (GCN) has demonstrated its powerful ability to process graph structure data, so many entity alignment models use GCN. GCN-Align [5] is a cross-language KG alignment framework, it uses GCN to convert two KGs into the same embedding space, and then calculates the distance between entity embeddings to completing the entity alignment. RDGCN [9] and AVR-GCN [6] improved the traditional GCN alignment framework. MuGNN [18] proposes a two-step method (KG completion and multi-channel graph neural network) for entity alignment. AliNet [8] introduces two-hop neighbors to reconcile heterogeneous neighborhood structures, and has achieved good results.

However, existing embedding-based entity alignment models mentioned above are still challenged that the neighborhood structure may be heterogeneous in different KGs. This may cause model to produce dissimilar entity embeddings for the corresponding entities, which will affect the accuracy of entity alignment. To tackle this problem, a straightforward idea is to mitigate the heterogeneity between KGs. AliNet [8] and MuGNN [18] above have noticed the problem. We are inspired by these works, to alleviate the heterogeneity, our idea is to utilize the latent space neighborhood in the KGs embedding space that can introduce distant neighbors to expand the topological neighborhood. In order to further consider the interdependence between entities that have been ignored in many existing researches, we convert the entity alignment into maximum bipartite matching problem to make global entity alignment. This problem can be further solved by the Hungarian algorithm. Our experiments with ablation studies on three entity alignment datasets prove the effectiveness of the proposed model. Latent space neighborhood information and global entity alignment decisions both contributes to the entity alignment performance improvement.

3   Problem Formulation

Formally, we represent a KG as a directed graph, where E represents the set of entities, R represents the set of relations, and T is the set of triples. Each triplet $(h, r, t) \in T$ indicates that the head entity h is connected to the tail entity t through the relation r. Entity alignment task is to automatically find more entity pairs that denote the same real-world identity in heterogeneous KGs. Entity alignment model is to take different KGs $G = (E, R, T)$ and $G' = (E', R', T')$ as input, and find a set of identical entities $S_e = \{(e, e') \in E \times E' \mid e \equiv e'\}$ representing the same object in the real-world. Those entity alignment seeds can be used as training data.

4   The Proposed Model

The goal of entity alignment is to discover the entity pairs representing the same object in heterogeneous KGs. In the paper, LatsEA uses HAKE [10] model to learn the latent space of the KG, then it aggregates the topological neighborhood and the latent space neighborhood of the central entity through vanilla GCN respectively. The average gating mechanism controls the topological neighborhood information and latent space neighborhood information aggregation to yield central entity embeddings. Then the entity similarity matrix is constructed by calculating the similarity between entity embeddings. To further consider the interdependence between entities and make global entity alignment decisions, we convert the entity similarity matrix into a bipartite graph, i.e., each row represents an entity in the source KG, and each column represents an entity in the target KG. So, the entity alignment problem is equivalent to maximum bipartite matching problem, which can be solved by Hungarian algorithm. The model architecture is illustrated in Fig. 1.

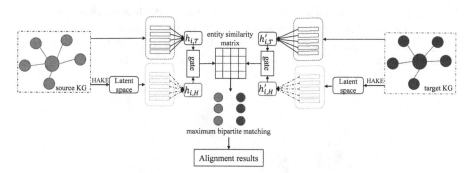

Fig. 1. Overview of the LatsEA. The red rectangle in the figure represents the topological neighborhood, and the green rectangle represents the latent space neighborhood. (Color figure online)

4.1   Basic Embedding Module

In our proposed model, in order to obtain a very meaningful entity embedding, the KG embedding model is required not only to model common relation patterns in the KG, but also to model semantic hierarchies. Because semantic hierarchies are very common in real-world applications. So, we choose Hierarchy-Aware Knowledge Graph Embedding (HAKE) [10] to learn the latent space of KG. HAKE maps KG into the polar coordinate system, the modulus part can model entities at different level, and the phase part can model entities at the same level of the hierarchy. Given a triple (h, r, t), the formula of HAKE as follows:

$$
\begin{aligned}
h_m \circ r_m &= t_m, \quad\quad where \quad h_m, t_m \in R^k, r_m \in R_+^k \\
(h_p + r_p) mod 2\pi &= t_p, \quad\quad where \quad h_p, r_p, t_p \in [0, 2\pi)^k
\end{aligned}
\tag{1}
$$

where h_m and h_p are generated by the modulus part and the phase part respectively. Therefore, the entity embeddings can be expressed as the concatenation of h_m and h_p. We calculate the cosine similarity between entity embeddings to generate the entity's latent space neighborhood, which will be described in next section. In practice, we train HAKE beforehand and freeze its parameters in the following process.

4.2 Topological Neighborhood and Latent Space Neighborhood

Existing entity alignment models mainly use the topology structure. But there is almost no way to use the structural information in latent space. The entity's topological neighborhood $N_T(u)$ is defined as follows:

$$
N_T(u) = \{v | v \in E, (u, v) \in R\}
\tag{2}
$$

where E is a set of entities, R is a set of relations. With a certain number of hops, if an entity has a path connected to the central entity, it will be part of the topological neighborhood. As shown in Fig. 2, the set of all blue circles build up the topological neighborhood when the number of hops is 2.

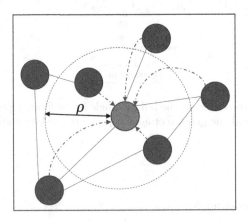

Fig. 2. An illustration of the latent space neighborhood and topological neighborhood. (Color figure online)

We introduce the latent space neighborhood to alleviate the heterogeneous. As shown in Fig. 2, if the distance between entity embeddings generated by HAKE is less than our predefined threshold ρ, the entity is the adjacent entity of the central entity in the latent space. These entities form the latent space neighborhood of the central entity. Therefore, the latent space neighborhood $N_H(u)$ can be expressed by the following formula:

$$N_H(u) = \{v|v \in E, d(e_u, e_v) < \rho\} \tag{3}$$

where ρ is a predefined distance threshold. e_u, e_v are embeddings of entity u and entity v, respectively. $d(e_u, e_v)$ is a function to calculate the distance between entity embeddings, we use the cosine similarity. The dotted circle in Fig. 2 represents the latent space neighborhood, the radius is the threshold ρ. The blue dotted line represents the topological neighborhood aggregation, and the green dotted line represents the latent space neighborhood aggregation.

4.3   Neighborhood Aggregation

The GCN was first proposed in ICLR [19], which deals specifically with graph-structured data and particularly powerful. Therefore, we use vanilla GCN to aggregate the topological neighborhood and latent space neighborhood respectively.

Topological Neighborhood Aggregation. The one-hop neighborhood is the most important structure information to learn the central entities' embeddings. The hidden representation of entity i at l^{th} layer is represented as $h_{i,T}^{(l)}$, it is obtained by aggregating its one-hop neighborhood, the calculation method is as follows:

$$h_{i,T}^{(l)} = \sigma\Big(\sum_{j \in N_T(i)} \frac{1}{a_i} W_T^{(l)} h_j^{(l-1)} \Big) \tag{4}$$

Where $N_T(\cdot)$ is a set of one-hop neighbors (including itself) of entity i, a_i is the normalization constant, $W_T^{(l)}$ is the weight matrix of the lth layer, $\sigma(\cdot)$ is an activation function, we use the hyperbolic tangent function (tanh). The input of each layer in GCN is a vertex feature matrix, it can encode nodes so that the vector representation of the entities contains the structural information in the graph.

Latent Space Neighborhood Aggregation. The latent space of the KG is learned through the HAKE [10] embedding model, which maps the KG to a continuous latent space. The HAKE embedding model is equivalent to a mapping function, which maps the nodes into a vector. The neighborhood in the latent space contains the entity in the dotted circle with radius ρ in Fig. 2. The definition of the latent space neighborhood is described in Sect. 4.2. It is a set of entities whose distance to the central entity is less than a given parameter ρ in the latent space. We use cosine similarity as the distance function $d(\cdot)$. Through the aggregation of latent space neighborhood, it is possible to capture long-range dependencies and alleviate the non-isomorphic neighborhood structures. Vanilla GCN is also used to encode entities in latent space neighborhood aggregation. Like the topological neighborhood aggregation, the hidden representation of the entity i in the lth layer is represented as $h_{i,H}^{(l)}$, which can be calculated using the following formula:

$$h_{i,H}^{(l)} = \sigma\left(\sum_{j \in N_H(i)} \frac{1}{a_i} W_H^{(l)} h_j^{(l-1)} \right) \tag{5}$$

Where $N_H(i)$ is a set of latent space neighbors (including itself) of entity i, a_i is the normalization constant, $W_H^{(l)}$ is the weight matrix of the lth layer.

4.4 Gated Topological and Latent Space Neighborhood Aggregation

After generating topological neighborhood information $h_{i,T}^{(l)}$ and latent space neighborhood information $h_{i,H}^{(l)}$ through GCN aggregation, they need to be aggregated to generate central entities embeddings $h_i^{(l)}$ in the lth layer. Because LSTM uses a gating mechanism to greatly alleviate the gradient disappearance and achieved relatively good results. Motivated by LSTM, we propose to use an average gating mechanism to aggregate topological neighborhood information and latent space neighborhood information. Different from AliNet [8], the average gating mechanism adds an average between different neighborhoods to reduce the influence of noise. First, the average vector $\bar{h}_i^{(l)}$ of topological neighborhood information $h_{i,T}^{(l)}$ and latent space neighborhood information $h_{i,H}^{(l)}$ is obtained at the lth layer, its calculation method is as follows:

$$\bar{h}_i^{(l)} = \frac{1}{n} \sum_{j \in T,H} h_{i,j}^l \tag{6}$$

Where n is the number of neighborhood types. The basic idea is that if the entity embeddings in a neighborhood is far from the average embedding, vector averaging can alleviate the noise introduced in a neighborhood. Therefore, the hidden representation $h_i^{(l)}$ of entity i at l^{th} layer can be calculated by the following formula:

$$h_i^{(l)} = GM(\bar{h}_i^{(l)}) + (1 - GM(\bar{h}_i^{(l)})) \cdot h_{i,T}^{(l)} \tag{7}$$

$$GM(\bar{h}_i^{(l)}) = \sigma(W\bar{h}_i^{(l)}) \tag{8}$$

where $GM(\cdot)$ is a gate that controls the aggregation of topological neighborhood information and latent space neighborhood information, this gate is as shown in (8). W is a weight matrix, so that the gate can be trained together with the model. We can find from formula (7) that when the latent space neighborhood brings more noise, its weight will be reduced, which can well alleviate the influence of the noise introduced by the latent space neighborhood and improve the accuracy of the model.

4.5   Global Entity Alignment Strategy

We calculate the similarity between entity embeddings generated by average gated topological and latent space neighborhood aggregation to generate the similarity matrix M. Many state-of-the-art models have determined entity alignment results in an independent manner. However, each entity alignment decision is interdependent. Source KG and target KG have a vertex set respectively, entity alignment is equivalent to finding as many matching entities as possible in two disjoint vertex sets. So, to make global entity alignment decisions, we formulate entity alignment as the maximum bipartite matching problem, which can be solved by Hungarian algorithm. To facilitate subsequent processing, we use 1 to subtract the similarity matrix M to get the matrix M^-. Therefore, if the element in M^- is smaller, the corresponding two entities are more similar. The complete process is shown in Algorithm 1, which has time complexity of $O(n^3)$ for entity alignment.

Algorithm 1: Hungarian algorithm

    Input:M^-
1   Subtracting the row minimum from each row;
2   Subtracting the col minimum from each row;
3   lineCount = 0;
4   **while** (lineCount <len(M^-)):
5   Cover all zeros with a minimum line_count;
6   **if** (lineCount == len(M^-)):
7   **break;**
8   **else:**
9   all uncovered elements in M^- subtract min uncovered element;
10   all elements in M^- that are covered twice add min uncovered element;
11   Find an optimal assignment in M^- with zeros cover;

Suppose there are three entities s_1, s_2, s_3 in the source KG, and three entities t_1, t_2, t_3 in the target KG. The example in Fig. 3 illustrates the algorithm.

To enable LatsEA to let the embeddings of aligned entities have small distance while those of unaligned entities have large distance, we use a set of pre-aligned entity pairs S_e as training data to train LatsEA. We minimizing the margin-based ranking loss function L in [9] to train model's parameters in an end-to-end way.

$$L = \sum_{(u,v)\in S_e} \sum_{(u',v')\in S'_e} [f(h_u, h_v) + \gamma - f(h_{u'}, h_{v'})]_+ \qquad (9)$$

Where $[x]_+ = max\{0, x\}$, S_e is the set of pre-aligned entity pairs, which is the positive sample during training. S'_e is the set of negative entity pairs yielded by corrupting (u, v). γ is margin hyper-parameters separating positive and negative entity pairs. $f(\cdot) = \| \cdot \|_2$ is the L2 distance metric function. We adopt Adam to minimize the loss function L. Then use the global entity alignment strategy to obtain the final entity alignment result.

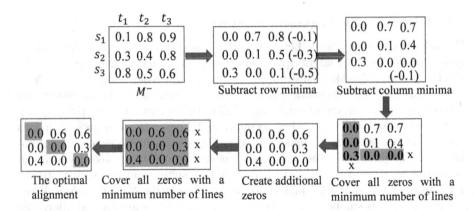

Fig. 3. Entity alignment as maximum bipartite matching problem using Hungarian algorithm.

5   Experiments

5.1   Datasets and Experiment Settings

Our experiments were conducted on DBP15K created by [17]. This dataset was generated based on the DBpedia multilingual KG and contains DBP_{ZH-EN} (Chinese-English), DBP_{JA-EN} (Japanese-English) and DBP_{FR-EN} (French-English). Each dataset contains 15,000 aligned entities for training and testing models. For detailed statistics, please refer to the original paper.

We investigate LatsEA's performance on entity alignment by comparing with three embedding-based entity alignment models and three GCN-based entity alignment models. They are MTransE [4], JAPE [17], AlignE [20], GCN [19], MuGCN [18] and AliNet (w/o rel. loss) [8]. Because of the AliNet [8] model only provides codes that does not optimize the relation loss, we only compare with AliNet (w/o rel. Loss) [8]. If LatsEA optimizes the relation loss proposed by AliNet, it will also perform better. We use grid search to find the appropriate value of the hyper-parameters. We choose hyper-parameters on the following possible values: margin γ in $\{0.5, 1.0, \ldots, 3.0\}$, learning rate in $\{0.0001, 0.001, 0.01, 0.1\}$, ρ in $\{0.80, 0.81, \ldots, 0.95\}$, the dimension of the hidden representation of each layer in $\{200, 300, 400, 500, 600\}$, the number of GCN layers in $\{1, 2, 3, 4\}$. The following hyper-parameters were used in the experiments. The γ is 1.5, the learning rate is 0.001, ρ is 0.93, the batch size is 1024. For each positive sample, 10 negative samples are collected. All GCN-based models stack two layers of GCN, and the dimension of the hidden representation of the three layers is 500, 400, and 300. We use $tanh(\cdot)$ as the activation function of neighborhood aggregation, and use the $ReLU(\cdot)$ as the activation function of gating mechanism. We use 70% of the seed alignments (10500 entity pairs for dbp15k) as the validation set and base on Hits@1 Performance termination training with a patience of 5 epochs. Following convention, we use Hits@1, Hits@10, and MRR to evaluate how the models perform. Higher Hits@1, Hits@10, and MRR scores indicate better performance.

5.2   Entity Alignment Results

Table 1 shows the performance of LatsEA and other entity alignment models. To show the fairness of the comparison, we reproduce the experimental results of MuGCN [18] and AliNet (w/o rel. Loss) [8] models under the same conditions. According to the experimental results, we found that LatsEA significantly outperforms the other entity alignment models on Hits@1 and MRR on the three datasets, the best results are highlighted in bold. For example, LatsEA achieves a Hits@1 score of 0.522, 0.539, and 0.538 on the three datasets, which are 0.017, 0.011, and 0.012 higher than the Hits@1 scores of AliNet (w/o rel. Loss) [8], respectively. Although LatsEA did not perform as well as MuGCN [18] on Hits@10, note that, the Hits@1 can better reflect the performance of the models. Because Hits@1 is equivalent to precision. LatsEA will perform better with global entity alignment strategy, and its results will be discussed in detail in Sect. 5.5. We think that these results have demonstrated LatsEA's effectiveness. The reason is that LatsEA uses the latent space neighborhood information to mitigate heterogeneous neighborhood structures.

Table 1. Results comparison on entity alignment (* marks the results obtained from AliNet)

Methods	$DBP15K_{ZH-EN}$			$DBP15K_{JA-EN}$			$DBP15K_{FR-EN}$		
	H@1	H@10	MRR	H@1	H@10	MRR	H@1	H@10	MRR
*MTransE	0.308	0.614	0.364	0.279	0.575	0.349	0.244	0.556	0.335
*JAPE	0.412	0.745	0.490	0.363	0.658	0.476	0.324	0.667	0.430
*AlignE	0.472	0.729	0.581	0.448	0.789	0.563	0.481	0.824	0.599
*GCN	0.487	0.790	0.559	0.507	0.805	0.618	0.508	0.808	0.628
MuGCN	0.493	**0.845**	0.611	0.501	**0.856**	0.620	0.498	**0.869**	0.622
AliNet	0.505	0.726	0.589	0.528	0.747	0.612	0.526	0.785	0.622
LatsEA	**0.522**	0.763	**0.613**	**0.539**	0.772	**0.625**	**0.538**	0.787	**0.632**

5.3   Effectiveness of Latent Space Neighborhood

Here, we set up two variants of LatsEA, LatsEA (To) only uses topological neighborhood information, LatsEA (Ls) only uses latent space neighborhood information. The entity alignment results are shown in Table 2. The best results are highlighted in bold. We observed that the latent space neighborhood brought improvement to entity alignment. This is because latent space neighborhood can introduce long-range neighbors to mitigate heterogeneous neighborhood structures. For example, on $DBP15K_{ZH-EN}$, the introduction of latent space neighborhood information can improve the entity alignment performance by about 2% on Hits@1, about 4% on Hits@10 and MRR. When LatsEA uses the latent space neighborhood information alone, the model performs poorly. This may be because the threshold ρ is large, which results in less useful information contained in the generated latent space neighborhood.

Table 2. Results on DBP15K w.r.t. different neighborhood

Methods	$DBP15K_{ZH-EN}$			$DBP15K_{JA-EN}$			$DBP15K_{FR-EN}$		
	H@1	H@10	MRR	H@1	H@10	MRR	H@1	H@10	MRR
LatsEA(Ls)	0.115	0.372	0.197	0.113	0.257	0.166	0.105	0.269	0.162
LatsEA(To)	0.504	0.728	0.578	0.528	0.757	0.616	0.535	**0.794**	0.632
LatsEA	**0.522**	**0.763**	**0.613**	**0.539**	**0.772**	**0.625**	**0.538**	0.787	**0.632**

5.4 Impact of Distance When Generating Latent Space Neighborhood

The purpose of introducing the latent space is to expand the neighborhood structure and reconcile the heterogeneity between KGs. We use grid search to find that the model performs better when $\rho = 0.93$. In order to deeply observe the influence of distance threshold ρ when generating latent space neighborhood, we evaluated the performance of LatsEA when the ρ is slightly higher than 0.93 or slightly lower than 0.93. The entity alignment results are shown in Table 3. Due to limited space, we only show the entity alignment results on the $DBP15K_{ZH-EN}$. The best results are highlighted in bold. It can be observed from the Table 3 that LatsEA performs best when the cosine similarity $\rho = 0.93$. We observe that the model's performance will declines as well when the ρ is larger. Because the larger ρ, latent space neighborhood will bring less useful information. Similarly, although the smaller ρ allows LatsEA to indirectly capture the more distant neighborhood information, such distant neighbors will also introduce more noise. Experiments results are consistent with our empirical analysis.

5.5 Effectiveness of Global Entity Alignment Strategy

Here, we examined the effectiveness of the proposed global entity alignment strategy. After generate the entity embeddings through our LatsEA, we obtain the similarity matrix of entities by calculating the similarities between these embeddings. In order to consider the interdependence between each entity alignment decision, we propose a global entity alignment strategy (GEAS). Concretely, we formulate entity alignment as the maximum bipartite matching problem, which

Table 3. Results on $DBP15K_{ZH-EN}$ w.r.t. different distance

Methods	$DBP15K_{ZH-EN}$		
	H@1	H@10	MRR
*GCN	0.487	**0.790**	0.559
LatsEA(ρ=0.92)	0.516	0.742	0.601
LatsEA	**0.522**	0.763	**0.613**
LatsEA(ρ=0.94)	0.519	0.717	0.606
LatsEA(ρ=0.95)	0.517	0.714	0.604

can be solved by Hungarian algorithm. Because the algorithm can make global optimal entity alignment decisions and ensure that each entity gets the optimal match. Therefore, we use Hits@1 to evaluate whether our proposed GEAS is effective. Hit@1 is equivalent to the accuracy of the model. The accuracy is shown in Table 4. We can observe from Table 4 that the performance on all datasets will be better with GEAS, revealing the effectiveness of considering the interdependence between entity alignment decisions.

Table 4. Results on DBP15K w.r.t. GEAS

Methods	$DBP15K_{ZH-EN}$	$DBP15K_{JA-EN}$	$DBP15K_{FR-EN}$
	Hits@1	Hits@1	Hits@1
LatsEA	0.522	0.539	0.538
LatsEA+GEAS	**0.551**	**0.582**	**0.593**

6   Conclusion

In this paper, we propose LatsEA for entity alignment, aiming at mitigating the heterogeneous among the topological neighborhood structures of counterpart entities. LatsEA introduces neighborhood structure in the latent space to capture distant neighbors' information, which can expand the topological neighborhood. It uses vanilla GCN to aggregate the information of adjacent entities in the topological neighborhood and latent space neighborhood, respectively. Then, it uses an average gating mechanism to aggregate topological neighborhood and latent space neighborhood information. In order to further consider the interdependence between entities and make global entity alignment decisions, we propose a global entity alignment strategy, i.e., formulate entity alignment as the maximum bipartite matching problem which is effectively solved by Hungarian algorithm. We evaluate LatsEA and global entity alignment strategy on three real-world entity alignment datasets, this result shows their effectiveness.

Because the contribution of distant neighbors to the central entity embeddings is different, in future work, we will introduce an attention mechanism to give different weights to neighbor entities with different contributions when aggregating latent space neighborhood information.

References

1. Ngomo, A.-C.N., Auer, S.: LIMES - a time-efficient approach for large-scale link discovery on the web of data. In: IJCAI 2011, pp. 1585–1590 (2011)
2. Pershina, M., Yakout, M., Chakrabarti, K.: Holistic entity matching across knowledge graphs. In: International Conference on Big Data 2015, pp. 1585–1590 (2015)
3. Bordes, A., Usunier, N., Garcia-Duran, A.: Translating embeddings for modeling multi-relational data. In: NIPS 2013, pp. 2787–2795 (2013)
4. Chen, M., Tian, Y., Yang, M., Zaniolo, C.: Multilingual knowledge graph embeddings for cross-lingual knowledge alignment. In: IJCAI 2017, pp. 1511–1517 (2017)
5. Wang, Z., Lv, Q., Lan, X., Zhang, Y.: Cross-lingual knowledge graph alignment via graph convolutional networks. In: EMNLP 2018, pp. 349–357 (2018)
6. Ye, R., Li, X., Fang, Y., Zang, H., Wang, M.: A vectorized relational graph convolutional network for multi-Relational network alignment. In: IJCAI 2019, pp. 4135—4141 (2019)
7. Xu, K., Wang, L., Yu, M., et al.: Cross-lingual knowledge graph alignment via graph matching neural network. In: ACL 2019, pp. 3156–3161 (2019)
8. Sun, Z., Wang, C., Hu, W., et al.: Knowledge graph alignment network with gated multi-hop neighborhood aggregation. In: AAAI 2020, pp. 222–229 (2020)
9. Wu, Y., Liu, X., Feng, Y., et al.: Relation-aware entity alignment for heterogeneous knowledge graphs. In: IJCAI 2019, pp. 5278–5284 (2019)
10. Zhang, Z., Cai, J., Zhang, Y., et al.: Learning hierarchy-aware knowledge graph embeddings for link prediction. In: AAAI 2020, pp. 3065–3072 (2020)
11. Wang, Z., Zhang, J., Feng, J., et al.: Knowledge graph embedding by translating on hyperplanes. In: AAAI 2014, pp. 1112–1119 (2014)
12. Lin, Y., Liu, Z., Sun, M., et al.: Learning entity and relation embeddings for knowledge graph completion. In: AAAI 2015, pp. 2181–2187 (2015)
13. Dettmers, T., Minervini, P., Stenetorp, P., et al.: Convolutional 2D knowledge graph embeddings. In: AAAI 2018, pp. 1811–1818 (2018)
14. Dai, Q., Nguyen, T.V., Nguyen, T.D., et al.: A capsule network-based embedding model for knowledge graph completion and search personalization. In: NAACL-HLT 2019, pp. 2180–2189 (2019)
15. Sun, Z., Deng, Z., Nie, J., et al.: RotatE: knowledge graph embedding by relational rotation in complex space. In: ICLR 2019 (2019)
16. Zhu, H., Xie, R., Liu, Z., et al.: Iterative entity alignment via joint knowledge embeddings. In: IJCAI 2017, pp. 4258–4264 (2017)
17. Sun, Z., Hu, W., Li, C.: Cross-lingual entity alignment via joint attribute-preserving embedding. In: International Semantic Web Conference 2017, pp. 628–644 (2017)
18. Cao, Y., Liu, Z., Li, C., et al.: Multi-channel graph neural network for entity alignment. In: ACL 2019, pp. 1452–1461 (2019)
19. Kipf, T.N., Welling, M.: Semi-supervised classification with graph convolutional networks. In: ICLR 2017 (2017)
20. Sun, Z., Hu, W., Zhang, Q., et al.: Bootstrapping entity alignment with knowledge graph embedding. In: IJCAI 2018, pp. 4396–4402 (2018)

NLP Applications

Few-Shot Charge Prediction with Multi-grained Features and Mutual Information

Han Zhang[1], Zhicheng Dou[2(✉)], Yutao Zhu[3], and Jirong Wen[4,5]

[1] School of Information, Renmin University of China, Beijing, China
zhanghanjl@ruc.edu.cn
[2] Gaoling School of Artificial Intelligence, Renmin University of China, Beijing, China
dou@ruc.edu.cn
[3] Université de Montréal, Montreal, Canada
[4] Beijing Key Laboratory of Big Data Management and Analysis Methods, Beijing, China
[5] Key Laboratory of Data Engineering and Knowledge Engineering, MOE, Beijing, China

Abstract. Charge prediction aims to predict the final charge for a case according to its fact description and plays an important role in legal assistance systems. With deep learning based methods, prediction on high-frequency charges has achieved promising results but that on few-shot charges is still challenging. In this work, we propose a framework with multi-grained features and mutual information for few-shot charge prediction. Specifically, we extract coarse- and fine-grained features to enhance the model's capability on representation, based on which the few-shot charges can be better distinguished. Furthermore, we propose a loss function based on mutual information. This loss function leverages the prior distribution of the charges to tune their weights, so the few-shot charges can contribute more on model optimization. Experimental results on several datasets demonstrate the effectiveness and robustness of our method. Besides, our method can work well on tiny datasets and has better efficiency in the training, which provides better applicability in real scenarios.

Keywords: Few-shot charge · Multi-grained · Mutual information

1 Introduction

Charge prediction aims to determine the final charge (e.g., *manslaughter, traffic offence*, or *theft*) for a case by analyzing the textual fact description of the defendants' behavior. As a subtask of legal judgment prediction (LJP), charge prediction plays an important role in legal assistance systems, thus has been widely applied in real scenarios. On the one hand, a charge prediction system can provide legal professionals, such as judges and lawyers, a quick and effective

© Springer Nature Switzerland AG 2021
S. Li et al. (Eds.): CCL 2021, LNAI 12869, pp. 387–403, 2021.
https://doi.org/10.1007/978-3-030-84186-7_26

reference to improve their work efficiency. On the other hand, it can also provide non-legal professionals with some simple and useful legal advice.

Automatic charge prediction has been studied for decades. Early studies focused on applying mathematical or statistical methods, such as counting the specific attributes (e.g., crime time and place) of the cases [7,9]. Later, researchers began to frame the charge prediction as a text classification problem and paid attention to designing manual features or extracting shallow features from fact descriptions to predict the charge [6,13,14]. However, these features rely heavily on human expertise and are specific for different types of cases, which limits their application to a larger range of domains. Recently, deep learning based methods have also achieved promising results on charge prediction due to their superiority on automatic feature extraction and combination [16,26,27,29].

However, the charge prediction is still a non-trivial problem. One of the challenges is **few-shot charges**, which is also the focus of this paper. In practice, the numbers of cases in different charges usually follow a long-tailed distribution, which means their case data are highly imbalanced. For example, in the real-world dataset Criminal [5], the most frequent ten charges (such as *theft* and *intentional injury*) cover around 78% cases, while the lowest frequent fifty charges (such as *scalping relics* and *tax-escaping*) cover less than 0.5% cases. Under this circumstance, deep learning based methods can hardly perform well because the training data are insufficient for these few-shot changes. Therefore, how to deal with the few-shot charges with limited cases is crucial for building a robust and effective charge prediction system.

To alleviate this problem, [5] introduced the attribute features of the law, which are shared by all charges, so as to transfer knowledge from high-frequency charges to low-frequency ones. [4] proposed a sequence enhanced capsule model and leveraged the focal loss to alleviate the few-shot problem. However, we find that they still have some limitations: (1) The introduced attribute features are artificial, and they are usually very discrete on the cases of the few-shot charges and thus contribute less to distinguishing them from other charges. For example, the charge *illegally granting loans* and *illegally absorbing public deposits* have the same characteristics on profit-making purpose and nonviolent crime, and they can only be distinguished by more fine-grained characteristics such as the defendant's affiliation. (2) Existing works are all optimized by the cross entropy or its similar variants, and they do not consider that the cross-entropy is easily affected by the prior distribution, which makes it difficult to classify few-shot charges. For example, charge A has 1,000 cases and few-shot charge B has only 5 cases, charge A contributes more to the cross entropy loss as the loss of each sample is directly added up.

To tackle these problems, in this work, we **first** propose using multi-grained features. We introduce a convolutional network (CNN) with multiple kernels to extract coarse-grained features and a bilinear CNN [11] to extract fine-grained features from the case descriptions. These two kinds of features are then fused by a capsule network. The attribute features provided by [5] are also leveraged as additional explicit knowledge. The fine-grained features, fused features, and attribute features are finally combined by a multi-layer perceptron for charge prediction. By this means, the representation of each description can be greatly enhanced and the

Dong has been relying on the private loan usury contract projects. Between 2012 and 2015, ... To raise funds and make a profit, Dong own directly or through others to high interest rates, ... borrow or take a stake in the way of sharing bonus to their friends and family and the society to absorb funds... As a private developer, ... he absorbs a deposit of X yuan illegally...	☑ Profit making ⎫ ⎬ Attributes ☑ No violence ⎭ ☑ Work Fine-grained

Case Fact Description

Charge

Illegally granting loans ✅ ✅

Illegally absorbing public deposits ✅ ✅ ✅

Fig. 1. An example of charge prediction. The two few-shot charges have the same attributes, thus the fine-grained features are necessary for distinguishing them.

classifier can obtain more information for predicting the charge. **Second**, we consider the prior probability distribution of different charges over the dataset, based on which we construct a mutual information loss function. The few-shot charges can thus be paid more attention during the training process. Finally, the whole model is optimized by both the charge prediction and attribute prediction tasks. Experimental results on a series of datasets with different sizes demonstrate the effectiveness and wide applicability of our method (Fig. 1).

2   Related Work

2.1   Charge Predication

Early work on charge prediction focused on quantitative and statistical methods. For example, [9] counted various facts in the case to predict the crime. [7] introduced a linear model and a nearest neighbor method to predict crime. With the success of machine learning in some areas, researchers begin to model the charge prediction problem as a text classification problem. The basic idea is to extract features from the case description and make predictions by machine learning methods, such as linear models, logistic model trees [12], or SVMs [21]. However, these methods are built on manual features, which heavily rely on human expertise and are hard to apply on large datasets with various charges.

Recently, deep learning based methods have achieved promising results on charge prediction or legal judgement prediction. For example, [22] proposed a hierarchical text matching model to predict the cases. [26] improved performance by introducing all laws into legal judge prediction task and building a relationship graph of all laws to distinguish the confusable cases. [15] used a seq2seq model with attention to predict the cause of the decision. [1] adopt gating mechanism to improve penalty prediction based on the charge. [19] applied multi-scale attention to deal with the cases of multiple defendants. To deal with the problem of few-shot charge prediction, [5] introduced the attribute characteristics (manually defined) as expertise knowledge into the charge prediction task, while [4] applied a focal loss and designed a capsule network.

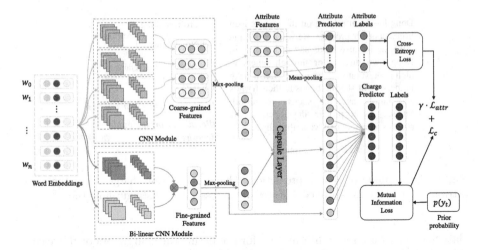

Fig. 2. The framework of MFMI.

Different from the existing work, we propose to extract fine-grained features automatically from the case descriptions to enhance the representation and design a mutual information loss function to take the prior distribution of different charges into account. Both strategies are helpful in improving the performance of few-shot charges.

2.2   Few-Shot Text Classification

In recent years, some studies have focused on text classification with few-shot samples. [2] proposed a prototype network structure based on mixed attention. Considering the domain differences of the text, [24] proposed the lifelong domain word embedding. [28] proposed to integrate a variety of measures in cross-domain few-shot learning. For invisible classes and cold start problems, [25] proposed an open world learning model to deal with invisible classes. [3] proposed a dynamic routing induction method to encapsulate abstract class representations.

3   Methodology

We propose a multi-task framework based on Multi-grained Features and Mutual Information (MFMI) for few-shot charge prediction. The structure of our MFMI is shown in Fig. 2. In general, MFMI considers three different kinds of features (namely coarse-grained features, fine-grained features, and attribute features) to represent a case and predict the charge based on the fused representations. To facilitate the few-shot charge prediction, we propose a loss function based on mutual information, with which the few-shot charges can contribute more on the model optimization, so the overall performance can be improved. Besides, similar to [5], we also add a supplementary task to predict whether the predefined attributes are contained in the fact, which is reported to be helpful in improving the performance. As a result, the whole model is trained with both our proposed mutual information loss on charge prediction and a cross-entropy loss on attribute prediction.

3.1 Task Formalization

Before introducing our model, we first give the formalization of the charge prediction task. Formally, assuming a fact description (text) containing n words as $X = (w_1, \cdots, w_n)$, where $w_i \in D$, and D is the fixed dictionary, the model is asked to predict (classify) the charge $y \in Y$ of the fact from the predefined charge set Y. Besides, as suggested by [5], we also predict the common attributes of the law articles as a supplementary task to boost the performance on few-shot charges prediction. It has the same input sequence X and aims at predicting the corresponding fact-findings of attributes $p = \{p_1, \cdots, p_k\}$ according to the fact. Here, k is the number of attributes, and $p_i \in \{0, 1\}$ is the label for a certain attribute. Generally, the charge prediction can be treated as a (multi-class) text classification task, while the attribute prediction can be regarded as a binary classification task.

3.2 Coarse-Grained Features

As illustrated in Fig. 2, the case description X is represented as a sequence of embeddings $\mathbf{X}$ by a looking-up operation on a pre-trained embedding table $\mathbf{E}$:

$$\mathbf{X} = [\mathbf{e}_1, \mathbf{e}_2, \cdots, \mathbf{e}_n], \quad \mathbf{e}_i = \text{Look-Up}(\mathbf{E}(w_i)), \tag{1}$$

where $\mathbf{X} \in \mathbb{R}^{n \times d}$, and $\mathbf{e}_i \in \mathbb{R}^d$ is the embedding of the i-th word w_i in the fact description.

To further enhance the representation of the word, we apply 1D CNNs with different kernels over the embedding sequence and compute the coarse-grained features as:

$$\mathbf{X}_t^C = \text{1D-CNN}(\mathbf{X}, s_t), \quad t \in [1, T], \tag{2}$$

where s_t is the kernel size of the t-th CNN and T is the number of CNNs. With different kernels, the semantic information in the consecutive words can be integrated together. For example, if $s_t = 2$, we can obtain a sequence of bigram representations. As these features can only reflect shallow semantic information, we call them coarse-grained features. Note that we apply zero padding during the convolution, thus $\mathbf{X}_t^C$ has a dimension of $n \times m$, where m is the number of output channels. Other neural networks, such as RNN [10], can also be applied to represent the word sequence, here we choose CNN because it can be computed in parallel and provide better efficiency in a real system.

3.3 Fine-Grained Features

In charge prediction, some charges usually have the same or similar fact descriptions. For example, the few-shot charge *illegally granting loans* and *illegally absorbing public deposits* are both nonviolent and have a profit-making purpose. It is hard to distinguish them based on only the coarse-grained features. Therefore, we propose to leverage a bilinear CNN module to extract fine-grained features for better distinguishing the charges. Bilinear CNN can integrate the information from two sub-CNNs by a bilinear transformation. In the original CNN structure, the features are usually refined through a max-pooling or mean-pooling operation, which can only take the first-order information into account. However, the pooling function of a bilinear CNN calculates the outer product of different feature channels. The outer product can capture the pairwise correlation between feature channels, providing a stronger characteristic representation than traditional CNNs.

Specifically, considering two 1D convolutional operations $f_A(\mathbf{X}, l)$ and $f_B(\mathbf{X}, l)$ computing at the position l of the embedding matrix $\mathbf{X}$, the bilinear transformation can be described as:

$$\text{Bilinear}(f_A, f_B) = f_A(\mathbf{X}, l)^\top f_B(\mathbf{X}, l). \tag{3}$$

As a result, if $f_A(\mathbf{X}, l)$ and $f_B(\mathbf{X}, l)$ have m_A and m_B channels respectively, the dimension of obtained features after bilinear transformation will be $m_A \times m_B$. Then, the features at all positions are aggregated by sum-pooling and produce an integrated feature representation $\Phi(\mathbf{X})$ as:

$$\Phi(\mathbf{X}) = \sum_{l \in L} \text{Bilinear}(f_A, f_B). \tag{4}$$

Finally, the output Φ of the bilinear transformation is flattened as a vector:

$$\mathbf{V} = \text{Flatten}(\Phi(\mathbf{X})). \tag{5}$$

To alleviate the overfitting problem [11], a signed square root and a L2 normalization are usually applied on $\mathbf{v}$ and output the fine-grained features $\mathbf{X}^F$:

$$\mathbf{V}' = \text{sign}(\mathbf{V})\sqrt{|\mathbf{V}|}, \quad \mathbf{X}^F = \mathbf{V}'/\|\mathbf{V}'\|_2. \tag{6}$$

3.4   Capsule Layer

A typical problem of CNNs is that they cannot obtain the relative position information among extracted features. To deal with this problem, we apply a capsule network [20] to integrate the spatial information and fuse the features for better representation. Compared with traditional neural networks, the basic unit of capsule network is the capsule, which conducts a series of operations on input vectors rather than scalars. The connections between capsules are implemented by a dynamic routing mechanism, which contains several affine transformations and nonlinear functions[1]. Here, we use Capsule($\cdot$) to denote a capsule layer.

---

[1] We suggest the reader to refer to the original paper [20] for the details of capsule network.

In particular, we use the capsule layer to aggregate the coarse-grained features $\mathbf{X}^C$ and fine-grained features $\mathbf{X}^F$ by:

$$\mathbf{X}^C = \text{Max-pooling}(\mathbf{X}_1^C, \cdots, \mathbf{X}_T^C), \quad \mathbf{X}^{\text{CAP}} = \text{Flatten}(\text{Capsule}(\mathbf{X}^C \oplus \mathbf{X}^F)), \tag{7}$$

where $\oplus$ is the concatenation operation. With such a capsule network, the spatial information of the features can be integrated into the refined representations.

3.5 Attribute Features

Inspired by [5], we also extract attribute features from the fact description. These attribute features are shared by various charges, thus can transfer knowledge from high-frequency charges to low-frequency ones.

Specifically, the attribute features are computed based on the coarse-grained features $\mathbf{X}_t^C$. First, we calculate attention weights $\mathbf{a} = \{\mathbf{a}_1, \cdots, \mathbf{a}_k\}$ for all attributes, where $a_i = [a_{i,1}, \cdots, a_{i,n}]$. For $\forall i \in [1, k]$ and $\forall j \in [1, T]$, $a_{i,j}$ is calculated by:

$$a_{i,j} = \frac{\exp(\tanh(\mathbf{W}^a \mathbf{X}_j^C)^\top \mathbf{u}_i)}{\sum_{t=1}^{T} \exp(\tanh(\mathbf{W}^a \mathbf{X}_t^C)^\top \mathbf{u}_i)}, \tag{8}$$

where $\mathbf{u}_i$ is the context vector of the i-th attribute (randomly initialized) to calculate how informative an element is to the attribute, and $\mathbf{W}^a$ is a weight matrix (parameter) shared by all attributes. Thereafter, we can obtain the fact-aware attribute features as:

$$\mathbf{X}_i^A = \sum_{t=1}^{T} a_{i,t} \mathbf{X}_t^C. \tag{9}$$

These features will be used for both charge prediction and attribute prediction (introduced in Sect. 3.7).

3.6 Aggregation and Prediction

To aggregate all obtained features, we concatenate them together and apply a multi-layer perceptron (MLP) to fuse them. Finally, the distribution Z_y of all crimes is calculated as:

$$Z_y = \text{MLP}(\mathbf{X}^F \oplus \mathbf{X}^{\text{CAP}} \oplus \mathbf{X}^A), \tag{10}$$

$$\mathbf{X}^A = \text{Mean-pooling}(\mathbf{X}_1^A, \cdots, \mathbf{X}_k^A). \tag{11}$$

3.7 Optimization

In general, our model is a multi-task learning framework. Both a charge prediction loss and an attribute prediction loss are used for optimizing our model.

Charge Prediction Loss Based on Mutual Information. As a multi-class classification problem, charge prediction is often optimized by a cross-entropy loss. Suppose that there are L categories, and the training data $(x, y) \sim \mathcal{T}$ have the distribution $p_\theta(y|x)$, the cross-entropy loss can be computed as:

$$\mathcal{L}_c = \mathbb{E}_{(x,y) \sim \mathcal{T}}[-\log p_\theta(y|x)], \tag{12}$$

$$p_\theta(y|x) = e^{Z_y} / \sum_{i=1}^{L} e^{Z_i}. \tag{13}$$

Directly applying cross-entropy loss into the charge prediction task is not suitable because the numbers of samples in different charge categories are extremely unbalanced. In this case, when we sample a batch of data during the training process, only a few low-frequency charges are contained, so they have less contribution for the optimization. As a result, the performance of the model on the few-shot charges is worse than that on others.

Essentially, whether a charge is few-shot or not is determined by its frequency in the training set, which is usually represented by the probability distribution. Therefore, the key point is how to integrate the prior probability distribution $p(y)$ of each charge into the loss function. When combining the cross entropy and the prior probability distribution, we find that the form fits the mutual information naturally [17]:

$$\log \frac{p_\theta(y|x)}{p(y)} \sim Z_y, \tag{14}$$

which is equivalent to:

$$\log p_\theta(y|x) \sim Z_y + \log p(y). \tag{15}$$

In other words, we integrate the prior probability distribution into the loss function by adding a term $\log p(y)$ to Eq. (12) as:

$$p_\theta(y|x) = \frac{e^{Z_y + \log p(y)}}{\sum_{i=1}^{L} e^{Z_i + \log p(i)}} = \left[1 + \sum_{i \neq y} \frac{p(i)}{p(y)} e^{Z_i - Z_y} \right]^{-1}. \tag{16}$$

The mutual information loss function is defined by taking Eq. (16) into Eq. (12).

Attribute Prediction Loss. As suggested by [5], we also add a supplementary task, namely, the attribute prediction, to improve the performance of our model on few-shot charge prediction. Specifically, we project the attribute features into the label space and use softmax function to get the final prediction result $\mathbf{p} = [p_1, \cdots, p_k]$ as:

$$\mathbf{z}_i = \mathrm{softmax}(\mathrm{MLP}(\mathbf{X}_i^A)), \quad p_i = \arg\max(\mathbf{z}_i), \tag{17}$$

Table 1. The statistics of all datasets.

Datasets	Criminal-S	Criminal-M	Criminal-L	Criminal-T
Train	61,589	153,521	306,900	6,860
Validation	7,755	19,250	38,429	2,730
Test	7,702	19,189	38,368	2,684

where p_i is the prediction result of the i-th attribute, and $\mathbf{z}_i$ is the predicted binary probability distribution. As each attribute is equally important in the model, we can easily calculate the attribute prediction loss by summing up the cross-entropy of all attributes:

$$\mathcal{L}_{attr} = -\sum_{i=1}^{k}\sum_{j=1}^{2}\hat{z}_{ij}\log(z_{ij}), \tag{18}$$

where $\hat{z}_i$ is the ground-truth label, and z_i is the predicted probabilities distribution.

Overall Loss. Finally, we combine $\mathcal{L}_c$ and $\mathcal{L}_{attr}$ to get the overall loss function $\mathcal{L}$ as follows:

$$\mathcal{L} = \mathcal{L}_c + \gamma \cdot \mathcal{L}_{attr}, \tag{19}$$

where γ is a hyper-parameter, which is used to adjust the weights of these two loss functions.

4   Experiments

4.1   Datasets

We conduct experiments on two real datasets Criminal [5] and CAIL [23].

Criminal: This dataset is for few-shot charges prediction. It contains three datasets with different sizes, denoted as Criminal-S (small), Criminal-M (medium), and Criminal-L (large), respectively. Each sample contains a fact description, a charge result, and attribute labels. The number of samples on these three datasets are as shown in Table 1. All datasets are divided into training, validation, and testing set with the ratio of 8:1:1.

In order to verify the performance of the model in terms of few-shot charges, all categories in the Criminal-S (small) dataset, are divided into three different classes according to their frequencies, where the charges with ≤10 cases are low-frequency charges and the charges with >100 cases are high-frequency charges.

In practical situations, the number of cases for each charge is usually small. In order to further test the robustness and performance of the models in a real world scenario, on the basis of Criminal-S (small), the number of cases of all charges in the training set is limited to less than 100. That is, all high-frequency charges become medium-frequency. This dataset is named as Criminal-T (tiny), the categories of samples in the dataset is the same as the above three datasets. The statistics of all datasets are shown in Table 1.

CAIL: This is a dataset for legal judgement prediction, which consists of three tasks (prediction of applicable law articles, charges, and term of penalty). The total number of samples is 101,619. The attribute information of each charge is not provided, thus we label it manually.

4.2 Baselines

On the Criminal dataset, we select two basic models and three state-of-the-art few-shot charge prediction models as baselines:

CNN and **LSTM:** These are two basic models for text classification. For CNN, we use {100, 200} as the number of output channels; while for LSTM, we set the size of the hidden states as {100, 200}.

Fact-Law Attention [16]: It improves the accuracy of charge prediction task by introducing additional legal articles and using attention mechanism to obtain most relevant legal article to enhance the representation capacity of fact description.

Secaps [4]: It uses a capsule network to improve the model's representation ability and manually adjusts the category weights to improve the accuracy of few-shot charges.

Attribute-Attention [5]: It constructs the features of artificial law attributes and predicts the charges and attributes simultaneously to improve performance.

On the CAIL dataset, we compare our model with three state-of-the-art models, which learn three legal judgement prediction tasks simultaneously, as introduced in Sect. 4.1:

Attribute-Attention-MTL: It has the same structure as the Attribute-Attention model, but is trained on three judgement prediction tasks.

TopJudge [29]: It leverages the dependencies of sub-tasks to improve the performance of legal judge prediction.

MPBFN [27]: It designs a multi-view forward prediction and backward validation framework to utilize the dependencies among sub-tasks of legal judge prediction.

4.3 Evaluation Metrics

Following existing studies [4,5], we adopt Accuracy (Acc.), Macro precision (MP), Macro Recall (MR), and Macro F1 (MF1) as the evaluation metrics. Among them, MR and MF1 are the preferred evaluation metrics for multi-class classification problems, especially for those with imbalance categories.

Table 2. Charge prediction results on the Criminal datasets. "†" indicates significant improvements ($p < 0.05$) compared with the best baseline.

Model	Criminal-S				Criminal-M				Criminal-L			
	Acc.	MP	MR	MF1	Acc.	MP	MR	MF1	Acc.	MP	MR	MF1
CNN-100	91.9	50.5	44.9	46.1	93.5	57.6	48.1	50.5	93.9	66.0	50.3	54.7
CNN-200	92.6	51.1	46.3	47.3	92.8	56.2	50.0	50.8	94.1	61.9	50.0	53.1
LSTM-100	93.5	59.4	58.6	57.3	94.7	65.8	63.0	62.6	95.5	69.8	67.0	66.8
LSTM-200	92.7	60.0	58.4	57.0	94.4	66.5	62.4	62.7	95.1	72.8	66.7	67.9
Fact-Law Att.	92.8	57.0	53.9	53.4	94.7	66.7	60.4	61.8	95.7	73.3	67.1	68.6
Secaps	**93.7**	67.8	66.3	65.8	94.7	**70.4**	68.3	68.2	**95.9**	77.2	73.5	73.7
Att. attention	93.4	66.7	69.2	64.9	94.4	68.3	69.2	67.1	95.8	75.8	73.7	73.1
MFMI	**93.7**	**69.3†**	**70.5†**	**68.2†**	**94.9**	70.2	**75.0†**	**71.0†**	**95.9**	**78.7†**	**77.4†**	**76.4†**

Table 3. Macro F1 values of various charges on the Criminal-S dataset. "†" indicates significant improvements ($p < 0.05$) compared with the best baseline.

Charge type	Low (<10)	Medium	High (>100)
# Charges	49	51	49
Secaps	52.4	59.2	85.9
Att. attention	49.7	60.0	85.2
MFMI	**55.9†**	**63.5†**	85.7

4.4 Experiment Settings

We adopt THULAC[2] for word segmentation on all fact descriptions in the datasets. The maximum length of the fact description is set as 500. The pre-trained embedding table is obtained by Word2Vec [18] with the dimension of 100. For CNN module, we use four kernels with the sizes of {2,4,8,16}, and the number of output channels is set as 64. For bilinear CNN module, we set the kernel sizes as {8,12}, and the number of output channels is also 64. As for the capsule layer, we set the number of capsule as the number of categories, and the dimension of each capsule is 32. The number of routings is 3. Adam [8] is applied as the optimizer with the learning rate of 1e–3. The settings of all baselines are consistent with their original papers.

4.5 Experimental Results

Experimental results on the Criminal datasets are shown in Table 2, Table 3 and Table 4. We can observe:

(1) In general, our MFMI achieves consistently better performance on all three sizes of datasets. This indicates that our method has wide applicability over various application scenarios (sufficient or insufficient data).

---

[2] https://github.cosm/thunlp/THULAC-Python.

Table 4. Charge prediction results on the tiny Criminal-T dataset. "†" indicates significant improvements ($p < 0.05$) compared with the best baseline.

	Acc.	MP	MR	MF1
Secaps	81.4	63.8	67.4	63.8
Att. attention	81.2	62.3	69.3	63.5
MFMI	**83.9**†	**63.7**	**70.7**†	**65.2**†

Table 5. Charge prediction results on the CAIL dataset.

	Acc.	MP	MR	MF1
TopJudge	82.10	**83.60**	78.42	79.05
MPBFN	82.14	82.28	80.72	80.72
Att. attention-MTL	83.65	80.84	82.01	81.55
MFMI	**84.20**	81.34	**82.74**	**81.65**

(2) Compared with accuracy, all methods perform poorly in terms of the Macro F1 metric. This is mainly because of the imbalance of training samples among different charges, and indicates the shortage of prediction for few-shot charges. However, our model achieves promising improvements (3.3%, 3.9%, and 3.3% absolutely on three datasets respectively), which demonstrates the robustness and effectiveness of our model. To further validate the performance on few-shot charges, we compare the results of our model with the two best baseline models on different frequency of charges. As shown in Table 3, we divide all charges into three classes based on the case number. The charges with less than 10 samples are regarded as low-frequency, while those with more than 100 samples are high-frequency. The rest charges are treated as medium-frequency. By this means, we obtain 49 low-frequency charges, 51 medium-frequency charges, and 49 high-frequency charges. From the results, we can see that our model significantly improves the performance (6.68% and 5.83% compared with the baseline models) on both low- and medium-frequency charges. These results clearly demonstrate the effectiveness of our model in dealing with few-shot charges.

(3) Specifically, on Criminal-S, MFMI achieves the best results in terms of all evaluation metrics. The value of Macro F1 is significantly improved by 3.6% compared with the previous best method. This proves the superiority of our method on small datasets. In practice, the number of samples for each charge is usually very small. To mimic the application in such real scenario, we build a tiny dataset Criminal-T (tiny) based on Criminal-S (small) and test the performance of MFMI and the other two best baselines. Criminal-T contains only 12,274 samples from the original Criminal dataset. The results are shown in Table 4. We can observe that MFMI achieves new state-of-the-art results on Accuracy, Macro Recall, and Macro F1. The improvements are significant. This proves

Table 6. Ablation results on the Criminal datasets. "†" indicates significant improvements ($p < 0.05$) compared with the best baseline.

Model	Criminal-S				Criminal-M				Criminal-L			
	Acc.	MP	MR	F1	Acc.	MP	MR	F1	Acc.	MP	MR	F1
MFMI	**93.7**	69.3	**70.5**†	**68.2**†	**94.9**	70.2	**75.0**†	**71.0**†	**95.9**	78.7	**77.4**	**76.4**†
w/o FF	92.1	63.1	68.8	63.6	93.6	66.4	73.7	68.1	94.0	70.1	76.9	72.3
w/ CE	**93.7**	**69.7**	67.0	67.1	94.7	**71.4**	69.8	69.4	95.8	**79.0**	73.1	74.8

Table 7. The time (seconds) taken per epoch.

Model	Criminal-T	Criminal-S
Att. attention	200	2,150
Secaps	272	2,200
MFMI	**19**	**150**

that our model is applicable in the real scenario and is capable of dealing with few-shot charges.

The experimental results on the CAIL dataset are shown in Table 5. It is worth noting that all baselines listed here are multi-task learning models, which are trained on three tasks (applicable law articles prediction, term of penalty prediction, and charge prediction). They use much more data and obtain more supervision signals. However, it is very interesting to see that our MFMI model achieves better results in terms of Accuracy, Macro Recall, and Macro F1. This implies that by leveraging fine-grained features and optimizing with the mutual information loss function, our method can make full use of the data and provide better performance.

4.6 Ablation Study

To investigate the effectiveness of fine-grained features and our proposed mutual information loss, we conduct an ablation study by removing them from the full model, respectively. The two variants are denoted as "w/o FF" and "w/ CE". The results are shown in Table 6. We can observe the performance degradation when removing either module. Specifically, when fine-grained features are not used, the performance of our model decreased significantly. The potential reason is that the fine-grained features can effectively improve the representation capability of our model. Thus the overall performance is improved. On the other hand, when replacing the mutual information loss with a normal cross-entropy (CE) loss, the accuracy has less change but the Macro F1 decreases significantly. By checking the results, we find that the performance on few-shot charges degrades more than the others. As the number of samples in few-shot charges is limited, the overall accuracy is less affected. This result demonstrates that the mutual information loss is effective on learning few-shot charges, which is consistent with our assumption.

4.7 Efficiency Analysis

We also compared the average time taken for each epoch during the training stage of our model and the two best few-shot charge prediction models. As shown in Table 7, our model spends less than a tenth of the time on both datasets compared to the baselines. This is because our model is based on several CNN modules, and the parallel training ability is far better than baselines with RNNs. In practice, our model with fast training speed is easier to deploy and apply.

5 Conclusion

In this work, we studied the problem of few-shot charge prediction. We proposed a multi-task learning framework, where both the charge prediction and attribute prediction are learned simultaneously. We extracted fine-grained and coarse-grained features to enhance the model's capability of representation, which are helpful for distinguishing the few-shot charges. Besides, we also proposed a loss function based on mutual information to enhance the learning on few-shot charges. Experimental results demonstrated the effectiveness of proposed methods on charge prediction, especially on the few-shot charges. Moreover, further experiments showed that our method has better scalability and efficiency. In the future, we will apply our method to other legal judgement prediction tasks and leverage the knowledge from other tasks to further improve the performance.

Acknowledgements. We thank all the anonymous reviewers for their insightful comments. This work was supported by National Natural Science Foundation of China No. 61872370 and No. 61832017, and Beijing Outstanding Young Scientist Program NO. BJJWZYJH012019100020098.

References

1. Chen, H., Cai, D., Dai, W., Dai, Z., Ding, Y.: Charge-based prison term prediction with deep gating network. In: Inui, K., Jiang, J., Ng, V., Wan, X. (eds.) Proceedings of the 2019 Conference on Empirical Methods in Natural Language Processing and the 9th International Joint Conference on Natural Language Processing, EMNLP-IJCNLP 2019, Hong Kong, China, 3–7 November 2019, pp. 6361–6366. Association for Computational Linguistics (2019). https://doi.org/10.18653/v1/D19-1667
2. Gao, T., Han, X., Liu, Z., Sun, M.: Hybrid attention-based prototypical networks for noisy few-shot relation classification. In: The Thirty-Third AAAI Conference on Artificial Intelligence, AAAI 2019, The Thirty-First Innovative Applications of Artificial Intelligence Conference, IAAI 2019, The Ninth AAAI Symposium on Educational Advances in Artificial Intelligence, EAAI 2019, Honolulu, Hawaii, USA, 27 January–1 February 2019, pp. 6407–6414. AAAI Press (2019). https://doi.org/10.1609/aaai.v33i01.33016407

3. Geng, R., Li, B., Li, Y., Ye, Y., Jian, P., Sun, J.: Few-shot text classification with induction network. CoRR abs/1902.10482 (2019). http://arxiv.org/abs/1902.10482

4. He, C., Peng, L., Le, Y., He, J., Zhu, X.: SEcaps: a sequence enhanced capsule model for charge prediction. In: Tetko, I.V., Kurková, V., Karpov, P., Theis, F.J. (eds.) Artificial Neural Networks and Machine Learning - ICANN 2019: Text and Time Series - 28th International Conference on Artificial Neural Networks, Munich, Germany, 17–19 September 2019, Proceedings, Part IV. LNCS, vol. 11730, pp. 227–239. Springer, Heidelberg (2019). https://doi.org/10.1007/978-3-030-30490-4_19

5. Hu, Z., Li, X., Tu, C., Liu, Z., Sun, M.: Few-shot charge prediction with discriminative legal attributes. In: Bender, E.M., Derczynski, L., Isabelle, P. (eds.) Proceedings of the 27th International Conference on Computational Linguistics, COLING 2018, Santa Fe, New Mexico, USA, 20–26 August 2018, pp. 487–498. Association for Computational Linguistics (2018). https://www.aclweb.org/anthology/C18-1041/

6. Katz, D.M., Bommarito, M.J., Blackman, J.: A general approach for predicting the behavior of the supreme court of the united states. PLOS ONE **12**(4), e0174698 (2017)

7. Keown, R.: Mathematical models for legal prediction. Comput./Law J. **2**, 829 (1980)

8. Kingma, D.P., Ba, J.: Adam: a method for stochastic optimization. In: Bengio, Y., LeCun, Y. (eds.) 3rd International Conference on Learning Representations, ICLR 2015, San Diego, CA, USA, 7–9 May 2015, Conference Track Proceedings (2015). http://arxiv.org/abs/1412.6980

9. Kort, F.: Predicting supreme court decisions mathematically: a quantitative analysis of the "right to counsel" cases. Am. Polit. Sci. Rev. **51**(01), 1–12 (1957)

10. Lai, S., Xu, L., Liu, K., Zhao, J.: Recurrent convolutional neural networks for text classification. In: Bonet, B., Koenig, S. (eds.) Proceedings of the Twenty-Ninth AAAI Conference on Artificial Intelligence, 25–30 January 2015, Austin, Texas, USA, pp. 2267–2273. AAAI Press (2015). http://www.aaai.org/ocs/index.php/AAAI/AAAI15/paper/view/9745

11. Lin, T., RoyChowdhury, A., Maji, S.: Bilinear CNN models for fine-grained visual recognition. In: 2015 IEEE International Conference on Computer Vision, ICCV 2015, Santiago, Chile, 7–13 December 2015, pp. 1449–1457. IEEE Computer Society (2015). https://doi.org/10.1109/ICCV.2015.170

12. Lin, W., Kuo, T., Chang, T., Yen, C., Chen, C., Lin, S.: Exploiting machine learning models for Chinese legal documents labeling, case classification, and sentencing prediction. Int. J. Comput. Linguist. Chin. Lang. Process. **17**(4) (2012). http://www.aclclp.org.tw/clclp/v17n4/v17n4a4.pdf

13. Liu, C., Chang, C., Ho, J.: Case instance generation and refinement for case-based criminal summary judgments in Chinese. J. Inf. Sci. Eng. **20**(4), 783–800 (2004). http://www.iis.sinica.edu.tw/page/jise/2004/200407_12.html

14. Liu, C., Hsieh, C.: Exploring phrase-based classification of judicial documents for criminal charges in Chinese. In: Esposito, F., Ras, Z.W., Malerba, D., Semeraro, G. (eds.) Foundations of Intelligent Systems, 16th International Symposium, ISMIS 2006, Bari, Italy, 27–29 September 2006, Proceedings. LNCS, vol. 4203, pp. 681–690. Springer, Heidelberg (2006). https://doi.org/10.1007/11875604_75

15. Liu, Z., Tu, C., Liu, Z., Sun, M.: Legal cause prediction with inner descriptions and outer hierarchies. In: Sun, M., Huang, X., Ji, H., Liu, Z., Liu, Y. (eds.) Chinese Computational Linguistics - 18th China National Conference, CCL 2019, Kunming, China, 18–20 October 2019, Proceedings. LNCS, vol. 11856, pp. 573–586. Springer, Cham (2019). https://doi.org/10.1007/978-3-030-32381-3_46

16. Luo, B., Feng, Y., Xu, J., Zhang, X., Zhao, D.: Learning to predict charges for criminal cases with legal basis. In: Palmer, M., Hwa, R., Riedel, S. (eds.) Proceedings of the 2017 Conference on Empirical Methods in Natural Language Processing, EMNLP 2017, Copenhagen, Denmark, 9–11 September 2017, pp. 2727–2736. Association for Computational Linguistics (2017). https://doi.org/10.18653/v1/d17-1289

17. Menon, A.K., Jayasumana, S., Rawat, A.S., Jain, H., Veit, A., Kumar, S.: Long-tail learning via logit adjustment. CoRR abs/2007.07314 (2020). https://arxiv.org/abs/2007.07314

18. Mikolov, T., Sutskever, I., Chen, K., Corrado, G.S., Dean, J.: Distributed representations of words and phrases and their compositionality. In: Burges, C.J.C., Bottou, L., Ghahramani, Z., Weinberger, K.Q. (eds.) Advances in Neural Information Processing Systems 26: 27th Annual Conference on Neural Information Processing Systems 2013. Proceedings of a Meeting Held 5–8 December 2013, Lake Tahoe, Nevada, United States, pp. 3111–3119 (2013). https://proceedings.neurips.cc/paper/2013/hash/9aa42b31882ec039965f3c4923ce901b-Abstract.html

19. Pan, S., Lu, T., Gu, N., Zhang, H., Xu, C.: Charge prediction for multi-defendant cases with multi-scale attention. In: Sun, Y., Lu, T., Yu, Z., Fan, H., Gao, L. (eds.) Computer Supported Cooperative Work and Social Computing - 14th CCF Conference, ChineseCSCW 2019, Kunming, China, August 16–18, 2019, Revised Selected Papers, Communications in Computer and Information Science, vol. 1042, pp. 766–777. Springer, Singapore (2019). https://doi.org/10.1007/978-981-15-1377-0_59

20. Sabour, S., Frosst, N., Hinton, G.E.: Dynamic routing between capsules. In: Guyon, I., et al. (eds.) Advances in Neural Information Processing Systems 30: Annual Conference on Neural Information Processing Systems 2017, 4–9 December 2017, Long Beach, CA, USA, pp. 3856–3866 (2017). https://proceedings.neurips.cc/paper/2017/hash/2cad8fa47bbef282badbb8de5374b894-Abstract.html

21. Sulea, O., Zampieri, M., Malmasi, S., Vela, M., Dinu, L.P., van Genabith, J.: Exploring the use of text classification in the legal domain. In: Ashley, K.D., et al. (eds.) Proceedings of the Second Workshop on Automated Semantic Analysis of Information in Legal Texts Co-located with the 16th International Conference on Artificial Intelligence and Law (ICAIL 2017), London, UK, 16 June 2017. CEUR Workshop Proceedings, vol. 2143. CEUR-WS.org (2017). http://ceur-ws.org/Vol-2143/paper5.pdf

22. Wang, P., Fan, Y., Niu, S., Yang, Z., Zhang, Y., Guo, J.: Hierarchical matching network for crime classification. In: Piwowarski, B., Chevalier, M., Gaussier, É., Maarek, Y., Nie, J., Scholer, F. (eds.) Proceedings of the 42nd International ACM SIGIR Conference on Research and Development in Information Retrieval, SIGIR 2019, Paris, France, 21–25 July 2019, pp. 325–334. ACM (2019). https://doi.org/10.1145/3331184.3331223

23. Xiao, C., et al.: CAIL2018: A large-scale legal dataset for judgment prediction. CoRR abs/1807.02478 (2018). http://arxiv.org/abs/1807.02478

24. Xu, H., Liu, B., Shu, L., Yu, P.S.: Lifelong domain word embedding via meta-learning. In: Lang, J. (ed.) Proceedings of the Twenty-Seventh International Joint Conference on Artificial Intelligence, IJCAI 2018, 13–19 July 2018, Stockholm, Sweden, pp. 4510–4516. ijcai.org (2018). https://doi.org/10.24963/ijcai.2018/627

25. Xu, H., Liu, B., Shu, L., Yu, P.S.: Open-world learning and application to product classification. In: Liu, L., et al. (eds.) The World Wide Web Conference, WWW 2019, San Francisco, CA, USA, 13–17 May 2019, pp. 3413–3419. ACM (2019). https://doi.org/10.1145/3308558.3313644

26. Xu, N., Wang, P., Chen, L., Pan, L., Wang, X., Zhao, J.: Distinguish confusing law articles for legal judgment prediction. In: Jurafsky, D., Chai, J., Schluter, N., Tetreault, J.R. (eds.) Proceedings of the 58th Annual Meeting of the Association for Computational Linguistics, ACL 2020, Online, 5–10 July 2020, pp. 3086–3095. Association for Computational Linguistics (2020). https://doi.org/10.18653/v1/2020.acl-main.280

27. Yang, W., Jia, W., Zhou, X., Luo, Y.: Legal judgment prediction via multi-perspective bi-feedback network. In: Kraus, S. (ed.) Proceedings of the Twenty-Eighth International Joint Conference on Artificial Intelligence, IJCAI 2019, Macao, China, 10–16 August 2019, pp. 4085–4091. ijcai.org (2019). https://doi.org/10.24963/ijcai.2019/567

28. Yu, M., et al.: Diverse few-shot text classification with multiple metrics. In: Walker, M.A., Ji, H., Stent, A. (eds.) Proceedings of the 2018 Conference of the North American Chapter of the Association for Computational Linguistics: Human Language Technologies, NAACL-HLT 2018, New Orleans, Louisiana, USA, 1–6 June 2018, Volume 1 (Long Papers), pp. 1206–1215. Association for Computational Linguistics (2018). https://doi.org/10.18653/v1/n18-1109

29. Zhong, H., Guo, Z., Tu, C., Xiao, C., Liu, Z., Sun, M.: Legal judgment prediction via topological learning. In: Riloff, E., Chiang, D., Hockenmaier, J., Tsujii, J. (eds.) Proceedings of the 2018 Conference on Empirical Methods in Natural Language Processing, Brussels, Belgium, 31 October–4 November 2018, pp. 3540–3549. Association for Computational Linguistics (2018). https://doi.org/10.18653/v1/d18-1390

Sketchy Scene Captioning: Learning Multi-level Semantic Information from Sparse Visual Scene Cues

Lian Zhou, Yangdong Chen, and Yuejie Zhang[✉]

School of Computer Science, Shanghai Key Laboratory of Intelligent Information Processing, Fudan University, Shanghai 200438, China
{16110240019,19110240010,yjzhang}@fudan.edu.cn

Abstract. To enrich the research about sketch modality, a new task termed Sketchy Scene Captioning is proposed in this paper. This task aims to generate sentence-level and paragraph-level descriptions for a sketchy scene. The sentence-level description provides the salient semantics of a sketchy scene while the paragraph-level description gives more details about the sketchy scene. Sketchy Scene Captioning can be viewed as an extension of sketch classification which can only provide one class label for a sketch. To generate multi-level descriptions for a sketchy scene is challenging because of the visual sparsity and ambiguity of the sketch modality. To achieve our goal, we first contribute a sketchy scene captioning dataset to lay the foundation of this new task. The popular sequence learning scheme, e.g., Long Short-Term Memory neural network with visual attention mechanism, is then adopted to recognize the objects in a sketchy scene and infer the relations among the objects. In the experiments, promising results have been achieved on the proposed dataset. We believe that this work will motivate further researches on the understanding of sketch modality and the numerous sketch-based applications in our daily life. The collected dataset is released at https://github.com/SketchysceneCaption/Dataset.

Keywords: Sketch · Sketchy scene · Dataset · Multi-level captioning · Sequence learning

1 Introduction

In recent years, sketch has emerged as one important data modality [5,24]. Compared to a natural image, a sketch only contains sparse and ambiguous visual information. Current works about sketch mainly focus on predicting one class label for a sketch, and such a label provides very limited semantic information [5]. Differently, the tasks about natural image are abundant, such as classification [3], captioning [1,22,23], and visual question answering [1]. What hinders the research about sketch is the lack of sketch datasets. Specifically, natural

© Springer Nature Switzerland AG 2021
S. Li et al. (Eds.): CCL 2021, LNAI 12869, pp. 404–418, 2021.
https://doi.org/10.1007/978-3-030-84186-7_27

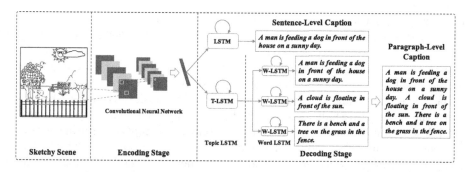

Fig. 1. An overview of the proposed Sketchy Scene Captioning framework. The visual attention models are not given for conciseness.

images are easy to be obtained and a lot of efforts have been put into annotating the images. By contrast, sketch is created by human and the generation of a sketch is time-consuming, which limits the volume of a sketch dataset and the visual details of the sketches in the dataset. Hence, most of current sketch datasets only contain sketches with a single object and the corresponding class label. Drawing inspiration from the task of natural image captioning, it is attractive to expand a sketch to a sketchy scene which contains several objects, and extend the class label to a sentence-level or even a paragraph-level description. In a word, a sketchy scene dataset with multi-level descriptions is in urgent need to promote the research about sketch.

One promising application with above extension is child education [16]. Specifically, with the wide popularity of tablet PC, it becomes common for a child to doodle on a touch screen. To interact with a child, a computer agent needs to understand what a child has drawn and give reasonable response to the child. For example, if a sketchy scene drawn by a child cannot match the sentence or paragraph given by the agent, the child is required to draw the sketchy scene again, which helps improve the drawing skill of the child. Another potential application is the assistance for the visually impaired people. With simple and sparse visual content, a sketchy scene can be easily turned into a concave-convex plate that can be read by a visually impaired person in a touch manner. With the corresponding caption transformed into human voice [25], the visually impaired person can feel what are depicted in the sketchy scene without the help of others. Other potential applications include large-scale sketchy scene retrieval via human language and automatic sketch management on the Web (e.g., to cluster the numerous sketchy scenes with similar topics).

Motivated by the observations above, we extend the task of sketch classification to Sketchy Scene Captioning, a task that aims to generate multi-level (i.e., sentence-level and paragraph-level) descriptions for a sketchy scene, as shown in Fig. 1. To the best of our knowledge, our work is the first attempt to generate multi-level and dense descriptions for a sketchy scene. Currently, to achieve the goal of sketchy scene captioning is very challenging for two reasons. **First,** compared to a natural image captioning dataset, only the generation of a sketchy scene is time-consuming, let alone the annotation of the sketchy scene. **Second**, a sketchy scene only contains very sparse visual cues. That is, an object is depicted only with some lines. In addition, the visual cues of a sketchy scene are also ambiguous. That is, the objects in different sketchy scenes have great variations in appearance, making it difficult to distinguish the objects. To overcome the above two challenges, we create a sketchy scene dataset with multi-level descriptions and achieve the goal of sketchy scene captioning using several sequence-learning-based models in the field of image captioning [13, 22, 23]. The contributions of this work are three-folds: 1) A new task termed Sketchy Scene Captioning is proposed to generate multi-level descriptions for a sketchy scene. This task can be treated as a new paradigm for comprehensive understanding of the sparse visual cues; 2) A sketchy scene captioning dataset is constructed based on *SketchyScene* dataset [26]. Currently, the new dataset contains 1,000 sketchy scenes with both the sentence-level and paragraph-level captions; and 3) Promising experimental results have been achieved on the newly collected dataset, demonstrating the potentials of Sketchy Scene Captioning. We hope this work could help motivate further researches on mining multi-level semantic information from sketchy scenes.

2   Related Works

In this section, we will briefly review the related works about sketch and image captioning. The differences between the prior works and ours will be discussed as well.

Current works about sketch mainly focus on Sketch Classification and Sketch-based Image Retrieval (SBIR). Sketch Classification is a task of recognizing what object is depicted in a sketch. SBIR aims to retrieve a natural image for a given query sketch. In recent years, great progresses have been made in the field of sketch. For example, Yu et al. [24] proposed Sketch-a-Net, a multi-scale and multi-channel deep neural network, to yield the sketch recognition performance surpassing that of humans on the *TU-Berlin* sketch dataset [5]. Sangkloy et al. [18] proposed the *Sketchy* dataset, which was the first large-scale collection of sketch-photo pairs for image retrieval. He et al. [7] proposed a deep visual-sequential fusion mechanism to model the visual and sequential patterns of the strokes of a sketch. Liu et al. [15] proposed a semantic-aware knowledge preservation method for sketch-based image retrieval. In spite of the above progresses, the related works about sketch classification are limited to assigning a class label to each sketch. In this paper, we go a step further to generate multi-level and dense descriptions for a sketchy scene.

Current methods for image captioning can be mainly divided into three categories, that is, template-based, retrieval-based, and sequence-learning-based. In the template-based method, the salient objects, their attributes, and the relations among objects in an image were first recognized, and a pre-defined template was then filled with the detected information to yield a full sentence [6]. The retrieval-based method first obtained the visually similar image with the query image, and then used the description of the retrieved image as the description of the query image [11]. However, these two methods could only generate relatively fixed sentences, relying on the given image-caption dataset. In the era of deep learning, sequence learning was adopted to adaptively generate a description for a natural image, where a Convolutional Neural Network (CNN) [8] encoder was used to encode the image into a high-level visual representation, and a Recurrent Neural Network (RNN) [20] decoder was adopted to "translate" the image representation into a sentence. Typically, Vinyals et al. [22] first proposed to use Inception [10] convolutional neural network as the encoder to convert an image into a fixed-length vector, and then use Long Short-Term Memory (LSTM) [9] neural network as the decoder to generate a caption for the image. Xu et al. [23] introduced two spatial visual attention mechanisms to help the model dynamically focus on the image regions corresponding to the word that was about to be generated. Besides, Krause et al. [13] designed a hierarchical LSTM model to generate a paragraph-level description for a natural image. Overall, current works on image captioning mainly focus on natural images. Differently, we explore the caption generation problem in the field of a different domain, that is, sketchy scene which only contains sparse and ambiguous visual information.

3   A New Dataset for Sketchy Scene Captioning

To the best of our knowledge, there is no available dataset for sketchy scene captioning. Hence, we need to first construct a sketchy scene dataset with sentence-level and paragraph-level descriptions. Next, we will describe how the dataset is collected in details.

3.1   *SketchyScene* Dataset without Descriptions

In the field of sketch, several sketch datasets, such as *TU-Berlin* [5] and *Sketchy* [18], have been proposed for sketch classification or cross-modal retrieval, and the sketches in these datasets are created by humans. However, each sketch in these datasets only contains one object with discrete class labels or together with the stroke orders. As a result, the related researches based on these datasets can only deal with single object, which indicates that to create a sketch dataset with annotations is very challenging. With single object in a sketch, these datasets cannot be used for captioning. To extend the research on sketch, Zou et al. [26] propose a brand new dataset called *SketchyScene* recently. The dataset consists of scene sketches where each scene sketch contains multiple objects. Each object in a scene sketch is assigned with one class label out of 45 categories. Because every scene has a corresponding natural or cartoon image for reference, all the sketchy scenes are supposed to be consistent with the real world. Besides, there is also segmentation information for each sketch. Because each scene sketch in *SketchyScene* is constructed by combining the separate instances of several categories, the volume of *SketchyScene* can grow relatively large, which ensures the

Fig. 2. Four sketchy scene examples from the *SketchyScene* dataset.

diversity of the sketchy scenes. Four examples from the *SketchyScene* dataset are shown in Fig. 2.

In spite of the advantages of *SketchyScene*, no sentence-level or paragraph-level descriptions are provided in *SketchyScene*. In this work, we choose to construct our sketchy scene captioning dataset based on *SketchyScene* with the following two reasons. **First,** *SketchyScene* provides realistic and diverse sketchy scenes, which makes the dataset suitable for sketchy scene captioning. It can be observed from Fig. 2 that the objects of each sketchy scene are quite diverse and the object arrangement in each scene is reasonable, making it meaningful to generate a sentence-level or even a paragraph-level description for the sketchy scene. **Second,** *SketchyScene* provides class labels for the objects in each sketchy scene, offering important hints for the annotators to give more accurate descriptions for each sketchy scene.

3.2   Description Collection for *SketchyScene* Dataset

We conduct the data collection in a manner of crowdsourcing. Hence, we first create a website to ease the annotation job. On the website, several annotation examples, a randomly picked target sketchy scene, and the category labels of the target sketchy scene are presented. To ensure the annotation quality, we only invite a number of graduate students in universities as volunteers who are well trained in English. We realize that some annotation instructions for volunteers are still needed to further improve the annotation quality. After analyzing a few initial annotated captions without instructions and being inspired by the proposed requirements when collecting *MSCOCO* [2] for image captioning, we summarize the following rules that the volunteers should obey when annotating a sketchy scene.

- Be faithful to the visual content of the presented sketchy scene. Do not describe anything unrelated to the sketchy scene (e.g., what may happen in the past or future).
- Do not describe what people may say in a sketchy scene.
- Do not give a name to a person or animal.
- Do not use any abbreviation in the descriptions.
- Try to use more specific words when possible. For example, use words such as "*girl*", "*boy*", "*woman*", and "*man*" instead of "*people*".

Sketchy Scene	Tags	Sentence-Level Caption	Paragraph-Level Caption
	cloud, cow, fence, flower, grass	A cow is standing on the grass.	A cow is standing on the ground. Some flowers and grass grow on the ground. There is a fence behind the cow. Two clouds are floating in the sky.
	car, cloud, cow, grass, house, mountain, road, tree	A car is on the road and a cow is eating grass beside the house.	There is a house beside the road. A cow is standing beside a tree and eating grass. A car is on the road and a mountain is behind the house.
	cat, cloud, dog, fence, flower, grass, people, sun, tree	A girl is standing on the ground with a dog and a cat in a sunny day.	A girl is standing in front of a fence. Some flowers and grass grow on the ground. A tree is standing near the girl. A dog and a cat is sitting under the tree. The sun is shining and two clouds are floating in the sky.

Fig. 3. Three representative examples of the collected multi-level captions for sketchy scenes.

- For sentence-level annotation, describe the sketchy scene with a brief summary, not necessary to include everything.
- For paragraph-level annotation, describe the sketchy scene as detailed as possible with all the given category labels.

With the settings above, we successfully collect 1,000 annotated scene sketches, where each scene sketch is associated with one sentence-level description and one paragraph-level description. Our website for data collection is still open for more annotations, and a new version of the dataset is expected to be released in the future. To share the idea of sketchy scene captioning and the collected dataset with other researchers timely, we currently use the collected 1,000 samples for exploration in this work.

3.3 Dataset Analysis

In this section, we will take a look at the newly collected dataset. Three representative examples are shown in Fig. 3. Column "Tags" shows the category labels corresponding to the objects in a sketchy scene, and these labels act as the guiding words for the volunteers when annotating a sketchy scene. The following two columns show the sentence-level and the paragraph-level descriptions of a sketchy scene, respectively. By analyzing the descriptions, we find that the annotators tend to take the most salient object as the subject of a sentence-level caption and describe its interactions with other possible objects in a sketchy scene. Differently, more objects and their interactions are described in a paragraph-level caption. Although the volunteers may not follow the instructions strictly, the quality of the captions is still good enough for our research.

We further conduct a quantitative analysis on the collected captions. For the sentence-level captions, there are 504 different words in total. The lengths of the sentence-level captions are concentrated in 10–20 words, and the distribution of caption length is roughly in line with the Gaussian distribution. Besides, most sentences have only 1–3 relations (e.g., verb and preposition) among objects, which means that the annotators tend to focus on the salient parts of a sketchy scene and ignore other details during the sentence-level annotation. For the paragraph-level captions, there are 681 different words in total. Most of the paragraph-level captions contain 3–6 sentences. The lengths of the paragraph-level captions are concentrated in 25–35 words, and the lengths of all the single sentences in the paragraph-level captions are concentrated in 6–14 words. It can be found that the sizes of the two vocabularies above are relatively small, which are caused by two reasons. **First**, compared to a natural image, a sketchy scene contains much less visual details (e.g., the color of an object). **Second**, there are only 45 object categories in the *SketchyScene* dataset and the annotators are required to use the given category labels when constructing a caption. Due to these two reasons, a sketch dataset cannot become as diverse as a natural image dataset, which is a stubborn problem in current research on sketch.

4   Multi-level Sketchy Scene Captioning Through Sequence Learning

In this work, the popular sequence-learning-based method is adopted for flexible sketchy scene captioning, as shown in Fig. 1. Our framework integrates Sketchy Scene Encoder for Deep Visual Features (i.e., encoding a sketchy scene at an abstract level to obtain a discriminative visual representation) and Sketchy Scene Decoder with Spatial Visual Attention (i.e., grasping more visual details of a sketchy scene while generating the description). It is a new attempt to generate multi-level descriptions for a sketchy scene through the sequence learning paradigm.

4.1   Sketchy Scene Encoder for Deep Visual Features

CNN is adopted as image encoder in sketchy scene captioning. Considering that a sketchy scene contains very sparse visual information, the outputs of different CNN layers are chosen as the visual features of a sketchy scene for comparison. The output of the fully-connected layer is a fixed-length vector that is denoted as v_{fc}. The output of a convolutional layer is a set of spatial feature vectors that are denoted as $v_{cv} = \{v_i | v_i \in \mathbb{R}^{d_v}, 1 \leq i \leq n\}$, where d_v is the feature dimension and n is the region number. These fine-grained features can be used for visual attention. The global representation of a sketchy scene v_0 is used to initialize the decoder and can be computed as:

$$v_0 = v_{fc} \quad \text{or} \quad (\sum_{i=1}^{n} v_i)/n, \; v_i \in v_{cv}. \tag{1}$$

4.2   Sketchy Scene Decoder with Spatial Visual Attention

Sentence-Level Decoder. The LSTM neural network is exploited as image decoder in sketchy scene captioning. The decoder generates a sentence $S = (s_0, \ldots, s_c, s_{c+1})$ conditioned on the input visual features ($\boldsymbol{v}_{fc}$ or $\boldsymbol{v}_{cv}$), where c is the length of the sentence, and s_0 and s_{c+1} denote the starting and ending tokens respectively. Each word in S is denoted as a one-hot vector. An embedding matrix is used to convert each word to a low-dimensional vector as follows:

$$x_t = \boldsymbol{W}_e s_t, 0 \leq t \leq c+1, \tag{2}$$

where $\boldsymbol{W}_e \in \mathbb{R}^{d_e \times V}$, d_e is the dimension of word embedding, and V is the vocabulary size. The inputs of the decoder at time step t ($1 \leq t \leq c+1$) include the embedding of the previous word x_{t-1} and a contextual vector z_t that is computed through the soft attention [23] as follows:

$$e_i^t = f_{att}(\boldsymbol{v}_i, \boldsymbol{h}_{t-1}), 1 \leq i \leq n, t \geq 1, \tag{3}$$

$$\alpha_i^t = \exp(e_i^t) / \sum_{k=1}^{n} \exp(e_k^t), \tag{4}$$

$$z_t = \sum_{i=1}^{n} \alpha_i^t \boldsymbol{v}_i, \tag{5}$$

where $\boldsymbol{h}_{t-1}$ is the hidden state of the decoder at time step $t-1$, and f_{att} is the soft visual attention function that is implemented as a fully-connected neural network. It should be noted that z_t exists only when a sketchy scene is encoded as a set of spatial feature vectors, otherwise the sketchy scene decoder is just a vanilla LSTM. Given the global visual representation $\boldsymbol{v}_0$ of a sketchy scene, the initial memory state $\boldsymbol{c}_0$ and the initial hidden state $\boldsymbol{h}_0$ can be obtained by feeding $\boldsymbol{v}_0$ into two separate fully-connected neural networks as follows:

$$c_0 = f_c(\boldsymbol{v}_0), \boldsymbol{h}_0 = f_h(\boldsymbol{v}_0), \tag{6}$$

where $tanh$ nonlinearity is adopted. It should be noted that the global visual representation $\boldsymbol{v}_0$ is only used to initialize the LSTM decoder. Given the visual features $\boldsymbol{F}$ (i.e., $\boldsymbol{v}_{fc}$ or $\boldsymbol{v}_{cv}$) of a sketchy scene, the LSTM decoder is learned by minimizing the negative logarithmic probability of the target sentence S as follows:

$$L_s = -\log P(S|\boldsymbol{F}) = -\sum_{t=1}^{c+1} \log P(s_t|s_0^{t-1}, \boldsymbol{F}), \tag{7}$$

where the whole conditional logarithmic probability can be decomposed into the multiplication of the logarithmic probability at each time step. The t-th word can be predicted by the output layer as follows:

$$P(s_t|s_0^{t-1}, \boldsymbol{F}) \propto \exp(\boldsymbol{E}_0(\boldsymbol{E}_1 \boldsymbol{h}_t + \boldsymbol{E}_2 \boldsymbol{z}_t)), \tag{8}$$

where $E_0 \in \mathbb{R}^{V \times d_m}$, $E_1 \in \mathbb{R}^{d_m \times d_m}$, $E_2 \in \mathbb{R}^{d_m \times d_v}$, and d_m is the number of the LSTM cell units.

Paragraph-Level Decoder. Considering that the average length of a paragraph is about 30 words, it is difficult for a single LSTM to generate such a long sequence with correct meanings. Thus, a hierarchical LSTM (H_LSTM) network [13] is exploited to generate a paragraph-level caption for a sketchy scene. The H_LSTM network consists of a topic LSTM and a word LSTM. The topic LSTM takes the visual features v_0 of a sketchy scene as input and generates a sequence of guiding signals $G = (g_0, \ldots, g_N, g_{N+1})$, where N is the number of topics, g_0 is the starting signal, $g_t = 1 (1 \leq t \leq N)$ indicates that a new sentence needs to be generated, and $g_{N+1} = 0$ indicates stopping generating the paragraph. Visual attention is also conducted by the topic LSTM. At time step t, the hidden state h_t^T and the contextual vector z_t^T of the topic LSTM are concatenated to obtain the topic vector that is used to guide the caption generation of the word LSTM. The word LSTM works similar to the sentence-level captioning decoder except that the visual features used to initialize it are replaced with the topic vector. The loss function of H_LSTM can be formulated as:

$$L_p = -\lambda_T \log P(G|v_{cv}) - \lambda_W \sum_{t=1}^{N} \log P(S_{W_t}|v_{cv}, h_t^T, z_t^T), \tag{9}$$

where λ_T and λ_W are the weighting coefficients of the topic LSTM loss and the word LSTM loss respectively, and S_{W_t} denotes the t-th sentence generated by the word LSTM.

5   Experiment and Analysis

5.1   Dataset and Preprocessing

We conduct the experiments on the newly collected dataset to verify the feasibility of the sketchy scene captioning task. The dataset is divided into training, validation, and testing sets with a ratio of 8:1:1, that is, 800 <sketchy scene, caption> pairs for training, 100 for validation, and 100 for testing. The words that appear at least 5 times in the training captions are kept, and a vocabulary of size 174 for sentence-level captions and another one of size 223 for paragraph-level captions are constructed. Each vocabulary includes a starting token "<start>", an ending token "<end>", and an unknown word token "<UNK>" for those words that appear less than 5 times in the training set.

5.2   Model Learning and Inference

In the experiments, the training samples are <sketchy scene, caption> pairs. That is, the captioning models are trained to generate a description for a sketchy scene. ResNet-101 [8] pre-trained on ImageNet [3] is used as the sketchy scene encoder. Because of the domain gap between a natural image and a hand-drawn sketch, the fine-tuning of the encoder is turned on during training. For each sketchy scene, the size of the output feature from the fully-connected layer before softmax operation is 1,000, and the size of the features from the convolutional layer before the last average pooling layer is 14 × 14 × 2,048. The number of LSTM cell units is set to 512. The dimension of word embedding is set to 512. Adam [12] is used as the model optimizer. Dropout [19] and early stopping are exploited to achieve model regularization. The BLEU [17] score on validation set is used for model selection. The initial learning rate is set to 0.0004 with a 0.5 decay ratio. The batch size is set to 40. Beam search is used for inference with a beam size of 5. The weighting coefficients λ_T and λ_W in Eq. (9) are both set to 1.

5.3   Quantitative Evaluation

We first verify the effectiveness of sentence-level sketchy scene captioning. Because the visual information of a sketchy scene is sparse and ambiguous, we explore how the representation of a sketchy scene affects the model performance by considering two factors: 1) visual feature ("*FC*" and "*CV*" denote the output features from the fully-connected layer and the convolutional layer, respectively); and 2) visual attention ("*ATT*" and "*NAT*" indicate that visual attention is turned on and off, respectively). It should be noted that visual attention is not conducted for the visual feature from the output of the fully-connected layer because a sketchy scene is simply encoded as a fixed-length vector in this case. The name of a model is denoted as the combination of the two factors above. The BLEU@n (**B@n**, n = 1, 2, 3, 4), METEOR (**M**) [4], ROUGE_L (**R**) [14], and CIDEr (**C**) [21] scores on testing set are reported in Table 1. These scores are computed by the *MSCOCO* captioning evaluation tool[1].

Table 1. The experimental results of sentence-level sketchy scene captioning on the new dataset.

Model	B@1	B@2	B@3	B@4	M	R	C
FC_NAT	25.6	16.2	11.0	7.8	10.3	31.1	39.0
CV_NAT	29.0	21.1	16.9	**14.4**	14.2	25.2	58.4
CV_ATT	**37.6**	**25.3**	**18.0**	13.0	**14.9**	**31.9**	**59.5**

Two aspects can be observed from Table 1. **First**, model *CV_NAT* performs better than model *FC_NAT* across all the metrics except **R**. This indicates that the fine-grained visual representation from the convolutional layer can better characterize the visual content of a sketchy scene compared to the visual representation from the fully-connected layer. The reason behind is that the visual information of a sketchy scene is sparse and the pooling operation before the

---

[1] https://github.com/tylin/coco-caption.

fully-connected layer causes too much information loss. Differently, the output from the convolutional layer can preserve more discriminative local features of a sketchy scene. With such discriminative details, the captioning model can better recognize the objects in a sketchy scene, which further helps the model infer the correct interactions among the objects in the sketchy scene. **Second**, model *CV_ATT* achieves higher scores than those of model *CV_NAT* across all the metrics except $B@4$, which is mainly due to the precise visual features produced by the attention mechanism. Specifically, the information loss of a sketchy scene still exists when its initial representation is obtained by averaging the spatial visual features. Meanwhile, its visual details are gradually forgotten by the captioning model as the process of caption generation goes on. However, with the help of visual attention, the captioning model can be guided with fine-grained visual details by focusing on the relevant regions when generating words, which helps alleviate the problem of forgetting.

We also conduct the experiments on paragraph-level sketchy scene captioning, and the results are reported in Table 2. It can be observed that the $B@n$, M, and R scores are comparable to the best results of the sentence-level captioning models, while the C score is worse. Considering the definitions of these metrics, our captioning model can generate a relatively fluent paragraph-level description with correct semantics. However, the captioning model may sometimes focus on the wrong key points of the sketchy scene, which are not consistent with those identified by a person to a certain degree.

Table 2. The experimental results of paragraph-level sketchy scene captioning on the new dataset.

Model	$B@1$	$B@2$	$B@3$	$B@4$	M	R	C
H_LSTM	43.6	28.4	19.8	14.3	17.0	33.4	30.8

5.4 Qualitative Evaluation

We first have an analysis on the sentence-level captioning results. As shown in Fig. 4, one generated sentence-level caption with the corresponding attention map sequence from model *CV_ATT* is given, where the attention maps highlight the regions that the captioning model learns to focus on at different time steps. It can be seen that what the caption describes matches the salient visual content of the sketchy scene quite well. Specifically, "*a big rabbit*" can be generated correctly when the model focuses on the object "*rabbit*". Meanwhile, the action "*crouching*" and the surroundings "*grass and flowers*" can be correctly recognized as well when the model focuses on the two sides of the sketchy scene. In addition, the model does not generate the description about the weather, such as "*on a cloudy day*", which may be due to the reason that the "*cloud*" is too small to be salient enough. It is worth noting that the model does not generate the description about the "*trees*" which occupy a large area of the sketchy scene.

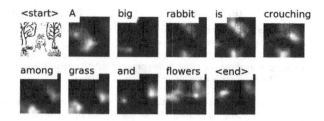

Fig. 4. A generated sentence-level caption with the corresponding attention map sequence.

As mentioned before, a sentence-level caption only describes the most salient parts of a sketchy scene. In the example, the *"rabbit"* has been treated as the salient object and the word *"rabbit"* may co-occur with *"grass"* and *"flowers"* more frequently in the dataset, and thus the *"trees"* are not treated as the salient objects by the captioning model.

Another three representative examples of the sentence-level captioning are given in Fig. 5. Generally, the salient objects in the selected sketchy scenes can be well identified except the objects *"woman"* and *"house"* in the first one, and this bad result may be caused by the imprecise visual representation of the sketchy scene. That is, the *"woman"* is occluded by the *"tree"* in front of her, making the captioning model fail to recognize the *"woman"* correctly. At the same time, the mistaken object *"chicken"* usually co-occurs with the object *"fence"*, and the *"house"* is then ignored by the model. It can be observed that the actions of the salient objects, that is, *"standing"*, *"driving"*, and *"playing"*, can be generated properly. The reason is that, the recognized objects co-occur with the actions frequently in the dataset, which helps the model generate the correct actions for the salient objects. In the first example, the mistaken object *"chicken"* is usually followed by the action *"standing"*, which makes the action still correct compared to the ground truth caption. Besides, the descriptions about the weather (i.e., *"on a sunny day"*) in all the examples are generated correctly. The reason may be that, the *"sun"* is usually located on the top of a sketchy scene in the dataset,

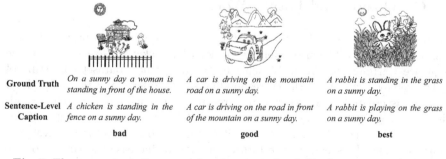

Ground Truth	*On a sunny day a woman is standing in front of the house.*	*A car is driving on the mountain road on a sunny day.*	*A rabbit is standing in the grass on a sunny day.*
Sentence-Level Caption	*A chicken is standing in the fence on a sunny day.*	*A car is driving on the road in front of the mountain on a sunny day.*	*A rabbit is playing on the grass on a sunny day.*
	bad	**good**	**best**

Fig. 5. Three representative examples of sentence-level sketchy scene captioning.

Ground Truth	Paragraph-Level Caption
A hen is guiding eight chicken to pass through the zebra crossing. Four cars are waiting at the edge of the zebra crossing. a police-man is sitting behind the cars. Tall buildings are located behind the police not far away.	*Four cars are parking on the road with a people standing behind them. A chicken with many UNK are UNK the UNK. Some houses stands in the distance.*
Several clouds are floating in the sky. A school bus is driving on the road. On one side of the road there are some girls and boys. On one side of the road there are some trees and a house. The trees are beside the house.	*A woman with a UNK on the road in front of the house. A car is driving on the road. Some trees are standing in the distance. Some grass are growing on the side of the road.*

Fig. 6. Two representative examples of paragraph-level sketchy scene captioning.

and thus the visual representation of the "*sun*" can be discriminative enough for the model to recognize the weather.

In the following, we will have an analysis on two representative examples of paragraph-level sketchy scene captioning, as shown in Fig. 6. It can be seen that both paragraphs are quite meaningful and describe many visual details of the sketchy scenes correctly. This indicates that the topic LSTM can give relatively correct guiding signals to the word LSTM and can stop the captioning process properly. Surprisingly, even the number of "*cars*" can be recognized correctly in the first example. These two examples show that it is promising to use a hierarchical LSTM model for paragraph-level sketchy scene captioning. There exist some problems in the results as well. For example, "*eight chicken*" is missed in the first example, and "*school bus*" and "*girls and boys*" are missed in the second example. Hence, how to learn a more discriminative sketchy scene representation and generate correct descriptions for the relatively small objects remains to be explored. Because no prior work about sketchy scene captioning exists, we cannot compare our results with other methods. However, the qualitative experimental results show that the generated multi-level captions for a sketchy scene by our models are quite meaningful, which proves that the proposed Sketchy Scene Captioning task is feasible and deserves further exploration.

6   Conclusion and Future Work

In this paper, a new task termed Sketchy Scene Captioning is proposed. This task aims to generate multi-level descriptions for a sketchy scene through the sequence learning paradigm. To achieve the goal, a new dataset consisting of 1,000 sketchy scenes with the corresponding sentence-level and paragraph-level captions is created. The experimental results show that our captioning models can recognize the main objects in a sketchy scene and the interactions among these objects. This proves that it is feasible to generate multi-level captions for a sketchy scene. In the future, we plan to increase the volume of the dataset and explore how to learn a better representation of a sketchy scene for caption generation. We hope this work can inspire further researches on the better understanding of sketchy scenes.

Acknowledgements. This work was supported by the National Natural Science Foundation of China (No. 61976057 and No. 61572140) and Shanghai Municipal R&D Foundation (No. 20511101203, No. 20511102702, No. 20511101403, No.19DZ2205700, and No. 2021SHZDZX0103).

References

1. Anderson, P., et al.: Bottom-up and top-down attention for image captioning and visual question answering. In: Proceedings of the Conference on Computer Vision and Pattern Recognition, pp. 6077–6086 (2018)
2. Chen, X., et al.: Microsoft COCO captions: data collection and evaluation server. CoRR abs/1504.00325 (2015). http://arxiv.org/abs/1504.00325
3. Deng, J., Dong, W., Socher, R., Li, L., Li, K., Li, F.: ImageNet: a large-scale hierarchical image database. In: Proceedings of the Conference on Computer Vision and Pattern Recognition, pp. 248–255 (2009)
4. Denkowski, M.J., Lavie, A.: Meteor universal: language specific translation evaluation for any target language. In: Proceedings of the Workshop on Statistical Machine Translation (WMT@ACL 2014), pp. 376–380 (2014)
5. Eitz, M., Hays, J., Alexa, M.: How do humans sketch objects? ACM Trans. Graph. **31**(4), 44:1–44:10 (2012)
6. Elliott, D., Keller, F.: Image description using visual dependency representations. In: Proceedings of the Conference on Empirical Methods in Natural Language Processing, pp. 1292–1302 (2013)
7. He, J., Wu, X., Jiang, Y., Zhao, B., Peng, Q.: Sketch recognition with deep visual-sequential fusion model. In: Proceedings of the ACM on Multimedia Conference, pp. 448–456 (2017)
8. He, K., Zhang, X., Ren, S., Sun, J.: Deep residual learning for image recognition. In: Proceedings of the Conference on Computer Vision and Pattern Recognition, pp. 770–778 (2016)
9. Hochreiter, S., Schmidhuber, J.: Long short-term memory. Neural Comput. **9**(8), 1735–1780 (1997)
10. Ioffe, S., Szegedy, C.: Batch normalization: accelerating deep network training by reducing internal covariate shift. In: Proceedings of the International Conference on Machine Learning, pp. 448–456 (2015)
11. Karpathy, A., Joulin, A., Li, F.: Deep fragment embeddings for bidirectional image sentence mapping. In: Proceedings of the Annual Conference on Neural Information Processing Systems, pp. 1889–1897 (2014)
12. Kingma, D.P., Ba, J.: Adam: a method for stochastic optimization. In: Proceedings of the International Conference on Learning Representations (2015). http://arxiv.org/abs/1412.6980
13. Krause, J., Johnson, J., Krishna, R., Fei-Fei, L.: A hierarchical approach for generating descriptive image paragraphs. In: Proceedings of the Conference on Computer Vision and Pattern Recognition, pp. 3337–3345 (2017)
14. Lin, C.Y.: ROUGE: a package for automatic evaluation of summaries. In: Proceedings of the Workshop on Text Summarization Branches Out (Post-Conference Workshop of ACL 2004), pp. 74–81 (2004)

15. Liu, Q., Xie, L., Wang, H., Yuille, A.L.: Semantic-aware knowledge preservation for zero-shot sketch-based image retrieval. In: Proceedings of the International Conference on Computer Vision, pp. 3661–3670 (2019)

16. Neshati, M., Fallahnejad, Z., Beigy, H.: On dynamicity of expert finding in community question answering. Inf. Process. Manage. **53**(5), 1026–1042 (2017)

17. Papineni, K., Roukos, S., Ward, T., Zhu, W.: BLEU: a method for automatic evaluation of machine translation. In: Proceedings of the Annual Meeting of the Association for Computational Linguistics, pp. 311–318 (2002)

18. Sangkloy, P., Burnell, N., Ham, C., Hays, J.: The sketchy database: learning to retrieve badly drawn bunnies. ACM Trans. Graph. **35**(4), 119:1–119:12 (2016)

19. Srivastava, N., Hinton, G.E., Krizhevsky, A., Sutskever, I., Salakhutdinov, R.: Dropout: a simple way to prevent neural networks from overfitting. J. Mach. Learn. Res. **15**(1), 1929–1958 (2014)

20. Sutskever, I., Vinyals, O., Le, Q.V.: Sequence to sequence learning with neural networks. In: Proceedings of the Annual Conference on Neural Information Processing Systems, pp. 3104–3112 (2014)

21. Vedantam, R., Zitnick, C.L., Parikh, D.: CIDEr: consensus-based image description evaluation. In: Proceedings of the Conference on Computer Vision and Pattern Recognition, pp. 4566–4575 (2015)

22. Vinyals, O., Toshev, A., Bengio, S., Erhan, D.: Show and tell: a neural image caption generator. In: Proceedings of the Conference on Computer Vision and Pattern Recognition, pp. 3156–3164 (2015)

23. Xu, K., et al.: Show, attend and tell: neural image caption generation with visual attention. In: Proceedings of the International Conference on Machine Learning, pp. 2048–2057 (2015)

24. Yu, Q., Yang, Y., Song, Y., Xiang, T., Hospedales, T.M.: Sketch-a-Net that beats humans. In: Proceedings of the British Machine Vision Conference, pp. 7.1–7.12 (2015)

25. Zou, C., Mo, H., Du, R., Wu, X., Gao, C., Fu, H.: LUCSS: language-based user-customized colourization of scene sketches. CoRR abs/1808.10544 (2018). http://arxiv.org/abs/1808.10544

26. Zou, C., et al.: SketchyScene: richly-annotated scene sketches. In: Proceedings of the European Conference on Computer Vision, pp. 438–454 (2018)

BDCN: Semantic Embedding Self-explanatory Breast Diagnostic Capsules Network

Dehua Chen[1], Keting Zhong[1], and Jianrong He[2(✉)]

[1] School of Computer Science and Engineering, Donghua University, Shanghai, China
chendehua@dhu.edu.cn
[2] Comprehensive Breast Health Center, Shanghai Ruijin Hospital, Shanghai, China

Abstract. Building an interpretable AI diagnosis system for breast cancer is an important embodiment of AI assisted medicine. Traditional breast cancer diagnosis methods based on machine learning are easy to explain, but the accuracy is very low. Deep neural network greatly improves the accuracy of diagnosis, but the black box model does not provide transparency and interpretation. In this work, we propose a semantic embedding self-explanatory Breast Diagnostic Capsules Network (BDCN). This model is the first to combine the capsule network with semantic embedding for the AI diagnosis of breast tumors, using capsules to simulate semantics. We pre-trained the extraction word vector by embedding the semantic tree into the BERT and used the capsule network to improve the semantic representation of multiple heads of attention to construct the extraction feature, the capsule network was extended from the computer vision classification task to the text classification task. Simultaneously, both the back propagation principle and dynamic routing algorithm are used to realize the local interpretability of the diagnostic model. The experimental results show that this breast diagnosis model improves the model performance and has good interpretability, which is more suitable for clinical situations.

Keywords: Semantic embedding · Capsule network · Interpretable · Breast tumor prediction

1 Introduction

Breast cancer is an important killer threatening women's health because of rising incidence. Early detection and diagnosis are the key to reduce the mortality rate of breast cancer and improve the quality of life of patients. Mammary gland molybdenum target report contains rich semantic information, which can directly reflect the results of breast cancer screening [1], and AI-assisted diagnosis of breast cancer is an important means. Therefore, various diagnostic models were born. Mengwan [2] used support vector machine (SVM) and Naive Bayes to classify morphological features with an accuracy of 91.11%. Wei [3] proposed

© Springer Nature Switzerland AG 2021
S. Li et al. (Eds.): CCL 2021, LNAI 12869, pp. 419–433, 2021.
https://doi.org/10.1007/978-3-030-84186-7_28

a classification method of breast cancer based on SVM, and the accuracy of the classifier experiment is 79.25%. These traditional AI diagnoses of breast tumors have limited data volume and low accuracy. Deep Neural Networks (DNN) enters into the ranks of the diagnosis of breast tumor. Wang [4] put forward a kind of based on feature fusion with CNN deep features of breast computer-aided diagnosis methods, the accuracy is 92.3%. Zhao [5] investigated capsule networks with dynamic routing for text classification, which proves the feasibility of text categorization. Existing models have poor predictive effect and lack of interpretation, which can not meet the clinical needs.

Based on the above pain points, we propose a semantic embedding self-explanatory Breast Diagnostic Capsules Network (BDCN), which diagnoses breast tumors based on the mammary gland molybdenum target report. Our contributions are as follows:

- Semantic segmentation algorithm is used to segment breast cancer lesions.
- Semantic tree is integrated into Bidirectional Encoder Representation from Transformers(BERT) pre-training to obtain word vectors.
- A capsule network with multi-head attention mechanism was proposed to predict breast tumors.
- Using Back Propagation to Realize Local Interpretation of BDCN Model.

2   Model

BDCN is based on semantic embedding, multi-head attention and capsule network to realize text diagnosis of breast cancer examination report. The overall architecture is shown in Fig. 1, which composed of three modules: the semantic segmentation layer of inspection report, the semantic tree knowledge embedding extraction word vector layer (Sem-Bert), and the capsule network assisted target feature multi-head attention representation classification layer (Muti-Cap).

The input is an original mammography report. We use segmentation algorithm to attain preprocessing and segmentation of lesions and the implementation details are shown in Sect. 2.1. Then we propose Sem-Bert method, which uses semantic tree to improve BERT and pre-training to obtain the word vector for the medical field. The details are discussed in Sect. 2.2. The word embedding $z_1, z_2, \cdots, z_N \in R^{d_z}$ is converted into the phrase embedding $w_1, w_2, \cdots, w_N \in R^{d_w}$ in one-dimensional convolution. Next, the feature capsule $fc_1, fc_2, \cdots, fc_M \in R^{d_f}$ is transformed into a prediction capsule $pc_j, j \in [1,3]$ based on the dynamic routing algorithm, and the result capsule $rc_j, j \in [1,3]$ is obtained by activating the network. Finally, the benign and malignant diagnosis of breast tumors was predicted, and the implementation details are shown in Sect. 2.3.

2.1   Check the Report Semantic Segmentation Layer

The main role of this layer is to achieve semantic segmentation of the report. The clinician can diagnose the breast tumor by analyzing the mammography report

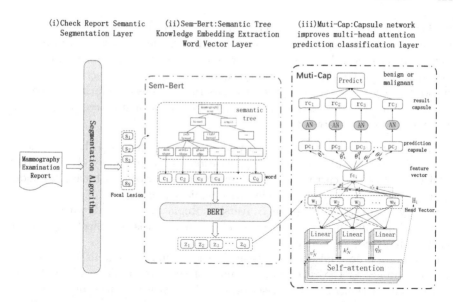

(i) Check Report Semantic Segmentation Layer   (ii) Sem-Bert: Semantic Tree Knowledge Embedding Extraction Word Vector Layer   (iii) Muti-Cap: Capsule network improves multi-head attention prediction classification layer

Fig. 1. Overall architecture of the BDCN.

to judge the pathological condition of the patient's breast and its surrounding tissue. However, a single report may contain multiple focal lesions in the same location or even in different gland background tissues, which may lead to lower prediction accuracy if generalized. Therefore, segmentation of breast report is an important link. The core steps of the segmentation algorithm are as follows:

- Rough segmentation. Divide the report into sentences, then according to different keywords such as skin, mass and axilla, they were divided into corresponding glandular background, focal lesions and axilla.
- Further subdivision. Most focal lesions mixed in the background part of the gland contain the keywords "nodules" or "densified shadow". The sentence of the background part of the gland is further subdivided to screen out the information of focal lesions and axillary parts. The sentences on the glandular background were further subdivided to screen for information on focal lesions and axillary lesions.
- Distinguish the left and right sides according to the the position described. According to "is divided into short sentences, when the "left" or "right" keyword appears, the short sentences are divided according to the position words; When "double" appears, it is divided into left and right sides.

2.2 Sem-Bert Layer

This layer is mainly to extract word vectors with BERT combined with semantic tree [6,7] and the process diagram of Sem-BERT to obtain word vectors is shown in Fig. 2. The Sem-Bert method constructs a semantic tree first. Semantic tree has obvious advantages of context hierarchy, so the construction of semantic tree can help to solve the problem of unreasonable word segmentation and context incoherence. According to the rule of "segment - organization description sentence - attribute description sentence", relational extraction is carried out, and dependency syntax is used to construct Extensible Markup Language (XML) semantic tree. The concrete content of report semantic tree construction includes the following five parts.

- Chinese word segmentation. Jieba word segmentation tool is the best choice to ensure the accuracy of word segmentation, and breast molybdenum target dictionary can be customized according to the knowledge of breast molybdenum target terminology combined with clinicians' guidance.
- Synonym conversion. Synonym conversion. Replacing the words with the same general meaning in medical science into unified words can effectively reduce the redundancy of semantic tree.
- Organization description sentence acquisition. Find the organization description sentence of the corresponding part of the paragraph, scan the report from left to right, and the description before each organization word A is encountered until the next new organization word B is encountered will be classified as the description of A.
- Organization segment subtree path acquisition. Each organization descriptive sentence is converted into an attribute description sentence and the attribute value is extracted. Taking the current organization as the root node, we find finer attributes through dependency syntax and extract the attribute value of each attribute in the organization word.
- Mammary report dictionary construction and XML transformation. The information extracted from different parts was added to the branches of the semantic tree, and the duplicated information was pruned to obtain the mammary gland report dictionary.

Taking the semantic tree as input, by configuring the BertConfig class, setting the Tokenizer word slicer, then numbering the words in the lexicon and converting them into dictionaries, selecting the word that makes the likelihood function increase the most, and selecting the word slicer according to the frequency. The structural diagram of the Sem-Bert method is shown in Fig. 2, taking the gland background description information as an example (circled by a red dotted line). Similar to Bert, the mammography report semantic tree needs to be converted into a sequence by means of token embedding, position embedding and segment embedding, while preserving its structural information. However, unlike traditional BERT, because the input is a semantic tree rather than a sequence of tokens, the positional embedding of the BERT input needs to be changed in order to preserve the structural information of the breast examination report. Position bedding is changed to level-position embedding and original-position embedding, marked by red and black numbers in Fig. 2, respectively. Level-position id represents the position of the same branch in the semantic tree, and gives each Token the same branch to scale the hierarchical order information of the semantic tree starting from 0. The Original-position id is represented in the same way as the

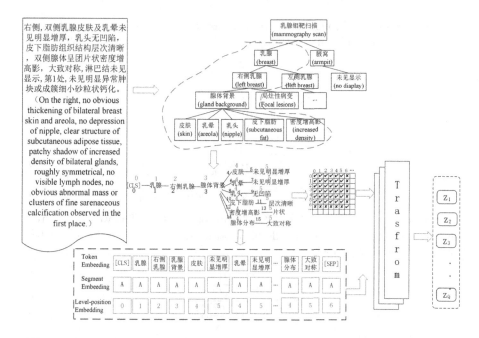

Fig. 2. The structure diagram of word vector was obtained by Sem-Bert method.

position id in BERT. However, in order to preserve the structural information, a relational matrix is introduced to record whether it is reachable under the same branch, reachable under the same branch, or not [20]. Lastly, multiple self-attentions in Transform are stacked with each other to code, and the final word vector $z_1, z_2, \cdots, z_N \in R^{d_z}$ is obtained by pre-training.

2.3   Muti-Cap Layer

This layer converts the word vectors of the pre-training layer into capsules, uses the capsule network to obtain the required prediction capsules, and combines effective information from multiple attention heads to achieve better classification. As one of the three most powerful semantic feature extractors, transform's self - attention mechanism is superior to CNN and neural network in word sense ambiguity resolution [8]. However, when the vector dimension is too high, the self-attention in each component represents different features, which leads to the fact that all attention can not fully capture the features [9]. Multi-head attention can learn the features of sequences from different aspects, which is helpful for the network to capture more abundant features. Each of these head vectors points to a feature capsule, performing attention independently. Long attention to extract the feature layer will get $w_1, w_2, \cdots, w_N \in R^{d_w}$ as its input in the layer. By way of generating M features capsule $fc_1, fc_2, \cdots, fc_M \in R^{d_f}$, we need to have M long since attention vector $h_1, h_2, \cdots, h_M \in R^{d_h}$, and for each Hi $H_i \in R^{d_f}, m \in [1, M]$ can obtain different key vectors k_n^i, value vectors v_n^i

and query vectors q_n^i through linear transformation, and then get more attention by the vector series. The generation process of the head vector is as follows:

$$\text{for } n = 1, 2, 3, \cdots, N \text{ do}$$

$$v_n^i = W_i^V w_n \tag{1}$$

$$k_n^i = W_i^K w_n \tag{2}$$

$$q_n^i = W_i^Q w_n \tag{3}$$

$$H_i = attention(q_n^i, k_n^i, v_n^i) \tag{4}$$

where $W_i^V \in R^{d_f \times d_w}, W_i^K \in R^{d_h \times d_w}$ are parameters. The weight of attention and the output feature capsule of multi-head attention can be calculated. The formulas are as following:

$$\alpha_n^i = \frac{exp(\frac{H_i^T k_n^i}{\sqrt{d_p}})}{\sum\limits_{n'} exp(\frac{H_i^T k_{n'}^i}{\sqrt{d_p}})} \tag{5}$$

$$fc_i = \sum_n \alpha_n^i v_n^i \tag{6}$$

The feature capsule is obtained by the sum of attention weights. As a parameter, the number of feature capsules is adjustable, which solves the problem that the intermediate parameters of three prediction capsules and M feature capsules are too large, and we need to normalize the attention weight:

$$\sum_n \alpha_n^i = 1 \tag{7}$$

With CNN classification, some important information will be lost in the operation of the aggregation layer, while capsule network represents a group of neurons by capsule, replacing the neuron output vector [10]. Therefore, the capsule network can effectively represent the location and semantics of features, and each upper capsule is the high-level semantics of the lower capsule. In addition, it can improve the information aggregation of multiple attention, so as to obtain more effective features and improve the ability of text representation. The structural schematic diagram is shown in Fig. 3.

The dynamic routing algorithm of the capsule network [11] is used to calculate the prediction capsule pc_i. Through multiple iterations of the routing process, Muti-Cap can determine the number of characteristic capsules flowing into the prediction capsules, which plays an significant role in the prediction capsules. Firstly, the feature capsule vector is calculated by inputting the feature capsule pc_i and multiplying the learning transformation matrix $\hat{W}_j$.

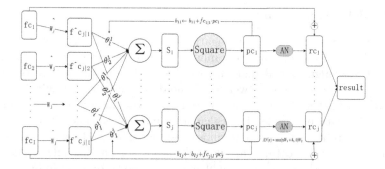

Fig. 3. Structure diagram of improved capsule network.

The dynamic routing weight θ_i^j is then determined by calculating the "routing softmax" of the initial logits b_{ij}, the formulas are as following:

$$\hat{fc}_{j|i} = \hat{W}_j fc_i + \hat{b}_{ij} \tag{8}$$

$$\theta_i^j = \frac{exp(b_{ij})}{\sum\limits_j exp(b_{ij})} \tag{9}$$

where $\hat{W}_j$ here is shared among each feature capsule to obtain the feature capsule vector, and b_{ij} is initialized to 0.

The final output is normalized activation by the function of squash [5] to make the whole model nonlinear and get the predicted capsule, as shown in Eqs. 10 and 11.

$$s_j = \sum_i \theta_i^j \hat{fc}_{j|i} \tag{10}$$

$$Squash(s_j) = \frac{\|s_j\|^2}{1 + \|s_j\|^2} \frac{s_j}{\|s_j\|} \tag{11}$$

where s_j is the original output capsule vector, which is the predicted capsule value before the square function is activated.

We introduce the activation network based on the dynamic routing algorithm in the traditional capsule network. The obtained prediction capsule was input into the activation network of linear transformation and ReLU activation function, and the final output was obtained by connecting with the residual fc_j feature capsule. The formula is shown in 12 and 13.

$$AN(x) = max(\mathbf{x}W_+c_1, 0)W_2 + c_2 \tag{12}$$

$$predict = fc_j + AN(pc_i) \tag{13}$$

In order to match the attention weight of multi head attention, we need to softmax the dynamic routing weight.

2.4   Model Interpretability

We use the principle of back propagation [12,17] to reach local interpretability. That is, a sample of mammography report to explain why the repors was diagnosed as benign or malignant breast cancer. The model defines two interpretable parameters: attention weight α_i^j and dynamic routing weight θ_i^j. Attention weight indicates whether attention right perform aimportant function in the formation of feature capsule. The calculation method is shown in Eq. 5 above. The dynamic routing weight determines the higher level classification capsule to which the current feature capsule will output it. The formula is described in the Eq. 9. As a classifier, the weight matrix of the traditional full-connection layer is fixed after training [13], which is not conducive to interpretation. However, in our model, all layers are fully connected. The capsule network part and the multi attention part are interdependent, which has an important impact on the research of local interpretation based on back propagation. As shown in the Fig. 4, prediction capsules are divided into three categories $j \in [1,3]$: benign, suspected malignant, and malignant breast cancer. By setting the parameter $P1$ of the capsule classification layer and the parameter $P2$ of the feature layer extracted by the multi-head self-attention.

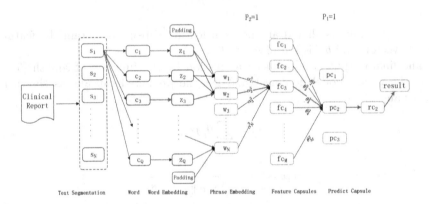

Fig. 4. Model interpretability process display diagram.

3   Experiment

In this section, we mainly discuss the validity and interpretability of our model. Firstly, the selection of methods in different stages is considered, such as the performance differences of Sem-Bert, BERT [19] and word2vec [21] in acquiring word vectors. Secondly, the model in this paper is compared with some unexplainable text classification models such as TextCNN [14] and LSTM [15]. Finally, we demonstrate the interpretability of our model through experiments.

3.1 Dataset

This dataset is the molybdenum mammography report of Shanghai Ruijin Hospital. We selected 1600 preoperative mammography data from 34 million original reports and included 2857 data after segmented pretreatment. The small sample dataset is classified into three categories, including benign, suspected malignant and malignant. The analysis of the dataset is shown in Table 1.

Table 1. Summary of mammography report dataset.

Dataset	Benign	Suspected of Malignant	Malignant
Training Set	1344	390	268
Test Set	734	58	63
Mammography	2078	448	331

3.2 Parameter Settings

The model was implemented by Tensorflow and trained by Adam Optimizer [16]. The multi-head attention part activation function uses a ReLU nonlinear function, the capsule network layer activation function Squash function. We set the number of class capsules is 3, the number of head vectors is 18.

3.3 Different Stage Selection

Evaluation Indexes. The main evaluation indexes of the experiment are Micro-Precision(Mi-P), Micro-Recall(Mi-R), Micro-F1-score(Mi-F1), Macro-Precision(Ma-P), Macro-Recall(Ma-R), Macro-F1-score(Ma-F1), Receiver Operating Characteristic(ROC).

Comparative Experiment. The main purpose of this section is to verify the validity of the BDCN model, including the advantages of Sem-Bert in obtaining word vectors and the classification accuracy of Muti-cap. In this experiment, six other models are selected as the baseline. Comparing experiment 'word2vec+Muti-Cap' and 'BERT+Muti-Cap' in mainly verifies whether texts in specific fields such as medical care will affect the acquisition of word vectors, thus affecting the prediction performance. The comparative experiments textCNN, textRNN, LSTM, and Capsule mainly verify the influence of different classification models on the prediction performance.

Experimental Results. Figure 5 shows the ROC curves of the model in this paper obtained according to different evaluation criteria under three classifications. For three classifications, the area under the curve and AUC values of micro and macro methods are different. In this paper, the dataset of benign, malignant and suspected malignant are large difference between three kinds of sample size, and there are obvious characteristics of malignant samples. So the ROC curve of class 2 malignant tag is more left, and the AUC area is larger. Meanwhile, the average (micro) AUC is larger than that of macro AUC. Since the samples in our dataset are unbalanced, so to give equal attention to the categories with small samples and those with large sample data. The ROC curves in subsequent comparative experiments were obtained by macro method.

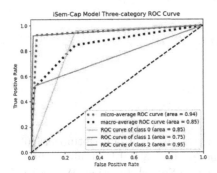

Fig. 5. ROC curve of three classifications of BDCN model

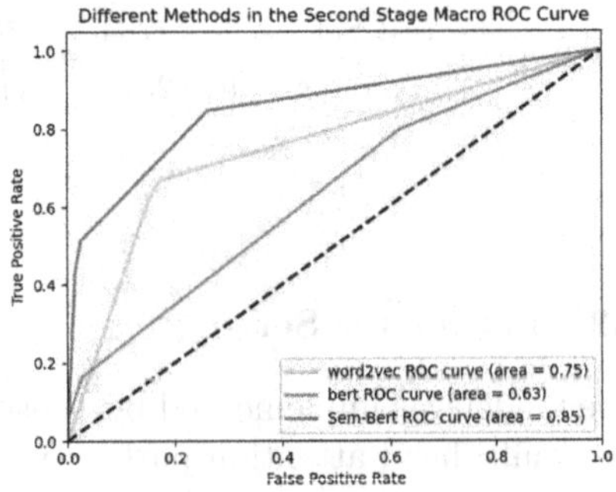

Fig. 6. ROC curves of word vectors are obtained by different methods

Comparing (1), (2) with (3) in Table 2 and Fig. 6, we can draw a conclusion: under the premise of using Muti-Cap method in the third stage, the evaluation index of the word vector method obtained by Sem-Bert in the second stage is higher, the ROC curve covers the other two comparative experiments, and the AUC value of the area under the ROC is significantly larger than the other two. Although there may be a loss of text information in the process of constructing semantic tree, the experiment effectively proves that the method of acquiring word vectors through semantic tree combined with BERT is more accurate than traditional word2vec and BERT method directly, which is very meaningful.

Table 3 compares the current mainstream deep learning models of text classification: CNN family textCNN, RNN family textRNN, LSTM family LSTM and capsule of Muti-Cap. The baselines are word2vec method to obtain the word vector, whereas in BDCN model, term vectors are obtained by Sem-Bert method. Comparing experiments (1),(2),(3) and (4) in Table 3, the capsule network from the view on the text classification task to task, effect and less textCNN this classic mode, but better results than simple LSTM, which shows that the attempt is meaningful. Moreover, through the improvement of the traditional capsule network and word vector acquisition method, our model has favorable performance

Table 2. Model selection at second stage

Model	Evaluation index					
	Mi-P(%)	Mi-R(%)	Mi-F1(%)	Ma-P(%)	Ma-R(%)	Ma-F1(%)
(1)word2vec+Muti-Cap	83.25	83.25	83.25	44.02	61.08	47.83
(2)BERT+Muti-Cap	87.37	87.37	87.37	78.26	42.37	45.56
(3)BDCN(our)	**91.58**	**91.58**	**91.58**	**75.95**	**79.73**	**77.14**

Table 3. Comparison to multimodal baselines

Model	Evaluation index					
	Mi-P(%)	Mi-R(%)	Mi-F1(%)	Ma-P(%)	Ma-R(%)	Ma-F1(%)
(1)textCNN	88.79	88.79	88.79	73.94	78.43	75.15
(2)textRNN	74.24	74.24	74.24	70.59	71.78	72.74
(3)LSTM	69.93	69.93	69.93	66.69	64.05	69.12
(4)Capsule	77.56	77.56	77.56	67.57	71.57	69.37
(5)BDCN(our)	**91.58**	**91.58**	**91.58**	**75.95**	**79.73**	**77.14**

results in each index. Experiments show that the selection of the second stage of this model has greatly improved the accuracy of the model.

Based on the back propagation principle, the BDCN can be interpreted to find the top three important weights in the dynamic routing weights according to the prediction results. Then the three important weights in the attention weights can be deduced in reverse, so as to find the three important features of judging the benign and malignant in this sample.

We plot a scatter plot and a bar chart, as shown in Fig. 7. They respectively explained the important features of benign, suspected malignant and malignant cases as well as the attention weight and routing weight. For example, according to the ranking of importance, the three important features that appear for the first time of each sample were selected. In Fig. 7(c), it can be seen that 'the shape of the burr', 'glandular degeneration' and 'lymph node display' are all important features of malignant tumors. In Fig. 7(a) 'roughly symmetrical glands', 'density increase', 'no depression of the nipple' and 'no lymph nodes displayed' are all important indicators of benign breast tumors. Suspected malignancy is a grading result of BI-RADS4, the more similar the characteristics are to malignant, the greater the probability of malignant tumors, and it is convenient for the doctor to remind the patient for further examination.

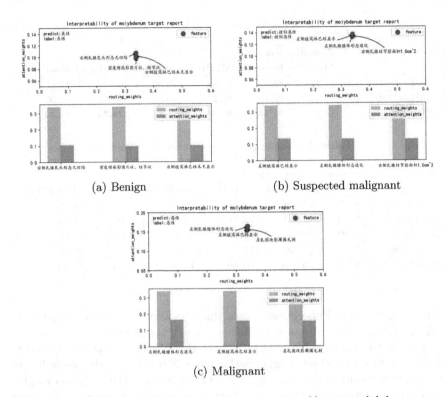

(a) Benign   (b) Suspected malignant

(c) Malignant

Fig. 7. Analysis of benign and malignant characteristics of breast molybdenum target

More intuitively show the local interpretability of the model, the extracted important features were labeled and displayed on the segmented breast molybdenum target samples. As shown in Fig. 7(a) 'no inverted nipples', 'no lymph nodes' and 'increased density'. Benign and malignant breast tumors can be determined by directly observing the words marked yellow, and the prediction accuracy of the BDCN is also verified from the side. On the basis of concise and accurate text report, we add the annotation of important words to make up for the lack of interpretation of text classification better.

predict: 良性
label: 良性
0022761342015103009042双侧乳腺皮肤及乳晕未见明显增厚，乳头无凹陷，皮下脂肪组织结构层次清晰，双侧腺体较丰富，呈团片状、结节状密度增高影，边缘膨隆，大致对称；淋巴结未见显示。右侧第2处：右乳钙化灶。

(a) Benign

predict: 疑似恶性
label: 疑似恶性
1003557542011021809060左乳下部皮肤略内凹，余双侧乳腺皮肤及乳晕未见明显增厚，乳头无凹陷，皮下脂肪组织结构层次清晰，双侧腺体部分退化，呈条索状改变，大致对称；淋巴结显示。左侧第1处：左乳内下象限深部偏高密度小结节，直径约1cm，边缘浅分叶，局部似见浸润毛刺。

(b) Suspected malignant

predict: 恶性
label: 恶性
0015488562011031709225左乳外侧皮肤增厚凹陷，乳头无凹陷，皮下脂肪组织结构层次清晰，双侧腺体部分退化，呈条索状改变，大致对称；淋巴结显示。左侧第1处：左乳外上可见团块影，周围可见毛刺，直径约为2.0cm，周围浸润性改变。

(c) Malignant

Fig. 8. Analysis of benign and malignant characteristics of breast molybdenum target

4 Conclusion

We proposed a semantically embedded self-interpreted breast diagnostic capsule network model. Semantic segmentation algorithm was used to segment the report, Sem-Bert method was used to obtain word vectors in medical field with hierarchical relationship, and capsule network with multiple attention was used to achieve prediction and classification of breast tumors. The validity of our model is better than other models in breast molybdenum target dataset. In addition, local self-interpretation method was used to provide intelligibility analysis, which was in line with doctors' clinical expectations. In the future, we will further study the global interpretability of the model and we hope to apply our technology to other diseases.

Acknowledgements. We would like to thank the anonymous reviewers for their helpful comments. This work was financially supported by the National Key R&D Program of China under Grant 2019YFE0190500.

References

1. Chinese Anti-Cancer Association: Committee of Breast Cancer Society.: Chinese Anti-Cancer Association Guidelines and Specifications for Diagnosis and Treatment of Breast Cancer 2019 Edition. Chin. J. Cancer (2019)
2. Mengwan, W., et al.: A Benign and malignant breast tumor classification method via efficiently combining texture and morphological features on ultrasound images. Comput. Math. Methods Med. Engl. (2020)
3. Wei, L., Yang, Y., Nishikawa, R.M.: Microcalcification classification assisted by content-based image retrieval for breast cancer diagnosis. Patt. Recogn. **42**(6), 1126–1132 (2009)
4. Wang, Z., Li, M., Wang, H., Jiang, H., Yao, Y., Zhang, H., Xin, J.: Breast cancer detection using extreme learning machine based on feature fusion with cnn deep features. IEEE Access **7**, 105146–105158 (2019)
5. Zhao, W., Ye, J., Yang, M., Lei, Z., Zhang, S., and Zhao, Z.: Investigating capsule networks with dynamic routing for text classification. arXiv preprint arXiv:1804.00538 (2018)
6. Chen, D., Huang, M., Li, W.: Knowledge-powered deep breast tumor classification with multiple medical reports. In: IEEE/ACM Transactions on Computational Biology and Bioinformatics, CA, USA (2019)
7. Jiang, D., He, J.: Tree framework with BERT word embedding for the recognition of Chinese implicit discourse relations. IEEE Access **8**, 162004–162011 (2020)
8. Long, J., Shelhamer, E., Darrell, T.: Fully convolutional networks for semantic segmentation. In: Proceedings of the IEEE Conference on Computer Vision and Pattern Recognition, pp. 3431–3440. Boston (2015)
9. Tang, G., Müller, M., Rios, A., Sennrich, R.: Why self-attention? a targeted evaluation of neural machine translation architectures. arXiv preprint arXiv:1808.08946 (2018)
10. Sabour, S., Frosst, N., Hinton, G. E.: Dynamic routing between capsules. In: Proceedings of the 31st International Conference on Neural Information Processing Systems. pp. 3859–3869 (2017)
11. Zhao, W., Peng, H., Eger, S., Cambria, E., Yang, M.: Towards scalable and reliable capsule networks for challenging NLP applications. arXiv preprint arXiv:1906.02829 (2019)
12. Wang, Z., Hu, X., Ji, S.: iCapsNets: towards interpretable capsule networks for text classification. arXiv preprint arXiv:2006.00075 (2020)
13. Zhang, W., Cai, L., Chen, M., Wang, N.: Progress in interpretability. In: International Conference on Mobile Computing, Applications, and Services, pp. 155-168. Springer, Cham (2019)
14. Rakhlin, A.: Convolutional Neural Networks for Sentence Classification. GitHub (2016)
15. Kalchbrenner, N., Danihelka, I., Graves, A.: Grid long short-term memory. arXiv preprint arXiv:1507.01526 (2015)
16. Bock, S., Weiß, M.: A proof of local convergence for the Adam optimizer. In: 2019 International Joint Conference on Neural Networks, IJCNN, Budapest (2019)
17. Aini, H., Haviluddin, H.: Crude palm oil prediction based on backpropagation neural network approac. Knowl. Eng, Data Sci. **2**(1), 1–9 (2019)
18. Cerda-Mardini, P., Araujo, V., Soto, A.: Translating natural language instructions for behavioral robot navigation with a multi-head attention mechanism. arXiv preprint arXiv:2006.00697 (2020)

19. Devlin, J., Chang, M. W., Lee, K., Toutanova, K.: Bert: pre-training of deep bidirectional transformers for language understanding. arXiv preprint arXiv:1810.04805 (2018)

20. Liu, W., et al..: K-bert: enabling language representation with knowledge graph. In: Proceedings of the AAAI Conference on Artificial Intelligence, Vol. 34: pp. 901–2908 (2020)

21. Mikolov, T., Chen, K., Corrado, G., Dean, J.: Efficient estimation of word representations in: vector space. arXiv preprint arXiv:1301.3781 (2013)

GCN with External Knowledge for Clinical Event Detection

Dan Liu, Zhichang Zhang$^{(\boxtimes)}$, Hui Peng, and Ruirui Han

College of Computer Science and Engineering, Northwest Normal University,
Lanzhou, China

Abstract. In recent years, with the development of deep learning and the increasing demand for medical information acquisition in medical information technology applications such as clinical decision support, Clinical Event Detection has been widely studied as its subtask. However, directly applying advances in deep learning to Clinical Event Detection tasks often produces undesirable results. This paper proposes a multi-granularity information fusion encoder-decoder framework that introduces external knowledge. First, the word embedding generated by the pre-trained biomedical language representation model (BioBERT) and the character embedding generated by the Convolutional Neural Network are spliced. And then perform Part-of-Speech attention coding for character-level embedding, perform semantic Graph Convolutional Network coding for the spliced character-word embedding. Finally, the information of these three parts is fused as Conditional Random Field input to generate the sequence label of the word. The experimental results on the 2012 i2b2 data set show that the model in this paper is superior to other existing models. In addition, the model in this paper alleviates the problem that "occurrence" event type seem more difficult to detect than other event types.

Keywords: Data augmentation · Pre-trained language model · Transformer · Clinical Event Detection · Electronic medical record

1 Introduction

Electronic medical records are an inevitable product of medical information. The use of natural language processing (NLP) technology to effectively detect clinical events in electronic medical records becomes crucial as the number of electronic medical records rapidly grows. It has been widely studied because of its potential help in constructing clinical event lines, medical Q & A, assisted diagnosis and other tasks. The task of Clinical Event Detection (CED) is to identify the boundary of the event in the electronic medical record and determine its type. The event detection to identify the boundary and determine type is usually considered as a sequence labeling task.

© Springer Nature Switzerland AG 2021
S. Li et al. (Eds.): CCL 2021, LNAI 12869, pp. 434–449, 2021.
https://doi.org/10.1007/978-3-030-84186-7_29

The emergence of deep learning has greatly improved the performance of the sequence labeling task model. Bidirectional long short-term memory network (BiLSTM) is widely employed in sequence labeling tasks owing to its high power to learn the contextual representation of words. Huang et al. [1] was the first to apply the bidirectional long short-term memory (BiLSTM) and the conditional random field (CRF) to sequence labeling tasks. But BiLSTM needs to be processed sequentially over time, it cannot be calculated in parallel. Instead, the Transformer not only advantage in modeling the long-range context, but also fully make use of the concurrence power of GPUs. Yan et al. [2] found that due to its position coding problem, the performance of Transformer in sequence labeling tasks is not as good as in other NLP tasks, and solves the position coding problem. However, due to the lack and particularity of clinical data, the performance of directly applying these technological advances to CED tasks is not ideal. For this reason, we need to use a large amount of medical information for tokens representation. In the past, Word2Vec [3] or Glove [4] was used to train a large number of unlabeled clinical texts to generate word embedding. There are two problems with this method. On the one hand, it is difficult to obtain a large amount of clinical data. On the other hand, the obtained data is not standardized, many mistakes, Various representations. Lee et al. [5] proposed a pre-trained biomedical language representation model for biomedical text mining (BioBERT) in 2019, whose performance on various biomedical text mining tasks largely surpassed BERT and previous advanced models. We use BioBERT for word embedding to solve the problem of poor model recognition performance caused by a large number of obscure professional terms in the medical field. In addition, the character-level features of words may show word features, for example, the beginning of "un" generally indicates negative characteristics. Therefore, adding character-level encoding will also have an impact on improving the performance of the model, and can solve the problem of Out Of Vocabulary(OOV). Lample et al. [6], Ma et al. [7] and Liu et al. [8] have added character-level coding to the model of the English NER task and proved its effectiveness. This article uses Convolutional Neural Network (CNN) as a character-level encoder for words [9].

After the Informatics Integrated Biology and Bedside Information (i2b2) sharing task was proposed in 2012 [10], we found that in order to better enable the event extraction task to better serve the later tasks such as disease diagnosis, the event types of the shared task also include "occurrence" type. However, the experimental results of the organizations that participated in the challenge in the past show that the "occurrence" type is more difficult to predict than other types of events, and it is not due to the small amount of data in this type. We analyzed the reasons and found a solution. (1) Most of the events in this type are nouns or verbs. We use the Part-of-Speech generated by Stanford CoreNLP tools as attention to help them identify. (2) The past practice always solves the problem of poor recognition based on the particularity of the medical field, while ignoring the language commonality between the medical field and the general field, and incidents of "occurrence" type is more inclined to event recognition in the general field. For this reason, this article introduces external knowledge to alleviates the problem that occurrence event type seem more difficult to detect than other event types.

At the same time, these external knowledge are helpful to the recognition of the event span and improve the overall performance of the model. In general, the contributions of this paper are as follows: (1) The pre-trained BioBERT language model is used for word-level coding, which effectively solves the problem of the lack and particularity of clinical data. (2) The Graph Convolutional Network (GCN) that introduces external knowledge alleviates the problem that occurrence event type seem more difficult to detect than other event types. (3) The experimental results show that the model in this paper is better than the previous optimal model.

2   Related Work

Constructing the clinical timeline is crucial to the patient diagnosis and treatment. The 2012 Informatics for Integrating Biology and the Bedside (i2b2) shared task [10] was the identification and linking of mentions of temporal expressions (TEs) (eg, dates, times, durations, and frequencies) and clinically relevant events (eg, patients problems, tests, treatments) in narratives. For this task, previous research on deep learning methods is mainly based on recurrent neural networks (RNN) [11,12], convolutional neural networks (CNN) [13,14], BiLSTM [15,16] and Attention [17] methods.

2.1   Clinical Event Detection

As a subtask of constructing the clinical timeline, Clinical Event Detection (CED) methods are mainly divided into the following two categories. (1) Method based on rules and machine learning: The method based on rules mainly summarizes relevant classification rules based on the experience and knowledge of knowledge engineers or domain experts, and then constructs corresponding rule templates as classifiers. Traditional machine learning methods are based on feature engineering. After 2012 i2b2 challenge task is proposed, The team involved in the task has adopted many different methods: rule-based , support vector machine (SVM) [18] conditional random field (CRF) [19], Markov Logic and some combination of these methods, the best performance is the CRF-based model proposed by Beihang University, Microsoft Research Asia, Beijing and Tsinghua University. Roberts et al. [20] used a combination of supervised, unsupervised and rule-based method, and the task ranked third. First, it uses the CRF classifier to identify event boundaries. Then use an independent SVM classifier for type detection. Kovacevic et al. [21] combined rules and machine learning and achieved F1 measure of 79.85%, it proposed the event CRF models were trained on relevant (type-specific) subsets of the training data and they all shared some feature groups. Although the rule-based method has high classification accuracy, it does not have the ability to learn from experience and is difficult to maintain. The rule making requires professional participation, time-consuming and labor-intensive, poor scalability, and it is difficult to promote and use in multiple fields. Traditional machine learning does not need to manually write rule templates, it can effectively solve the problems in rule-based methods. But, it is time consuming to extract features, and its consumption tends to grow as the size of the data set becomes larger, which is prone to dimensional disasters. (2) Methods based on deep learning: The emergence of deep learning greatly reduces the difficulty of obtaining text features. Zhu et al. [22] proposed a bidirectional LSTM-CRF model is trained for clinical concept extraction using the contextual word embedding model, it achieved the best performance among reported baseline models on the i2b2 2010 challenge dataset and the result is higher than this article, this is due to the dataset of this article has added three new event types three: evidential, occurrence and clinical department, in particular, the evidential and occurrence event types seem more difficult to detect than other event types [10]. Recently, research on the 2012 i2b2 data set has decreased, but the NER task has been widely studied. The LSTM and CRF models greatly improve the performance of the NER task [23]. Transformer is widely used in NER tasks due to its parallelism and advantages in modeling long-range context [2]. Chen et al. [24] proposed a simple but effective CNN-based network for NER, Gated Relation Network (GRN), which is more capable than common CNNs in capturing long-term context. Lin et al. [25] also used Self Attention when solving NER tasks. Graph Neural Networks (GNNs) are also widely used in NER tasks [26,28].

Table 1. Stanford CoreNLP semantic analysis example.

Resolve name	Content
Sentence	She had a CT scan
Part-of-Speech	[('She', 'PRP'), ('had', 'VBD'), ('a', 'DT'), ('CT', 'NN'), ('scan', 'VB')]

Constituency Parsing	(ROOT
	  (S
	    (NP (PRP She))
	    (VP (VBD had)
	      (S
	        (NP (DT a) (NN CT))
	        (VP (VB scan))))))

Dependency Parsing	[('ROOT', 0, 2), ('nsubj', 2, 1), ('det', 4, 3), ('nsubj', 5, 4), ('ccomp', 2, 5)]

2.2 Stanford CoreNLP

In this section, we will briefly introduce the semantic parsing tool Stanford CoreNLP used in this article. It is a natural language processing toolkit. CoreNLP enables users to derive linguistic annotations for text, including token and sentence boundaries, Part-of-Speech, named entities, numeric and time values, dependency and constituency parses, coreference, sentiment, quote attributions, and relations. CoreNLP currently supports 6 languages: Arabic, Chinese, English, French, German, and Spanish. This article mainly uses its Part-of-Speech, dependency and constituency parses functions. Since this tool is suitable for sentences in the general domain, that is, the model used by the tool is trained on a large number of general domain corpora, so this article calls it the introduction of external knowledge. Taking the sentence "She had a CT scan" in the medical field as an example, The Part-of-Speech, dependency and constituency parses of the sentences parsed by the Stanford CoreNLP tool are shown in Table 1. When we train the model, we need to encode the information in Table 1 into a matrix form, which will be described in detail in a later part.

2.3 BioBERT

In this section, we briefly introduce the pre-trained language model (BioBERT) used in this article. Direct application of NLP advancements to clinical text mining often yields unsatisfactory results due to a word distribution shift from general domain corpora to clinical corpora. Lee et al. [5] investigate how the recently introduced pre-trained language model BERT can be adapted for biomedical corpora, and proposed a domain-specific language representation model pre-trained on large-scale biomedical corpora (BioBERT) to solve this problem. BioBERT initializes weights from BERT, which is pre-trained on the English Wikipedia and Books Corpus general domain corpus. Then, BioBERT is pre-trained on PubMed abstract and PMC full-text article biomedical corpus.

3   Model

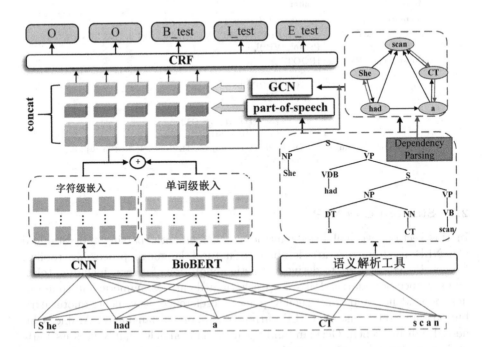

Fig. 1. The framework of our model.

The model of this article mainly includes four parts: character-word embedding module, Part-of-Speech attention module, semantic GCN module, CRF decoding module. Figure 1 shows an overview of our model, where CNN is a character-level encoder for words, BioBERT is a word-level encoder for words, and the semantic analysis tool uses Stanford CoreNLP. First, input a sentence to generate character-level embedding of words through CNN, and generate word-level embedding of words through biomedical pre-training language model BioBERT, the two are spliced together. And then perform Part-of-Speech attention coding for character-level embedding, perform semantic graph convolutional network coding for the spliced character-word embedding. Finally, the information of these three parts is fused as CRF input to generate the sequence label of the word. We will introduce each part in detail in the following chapters.

3.1   Character-Word Embedding

Given a sequence of N tokens $H = [H_1, H_2, ..., H_N]$, For each token H_i, We first splicing word-level and character-level embedding $H_i = [C_i; W_i]^T$, where C_i is character-level embedding, W_i is word-level embedding. And provide it to subsequent modules. So a sentence can be expressed as $H = [[C_1; W_1]^T; [C_2; W_2]^T; ...; [C_N; W_N]^T]^T$, $H \in R^{N \times (cd+wd)}$, where cd is the character-level encoding dimension, wd is the word-level encoding dimension. The following describes the details of word-level embedding and character-level embedding of tokens in detail.

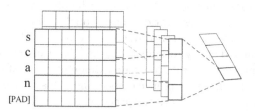

Fig. 2. The architecture of character-level embedding

Character-Level Embedding: For a token character-level embedding C_i, As shown in Fig. 2, for the character sequence of the token i in a sentence $x_i = x_i^1, x_i^2, ..., x_i^{cd}$, ($cd$ represents the number of characters in the longest word, namely, the maximum length of the word). We vectorize it and use the character embedding method to get the vector representation of each character $c^i = e^c(x^i)$, Then the character-level embedding of the word is expressed as a matrix $c = [c^1; c^2; ...; c^{wl}]^T$, $c \in R^{wl \times cd}$, Let's perform a convolution operation. Assuming that there are cd convolution kernels, the formula for the n-th convolution operation of the m-th convolution kernel is as follows: $h^{mn} = w \cdot c^{n:n+k-1} + b$, The size of the sliding window contains k characters, which is represented by the symbol $c^{n:n+k-1}$, w represents the convolution kernel, each time the feature is obtained by sliding k characters h^{mn}, that is, the red box and the yellow box in the Fig. 2. The m-th convolution kernel generates feature vectors for all characters sliding $h^m = [h^{m1}; h^{m2}; ...; h^{m(wl-k+1)}]^T$, So, the character-level features of the words generated by a set of convolution kernels are $h = [h^1; h^2; ...; h^{cd}]$, Then perform maximum pooling to get the character-level representation of the token $C_i = [max(h^1), max(h^2), ..., max(h^{cd})]$, $C_i \in R^{N \times cd}$.

Word-Level Embedding: In order to solve the problem of the particularity of clinical data, this article uses BioBERT (a domain-specific language representation model pre-trained on large-scale biomedical corpora). BioBERT initializes weights from BERT, which is pre-trained on the English Wikipedia and Books Corpus general domain corpus. Then, BioBERT is pre-trained on PubMed abstract and PMC full-text article biomedical corpus. Specifically, it uses pre-trained BioBERT on PubMed for 1M steps model, this version as BioBERT v1.1 (+PubMed). Fine-tuned based on this model, the BioBERT model takes in word sequence $s = w_1, w_2...w_N$ (N represents the maximum sentence length), Calculate the output of the BioBERT layer using the equations below:

$$H^w = BioBERT(s) \tag{1}$$

where $H^w \in R^{l \times wd}$, wd is the word-level embedding dimension, the model limits it to a multiple of 768. This article is the last BioBERT layer, so the size is 768.

Part-of-Speech Attention. We found that most event words are basically verbs or nouns, so Part-of-Speech features are helpful to event recognition. In order to learn sentence representations based on Part-of-Speech attention, we follow the self-attention method introduced by Lin et al. [27], this method has also been used in named entity recognition tasks [25]. It uses attention to convert the sentence into multiple vectors to extract different parts of the sentence, and uses a matrix to represent the sentence embedding. This article replaces the self-attention of the sentence with Part-of-Speech attention. The specific calculation formula is as follows:

$$Pos = SoftMax(W_2 tanh(W_1 P^T)) \tag{2}$$

$$H' = Pos \cdot C \tag{3}$$

Where P represents the Part-of-Speech matrix corresponding to the sentence, we use one hot to represent, $P \in R^{N \times 37}$, 37 is the number of all types of Part-of-Speech of the Stanford CoreNLP tool, and C represents the character-level embedding of the word. W_1 and W_2 are two trainable parameters for calculating the Part-of-Speech attention score vector Pos. The W_2 first dimension is fixed to N for information fusion later. We get a matrix of weighted sums of token vectors for a Part-of-Speech attention H', $H' \in R^{N \times cd}$.

3.2 Semantic GCN

Since the original input sentences are plain texts without inherent graphical structure, we first construct graphs based on the sequential information of texts and the semantic information in sentences parsed using Stanford CoreNLP tool. Then, we apply GCN [28–30] which propagates information between neighboring nodes in the graphs, to extract events.

Graph Construction: We create three kinds of information fusion people graphs for each sentence, each graph is defined as $G = (V, E)$, where V is the node set (word) and E is the edge set. Figure 3 shows the structure of the sentence "She had a CT scan." The process of constructing a graph is divided into three steps.

- Adjacent graph: For each pair of adjacent words in the sentence, we add one directed edge from the left word to the right one, allowing local contextual information to be utilized.
- Constituency parsing graph: According to the constituency parsing obtained by the Stanford CoreNLP tool, we connect the leftmost node of all subtrees to the last edge node according to the established tree structure.
- Dependency parsing graph: The dependency parsing obtained by the Stanford CoreNLP tool, we build a dependency parsing graph based on the dependencies between words.

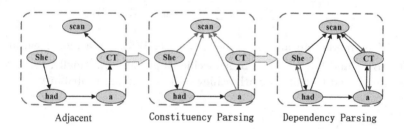

Fig. 3. Semantic analysis graph.

Bi-GCN: In order to consider both incoming and outgoing features for each node, we use Bi-GCN to extract graph features [31,32]. Given a graph $G = (V, E)$, and the word representation $H = [H_1, H_1, ..., H_N]^T$, the graph feature $H'' \in R^{N \times 2d_f}$ learned from Bi-GCN is expressed as follows, we use them the same as Luo et al. [28]:

$$\overrightarrow{f_i} = ReLU(\sum_{e_{ij} \in E} \overrightarrow{W_f} H_j + \overrightarrow{b_f}) \tag{4}$$

$$\overleftarrow{f_i} = ReLU(\sum_{e_{ji} \in E} \overleftarrow{W_f} H_j + \overleftarrow{b_f}) \tag{5}$$

$$H'' = [\overrightarrow{f_i}; \overleftarrow{f_i}] \tag{6}$$

where $W_f \in R^{2d_f \times (cd+wd)}$ and $b_f \in R^{d_f}$, d_f are trainable parameters, it is the hidden size of GCN, $ReLU$ is the non-linear activation function. e_{ij} represents the edge outgoing from token H_i, and e_{ji} represents the edge incoming to token H_i.

3.3 CRF Decoding

The character-word embedding representation, Part-of-Speech attention coding representation and semantic graph convolutional coding representation of the word are spliced together to obtain a matrix $\hat{H} = [H; H'; H''], \hat{H} \in R^{N \times (2cd+wd+2d_f)}$. It is input to the CRF layer to predict the corresponding tag sequence. The probability of a label sequence $Y = y_1, y_2 \ldots y_l$ is

$$P(y|H) = \frac{\exp(\sum_i (W^{y_i}_{CRF} h_i + b^{(y_{i-1}, y_i)}_{CRF}))}{\sum_{y'} \exp(\sum_i (W^{y_i}_{CRF} h_i + b^{(y_{i-1}, y_i)}_{CRF}))} \tag{7}$$

Where y' represents an arbitrary label sequence, $W^{y_i}_{CRF}$ is a model parameter specific to y_i, and $b^{(y_{i-1}, y_i)}_{CRF}$ is a bias specific to y_{i-1} and y_i. Finally, the Viterbi Algorithm is used to find the path achieves the maximum probability.

4   Experiment

4.1 Dataset

To evaluate our proposed model, we experiment on 2012 i2b2 challenge dataset, the training corpus consists of 190 electronic medical records, which contains 2250 sentences (The number after adjusting the sentence length), and the test corpus of 120 electronic medical records, which contains 1741 sentences (The number after adjusting the sentence length), event types include clinical department, evidential, "occurrence", problem, test, treatment (Grouin et al., 2013). The 2012 i2b2 challenge dataset does not have development set, this article divides the test set into a test set and a development set at a ratio close to 1:1. Among them, there are 821 sentences in the development set, 920 sentences in the test set.

4.2 Evaluation Metrics

This paper CED is a sequence labeling task, standard precision (P), recall (R) and F1-score (F) are used as evaluation metrics. In order to prove the effectiveness of the model, this article also uses the same evaluation metrics as 2012 i2b2 challenge (Span F1-score and Type accuracy) [10]. We used the Span F1-score, the harmonic mean of precision and recall of the predict output span against the gold standard span to evaluate event span detection performance. The calculation of Span F1-score is the same as the calculation of the F1-score evaluation metrics. It is worth noting that the Span F-score is lenient matching (predict event span overlap with the gold standard span). For event types, we calculated classification accuracy, that is, the percentage of correctly identified event types for the events whose spans are detected correctly, the specific calculation process is as follows:

$$P = \frac{|pred.type \cap glod.type|}{|pred.span \cap glod.span|} \tag{8}$$

where, "pred.type" means predict type output, "gold.type" means gold standard type, " pred.span" means predict span output, "gold.span" means gold standard span.

4.3 Settings

In the process of data preprocessing, in order to solve the large gap in sentence length, we split the sentence according to several punctuation marks. Such as "," and ";" etc., these punctuation marks can be used to break the sentence, the sentence is still complete. Then join several short sentences that are adjacent but whose total length does not exceed the maximum length. Because some special characters have their own characteristics, this article will deal with them to increase accuracy, such as "'s", splice it directly to the previous word, and we replace all digits with "0". In the experiment based on the CED task. We use the BIOES tag schema. For character-level embedding, we set randomly initialized character embedding size to 30. For word-level embedding we only take the last layer, the dimension is 768. Since the Part-of-Speech attention coding representation and the semantic map convolution representation are obtained from the introduced external knowledge, in order to avoid negative effects and reduce the problem of error propagation. We still have to focus on coding in the medical field, supplemented by coding with external knowledge. Therefore, the dimension of the trainable parameters represented by the semantic graph convolutional coding needs to be as small as possible than the word-level coding dimension of words, we set it to 120. The batch size for training is 16, epochs is 100, We use SGD and 0.9 momentum to optimize the model. During the optimization, we use the triangle learning rate where the learning rate rises to the pre-set learning rate (0.0008) at the first 1% steps and to 0 in the left 99% steps [33].

4.4   Evaluation on CED

We compare proposed model with the latest model on the 2012 i2b2 challenge dataset. In addition, we will also apply the latest model for NER to the data set of this article for comparative experiments. The 2012 i2b2 challenge test results are shown in Table 2 and Table 3.

The overall results of the model using P, R and F evaluation metrics are shown in Table 2. In the first block, the model combines rule-based and machine learning approaches that rely on morphological, lexical, syntactic, semantic, and domain specific features. In the second block, we give the model performance based on BiLSTM, the model uses TENER for character level embedding and the Glove 100d pre-trained embedding for word level embedding. In the third block of, we give the model performance based on Transformer, the model uses TENER for character level embedding and the ELMo for word level embedding. In the fourth part, we give the performance of another model experiment we did. The model uses TENER for character-level embedding, BioBERT for word-level embedding, and TENER for final encoding. In addition, the model also has data enhancements. In the last block, we give the experimental result of our proposed model. We can observe that our proposed model outperforms other models. It improves the F1-score from 79.85% to 80.55% on overall performance. Compared with the last experiment, our model improves the F1-score from 80.26% to 80.55%.

Table 2. Results of Precision, Recall and F1-score metrics.

Model	Precision	Recall	F1-score
Rule-based and machine learning [21]	0.8147	0.7805	0.7985
BiLSTM_CRF	0.7484	0.5272	0.6186
ELMo_TENER [2]	0.7454	0.7805	0.7626
BioBERT_TENER_Data Augmentation	0.8101	0.7953	0.8026
Ours	**0.8152**	**0.7961**	**0.8055**

For the results use span F1-score and type accuracy evaluation metrics which is depicted in Table 3, The first three block are the results of the top three participating in the 2012 i2b2 challenge, the forth block used a combination of supervised, unsupervised and rule-based method, and the task ranked third. First, it identifies event boundaries with a CRF classifier. Then it detects type using separate SVM classifiers. The fifth block, we give the model performance based on Transformer, the Glove 100d pre-trained embedding for word level embedding and model uses TENER for character level embedding, the TENER for word-character level embedding. In the sixth part, we give the performance of another model experiment we did. The model uses TENER for character-level embedding, BioBERT for word-level embedding, and TENER for final encoding. In addition, the model also has data enhancements. In the last block, we give the experimental result of our proposed model. We can observe that our proposed model improves Span F1-score from 91.66% to 91.68%, but our method improves the Type accuracy score from 86% to 92.76%. Compared with the last experiment, our model improves the Span F1-score from 90.33% to 91.68%.

Table 3. Results of Span F1-score and Type accuracy metrics.

Model	Span F1-score	Type accuracy
Beihang University et al. (CRF) [10]	0.9166	0.8600
Vanderbilt University (CRF _SVM) [10]	0.9000	0.84.00
The University of Texas (CRF_SVM) [10]	0.8900	0.8000
Supervised, unsupervised and rule-based [20]	0.8933	0.8045
TENER (the Glove 100d) [2]	0.7424	0.7505
BioBERT_TENER_Data Augmentation	0.9033	**0.9300**
Ours	**0.9168**	0.9276

4.5 Ablation Study

We examine the contributions of four main components, namely, BioBERT word-level embedding, CNN character-level embedding, Part-of-Speech attention coding, and semantic GCN coding. The experimental results are shown in Table 4, Where "-" means remove component, "→" means replace component.

We can observe that BioBERT word-level embedding, CNN character-level embedding, Part-of-Speech attention coding and semantic GCN coding improved the performance of the model to varying degrees. Especially, BioBERT and semantic GCN coding representation. BioBERT has been pre-trained on a large amount of biomedical data and has lots of biomedical information. For semantic GCN coding, because the external knowledge it uses is a tool that is trained on a large amount of general predictive data. The understanding of sentence structure has played a very good help, and has a certain enlightening effect on determining the boundary of the event. Therefore, it is helpful to improve the performance of the model. The character-level embedding has less impact on the result, so analyze the reason is that the word-level encoding dimension is much smaller than the 768 of the BioBERT word-level encoding, but if the character-level encoding is also adjusted to a large value, the training efficiency of the model will be very low. For the Part-of-Speech attention module, it is helpful to the model in judging whether it is an event or not, but because its dimension is smaller than that of other information, Has little effect. Experimental results show, these four components can help the model learn medical text information better.

Table 4. Results of ablation study.

Model	Precision	Recall	F1-score
Ours	**0.8152**	**0.7961**	**0.8055**
BioBERT→Glove 100d	0.7308	0.5849	0.6498
-CNN	0.8025	0.7892	0.7958
-Part-of-Speech	0.7553	0.7806	0.7677
-GCN	0.7547	0.6120	0.6759

4.6 Type Analysis

The evaluation task challenged by i2b2 shows that event detection in 2012 seems more challenging than i2b2 in 2010. This is due to the addition of three new event types: "evidence", "occurrence" and "clinical department". In particular, the types of "evidence" and "occurrence" seem to be more difficult to detect than other types. The results of our last experiment also found that the "evidential" and "occurrence" event types seem more difficult to detect than other event types. Especially occurrence EVENT type, there is enough data volume, but the result is much lower than other types. The experimental results of our previous model and the model in this article on different types are shown in the Table 5, where the "Last" represents the result of our last experiment, where the "Ours" represents the experimental results of this article. We found that due to the "occurrence" type, it is more biased towards the general field. In the model in this article, we have introduced external knowledge, so its F value has been greatly improved compared to the previous model and the model in our last experiment, while the performance of other types has not significantly decreased.

Table 5. Results of different event types.

Event type	F1-score (Last)	F1-score (Ours)	Train (Dataset)	Test (Dataset)
Clinical department	0.8164	0.8153	0.0605	0.0539
Evidential	0.7519	0.7532	0.0449	0.0438
Occurrence	0.6632	**0.6910**	0.1995	0.1838
Problem	0.8375	0.8367	0.3050	0.3170
Test	0.8409	0.8393	0.1576	0.1599
Treatment	0.8420	0.8345	0.2325	0.2417

5   Conclusion

This paper proposes a multi-granularity information fusion encoder-decoder framework. This framework uses the pre-trained language model (BioBERT) to generate word-level features, and solves the problem of poor model recognition performance caused by obscure professional terms in electronic medical records. The Graph Convolutional Network that introduces external knowledge improves the performance of the model in identifying "occurrence" type. Further improve the overall performance of the model. Experiments on the 2012 i2b2 challenge dataset show that our model achieves superior performance than other existing models.

References

1. Huang, Z., Xu, W., Yu, K.: Bidirectional LSTM-CRF models for sequence tagging. arXiv preprint arXiv:1508.01991 (2015)
2. Yan, H., Deng, B., Li, X., et al.: TENER: adapting transformer encoder for named entity recognition. arXiv preprint arXiv:1911.04474 (2019)
3. Mikolov, T., Chen, K., Corrado, G., et al.: Efficient estimation of word representations in vector space. arXiv preprint arXiv:1301.3781 (2013)
4. Pennington, J., Socher, R., Manning, C.: Glove: global vectors for word representation. In: Proceedings of the 2014 Conference on Empirical Methods in Natural Language Processing, pp. 1532–1543 (2014)
5. Lee, J., Yoon, W., Kim, S., et al.: BioBERT: a pre-trained biomedical language representation model for biomedical text mining. Bioinformatics **36**(4), 1234–1240 (2020)
6. Lample, G., Ballesteros, M., Subramanian, S., et al.: Neural architectures for named entity recognition. In: Proceedings of the 2016 Conference of the North American Chapter of the Association for Computational Linguistics: Human Language Technologies, pp. 260–270. Association for Computational Linguistics, San Diego (2016)
7. Ma, X., Hovy, E.: End-to-End sequence labeling via bi-directional LSTM-CNNs-CRF. In: Proceedings of the 54th Annual Meeting of the Association for Computational Linguistics (Volume 1: Long Papers), pp. 1064–1074. Association for Computational Linguistics, Berlin (2016)
8. Liu, L., Shang, J., Ren, X., et al.: Empower sequence labeling with task-aware neural language model. Proc. AAAI Conf. Artif. Intell. **32**(1), 5253–5260 (2018)
9. Chiu, J., Nichols, E.: Named entity recognition with bidirectional LSTM-CNNs. Trans. Assoc. Comput. Ling. **4**, 357–370 (2016)
10. Grouin, C., Grabar, N., Hamon, T., et al.: Eventual situations for timeline extraction from clinical reports. J. Am. Med. Inform. Assoc. **20**(5), 820–827 (2013)
11. Fries, J.: Brundlefly at SemEval-2016 Task 12: recurrent neural networks vs. joint inference for clinical temporal information extraction. In: Proceedings of the 10th International Workshop on Semantic Evaluation(SemEval-2016), pp. 1274–1279. Association for Computational Linguistics, San Diego (2016)
12. Cheng, F., Miyao, Y.: Classifying temporal relations by bidirectional LSTM over dependency paths. In: Proceedings of the 55th Annual Meeting of the Association for Computational Linguistics (Volume 2: Short Papers), pp. 1–6. Association for Computational Linguistics, Vancouver (2017)
13. Dligach, D., Miller, T., Lin, C., et al.: Neural temporal relation extraction. In: Proceedings of the 15th Conference of the European Chapter of the Association for Computational Linguistics: Volume 2, Short Papers, pp. 746–751. Association for Computational Linguistics, Valencia (2017)

14. Li, P., Huang, H.: UTA DLNLP at SemEval-2016 Task 12: deep learning based natural language processing system for clinical information identification from clinical notes and pathology reports. In: Proceedings of the 10th International Workshop on Semantic Evaluation (SemEval-2016), pp. 1268–1273. Association for Computational Linguistics, San Diego (2016)

15. Tourille, J., Ferret, O., Neveol, A., et al.: Neural architecture for temporal relation extraction: a BI-LSTM approach for detecting narrative containers. In: Proceedings of the 55th Annual Meeting of the Association for Computational Linguistics (Volume 2: Short Papers), pp. 224–230 (2017)

16. Lin, C., Miller, T., Dligach, D., et al.: Self-training improves recurrent neural networks performance for temporal relation extraction. In: Proceedings of the Ninth International Workshop on Health Text Mining and Information Analysis, pp. 165–176. Association for Computational Linguistics, Brussels (2016)

17. Zhao, S., Li, L., Lu, H., et al.: Associative attention networks for temporal relation extraction from electronic health records. J. Biomed. Inform. **99**(103309) (2019)

18. Cortes, C., Vapnik, V.: Support-vector networks. In: Proceedings of the Twelfth International Conference on Machine Learning, vol. 20, no. 3, pp. 273–297 (1995)

19. Lafferty, J., McCallum, A., Pereira, F.: Conditional random fields: probabilistic models for segmenting and labeling sequence data. In: Proceedings of the Eighteenth International Conference on Machine Learning, pp. 282–289 (2001)

20. Roberts, K., Rink, B., Harabagiu, S.: A flexible framework for recognizing events, temporal expressions, and temporal relations in clinical text. J. Am. Med. Inform. Assoc. **20**(5), 867–875 (2013)

21. Kovacevic, A., Dehghan, A., Filannino, M., et al.: Combining rules and machine learning for extraction of temporal expressions and events from clinical narratives. J. Am. Med. Inform. Assoc. **20**(5), 820–827 (2013)

22. Zhu, H., Paschalidis, I., Tahmasebi, A.: Clinical concept extraction with contextual word embedding. J. Am. Med. Inform. Assoc. **26**(11), 1297–1304 (2019)

23. Akhundov, A., Trautmann, D., Groh, G.: Sequence labeling: a practical approach. arXiv preprint arXiv:1808.03926 (2018)

24. Chen, H., Lin, Z., Ding, G., et al.: GRN: gated relation network to enhance convolutional neural network for named entity recognition. In: Proceedings of the AAAI Conference on Artificial Intelligence (2019). https://doi.org/10.1609/aaai.v33i01.33016236

25. Lin, Y., Lee, D., Shen, M., et al.: TriggerNER: learning with entity triggers as explanations for named entity recognition. In: Proceedings of the 58th Annual Meeting of the Association for Computational Linguistics, pp. 8503–8511. Association for Computational Linguistics, Online (2020)

26. Liu, P., Chang, S., Huang, X., et al.: Contextualized non-local neural networks for sequence learning. In: Proceedings of the AAAI Conference on Artificial Intelligence, vol. 33, no. 1, pp. 6762–6769 (2019)

27. Lin, Z., Feng, M., Santos, C.: A structured self-attentive sentence embedding. In: International Conference on Learning Representations (ICLR) (2017)

28. Luo, Y., Zhao, H.: Bipartite flat-graph network for nested named entity recognition. In: Proceedings of the 58th Annual Meeting of the Association for Computational Linguistics, pp. 6408–6418. Association for Computational Linguistics, Online (2020)

29. Kipf, T., Welling, M.: Semi-supervised classification with graph convolutional networks. In: International Conference on Learning Representations (ICLR) (2017)

30. Qian, Y., Santus, E., Jin, Z., et al.: Graphie: a graph-based framework for information extraction. In: Proceedings of the 2019 Conference of the North American Chapter of the Association for Computational Linguistics: Human Language Technologies, Volume 1 (Long and Short Papers), pp. 751–761. Association for Computational Linguistics, Minneapolis (2019)
31. Marcheggiani, D., Titov, I.: Encoding sentences with graph convolutional networks for semantic role labeling. In: Proceedings of the 2017 Conference on Empirical Methods in Natural Language Processing, pp. 1506–1515. Association for Computational Linguistics, Copenhagen (2017)
32. Fu, T., Li, P., Filannino, M., Ma, W.: GraphRel: modeling text as relational graphs for joint entity and relation extraction. In: Proceedings of the 57th Annual Meeting of the Association for Computational Linguistics, pp. 1409–1418. Association for Computational Linguistics, Florence (2019)
33. Smith, L.: Cyclical learning rates for training neural networks. In: 2017 IEEE Winter Conference on Applications of Computer Vision (WACV), pp. 464–472. IEEE, Santa Rosa (2017)

A Prompt-Independent and Interpretable Automated Essay Scoring Method for Chinese Second Language Writing

Yupei Wang[2] and Renfen Hu[1(✉)]

[1] Institute of Chinese Information Processing, Beijing Normal University, Beijing, China
irishu@mail.bnu.edu.cn
[2] School of Science, Beijing Jiaotong University, Beijing, China

Abstract. With the increasing popularity of learning Chinese as a second language (L2), the development of an automated essay scoring (AES) method specially for Chinese L2 essays has become an important task. To build a robust model that could easily adapt to prompt changes, we propose 90 linguistic features with consideration of both language complexity and correctness, and introduce the Ordinal Logistic Regression model that explicitly combines these linguistic features and low-level textual representations. Our model obtains a high QWK of 0.714, a low RMSE of 1.516 and a considerable Pearson correlation of 0.734. With a simple linear model, we further analyze the contribution of the linguistic features to score prediction, revealing the model's interpretability and its potential to give writing feedback to users. This work provides insights and establishes a solid baseline for Chinese L2 AES studies.

Keywords: Automated essay scoring · Chinese · Second language

1 Introduction

Automated Essay Scoring (AES) is one of the most important Natural Language Processing (NLP) applications in the field of education [15,18], and has been widely used in standardized language tests [2,3]. However, existing works mainly focus on the scoring of English essays [1,29,36] or Chinese essays by native speakers and minority learners [6,19,27]. Although Chinese second language (L2) acquisition has enjoyed an increasing boom in recent decades, the AES system designed for Chinese L2 writing has received much less attention.

Meanwhile, existing AES methods face two important challenges. Firstly, the scoring models are mostly built in a prompt-dependent style, i.e. training and

Supported by the National Social Science Fund of China (No. 18CYY029), the National Natural Science Fund of China (No. 62006021), and the National Training Program of Innovation and Entrepreneurship for Undergraduates (No. 202110004069).

© Springer Nature Switzerland AG 2021
S. Li et al. (Eds.): CCL 2021, LNAI 12869, pp. 450–470, 2021.
https://doi.org/10.1007/978-3-030-84186-7_30

testing for each specific prompt. It requires to collect prompt-specific data, yielding great costs in dataset construction [2]. Besides, the built models are of weak generalization capabilities and cannot be used to score essays of other prompts. Secondly, although neural network methods have achieved great success in NLP tasks, the gains in neural AES systems are far from being satisfactory. For example, Mayfield and Alan [17] find that fine-tuning BERT produces similar performance to classical models at significant additional cost. Apart from the costs, the deep neural models are also weak in interpretability of the results. However, it is a very important property for AES users who expect to get feedback on the writing, not just a score [14,15,31].

To solve the above problems, this paper proposes a prompt-independent and interpretable AES method for Chinese L2 writing. Specifically, we build prompt-independent models that could make full use of L2 writing data, and make predictions without the prompt limitations. For interpretability considerations, we extract 90 linguistic indices on accounting of the usage of characters, words, clauses, collocations, dependency structures, syntactic constructions which are emphasized in Chinese L2 acquisition, and 5 indices that address different types of writing errors. Furthermore, we integrate these linguistic and correctness indices into text representations, and introduce the Ordinal Logistic Regression (OLR) model to the AES task for Chinese second language writing. Our model achieves a high quadratic weighted Kappa (QWK) score of 0.714, a low Root Mean Square Error (RMSE) of 1.516, and a high Pearson coefficient of 0.734, performing much better than the classical machine learning models and neural network baselines.

The contribution of this paper is two-fold. (1) Instead of building prompt-specific essay scoring models, it presents a generic model that could make full use of writing data, and score general narrative and argumentative essays. (2) By integrating various dimensions of linguistic features which are emphasized in Chinese L2 acquisition, the models are both effective and interpretable when making predictions. The source code of our method is publicly available[1].

2   Related Work

2.1   Prompt-Specific vs. Prompt-Independent

Most existing AES methods are built as a prompt-specific style, i.e. training and testing with prompt-specific data [8,29,31] or relying on prompt-specific features [2]. They can sometimes achieve better results than those trained regardless of prompts. However, the application of the models are limited to specific topics and they could not make full use of the data. In addition, it will be costly and time-consuming to obtain training data each time when a new prompt is introduced. Two approaches have been developed to improve the situation. One is to directly use all the data [1], ignoring the differences of the prompts. Another method is domain adaptation [5,8,24], which could make better use of available essays in all prompts and make the model robust to the change of prompts.

[1] https://github.com/iris2hu/L2C-rater.

2.2   Interpretability and Feedback

Many success have been achieved by holistic scoring of essays. However, this method faces challenges in providing effective feedback to the students due to its poor interpretability, especially for those neural models [1,8,29]. Taghipour and Ng examine the score variations for three essays after processing each word by the neural network, and find that the model is able to learn essay length and essay content [29]. Alikaniotis et al. visualize the "quality" of the word vectors [1]. However, these methods could only give very shallow explanations of the model behaviors, and are not able to give end user feedback.

A mainstream approach to solve this problem is to score the essays from different dimensions, such as coherence, argument strength, prompt adherence and organization [20–23]. Although it can help the students to understand the shortcomings of their essays, more detailed feedback is still welcome. Woods et al. developed a model-driven sentence selection approach, which can give students sentence-level advice in detail [31]. Ke et al. identify a set of attributes that can explain an argument's persuasiveness and annotate each argument in corpus with the values of these attributes [14].

2.3   Automated Essay Scoring of Chinese Essays

The research on Chinese AES has received much less attention compared with English AES, and lots of work focus on essays of native Chinese speakers. However, the characteristics of L2 essays are quite different from those by native speakers [4,33]. A series of works show that linguistic complexity features are quite effective in measuring the quality of Chinese L2 writing [13,30,32,34]. However, most of these studies only examine their method with a small set of essays on limited prompts. The effectiveness of these features on large-scale datasets remains to be discussed, and their roles in AES systems are also worthy of further exploration. Motivated by previous works, this paper proposes a prompt-independent AES approach for Chinese L2 writing, which integrates a wide scope of linguistic complexity features to enhance the interpretability of the models.

3   The Proposed Method

3.1   The Interpretable Representations of Essay Features

We extracted three types of interpretable features to represent the L2 essays, including linguistic complexity, writing errors and various dimensions of textual features. We measure the linguistic complexity with consideration of the diversity and sophistication of characters, words, clauses, collocations, dependency structures and syntactic constructions. Regarding the writing errors, we build punctuation, character, vocabulary, sentence and discourse level indices. In terms of textual features, we introduce characters, words, ngrams and part-of-speeches.

Linguistic Complexity Features. The effectiveness of linguistic complexity features has been well addressed in predicting the Chinese L2 writing quality [13,30,33]. In addition, they could provide direct feedback to the users on accounting of the usages of different linguistic units, which is highly explainable. Therefore, this paper designs and constructs a comprehensive set of linguistic complexity measures of Chinese L2 writing. These measures are integrated into the representations of L2 essays.

It should be noted that when designing the feature set, it is not applicable to directly transfer the ones that work in English AES or AES systems for Chinese native speakers to Chinese L2 AES, because Chinese has a lot of language-specific features that are emphasized in second language acquisition. Hu pointed out that indices based on language-specific features have stronger predictive power and higher efficiency in predicting the L2 writing scores [11]. Hence the linguistic complexity feature set for Chinese L2 AES should take into account both the language-independent and language-specific features. In this paper, we build 90 linguistic indices of writing quality from the following dimensions. A full list of the indices and their descriptions can be seen in the Appendix A.

Chinese Characters and Vocabulary. We build four indices in this dimension, including the number of Chinese characters, the number of Chinese words, lexical diversity and lexical sophistication. The lexical diversity index is computed as the root type token ratio (RTTR) of words. The lexical sophistication is built as the ratio of sophisticated words. In this study, we identify the words of HSK-5 level, HSK-6 level and out of the HSK vocabulary as the sophisticated words.

Sentences and Clauses. Seven indices are proposed to measure the sentence and clausal complexity, including the mean length of sentences, the mean length of clauses, the mean length of T-units, number of clauses per sentence, number of T-units per sentence, the mean depth of the dependency trees and the max depth of the dependency trees.

Collocations and Bigrams. We introduce 21 collocation-based indices and two bigram-based indices in this dimension. First, eight types of collocations are considered by following Hu and Xiao's work [12], including Verb-Object (VO), Subject-Predicate (SP), Adjective-Noun (AN), Adverb-Predicate (AP), Classifier-Noun (CN), Preposition-Postposition (PP), Preposition-Verb (PV) and Predicate-Complement (PC), where the former four (VO, SP, AN, AP) are universal collocation types that exist in different languages, while the later four (CN, PP, PV, PC) are language-specific types that have been greatly emphasized in Chinese second language acquisition. Similar to lexical diversity, the collocation diversity is built as the RTTRs of different types of collocations, including all the collocations, language-specific collocations, language-independent collocations, and each type of the collocations, resulting in 11 diversity indices. Besides, to measure the collocation sophistication, we introduce the ratio of low frequency collocations and language-specific collocations by following Hu's work[2] [11].

---

[2] https://github.com/iris2hu/Chinese-collocation-complexity.

Also, the ratio of each type of collocations is computed. To cover more language usages, we implement the bigram diversity and sophistication as well by considering the bigrams as a specific type of collocations.

Dependency Structures. The eight types of collocations are extracted from dependency parsing trees with rule-based methods [12]. Although they can well reflect the important knowledge in Chinese L2 acquisition, there are still two problems. One is that they only target at a part of the syntactic relations, hence lacking a whole picture of the syntactic structures. Another is that the collocation diversity and sophistication are not able to measure the fine-grained phrasal complexity underlying the structures, e.g. the number and length of the modifiers. To address the above two questions, this paper proposes 41 dependency based indices that measure the distance, diversity and ratio of all the dependency triples. In this work, we use the LTP dependency parser[3] and 13 dependency relations are considered when building the corresponding indices.

Constructions. The acquisition of grammatical constructions is one of the most important aspects of Chinese L2 teaching and learning [16,28,37]. Both the standardized language test developers and textbook editors make great efforts in designing appropriate construction lists for certain levels of learners.

Consider the importance of construction knowledge in Chinese L2 acquisition, this paper proposes to measure the density and ratio of constructions with regarding to their levels. Specifically, we employ the construction list from the General Syllabus of International Chinese Teaching [9] which include 62 constructions of five levels. After automatic recognition of the constructions, we build 15 indices to reflect the density and ratio of different levels of constructions.

Writing Error Features. In addition to the linguistic complexity, the correctness of L2 production also plays an important role in automated essay scoring or speech rating. Therefore, we adopt five indices of writting errors, i.e. the number of punctuation errors, Chinese character errors, word level errors, sentence level errors and discourse level errors with reference to the annotation in HSK Dynamic Composition Corpus[4].

Multi-granularity Text Features. The high correlation between lexical complexity and writing scores has been witnessed in many studies [19,30]. However, it is still beneficial to further retain the full picture of the textual features.

To represent a text, we extract character, word and part-of-speech unigrams, bigrams and trigrams as features since they could reflect multi-granularity language usages, and could be more explainable than neural representations e.g. word embeddings. We use the tf–idf weighted representations of these features, and each essay can be represented as a text vector:

---

[3] https://github.com/HIT-SCIR/ltp.
[4] http://hsk.blcu.edu.cn/.

$$TextVec = (tfidf_1, tfidf_2, ..., tfidf_N) \tag{1}$$

Where N denotes the total number of unique language units that appear in the corpora.

For the above linguistic complexity, writing error and multi-granularity text features, we conduct preliminary experiments to make feature selection and combination. The detailed process will be introduced in the Experiment section.

3.2 The Ordinal Logistic Regression Model

Automated essay scoring is mainly built as classification or regression tasks. Although the classification models can achieve good results, they treat each score as an independent category, hence losing the ordering information. While for linear regression, it may suffer from violations of modeling assumptions because of the small, discrete, range of possible scores [31].

To address the above questions, this paper proposes to use the Ordinal Logistic Regression (OLR) model in Chinese L2 AES since the OLR method is an effective classification method for ordinal categories [25]. In the classification problem with ordinal classes, the loss of mis-predicting a certain category into different categories should not be the same, e.g. predicting 2 as 3 vs. predicting 2 as 6. The traditional classification loss functions need to be improved to adapt to this relationship [26]. Woods et al. firstly introduce this method into English AES study and achieve impressive results [31]. Inspired by their work, this paper introduces the OLR models into Chinese L2 AES and compares its effectiveness to multiple classical machine learning and neural baselines.

A practical loss of ordinal classification is threshold-based, which is specifically divided into Immediate-threshold loss and All-threshold loss. The latter, which we actually use, is more general than the former. All-threshold loss are represented as (2):

$$\text{Loss}_{\text{AT}}(z) = \sum_{k=1}^{l-1} f\left(s(k;i)\left(\theta_k - z\right)\right) \quad s(k;i) = \begin{cases} -1, k < i \\ +1, k \geq i \end{cases} \tag{2}$$

where z is a specific predicted value, (θ_{i-1}, θ_i) refers to the "correct" segment, and $f(\cdot)$ could be any kind of loss function for multiclass classification problem.

In this study, we employ a very large feature space, requiring regularization to alleviate possible over-fitting. The Regularized Logistic Regression (RLR) minimization objective is defined as

$$\text{Loss}_{\text{RLR}} = \sum_{i=1}^{N} \log\left(1 + \exp\left(-y_i \cdot \mathbf{x}_i^T \mathbf{w}\right)\right) + \frac{\lambda}{2}\mathbf{w}^T\mathbf{w} \tag{3}$$

Defining $h(z) := \log(1 + \exp(z))$, bringing $h(\cdot)$ into $Loss_{AT}(\cdot)$ as $f(\cdot)$, and summing $Loss_{AT}(\cdot)$ of all training examples, we have the minimization objective for the All-threshold version of Ordinal Logistic Regression (OLR-AT)

$$\text{Loss}_{\text{ATL}} = \sum_{i=1}^{N} \left[\sum_{k=1}^{y_i-1} h\left(\theta_k - \mathbf{x}_i^T \mathbf{w}\right) + \sum_{k=y_i}^{l-1} h\left(\mathbf{x}_i^T \mathbf{w} - \theta_k\right) \right] + \frac{\lambda}{2}\mathbf{w}^T\mathbf{w} \quad (4)$$

where label $k \in \{1, \ldots, l\}$ corresponds to the segment (θ_{k-1}, θ_k). θ_0 and θ_l denotes $-\infty$ and $+\infty$ respectively. $\{\mathbf{x}_1, \ldots, \mathbf{x}_n\}, \mathbf{x}_i \in \mathbb{R}^d$ are training examples while $\{y_1, \ldots, y_n\}, y_i \in \{1, \ldots, l\}$ are their labels.

4   Experiments

4.1   Dataset and Preprocessing

In the experiments, we use the essay data from HSK Dynamic Composition Corpus. HSK is a standardized test of Chinese language proficiency for non-native Chinese speakers. The essays are rated from 40 points to 95 points with five as an interval, yielding 12 different categories. The mean score is 69.499 and the Standard Deviation is 10.980. We use the 10277 argumentative and narrative essays (over 3.7 million Chinese characters) from the corpus to train and test the AES model. For a reliable evaluation, we conduct 5-fold cross validation. First, 1477 essays are randomly selected as the test set, and the remaining 8800 essays are split into five groups. Each time four groups are used for training and the left one is used as the development set, which helps to find the optimal parameters. Therefore, all the experiments are conducted five times and the average results on test set is reported.

As the essays in the corpus are manually labeled with different types of writing errors. After retrieving the writing error indices, we carefully remove the annotation tags and transform the essays to their original states, i.e. the original texts written by the test takers. Then we use the method proposed in Sect. 3.1 to obtain the 90 linguistic complexity indices. For multi-granularity text representations, we use `jieba` to conduct word segmentation and POS tagging, and the TfidfTransformer in `scikit-learn` to get the feature weights.

4.2   Feature Selection

To examine the predictive power of different types of linguistic complexity and writing error indices, we conduct step-wise linear regression in each dimension, and the result can be seen in Table 1. It suggests that all of the six dimensions of indices could explain the score variances to some extent, where the indices built upon Chinese characters and vocabulary, collocations and bigrams, and dependency structures have stronger predictive power than the indices in other dimensions.

Table 1. Step-wise regression results in each dimension. The numbers in brackets denote the number of indices entered and remained in the step-wise regression respectively.

Dimension	R	R^2
Chinese characters and vocabulary (4, 3)	0.648	0.420
Sentences and clauses (7, 4)	0.197	0.039
Collocations and bigrams (23, 8)	0.587	0.345
Dependency structures (41, 16)	0.610	0.372
Constructions (15, 9)	0.248	0.061
Writing Error Features (5, 4)	0.254	0.065

Before building the essay scoring model, we make feature selection of the proposed linguistic indices to avoid multicollinearity problem. For the 90 linguistic complexity indices, we select 33 indices with the step-wise regression method. After integrating the five writing error features, the step-wise regression model yields 31 effective features. In the following experiments, these two feature sets are used as the `ling` and `ling+err` settings. The selected features can be seen in Appendix A. For the multi-granularity textual features, we examine different feature combinations in preliminary experiments and find that the combination of word unigrams and pos features could achieve optimal performance with efficient feature space, thus they are used as the `text` setting.

4.3   Models, Parameters and Evaluation Metrics

For the text representations, the min term frequency is set to 10. For OLR-AT model, the penalty coefficient λ is set to 1.0. To make a comparison, we build two types of baselines in the experiments, including regression-based and tree-based machine learning models that use the same input features as our OLR method, and an effective neural AES model introduced by Taghipour and Ng [29] which extracts features automatically and implicitly.

Linear Regression. Linear Regression (LiR) refers to the process of fitting a multi-dimensional linear function to all data points as much as possible. Adding the L1 or the L2 regular term to the cost function yields two variants, i.e. LASSO and Ridge Regression.

Logistic Regression. Logistic Regression (LoR) is a generalized linear model for binary classification. In multi-class scenario, it can be implemented with a One vs. Rest scheme. In our experiment, we set the maximum iteration threshold to a large value 1000 to ensure that the algorithm converges as much as possible.

Random Forest Regression. Random Forest (RF) is based on bagging mechanism and contains multiple decision trees generated in parallel. Each decision tree randomly selects a part of the feature vector for training, and the output is the average results of the trees. In the experiments, the maximum tree depth of the Random Forest Regression is set to 40.

XGBoost Regression. XGBoost is an improved version of the gradient boosting algorithm GDBT. In the experiments, the maximum tree depth is constrained to 3, the number of estimators is constrained to 300, the learning rate is set to 0.05, and the gamma is set to 5.

CNN+LSTM. The CNN+LSTM architecture is a classical neural baseline for English AES task [29]. Here we introduce this model to Chinese L2 AES. We use 300-dim Chinese word vectors[5] pre-trained on Sogou news corpus. We train the network for 20 epochs and the batch size is 32. The training is stopped when the model does not make further improvement after 1000 batches of training. The vocabulary size is set as 20000. Other settings align with Taghipour and Ng [29].

Att-BLSTM. The Att-BLSTM architecture was first proposed for relation classification task [38]. Here we adjust its output from a vector to a scalar so that it can be used for regression task. The other settings, e.g. the use of word embeddings, epochs and batch size, are consistent with those in CNN+LSTM. The model parameters align with Zhou et al. [38].

There are many metrics that can measure the correlation and consistency between the outputs of the AES system and the scores of human experts [35]. In this work we employ three of them: Quadratic Weighted Kappa (QWK), Root Mean Square Error (RMSE) and Pearson coefficient (Pears.). QWK is widely adopted for evaluating AES methods [1,27,29,31], RMSE is a standard way to measure the error of models, while Pearson coefficient could reflect scoring consistency.

4.4   Results

In the experiments, the machine learning methods use four different feature sets as described above: ling, ling+err, ling+text and ling+err+text. When using the combination of linguistic and text features, we concatenate the feature matrices. The CNN+LSTM and Att-BLSTM baselines employ two settings by initializing the word vectors randomly or with the pre-trained Sogou embeddings. Table 2 shows the results of our OLR-AT model and other baselines.

---

[5] https://github.com/Embedding/Chinese-Word-Vectors.

Table 2. Results of Chinese L2 AES. The **bold** denotes the best result under the same feature setting.

Method	Mode	QWK	RMSE	Pears.	Mode	QWK	RMSE	Pears.
LiR	ling	0.640	**1.636**	**0.679**	ling+text	0.269	3.576	0.299
	ling+err	**0.668**	**1.585**	**0.702**	ling+err+text	0.276	3.557	0.307
LoR	ling	0.598	1.813	0.620	ling+text	0.641	1.720	0.663
	ling+err	0.640	1.715	0.661	ling+err+text	0.663	1.667	0.681
RFR	ling	0.625	1.657	0.668	ling+text	0.652	1.603	0.694
	ling+err	0.655	1.601	0.695	ling+err+text	0.667	1.575	0.706
XGBR	ling	0.576	1.690	0.652	ling+text	0.587	1.676	0.659
	ling+err	0.613	1.625	0.687	ling+err+text	0.621	1.616	0.690
CNN+LSTM	Random	0.496	1.845	0.551	Sogou	0.504	1.831	0.560
Att-BLSTM	Random	0.520	1.825	0.568	Sogou	0.531	1.812	0.578
OLR-AT	ling	**0.644**	1.650	0.674	ling+text	**0.697**	**1.554**	**0.718**
	ling+err	0.666	1.616	0.691	ling+err+text	**0.714**	**1.516**	**0.734**

It can be seen that the OLR-AT model on `ling+err+text` feature setting achieves the best performance overall, suggesting the effectiveness of the OLR model and the use of feature combinations. The different models and feature settings also yield different results. We make comparisons of them as below.

Feature Settings. The OLR-AT and machine learning methods all integrate four feature settings. First, `ling+err` brings consistent improvements to `ling` after integrating the error information. This echoes the emphasis on writing error information in HSK standards [7]. Second, except for Linear Regression (LiR), all models obtain the best results under `ling+err+text`. It is worth noting that the very simple Linear Regression model achieves almost the best results under `ling` and `ling+err`. It indicated that LiR, as a simple, effective and interpretable model, might be weak in dealing with high-dimension feature space. To solve this problem, we could introduce parameter regularizations, which will be further explored in the Discussion section.

Models. From a model point of view, we firstly notice that the neural baseline CNN+LSTM does not achieve comparable results of OLR-AT and other machine learning methods, suggesting that it is not applicable to directly transfer the method that work in English to Chinese[6]. Similarly, the neural model Att-BLSTM, which performs well in other tasks such as relation classification, does not obtain competitive results either. It is worth noting that the OLR-AT model surpasses almost all other models. Also, after adding text features to `ling+err`, the performance of OLR-AT improves by 7.2%, compared with 3.6% of Logistic Regression, 1.8% of Random Forest and 1.3% of XGBoost.

---

[6] According to [29], the CNN+LSTM model achieves a high QWK on the English ASAP dataset: 0.717 (AES and Rater1) and 0.710 (AES and Rater2). The QWK of two human raters is 0.754.

Since the HSK dataset does not release scores of different human raters, we are not able to compare the Chinese AES results to human performance. As a reference, the English ASAP (Automated Student Assessment Prize) dataset reported the average between-rater QWK as 0.754[7]. Given our best model (OLR-AT under `ling+err+text`) achieves a QWK of 0.714, it indicates that our method could be a solid work for Chinese L2 AES task. Next, we will make further discussion of the models' errors and shed some light on future work. In addition, we will explore the improved Linear Regression model since it is a simple yet the most explainable model which has the potential to offer users feedback.

5   Discussion

5.1   Analysis on Confusion Matrix

To illustrate the models' behaviors, Fig. 1 shows the confusion matrix of the OLR-AT model under `ling+err+text`.

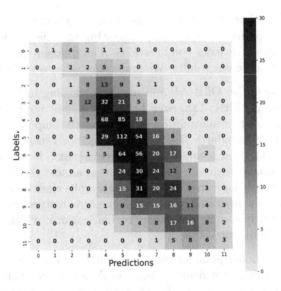

Fig. 1. Confusion matrix of OLR-AT results

It can be seen that the darker blocks distributed around the main diagonal, indicating a high QWK of the model. However, there are still some outliers deviating from the main diagonal. We manually checked the essays that are scored too high or too low (±2 classes), and find the errors are mainly due to the following reasons:

---

[7] ASAP is the most popular dataset in English AES studies, and the dataset can be downloaded at Kaggle.

- For essays with high predicted scores, they typically have a high proficiency of language uses, but the contents deviate from their prompts, or (for argumentative essays) are lack of organization when expressing opinions. Existing feature sets and algorithms cannot detect these writing flaws.
- For essays with low predicted scores, we find some rating exceptions by the human raters, e.g. giving high scores to unfinished essays. Since the length of the essay is an important feature in our model, the AES scores lower than the human raters. Therefore, on the whole, the algorithm's ability to capture current features is in place, and the evaluation effect is relatively reliable.

From the above analysis, we find that it would be helpful to further introduce prompt-essay relevance measures, as well as the discourse level indices e.g. cohesion and coherence. We will conduct the research from these aspects in future work.

5.2 Revisiting Linear Regression: Interpretability and Potential of Providing Feedback

Why does the result of Linear Regression drop so significantly after introducing the `text` representations? We speculate that the reason lies in the high-dimension and sparse feature space of text representations, which could easily lead to over-fitting of linear model. To verify this, we implement Linear Regression under `text` setting, and compare it with the results of `ling+text` and `ling+err+text` as shown in Table 3. It can be easily seen that the `text` only setting has a low performance, and integrating text features to linguistic features does not make improvements to `ling` and `ling+err` settings.

Table 3. The results of Linear Regression with different feature sets.

Mode	QWK	RMSE	Pears.
text	0.207	3.787	0.232
ling+text	0.269	3.576	0.299
ling+err+text	0.276	3.557	0.307

Further, we use one of the variants of linear regression - Ridge Regression, that is, adding an L2 regular term to the objective function of Linear Regression. Ridge regression loses unbiasedness in exchange for high numerical stability [10]. The results are shown in Table 4. Ridge Regression greatly alleviated the over-fitting phenomenon. Different from Linear Regression, Ridge makes clear improvements after integrating text features, surpassing the tree-based methods even. It is close to OLR-AT in terms of QWK, and even scores slightly better than OLR-AT on RMSE and Pearson correlation coefficient. The reason is that the OLR-AT method focuses on correct classification, while Ridge aims to minimize the sum of squares of deviations under the constraint of regular terms.

Table 4. The comparison of Linear Regression and Ridge Regression

Method	Mode	QWK	RMSE	Pears.	Mode	QWK	RMSE	Pears.
LiR	ling	0.640	1.636	0.679	ling+text	0.269	3.576	0.299
	ling+err	0.668	1.585	0.702	ling+err+text	0.276	3.557	0.307
Ridge	ling	0.636	1.640	0.676	ling+text	0.694	1.538	0.723
	ling+err	0.667	1.585	0.702	ling+err+text	0.709	1.510	0.735

The power of Ridge Regression, a simple linear model, inspires us to explore the interpretation the results and the possibility to provide feedback to L2 students. Note that Linear Regression under `ling+err` setting achieves better performance than that of OLR-AT, indicating that the linear relationship does exist between low-dimension linguistic features and the writing scores, and the linguistic features explain a large amount of score variances. Thus, studying how linear model uses linguistic features to score essays helps understand what an excellent essay should be like in a model perspective. Figure 2 shows 31 box plots of the selected `ling+err` features, showing the effect of each feature on essay scores.

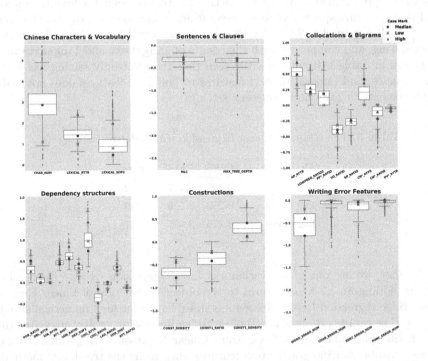

Fig. 2. Effect plots of 31 selected features in `ling+err` setting. We use product of linear model coefficients and feature values $\omega_i x_i$ to convey the effect, where ω_i denotes the coefficient of feature i from the linear model, and x_i is the corresponding feature value.

In addition, we choose three essays of high score (95 points), medium score (65 points), and low score (45 points) respectively and mark their effect value in the box plot to show how each feature contributes to their final scores. The three example essays can be seen in Appendix B. From the effects in Fig. 2, we can clearly see the pros and cons of each essay, and provide corresponding feedback as below:

- *Essay of high score (the green triangle).* The student can write a long essay and use diverse and sophisticated words. In terms of syntactic usages, the essay employs sufficient syntactic constructions, but the use of language-specific structures is limited. Considering the correctness, the student can write Chinese characters with a high accuracy, but there are still some word, sentence and punctuation errors.
- *Essay of medium score (the blue circle).* The essay is of medium length and lexical diversity. The vocabulary used in the essay is relatively simple. From the syntactic view, the student is able to produce language-specific structures and idiomatic expressions skillfully. The student can use most words correctly, but the essay still contains some character, sentence and discourse errors.
- *Essay of low score (the red cross).* The essay is short with a limited word and collocation vocabulary, but the student is able to use some sophisticated words and elementary level constructions. Also, The student can produce correct and fluent text with relatively few mistakes.

6   Conclusion and Future Work

In this paper, we propose a prompt-independent and interpretable AES method for Chinese L2 writing. We build explainable representations of both the linguistic and text features, and the threshold-based Ordinal Logistic Regression model is introduced to our Chinese L2 AES task. The result on OLR-AT model under `ling+err+text` setting obtains a high QWK score of 0.714, a low RMSE of 1.516, and a high Pearson coefficient of 0.734. Further, we find that with our method and the feature set, the model is explainable and has the potential to offer users feedback on the writing. This work provides insights and a solid baseline for AES studies of Chinese L2 writing.

At present, our method has integrated linguistic complexity, writing correctness and text features of essays. It still needs further study on developing prompt-essay relevance measures, as well as the discourse level indices e.g. cohesion and coherence. Just as noteworthy, in Table 2, the CNN+LSTM architecture seems to perform poorly, far inferior to its excellent performance on the English ASAP dataset. This explains to a certain extent how different the scoring standards for Chinese L2 essays and those for native English speakers. It should be pointed out that we are not denying the possibility of applying neural networks to Chinese L2 AES task. Since CNN+LSTM structure itself is relatively simple, it cannot fully detect the features that a Chinese L2 AES task needs. As a model that has not been specially adjusted for the Chinese AES task, Att-BLSTM's improvement over CNN+LSTM has shown that neural networks have potential to achieve better results. Thus in future work, neural networks with stronger learning abilities, together with a good interpretation method may play important roles in this task.

Appendix

A   The Linguistic Complexity and Writing Error Features

We include the detailed list of 90 linguistic complexity and 5 writing error features in Table 5. As described in the main paper, the feature selection module yields two feature sets, i.e. `ling` (33 indices, denoted as ◇) and `ling+err` (31 indices, denoted as ♣).

Table 5. The feature sets in this study.

ID	Feature	Description
Chinese characters and vocabulary		
1	CHAR_NUM ◇ ♣	Number of Chinese characters
2	WORD_NUM ◇	Number of words
3	LEXICAL_RTTR ◇ ♣	Root type token ratio (RTTR) of words
4	LEXICAL_SOP2 ◇ ♣	Root ratio of sophisticated words
Sentences and clauses		
5	MLS	Mean length of sentences
6	MLC ◇ ♣	Mean length of clauses
7	MLTU	Mean length of T-units
8	NCPS	Number of clauses per sentence
9	NTPS	Number of T-units per sentence
10	MEAN_TREE_DEPTH	Mean depth of syntactic trees
11	MAX_TREE_DEPTH ◇ ♣	Max depth of syntactic trees
Collocations and bigrams		
12	COLL_RTTR	RTTR of all the collocations
13	UNIQUE_RTTR	RTTR of Chinese unique collocations
14	GENERAL_RTTR ◇	RTTR of language-independent collocations
15	UNIQUE_RATIO2 ◇	Ratio of Chinese unique collocations
16	LOWFREQ_RATIO2 ◇ ♣	Ratio of sophisticated collocations
17	VO_RATIO ♣	Ratio of verb-object collocations
18	VO_RTTR ◇	RTTR of verb-object collocations
19	SP_RATIO ♣	Ratio of subject-predicate collocations
20	SP_RTTR ◇	RTTR of subject-predicate collocations
21	AN_RATIO	Ratio of adjective-noun collocations
22	AN_RTTR	RTTR of adjective-noun collocations

(continued)

Table 5. (*continued*)

ID	Feature	Description
23	AP_RATIO ◇	Ratio of adverb-predicate collocations
24	AP_RTTR ♣	RTTR of adverb-predicate collocations
25	CN*_RATIO ◇ ♣	Ratio of classifier-noun collocations
26	CN*_RTTR ◇ ♣	RTTR of classifier-noun collocations
27	PP*_RATIO ♣	Ratio of preposition-postposition collocations
28	PP*_RTTR ◇	RTTR of preposition-postposition collocations
29	PV*_RATIO	Ratio of preposition-verb collocations
30	PV*_RTTR ◇ ♣	RTTR of preposition-verb collocations
31	PC*_RATIO	Ratio of predicate-complement collocations
32	PC*_RTTR	RTTR of predicate-complement collocations
33	BIGRAM_RTTR ◇	RTTR of bigrams
34	BIGRAM_SOP2 ◇	Root ratio of sophisticated bigrams
Dependency structures		
35	DEP_RTTR	RTTR of dependency triples
36	DEP_SOP2 ◇ ♣	Root ratio of sophisticated dependency triples
37	HED_RTTR	RTTR of HED dependency triples
38	HED_RATIO	Ratio of HED dependency triples
39	COO_RTTR ◇ ♣	RTTR of COO dependency triples
40	COO_RATIO ◇ ♣	Ratio of COO dependency triples
41	SBV_RTTR	RTTR of SBV dependency triples
42	SBV_RATIO ◇	Ratio of SBV dependency triples
43	ADV_RTTR	RTTR of ADV dependency triples
44	ADV_RATIO	Ratio of ADV dependency triples
45	ATT_RTTR ◇	RTTR of ATT dependency triples
46	ATT_RATIO ♣	Ratio of ATT dependency triples
47	VOB_RTTR	RTTR of VOB dependency triples
48	VOB_RATIO ◇ ♣	Ratio of VOB dependency triples
49	FOB_RTTR	RTTR of FOB dependency triples
50	FOB_RATIO	Ratio of FOB dependency triples
51	POB_RTTR	RTTR of POB dependency triples
52	POB_RATIO ◇ ♣	Ratio of POB dependency triples
53	IOB_RTTR ♣	RTTR of IOB dependency triples
54	IOB_RATIO	Ratio of IOB dependency triples
55	DBL_RTTR ♣	RTTR of DBL dependency triples
56	DBL_RATIO ◇	Ratio of DBL dependency triples
57	RAD_RTTR	RTTR of RAD dependency triples
58	RAD_RATIO	Ratio of RAD dependency triples
59	CMP_RTTR	RTTR of CMP dependency triples
60	CMP_RATIO	Ratio of CMP dependency triples

(*continued*)

Table 5. (*continued*)

ID	Feature	Description
61	LAD_RTTR	RTTR of LAD dependency triples
62	LAD_RATIO ♣	Ratio of LAD dependency triples
63	COO_DIST	Mean distance of COO dependency triples
64	SBV_DIST	Mean distance of SBV dependency triples
65	ADV_DIST	Mean distance of ADV dependency triples
66	ATT_DIST ◇ ♣	Mean distance of ATT dependency triples
67	VOB_DIST ◇ ♣	Mean distance of VOB dependency triples
68	FOB_DIST	Mean distance of FOB dependency triples
69	POB_DIST ◇	Mean distance of POB dependency triples
70	IOB_DIST ◇	Mean distance of IOB dependency triples
71	DBL_DIST	Mean distance of DBL dependency triples
72	RAD_DIST	Mean distance of RAD dependency triples
73	CMP_DIST	Mean distance of CMP dependency triples
74	LAD_DIST	Mean distance of LAD dependency triples
75	MEAN_DIST	Mean distance of all the dependency triples
Constructions		
76	CONST_DENSITY ◇ ♣	Number of constructions/N_char
77	CONST1_RATIO ◇ ♣	Ratio of level-1 constructions
78	CONST1_DENSITY ◇ ♣	Number of level-1 constructions/N_char
79	CONST2_RATIO	Ratio of level-2 constructions
80	CONST2_DENSITY	Number of level-2 constructions/N_char
81	CONST3_RATIO	Ratio of level-3 constructions
82	CONST3_DENSITY	Number of level-3 constructions/N_char
83	CONST4_RATIO	Ratio of level-4 constructions
84	CONST4_DENSITY	Number of level-4 constructions/N_char
85	CONST5_RATIO	Ratio of level-5 constructions
86	CONST5_DENSITY	Number of level-5 constructions/N_char
87	CONST_LOW_RATIO	RATIO of low level constructions (level 1–2)
88	CONST_HIGH_RATIO	RATIO of high level constructions (level 4–5)
89	CONST_LOW_DENSITY	Number of low level constructions/N_char
90	CONST_HIGH_DENSITY	Number of high level constructions/N_char
Writing errors		
91	PUNC_ERROR_NUM ♣	Number of punctuation errors
92	CHAR_ERROR_NUM ♣	Number of character level errors
93	WORD_ERROR_NUM ♣	Number of word level errors
94	SENT_ERROR_NUM ♣	Number of sentence level errors
95	DISCOURSE_ERROR_NUM	Number of discourse level errors

B The Example Essays

Essay 1 (high, 95 points)

如何看待"安乐死"?

How should we view euthanasia?

在二十世纪的今天，"安乐死"已不是什么新话题。在不同的国家，已曾有人要求、助他人安乐死；只是每个国家的法律有，因此对助他人求安乐死的人的对待亦有所不同。比方说，根据香港的法律，自和人均属违法。法院不判犯人死刑；自的人如果自不遂，救後活过来，理论上是要坐牢的。

法律归法律，竟生与死是人一生的大事，不能绝对由法律。代人崇尚自由，什么都讲求选择——生育女、职业、居住地、个人格……偏偏在出生件事情上，没有一个人可以选择。那么，人最少应该可以选择何时死亡了吧？

讨论"安乐死"的基础，应该是人对生命的尊重。要不然，有了安乐死为後盾，人可以在痛苦中言死去，而存心作奸犯科的，更大有理由害命。

我为"安乐死"有其可取之处，但得以可为大前提。首先，只有身患重病，到了末期段，亦明知没有医治方法的病人，有权选择"安乐死"。第二，这个选择必须由病人在清醒的下自作出，最好有书面证明。第三，关于病情的判断，应有最少位医生签署证明作实。有些人或还会加上第四个条件，就是只准许以「被动」方式施行"安乐死"，即拔掉维生器具；不能施行「主动」方式的"安乐死"，即注射毒药等。

"安乐死"可以免除一些受痛苦煎熬的病人的困苦，使他们仍有一点尊严地去世，有其可取之处；但防止有人滥用"安乐死"，亦致为重要。

Essay 2 (medium, 65 points)

如何解决"代沟"问题?

How to solve the generation gap problem?

从古代到在，代沟问题是在人们的生活上常存的。着社会的发展速度加快，代沟问题的程度也很深了。那么我们如何解决这个问题呢？

对我的意见来说，最好的解决方法就是增加两代之间的对话。为了增加对话呢，需要两代互相的努力。首先找个共同的话题开始慢慢得增加在一起的时间。

举例子说，我常用电脑的。无论工作还是玩儿，都来电脑做的。但是我妈对我这个样子太不满意了。这也是一种代沟吧。妈妈是对电脑外行，但孩子每天用电脑所以她对我不懂的地方也越来越多。她还不满意吗？我妈跟我经过一段时间的对话，我才了解了她的心情是如何。然后我开始教妈电脑。在呢我妈跟我一起用电脑。如果我对妈妈的态度只是不满意没有跟她说话的话，我解决不了这个问题。我感觉得这样的办法对孩子也有教育方面的好处。

总而言之，我想代沟的问题呢，应该通过两代互相的努力增加对话时间的时候，可以解决的。

Essay 3 (high, 45 points)

我看流行歌曲

My opinion on popular songs

我非常喜欢流行歌曲，因为流行歌曲不但动听，而且可以表达自己的想法和感情。比如说：周杰伦、王力宏、田震、那英、周传雄等等他们都是发自自己内心再唱歌。

还有那些作词、作曲的人都用音乐来表达自己或其他人的情感，对社会赞赏和不满。

流行歌曲里大部分都情歌，一个人想对自己喜欢的人告白，用歌曲是最好不过的了。

希望人人都喜欢流行歌曲。

References

1. Alikaniotis, D., Yannakoudakis, H., Rei, M.: Automatic text scoring using neural networks. In: Proceedings of the 54th Annual Meeting of the Association for Computational Linguistics (Volume 1: Long Papers), pp. 715–725 (2016)
2. Attali, Y., Burstein, J.: Automated essay scoring with e-rater® v. 2. J. Technol. Learn. Assess. **4**(3) (2006)
3. Burstein, J., Chodorow, M.: Automated essay scoring for nonnative English speakers. In: Computer Mediated Language Assessment and Evaluation in Natural Language Processing (1999)
4. Cao, X., Deng, S.: The contrastive analysis of the writing performance in Chinese as L1 and L2: based on the Chinese compositions on the same topic among Chinese senior elementary students and Vietnam senior university students. TCSOL Stud. (huáwén jiaoxué yu yánjiu) (2), 39–46 (2012)
5. Cao, Y., Jin, H., Wan, X., Yu, Z.: Domain-adaptive neural automated essay scoring. In: Proceedings of the 43rd International ACM SIGIR Conference on Research and Development in Information Retrieval, pp. 1011–1020 (2020)
6. Chang, T.H., Lee, C.H.: Automatic Chinese essay scoring using connections between concepts in paragraphs. In: 2009 International Conference on Asian Language Processing, pp. 265–268. IEEE (2009)
7. Dan, N.: A historical account of the assessment criteria for HSK writing. J. Yunnan Normal Univ. (Teaching and Research on Chinese as a Foreign Language Edition) (Yúnnán shifàn dàxué xuébào (duìwài hànyu jiaoxué yu yánjiu ban)) **7**(6) (2009)
8. Dong, F., Zhang, Y., Yang, J.: Attention-based recurrent convolutional neural network for automatic essay scoring. In: Proceedings of the 21st Conference on Computational Natural Language Learning (CoNLL 2017), pp. 153–162 (2017)
9. Hanban: General Course Syllabus for International Chinese Language Teaching. Foreign Language Teaching and Research Press, Beijing (2009)
10. Hoerl, A.E., Kennard, R.W.: Ridge regression: biased estimation for nonorthogonal problems. Technometrics **12**(1), 55–67 (1970)
11. Hu, R.: On the relationship between collocation-based syntactic complexity and Chinese second language writing. Appl. Linguist. (yuyán wénzì yìngyòng) (1), 132–144 (2021)
12. Hu, R., Xiao, H.: The construction of Chinese collocation knowledge bases and their application in second language acquisition. Appl. Linguist. (yuyán wénzì yìngyòng) (1), 135–144 (2019)
13. Huang, Z., Xie, J., Xun, E.: Study of feature selection in HSK automated essay scoring. Comput. Eng. Appl. (jìsuànji gongchéng yu yingyòng) (6), 118–122+126 (2014)

14. Ke, Z., Carlile, W., Gurrapadi, N., Ng, V.: Learning to give feedback: modeling attributes affecting argument persuasiveness in student essays. In: IJCAI, pp. 4130–4136 (2018)
15. Ke, Z., Ng, V.: Automated essay scoring: a survey of the state of the art. In: IJCAI, pp. 6300–6308 (2019)
16. Lu, J.: The grammatical teaching in Chinese second language teaching. Lang. Teach. Res. (yuyán jiaoxué yu yánjiu) (3), 1–8 (2000)
17. Mayfield, E., Black, A.W.: Should you fine-tune BERT for automated essay scoring? In: Proceedings of the Fifteenth Workshop on Innovative Use of NLP for Building Educational Applications, pp. 151–162 (2020)
18. Page, E.B.: Grading essays by computer: progress report. In: Proceedings of the Invitational Conference on Testing Problems, pp. 87–100 (1966)
19. Peng, X., Ke, D., Chen, Z., Xu, B.: Automated Chinese essay scoring using vector space models. In: 2010 4th International Universal Communication Symposium, pp. 149–153. IEEE (2010)
20. Persing, I., Davis, A., Ng, V.: Modeling organization in student essays. In: Proceedings of the 2010 Conference on Empirical Methods in Natural Language Processing, pp. 229–239 (2010)
21. Persing, I., Ng, V.: Modeling thesis clarity in student essays. In: Proceedings of the 51st Annual Meeting of the Association for Computational Linguistics (Volume 1: Long Papers), pp. 260–269 (2013)
22. Persing, I., Ng, V.: Modeling prompt adherence in student essays. In: Proceedings of the 52nd Annual Meeting of the Association for Computational Linguistics (Volume 1: Long Papers), pp. 1534–1543 (2014)
23. Persing, I., Ng, V.: Modeling argument strength in student essays. In: Proceedings of the 53rd Annual Meeting of the Association for Computational Linguistics and the 7th International Joint Conference on Natural Language Processing (Volume 1: Long Papers), pp. 543–552 (2015)
24. Phandi, P., Chai, K.M.A., Ng, H.T.: Flexible domain adaptation for automated essay scoring using correlated linear regression. In: Proceedings of the 2015 Conference on Empirical Methods in Natural Language Processing, pp. 431–439 (2015)
25. Rennie, J.D.: Ordinal logistic regression (2005). http://people.csail.mit.edu/jrennie/writing/olr.pdf
26. Rennie, J.D., Srebro, N.: Loss functions for preference levels: regression with discrete ordered labels. In: Proceedings of the IJCAI Multidisciplinary Workshop on Advances in Preference Handling, vol. 1. Citeseer (2005)
27. Song, W., Zhang, K., Fu, R., Liu, L., Liu, T., Cheng, M.: Multi-stage pre-training for automated Chinese essay scoring. In: Proceedings of the 2020 Conference on Empirical Methods in Natural Language Processing (EMNLP), pp. 6723–6733 (2020)
28. Sun, D.: Two theoretical issues on the studies of the pedagogic grammar of TCSL. Lang. Teach. Res. (yuyán jiaoxué yu yánjiu) (2), 30–39 (2016)
29. Taghipour, K., Ng, H.T.: A neural approach to automated essay scoring. In: Proceedings of the 2016 Conference on Empirical Methods in Natural Language Processing, pp. 1882–1891 (2016)
30. Wang, Y.: The correlation between lexical richness and writing score of CSL learner–the multivariable linear regression model and equation of writing quality. Appl. Linguist. (yyán wénzì yìngyòng) (2), 93–101 (2017)
31. Woods, B., Adamson, D., Miel, S., Mayfield, E.: Formative essay feedback using predictive scoring models. In: Proceedings of the 23rd ACM SIGKDD International Conference on Knowledge Discovery and Data Mining, pp. 2071–2080 (2017)

32. Wu, J.: The research of indices of the grammatical complexity in South Korean native speakers' Chinese writing and its relationship with writing quality. Linguist. Sci. (yuyán kexué) (5), 66–75 (2018)

33. Wu, J., Wei, Z., Lu, D.: Assessing Chinese L2 writing quality on basis of language features and content quality. Chin. Teach. World (shìjiè hànyu jiaoxué) **33**(2), 130–144 (2019)

34. Wu, P., Xing, H.: The influence of content, lexical and discourse features on the quality of CSL learners' writing. Lang. Teach. Res. (yuyán jiaoxué yu yánjiu) (2), 24–32 (2020)

35. Yannakoudakis, H., Cummins, R.: Evaluating the performance of automated text scoring systems. In: Proceedings of the Tenth Workshop on Innovative Use of NLP for Building Educational Applications, pp. 213–223 (2015)

36. Yannakoudakis, H., Briscoe, T., Medlock, B.: A new dataset and method for automatically grading ESOL texts. In: Proceedings of the 49th Annual Meeting of the Association for Computational Linguistics: Human Language Technologies, pp. 180–189 (2011)

37. Zhao, J.: Pedagogical grammar of Chinese as a second language: combination of grammar framework and fragmentation. Lang. Teach. Res. (yuyán jiaoxué yu yánjiu) (2), 1–10 (2018)

38. Zhou, P., et al.: Attention-based bidirectional long short-term memory networks for relation classification. In: Proceedings of the 54th Annual Meeting of the Association for Computational Linguistics (Volume 2: Short Papers), pp. 207–212 (2016)

A Robustly Optimized BERT Pre-training Approach with Post-training

Zhuang Liu[1(✉)], Wayne Lin[2], Ya Shi[3], and Jun Zhao[4]

[1] Dongbei University of Finance and Economics, Dalian, China
liuzhuang@dufe.edu.cn
[2] University of Southern California, LA, USA
[3] Union Mobile Financial Technology, Beijing, China
[4] IBM Research, Beijing, China

Abstract. In the paper, we present a *'pre-training'*+*'post-training'*+*'fine-tuning'* three-stage paradigm, which is a supplementary framework for the standard *'pre-training'*+*'fine-tuning'* language model approach. Furthermore, based on three-stage paradigm, we present a language model named PPBERT. Compared with original BERT architecture that is based on the standard two-stage paradigm, we do not fine-tune pre-trained model directly, but rather post-train it on the domain or task related dataset first, which helps to better incorporate task-awareness knowledge and domain-awareness knowledge within pre-trained model, also from the training dataset reduce bias. Extensive experimental results indicate that proposed model improves the performance of the baselines on 24 NLP tasks, which includes eight GLUE benchmarks, eight SuperGLUE benchmarks, six extractive question answering benchmarks. More remarkably, our proposed model is a more flexible and pluggable model, where post-training approach is able to be plugged into other PLMs that are based on BERT. Extensive ablations further validate the effectiveness and its state-of-the-art (SOTA) performance. The open source code, pre-trained models and post-trained models are available publicly.

Keywords: BERT · Pre-training · Post-training · SQuAD · GLUE · SuperGLUE

1 Introduction

Recently, the introduction of pre-trained language models (PLMs), including GPT [18], BERT [3], and ELMo [17], among many others, has achieved tremendous success to the natural language processing (NLP) research. Typically, the basic structure of such a model consists of two successive stages, one step during the pre-training phase and another step during the fine-tuning phase. During the pre-training phase it pre-trains on unsupervised dataset firstly, then during the fine-tuning phase it fine-tunes on downstream supervised NLP tasks. Up to now, these models obtained the best performance on various NLP tasks. Some of the most

S. Li et al. (Eds.): CCL 2021, LNAI 12869, pp. 471–484, 2021.
https://doi.org/10.1007/978-3-030-84186-7_31

prominent examples are BERT, and BERT based SpanBERT [5], ALBERT [8]. These PLMs are trained on the large unsupervised corpus through some unsupervised training objectives. However, it is not obvious that the model parameters which is obtained during unsupervised pre-training phase can be well-suited to support the this kind of transfer learning. Especially during the fine-tuning phase, for the target NLP task only a small amount of supervised text data is available, fine-tuning the pre-trained model are potentially brittle. And for the pre-trained model, supervised fine-tuning requires substantial amounts of task-specific supervised training dataset, not always available. For example, in GLUE benchmark [25], Winograd Schema dataset [9] have only 634 training data, too small for fine-tuning natural language inference (NLI) task. Moreover, although PLMs, such BERT, can learn contextualized representations across many NLP tasks (to be task-agnostic), which leverages PLMs alone still leaves the domain-specific challenges unresolved (BERT are trained on general domain corpora only, and capture a general language knowledge from training dataset, but lack domain or task-specific data severely). For example, in financial domain, they often contain unique vocabulary information, such as stock, bond type, and the sizes of labeled data are also very small (even only few hundreds of samples). In the paper, to overcome the aforementioned issues, we proposed a novel three-stage BERT (called PPBERT) architecture, in which we add a second stage of training, that is *post-training*, to improving the original BERT architecture model.

Typically there are two directions to pursue new state-of-art in the post pre-trained PLMs era. One is to construct novel neural network architecture model based on PLMs, like BERTserini [26] and BERTCMC [15]. Other approach is to optimize pre-training, like GPT 2.0 [18], MT-DNN [10], SpanBERT [5], and ALBERT [8]. In the paper, we present another novel method to improve the PLMs. We present a *'pre-training'+'post-training'+'fine-tuning'* three-stage paradigm and further present a language model named PPBERT. Compared with original BERT architecture that is based on the standard *'pre-training'+'fine-tuning'* PLMs approach, we do not fine-tune pre-trained models directly, but rather **post-train** them on the domain or task related training dataset first, which helps to better incorporate task-awareness knowledge and domain-awareness knowledge within pre-trained model, also in the training dataset can reduce bias. More specifically, our framework involves three sequential stages: pre-training stage using on large-scale corpora (see Subsect. 2.1), post-training stage using the task or domain related datasets via multi-task continual learning method (see Subsect. 2.2), and fine-tuning stage using target datasets, even with little labeled samples or without labeled samples (see Subsect. 2.3). Thus, PPBERT can benefits from the regularization effect since it leverages cross-domain or cross-task data, which helps model generalize better with limited data and adapt to new domains or tasks better.

Sum up, on a wide variety of tasks our proposed post-training process outperforms existing BERT benchmark, and achieved better performance on small dataset and domain-specific tasks in particular substantially. Specifically, we compared our model with BERT baselines on GLUE and SuperGLUE benchmark tasks and consistently significantly outperform BERT on all of 16 tasks

(8 GLUE tasks and 8 SuperGLUE tasks), increasing by the GLUE average score of 87.02, showing an absolute improvement of 2.97 over BERT; showing an absolute improvement of 5.55, pushing the SuperGLUE to 74.55. More remarkably, our model is a more flexible and pluggable. The post-training appoach can be straight plugged into other PLMs based on BERT. In our ablation studies, we plug the post-training strategy into original BERT (i.e., PPBERT) and its variant, ALBERT (called PPALBERT), respectively. Our approaches advanced the SOTA results for five popular question answering datasets, surpassing the previous pre-trained models by at least 1 point in absolute accuracy. Moreover, through further ablation studies, the best model obtains SOTA results on small datasets (1/20 training set). All of these clearly demonstrate our proposed three-stage paradigms exceptional generalization capability via post-training learning.

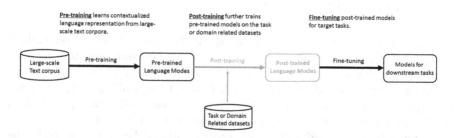

Fig. 1. An illustration of the architecture for our PPBERT, which is a '*pre-training*'-'*post-training*'-then-'*fine-tuning*' three-stage BERT. Compared with standard BERT architecture that has the two-stage '*pre-training*'-then-'*fine-tuning*', we do not directly fine-tune pre-trained models, but rather add a second stage of training (called '*post-training*'). More specifically, during the pre-training stage, we first on the large-scale dataset conduct unsupervised pre-training, and then during the post-training stage post-train pre-trained models on the task or domain related dataset, and last during the fine-tuning stage conduct fine-tuning on downstream supervised NLP tasks.

2   The Proposed Model: PPBERT

As shown in Fig. 1, the standard BERT is built based on two-stage paradigm architecture, '***pre-training***'+'***fine-tuning***'. Compared traditional pre-training methods, PPBERT does not fine-tune the pre-trained model directly after pre-training, but rather continues to post-train the pre-trained model on the task or domain related corpus, helping to reduce bias. During post-training processing our proposed PPBERT framework can continuously update pre-trained model. The architecture of our PPBERT architecture is shown in Fig. 1.

2.1   Pre-training

The training procedure of our proposed PPBERT has 2 processing: pre-training stage and post-training stage. As BERT outperforms most existing models, we do not intend to re-implement it but focus on the second training stage: Post-training. The pre-training processing follows that of the BERT model. We first use original BERT and further adopt a joint post-training method to enhance BERT. Thus, our proposed PPBERT is more flexible and pluggable, where post-training approach is able to be plugged into other language models based on BERT, such as ALBERT [8], SpanBERT [5], not only applied to original BERT.

2.2 Post-training

Compared with original BERT architecture that has two-stage paradigm, '*pre-training*'+'*fine-tuning*', we do not fine-tune pre-trained model, but rather first *post-train* the model on the task or domain related training dataset directly. We add a second training stage, that is '*post-training*' stage, on an intermediate task before target-task fine-tuning.

Training Details. In the post-training stage, its aims to train the pre-trained model on the task or domain related annotated data continuously, to learn task knowledge or domain knowledge from different post-training tasks by keeping updating the pre-trained model. Thus, it brings a big challenge: How to train these post-training tasks in a continual way, and more efficiently post-train a new task without forgetting the knowledge that is learned before.

Inspired by [2,22] and [16], which show Continual Learning can train the model with several tasks in sequence, but we find that, standard Continual Learning method trains the model with only one task at each time with the demerit that it is easy to forget the knowledge previously learned. Also concurrently, inspired by [10,12] and [4,13], which show Multi-task Learning can allow the use of different training corpus to train sub-parts of neural networks, but we find that, although Multi-task Learning could train multiple tasks at the same time, it is necessary that all customized pre-training tasks are prepared before the training could proceed. So this method takes as much time as continual learning does, if not more. So we present a multi-task continual learning method to tackle with this problem. More specifically, whenever a new post-training task comes, the multi-task continual learning method first utilizes the parameters that is previously learned to initialize the model, and then simultaneously train the newly-introduced task together with the original tasks, which will make sure that the learned parameters can encode the knowledge that is previously learned. More crucially, during post-training we allocate each task K training iterations, and then further assign these K iterations for each task to different stages of training. Also concurrently, instead of updating parameters over a batch, we divide a batch into more sub-batches and accumulate gradients on those sub-batches before parameter updates, which allows for a smaller sub-batch to be consumed in each iteration, more conducive to iterating quickly by using distributed training. As a result, proposed PPBERT can continuously update pre-trained model using the multi-task continual learning method. So we can guarantee the efficiency of our post-training without forgetting the knowledge that is previously trained.

Post-training Datasets. As discussed above, fine-tuning processing has main challenges, on the target task directly, as follows: **i)** during the fine-tuning phase, there is only a small amount of supervised training data, fine-tuning the pre-trained model are potentially brittle; **ii)** for the pre-trained model, its supervised fine-tuning requires substantial amounts of task-specific supervised training dataset, limited and indirect, not always available; **iii)** leveraging BERT alone leaves the domain or task-specific questions unresolved. To enhance the performance of pre-trained model, we need to effectively fuse task knowledge (from related NLP tasks supervised data) or domain knowledge (from related in-domain supervised data). As a common NLP task, Questions and Answers (QA), to get the answer based on a question, requires reasoning on facts relevant to the given question and deep semantic understanding of document. Thus, a large-scale QA supervised corpus can benefit most NLP tasks. Similarly, NLI task (a.k.a. RTE) and sentiment analysis (SA) are also two important and basic tasks for natural language understanding. Eventually, we use QA dataset (CoQA), NLI dataset (SNLI) and SA dataset (YELP) as post-training datasets. We post-train our model on CoQA, SNLI and YELP data simultaneously.

In this work, for generality and wide applicability of our proposed PPBERT, we use only CoQA, SNLI and YELP as post-training datasets. Note that, because PPBERT adopts the effective multi-task continual learning training method (Sect. 2.2), its post-training datasets are easily scalable, which is meant to be combined further with other datasets, including domain specific data.

2.3   Fine-Tuning

In fine-tuning processing, we first initialize PPBERT model with the post-trained parameters, and then use supervised dataset from specific tasks to further fine-tune. In general, for each downstream task, after being fine-tuned it has its own fine-tuned models.

3   Experiments

3.1   Tasks

To evaluate our proposed approach, we use a comprehensive experiment tasks, as follows:

i) in Sect. 3, eight tasks in the GLUE benchmark [25] and eight tasks in the SuperGLUE benchmark [24];

ii) in Sect. 4, five question answering tasks, two natural language inference tasks and two tasks in domain adaptation, financial sentiment analysis and financial question answering.

We expect that these NLP tasks will benefit from proposed '*pre-training*'+'*post-training*'+'*fine-tuning*' three-stage paradigm particularly.

3.2   Datasets

This subsection briefly describes the datasets.

GLUE. The General Language Understanding Evaluation (GLUE) benchmark [25] is a collection of eight datasets to evaluate NLU tasks. GLUE[1] consists of a series of NLP task datasets (See Table 1), including: Corpus of Linguistic Acceptability (CoLA), Multi-genre Natural Language Inference (MNLI), Recognizing Textual Entailment (RTE), Quora Question Pairs (QQP), Semantic Textual Similarity Benchmark (STS-B), Stanford Sentiment Treebank (SST-2), Question Natural Language Inference (QNLI), Microsoft Research Paraphrase Corpus (MRPC).

Table 1. Summary of the GLUE benchmark.

Corpus	Task	#Train	#Dev	#Test	Metrics
CoLA	Acceptability	8.5k	1k	1k	Matthews corr
STS-B	Similarity	7k	1.5k	1.4k	Pearson/Spearman corr
QQP	Paraphrase	364k	40k	391k	Accuracy/F1
MRPC	Paraphrase	3.7k	408	1.7k	Accuracy/F1
SST-2	Sentiment	67k	872	1.8k	Accuracy
QNLI	QA/NLI	108k	5.7k	5.7k	Accuracy
MNLI	NLI	393k	20k	20k	Accuracy
RTE	NLI	2.5k	276	3k	Accuracy

Notes: The details of GLUE benchmark. The #Train, #Dev and #Test denote the size of the training set, development set and test set of corresponding corpus respectively.

SuperGLUE. Similar to GLUE, the SuperGLUE benchmark [24] is a new benchmark that is more difficult language understanding task datasets[2], including: BoolQ, CommitmentBank (CB), Choice of Plausible Alternatives (COPA), Multi-Sentence Reading Comprehension (MultiRC), Reading Comprehension with Commonsense Reasoning (ReCoRD), Recognizing Textual Entailment (RTE), Words in Context (WiC), Winograd Schema Challenge (WSC).

SQuAD. The Stanford Question Answering Dataset (SQuAD) is one of the most popular machine reading comprehension challenges datasets. SQuAD is a typical extractive machine reading comprehension task, including a question and a paragraph of context. Its aim is to give a text span extracted from the document based on the given question. SQuAD consists of two versions: SQuAD [20] (in this version, the provided document always contains an final answer) and SQuAD v2.0 [19] (in this version, some questions are not answered from the provided document).

---

[1] https://gluebenchmark.com/.
[2] https://super.gluebenchmark.com/.

Financial Datasets. To better demonstrate the generality of our post-training approach, we further perform domain adaptation experiments on two financial tasks, FiQA sentiment analysis (SA) dataset and FiQA question answering (QA) dataset. As part of the companion proceedings for WWW'18 conference, [14] released two very small financial datasets (FiQA).

Additional Benchmarks. As shown in Table 6, we present additional datasets for extractive question answering tasks, including RACE [7], NewsQA [23], TrivaQA [6], HotpotQA [28]. More details are provided in the supplementary materials.

Table 2. The overall performance of PPBERT and the comparison against BERT models on GLUE benchmark.

| | | BASE model | | LARGE model | | | |
| | | Test Set | | Dev Set | | Test Set | |
Task	Human Perf.	BERT[†]	PPBERT[‡]	BERT[†]	PPBERT[‡]	BERT[†]	PPBERT[‡]
CoLA	66.4	52.1	52.3	60.6	61.3	60.5	61.1
SST-2	97.8	93.5	94.6	93.2	95.7	94.9	95.7
MRPC	86.3/80.8	84.8/88.9	85.7/89.2	88.0	89.6	85.4/89.3	87.2/90.2
STS-B	92.7/92.6	87.1/85.8	87.6/86.5	90.0	91.3	87.6/86.5	90.5/89.8
QQP	59.5/80.4	89.2/71.2	88.8/73.0	91.3	92.2	89.3/72.1	90.6/73.9
MNLI	92.0/92.8	84.6/83.4	85.9/85.1	86.6	88.7	86.7/85.9	88.3/88.4
QNLI	91.2	90.5	92.2	92.3	93.8	92.7	93.7
RTE	93.6	66.4	72.3	70.4	84.2	70.1	80.3
(Avg)	85.94	80.00	81.53 (1.53 ↑)	84.05	87.02 (2.97 ↑)	82.45	85.03 (2.58 ↑)

Notes: The results on GLUE benchmark [25], where the results on test set are scored by the GLUE evaluation server and the results on dev set are the median of three experimental results. The metrics for these tasks are shown in Table 1. Purple-colored texts indicate the results on par with or pass human performance. [‡] indicates our proposed model. [†] indicates original model BERT [3].

Table 3. Results on SuperGLUE benchmark.

Single Model	BoolQ	CB	COPA	MultiRC	ReCoRD	RTE	WiC	WSC	(Avg)
Human Perf.[§]	89.0	95.8/98.9	100.0	81.8/51.9	91.7/91.3	93.6	80.0	100.0	89.79
BERT[§]	77.4	75.7/83.6	70.6	70.0/24.1	72.0/71.3	71.7	69.6	64.4	69.00
PPBERT (ours)	80.3	81.4/86.9	74.2	76.5/40.7	78.7/77.5	77.4	72.9	68.7	74.55

Notes: All results are based on a 24-layer architecture (LARGE model). PPBERT results on the development set are a median over three runs. Model references: [§]: ([24]).

3.3 Experimental Results

We evaluate the proposed PPBERT on two popular NLU benchmarks: GLUE and SuperGLUE. We compare PPBERT with standard BERT model and demonstrate the effectiveness of with '*post-training*'.

GLUE Results. We evaluated performance on GLUE benchmark, with the large models and the base models of each approach. We reports the results of each method on the development dataset and test dataset. The detailed experimental results on GLUE are presented in Table 2. As illustrated in the BASE models columns of Table 2, PPBERT$_{BASE}$ achieves an average score of 81.53, and outperforms standard BERT$_{BASE}$ on all of the 8 tasks. As shown, in test dataset parts of LARGE models sections in Table 2, PPBERT$_{LARGE}$ outperform BERT$_{LARGE}$ on all of the 8 tasks and achieves an average score of 85.03. We also observe similar results in the dev set column, achieveing an average score of 87.02 on the dev set, a 2.97 improvement over BERT$_{LARGE}$. From this data we can see that PPBERT$_{LARGE}$ matched or even outperformed human level.

SuperGLUE Results. Table 3 shows the performances on 8 SuperGLUE tasks. As shown in Table 3, it is apparent that PPBERT outperforms BERT on 8 tasks significantly. The main gains from PPBERT are in the MultiRC (+6.5) and in ReCoRD (+6.7), both accounting for the rise in PPBERT's GLUE score. Also, as Table 3 shows, there is a huge gap between human performance (89.79) and the performance of PPBERT (74.55).

Overall Trends. Table 2 and Table 3 respectively show our results on GLUE and SuperGLUE with and without '*post-training*'. As shown, we compare proposed method to standard BERT benchmarks on 16 baseline tasks, and find on every task our proposed PPBERT outperforms BERT. Since in pre-training phase PPBERT has the same architecture and pre-training objective as standard BERT, the main gain is attributed to '*post-training*' in post-training phase. If we consider the gains, especially PPBERT is better at natural language inference and question answering tasks, and is not good at syntax-oriented task. In GLUE benchmark (we also observe similar results in SuperGLUE), for example, i) for the question answering tasks (QNLI, MultiRC, ReCoRD) and the natural language inference tasks (MNLI and RTE), we achieves significant accuracy gain of at least 1 point improvement. ii) for sentiment task (SST-2), although we observe a smaller gain (+0.8), it is mainly because the accuracy has been already high, a reasonable score (obtained a accuracy score of 95.7); iii) for simple sentence task, we observe the smallest gain (+0.2) on all tasks in the syntax-oriented (CoLA) task. Besides, this mirrors results also reported in [1], who show that few pre-training tasks other than language modeling offer any advantage for CoLA. iv) for MRPC and RTE tasks, as shown in Table 2 and Table 3, what is interesting in the results is that we find consistent improvements after post-training This reveals that the learned PPBERT representation by '*pre-training*'+'*post-training*' allows much more effective domain adaptation than the BERT representation by '*pre-training*' only.

4 Ablation Study and Analyses

4.1 Cooperation with Other Pre-trained LMs

Our proposed PPBERT is a more flexible and pluggable, where post-training approach can be plugged into other PLMs based on BERT, not only applied to original BERT model. We further validate the performance of PPBERT when '*post-training*' approach on different pre-trained LMs. We compare post-training by plugging it into original BERT (i.e., PPBERT) and and its variant, ALBERT (called PPALBERT) pre-trained LMs, respectively. Also, we further post-train the most recent proposed PPALBERT with one additional QA dataset (SearchQA), and call it PPALBERT$_{\text{LARGE}}$-QA.

Comparisons to SOTA Models. We evaluate our models on the popular SQuAD benchmark (Sect. 3.2). Performance of each model is evaluated on the two standard metric values: F1 score and exact match (EM) score. F1 score measures the precision and recall, and less strict than then EM score. EM score measures whether the model output exactly matches the ground answers.

Table 4. Comparison with state-of-the-art results on the Dev set of SQuAD.

Single Model	SQuAD1.1 EM/F1	SQuAD2.0 EM/F1
Human Perf.	82.3/91.2	86.8/89.5
ALBERT$_{\text{BASE}}$ [8]	82.1/89.3	76.1/79.1
BERT$_{\text{LARGE}}$ [3]	84.1/90.9	79.0/81.8
XLNet$_{\text{LARGE}}$ [27]	89.0/94.5	86.1/88.8
RoBERTa$_{\text{LARGE}}$ [11]	88.9/94.6	86.5/89.4
ALBERT$_{\text{LARGE}}$ [8]	<u>89.3/94.8</u>	<u>87.4/90.2</u>
PPBERT$_{\text{LARGE}}$ (ours)	85.2/92.1	82.2/84.8
PPALBERT$_{\text{LARGE}}$ (ours)	89.6/95.0	87.6/90.4
PPALBERT$_{\text{LARGE}}$-QA (ours)	**89.7/95.1**	**87.7/90.5**

Notes: Results on SQuAD 1.1/2.0 development dataset. Best scores are in bold texts, and the previous best scores are underlined.

Table 4 details performance gains when exploiting each of the three post-trained LMs on SQuAD datasets (two versions, respectively). As shown in Table 4, on the SQuAD dev dataset (version 1.1), compared with BERT baseline, adding post-training stage improves the EM by 1.1 points ($84.1 \rightarrow 85.2$), and F1 1.2 points ($90.9 \rightarrow 92.1$). Similarly, PPALBERT$_{\text{LARGE}}$ also outperforms ALBERT$_{\text{LARGE}}$ baseline, by 0.3 EM and 0.2 F1. Especially, PPALBERT$_{\text{LARGE}}$-QA using further post-training relatively improves 0.1 EM and 0.1 F1 over PPALBERT$_{\text{LARGE}}$, respectively. We also observe similar results on SQuAD v2.0 development set. The most recent proposed PPALBERT sets a new state-of-the-art, achieving 87.7 EM and 90.5 F1.

Performance on Other QA and NLI Tasks. Furthermore, extensive experiments on six NLP tasks about semantic relationship are conducted, including two natural language inference benchmarks (QNLI and MNLI-m, both from GLUE), and four extractive question answering benchmarks (TriviaQA, RACE, HotpotQA and NewsQA). All benchmarks except RACE, we use the same fine-tuning method as SQuAD. Different from others, RACE is a multiple-choice QA dataset. The experimental results for PPALBERT are shown in Table 5. As depicted in Table 5, both PPALBERT$_{\text{LARGE}}$ and PPALBERT$_{\text{LARGE}}$-QA achieve state-of-the-art accuracy across all settings. Overall, as expected, only utilizing '*pre-training*' is inferior to our proposed '*pre-training*'-then-'*post-training*' method. The experimental results (Sect. 4.1 and Sect. 4.1) described above, indicate that our two stage training paradigm is very flexible, and proposed post-training appoach could be easily plugged into other PLMs. More remarkably, we achieve new SOTA performances on existing baselines.

Table 5. Performance on six QA and NLI tasks.

Single Model	NewsQA	TrivaQA	HotpotQA	RACE	QNLI	MNLI-m
BERT$_{\text{LARGE}}$[†]	68.8	77.5	78.3	72.0	92.3	86.6
SpanBERT$_{\text{LARGE}}$[†]	73.6	83.6	83.0	-	93.3	87.0
RoBERTa$_{\text{LARGE}}$ [‡]	-	-	-	83.2	94.7	90.2
ALBERT$_{\text{LARGE}}$ [§]	-	-	-	86.5	95.2	90.4
PPALBERT$_{\text{LARGE}}$ (ours)	74.6	84.3	83.4	86.7	95.6	90.7
PPALBERT$_{\text{LARGE}}$-QA (ours)	**74.8**	**84.5**	**83.5**	**86.8**	**95.9**	**90.9**

Notes: The details of NewsQA, TrivaQA, HotpotQA and RACE are shown in Table 6. QNLI and MNLI-m are from GLUE. Model references: [†]: ([5]), [‡]: ([11]), [§]: ([8]).

Table 6. The details of QA datasets.

Dataset	Lang.	#Query	#Documents	Query	Documents	Answer type
SQuAD 1.1[†]	EN	100K	536	CS	Wiki	Span of words
SQuAD 2.0[‡]	EN	150K	500	CS	Wiki	Span of words
NewsQA [23]	EN	100K	10K	CS	CNN	Span of words
HotpotQA [28]	EN	78K	113k	CS	Wiki	Span/substring of words
TrivaQA [6]	EN	40K	660K	TW	Wiki./Web doc	Span/substring of words
RACE [7]	EN	870K	50K	EE	EE	Multiple-choice
CoQA [21]	EN	127K	8K	CS	QA Dialog	Span/substring of words

Notes: CS denotes Crowdsourced. TW denotes Trivia websites. EE denotes English exam. Model references: [†]: ([20]), [‡]: ([19]).

5 Conclusion

In the paper, we present a *'pre-training'*+*'post-training'*+*'fine-tuning'* three-stage paradigm and a language model named PPBERT based on the three-stage paradigm, which is a supplementary framework for the standard *'pre-training'*+*'fine-tuning'* two-stage architecture. Our proposed three-stage paradigm helps to incorporate task-awareness knowledge and domain knowledge within pre-trained model, also reduce the bias in the training corpus. PPBERT can benefits from the regularization effect since it leverages cross-domain or cross-task data, which helps model generalize better with limited data and adapt to new domains or tasks better. With the latest PLMs as baseline and encoder backbone, PPBERT is evaluated on 24 well-known benchmarks, which outperformS strong baseline models and obtains new SOTA results. We hope this work can encourage further research into the language models training, and the future works involve the choice of other transfer learning sources such as CV etc.

Acknowledgements. We would like to thank the reviewers for their helpful comments and suggestions to improve the quality of the paper. The authors gratefully acknowledge the financial support provided by the Basic Scientific Research Project (General Program) of Department of Education of Liaoning Province, the University-Industry Collaborative Education Program of the Ministry of Education of China (No.202002037015).

References

1. Bowman, S.R., Pavlick, E., Grave, E.: Looking for Elmo's friends: sentence-level pretraining beyond language modeling. CoRR abs/1812.10860 (2018)
2. Chen, Z., Liu, B.: Lifelong Machine Learning. Synthesis Lectures on Artificial Intelligence and Machine Learning, 2nd edn. Morgan & Claypool Publishers, Williston (2018). https://doi.org/10.2200/S00832ED1V01Y201802AIM037

3. Devlin, J., Chang, M., Lee, K., Toutanova, K.: BERT: pre-training of deep bidirectional transformers for language understanding. In: Burstein, J., Doran, C., Solorio, T. (eds.) Proceedings of the 2019 Conference of the North American Chapter of the Association for Computational Linguistics: Human Language Technologies, NAACL-HLT 2019, Minneapolis, MN, USA, 2–7 June, 2019, Volume 1 (Long and Short Papers), pp. 4171–4186. Association for Computational Linguistics (2019). https://doi.org/10.18653/v1/n19-1423

4. Hou, M., Chen, X., Huang, S., Xie, S., Zhou, G.: Generalizing deep multi-task learning with heterogeneous structured networks. In: Proceedings of ICLR (2020)

5. Joshi, M., Chen, D., Liu, Y., Weld, D.S., Zettlemoyer, L., Levy, O.: Span-BERT: improving pre-training by representing and predicting spans. CoRR abs/1907.10529 (2019). http://arxiv.org/abs/1907.10529

6. Joshi, M., Choi, E., Weld, D.S., Zettlemoyer, L.: Triviaqa: a large scale distantly supervised challenge dataset for reading comprehension. In: Barzilay, R., Kan, M. (eds.) Proceedings of the 55th Annual Meeting of the Association for Computational Linguistics, ACL 2017, Vancouver, Canada, 30 July – 4 August, Volume 1: Long Papers, pp. 1601–1611. Association for Computational Linguistics (2017). https://doi.org/10.18653/v1/P17-1147

7. Lai, G., Xie, Q., Liu, H., Yang, Y., Hovy, E.H.: RACE: large-scale reading comprehension dataset from examinations. In: Palmer, M., Hwa, R., Riedel, S. (eds.) Proceedings of the 2017 Conference on Empirical Methods in Natural Language Processing, EMNLP 2017, Copenhagen, Denmark, 9–11 September, 2017, pp. 785–794. Association for Computational Linguistics (2017). https://doi.org/10.18653/v1/d17-1082

8. Lan, Z., Chen, M., Goodman, S., Gimpel, K., Sharma, P., Soricut, R.: ALBERT: A lite BERT for self-supervised learning of language representations. In: 8th International Conference on Learning Representations, ICLR 2020, Addis Ababa, Ethiopia, April 26–30, 2020. OpenReview.net (2020), https://openreview.net/forum?id=H1eA7AEtvS

9. Levesque, H.J., Davis, E., Morgenstern, L.: The winograd schema challenge (2012)

10. Liu, X., et al.: The microsoft toolkit of multi-task deep neural networks for natural language understanding. In: Celikyilmaz, A., Wen, T. (eds.) Proceedings of the 58th Annual Meeting of the Association for Computational Linguistics: System Demonstrations, ACL 2020, Online, 5–10 July, 2020, pp. 118–126. Association for Computational Linguistics (2020). https://doi.org/10.18653/v1/2020.acl-demos.16

11. Liu, Y., et al.: Roberta: a robustly optimized BERT pretraining approach. CoRR abs/1907.11692 (2019). http://arxiv.org/abs/1907.11692

12. Liu, Z., Huang, D., Huang, K., Li, Z., Zhao, J.: Finbert: a pre-trained financial language representation model for financial text mining. In: Proceedings of the Twenty-Ninth International Joint Conference on Artificial Intelligence, IJCAI 2020, 5–10 January, 2021, Yokohama, Japan, pp. 4513–4519 (2020)

13. Liu, Z., Huang, K., Huang, D., Liu, Z., Zhao, J.: Dual head-wise coattention network for machine comprehension with multiple-choice questions. In: d'Aquin, M., Dietze, S., Hauff, C., Curry, E., Cudré-Mauroux, P. (eds.) CIKM 2020: The 29th ACM International Conference on Information and Knowledge Management, Virtual Event, Ireland, 19–23 October, 2020, pp. 1015–1024. ACM (2020). https://doi.org/10.1145/3340531.3412013

14. Maia, M., et al. (eds.): Proceedings of WWW. ACM (2018). https://doi.org/10.1145/3184558

15. Ohsugi, Y., Saito, I., Nishida, K., Asano, H., Tomita, J.: A simple but effective method to incorporate multi-turn context with BERT for conversational machine comprehension. CoRR abs/1905.12848 (2019). http://arxiv.org/abs/1905.12848
16. Parisi, G.I., Kemker, R., Part, J.L., Kanan, C., Wermter, S.: Continual lifelong learning with neural networks: a review. Neural Networks. **113**, 54–71 (2019). https://doi.org/10.1016/j.neunet.2019.01.012
17. Peters, M.E., et al.: Deep contextualized word representations. In: Walker, M.A., Ji, H., Stent, A. (eds.) Proceedings of the 2018 Conference of the North American Chapter of the Association for Computational Linguistics: Human Language Technologies, NAACL-HLT 2018, New Orleans, Louisiana, USA, 1–6 June, 2018, Volume 1 (Long Papers), pp. 2227–2237. Association for Computational Linguistics (2018). https://doi.org/10.18653/v1/n18-1202
18. Radford, A., Narasimhan, K., Salimans, T., Sutskever, I.: Improving language understanding by generative pre-training. In: Proceedings of Technical Report, OpenAI (2018). https://github.com/openai/finetune-transformer-lm
19. Rajpurkar, P., Jia, R., Liang, P.: Know what you don't know: unanswerable questions for squad. In: Gurevych, I., Miyao, Y. (eds.) Proceedings of the 56th Annual Meeting of the Association for Computational Linguistics, ACL 2018, Melbourne, Australia, 15–20 July, 2018, Volume 2: Short Papers, pp. 784–789. Association for Computational Linguistics (2018). https://doi.org/10.18653/v1/P18-2124, https://www.aclweb.org/anthology/P18-2124/
20. Rajpurkar, P., Zhang, J., Lopyrev, K., Liang, P.: Squad: 100,000+ questions for machine comprehension of text. In: Su, J., Carreras, X., Duh, K. (eds.) Proceedings of the 2016 Conference on Empirical Methods in Natural Language Processing, EMNLP 2016, Austin, Texas, USA, 1–4 November, 2016, pp. 2383–2392. The Association for Computational Linguistics (2016). https://doi.org/10.18653/v1/d16-1264
21. Reddy, S., Chen, D., Manning, C.D.: CoQA: a conversational question answering challenge. Trans. Assoc. Comput. Linguist. **7**, 249–266 (2019). https://transacl.org/ojs/index.php/tacl/article/view/1572
22. Sun, Y., Wang, S., Li, Y.: ERNIE: enhanced representation through knowledge integration. CoRR abs/1904.09223 (2019). http://arxiv.org/abs/1904.09223
23. Trischler, A., et al.: Newsqa: a machine comprehension dataset. In: Proceedings of the 2nd Workshop on Representation Learning for NLP (2017)
24. Wang, A., et al.: SuperGLUE: a stickier benchmark for general-purpose language understanding systems. In: Wallach, H.M., Larochelle, H., Beygelzimer, A., d'Alché-Buc, F., Fox, E.B., Garnett, R. (eds.) Advances in Neural Information Processing Systems 32: Annual Conference on Neural Information Processing Systems 2019, NeurIPS 2019, 8–14 December, 2019, Vancouver, BC, Canada, pp. 3261–3275 (2019). https://proceedings.neurips.cc/paper/2019/hash/4496bf24afe7fab6f046bf4923da8de6-Abstract.html
25. Wang, A., Singh, A., Michael, J., Hill, F., Levy, O., Bowman, S.R.: GLUE: a multitask benchmark and analysis platform for natural language understanding. In: 7th International Conference on Learning Representations, ICLR 2019, New Orleans, LA, USA, 6–9 May, 2019. OpenReview.net (2019). https://openreview.net/forum?id=rJ4km2R5t7

26. Yang, W., et al.: End-to-end open-domain question answering with BERTserini. In: Ammar, W., Louis, A., Mostafazadeh, N. (eds.) Proceedings of the 2019 Conference of the North American Chapter of the Association for Computational Linguistics: Human Language Technologies, NAACL-HLT 2019, Minneapolis, MN, USA, 2–7 June, 2019, Demonstrations, pp. 72–77. Association for Computational Linguistics (2019). https://doi.org/10.18653/v1/n19-4013

27. Yang, Z., Dai, Z., Yang, Y., Carbonell, J.G., Salakhutdinov, R., Le, Q.V.: XLnet: generalized autoregressive pretraining for language understanding. In: Wallach, H.M., Larochelle, H., Beygelzimer, A., d'Alché-Buc, F., Fox, E.B., Garnett, R. (eds.) Advances in Neural Information Processing Systems 32: Annual Conference on Neural Information Processing Systems 2019, NeurIPS 2019, 8–14 December, 2019, Vancouver, BC, Canada, pp. 5754–5764 (2019). https://proceedings.neurips.cc/paper/2019/hash/dc6a7e655d7e5840e66733e9ee67cc69-Abstract.html

28. Yang, Z., et al.: Hotpotqa: a dataset for diverse, explainable multi-hop question answering. In: Riloff, E., Chiang, D., Hockenmaier, J., Tsujii, J. (eds.) Proceedings of the 2018 Conference on Empirical Methods in Natural Language Processing, Brussels, Belgium, 31 October – 4 November, 2018, pp. 2369–2380. Association for Computational Linguistics (2018). https://doi.org/10.18653/v1/d18-1259

Author Index